# Perspectives on Theoretical Chemistry

Christopher J. Cramer • Donald G. Truhlar
Editors

# Perspectives on Theoretical Chemistry

Five Decades of Theoretical Chemistry Accounts
and Theoretica Chimica Acta

*With contributions from*

Christopher J. Cramer • Gino A. DiLabio • Filipp Furche
Sophya Garashchuk • Peter M. W. Gill • Hua Guo • So Hirata
Brian K. Kendrick • Hans Lischka • Wenjian Liu
Fernando R. Ornellas • Irina Paci • Kirk A. Peterson
Markus Reiher • Jeffrey R. Reimers • Manuel Smeu
Seiichiro Ten-no • Diego Troya • Donald G. Truhlar
Christoph van Wüllen • Dong H. Zhang

*Editors*
Prof. Christopher J. Cramer
Department of Chemistry and
Supercomputing Institute
University of Minnesota
Minneapolis, MN
USA

Prof. Donald G. Truhlar
Department of Chemistry and
Supercomputing Institute
University of Minnesota
Minneapolis, MN
USA

Originally Published in Theor Chem Acc, Volume 131 (2012), Number 1
© Springer-Verlag Berlin Heidelberg 2012

ISBN 978-3-642-28444-1
Springer Heidelberg New York Dordrecht London

Library of Congress Control Number: 2012934815

© Springer-Verlag Berlin Heidelberg 2012
This work is subject to copyright. All rights are reserved by the Publisher, whether the whole or part of the material is concerned, specifically the rights of translation, reprinting, reuse of illustrations, recitation, broadcasting, reproduction on microfilms or in any other physical way, and transmission or information storage and retrieval, electronic adaptation, computer software, or by similar or dissimilar methodology now known or hereafter developed. Exempted from this legal reservation are brief excerpts in connection with reviews or scholarly analysis or material supplied specifically for the purpose of being entered and executed on a computer system, for exclusive use by the purchaser of the work. Duplication of this publication or parts thereof is permitted only under the provisions of the Copyright Law of the Publisher's location, in its current version, and permission for use must always be obtained from Springer. Permissions for use may be obtained through RightsLink at the Copyright Clearance Center. Violations are liable to prosecution under the respective Copyright Law.
The use of general descriptive names, registered names, trademarks, service marks, etc. in this publication does not imply, even in the absence of a specific statement, that such names are exempt from the relevant protective laws and regulations and therefore free for general use.
While the advice and information in this book are believed to be true and accurate at the date of publication, neither the authors nor the editors nor the publisher can accept any legal responsibility for any errors or omissions that may be made. The publisher makes no warranty, express or implied, with respect to the material contained herein.

Printed on acid-free paper

Springer is part of Springer Science+Business Media (www.springer.com)

# Contents

**Fifty years of TCA** .......................................................................................... 1
Christopher J. Cramer

**Computational modeling of extended systems** ...................................................... 7
Gino A. DiLabio

**Nanoparticle morphology and aspect ratio effects in Ag/PVDF nanocomposites** ............ 9
Christopher K. Rowan, Irina Paci

**Toward ab initio refinement of protein X-ray crystal structures:
interpreting and correlating structural fluctuations** ........................................ 21
Olle Falklöf, Charles A. Collyer, Jeffrey R. Reimers

**Conduction modulation of $\pi$-stacked ethylbenzene wires on Si(100)
with substituent groups** .................................................................................. 37
Manuel Smeu, Robert A. Wolkow, Hong Guo

**Theoretical characterization of reaction dynamics in the gas phase
and at interfaces** ............................................................................................. 45
Hua Guo

**Calculation of the zero-point energy from imaginary-time quantum
trajectory dynamics in Cartesian coordinates** .................................................. 49
Sophya Garashchuk

**Time-dependent wave packet propagation using quantum hydrodynamics** ............ 59
Brian K. Kendrick

**Dynamics of collisions of hydroxyl radicals with fluorinated
self-assembled monolayers** .............................................................................. 79
Diego Troya

**A full-dimensional time-dependent wave packet study of the
$OH + CO \rightarrow H + CO_2$ reaction** ............................................................ 91
Shu Liu, Xin Xu, Dong H. Zhang

**Electronic structure theory: present and future challenges** ................................ 99
So Hirata

**Electron correlation methods based on the random phase approximation** ............ 103
Henk Eshuis, Jefferson E. Bates, Filipp Furche

v

**Uniform electron gases** .......................... 121
Peter M.W. Gill, Pierre-François Loos

**Explicitly correlated wave functions: summary and perspective** .................... 131
Seiichiro Ten-no

**Theoretical chemistry: current applications to photochemistry
and thermochemistry** ........................ 143
Fernando R. Ornellas

**Electronically excited states and photodynamics: a continuing challenge** .................... 147
Felix Plasser, Mario Barbatti, Adélia J.A. Aquino, Hans Lischka

**Chemical accuracy in ab initio thermochemistry and spectroscopy:
current strategies and future challenges** ........................ 161
Kirk A. Peterson, David Feller, David A. Dixon

**Negative energy states in relativistic quantum chemistry** ........................ 181
Christoph van Wüllen

**Fully relativistic theories and methods for NMR parameters** ........................ 187
Yunlong Xiao, Qiming Sun, Wenjian Liu

**Exact decoupling of the relativistic Fock operator** ........................ 205
Daoling Peng, Markus Reiher

# EDITORIAL

# Fifty years of TCA

**Christopher J. Cramer**

Published online: 13 January 2012
© Springer-Verlag 2012

Most of us probably do not remember the appearance of Volume 1, Number 1 of *Theoretica Chimica Acta* (TCA); it was in 1962. The present editorial board members were all very young at that time, if they had even yet been born. TCA was the first journal devoted *exclusively* to theoretical chemistry. Since then many other theoretical chemistry journals have debuted, including *Int. J. Quantum Chem.*, *J. Comput. Chem.*, *Computational and Theoretical Chemistry* (formerly *Theochem*), and *J. Chem. Theory Comput.* Meanwhile, in 1996, TCA changed its name from the Latin *Theoretica Chimica Acta* to the English *Theoretical Chemistry Accounts*.

The golden jubilee of the first theoretical chemistry journal provides an occasion both to look back and to look forward. When TCA was inaugurated, quantum mechanics was 37 years old. Now, it is 87 years old. When TCA was inaugurated, *J. Chem. Phys.* was 29 years old, *Phys. Rev.* as a journal of the American Physical Society was 49 years old, and *J. Phys. Chem.* was 66 years old. In the most recent impact factors (those for 2010), *J. Chem. Phys.*, TCA, and *Phys. Rev. A* all have impact factors of 2.9, and *J. Phys. Chem. A* has an impact factor of 2.7. Of the 49 Nobel Prizes in chemistry since 1962 (the 2011 Prize has not been announced at the time we are putting this to press), theoretical chemists have shared or won six: Lars Onsager, William Lipscomb, Ilya Progogine, Kenichi Fukui and Roald Hoffman (shared), Rudy Marcus, and Walter Kohn

and John Pople (shared). We feel safe predicting that theoretical chemists will take home more than six of the next 49.

Let's look back in two ways. First, we wish to thank all those who have contributed to TCA through editorial service or editorial advisory service. Second, we wish to single out the most highly cited papers in the history of TCA. Then, we will look forward by means of the rest of this special issue.

## 1 Editorial service

Here is a list of those who have provided editorial service to TCA and the capacities in which they served. On behalf of the entire theoretical chemistry community, we thank them for their service.

***Honorary Editor***
Klaus Ruedenberg

***Editor-in-Chief***
Cramer, Christopher J.
Hartmann, Hermann
Ruedenberg, Klaus
Truhlar, Donald G.

***Editor or Associate Editor***
Baer, Roi
Cársky, Petr
Chandler, David
Cramer, Christopher J.
Davidson, Ernest R.
Handy, Nicholas C.
Iwata, Suehiro
Jortner, Joshua

Published as part of the special collection of articles celebrating the 50th anniversary of Theoretical Chemistry Accounts/Theoretica Chimica Acta.

C. J. Cramer (✉)
Department of Chemistry, University of Minnesota,
207 Pleasant Street SE, Minneapolis, MN 55455-0431, USA
e-mail: cramer@umn.edu

Kutzelnigg, Werner
Michl, Josef
Miller, William H.
Morokuma, Keiji
Roos, Björn O.
Salahub, Dennis R.
Schaefer III, Henry F.
Truhlar, Donald G.
Wu, Yundong

**Chief Advisory Editor**
Truhlar, Donald G.

**Editorial Board**
Ahlrichs, Reinhart
Almlöf, Jan E.
Ángyán, János G.
Apeloig, Yitzhak
Aquilanti, Vincenzo
Åqvist, Johan
Baerends, Evert J.
Bagchi, Biman A.
Baldridge, Kim K.
Ballhausen, Carl J.
Barone, Vincenzo
Bartlett, Rodney J.
Basilevsky, Mikhail V.
Bauschlicher, Charles W.
Becke, Axel D.
Berne, Bruce J.
Bertrán, Juan
Botschwina, Peter
Brooks, Charles
Brown, Ronald D.
Bunge, Carlos F.
Cársky, Petr
Case, David A.
Catlow, Richard
Champagne, Benoit
Chelikowsky, James R.
Child, Mark S.
Clary, David C.
Collins, Michael A.
Davidson, Ernest R.
DiLabio, Gino A.
Dykstra, Clifford E.
Eisenstein, Odile
Elber, Ron
Field, Martin J.
Freed, Karl F.
Frenkel, Daan
Frenking, Gernot
Fried, Joel R.

Fukui, Kenichi
Gagliardi, Laura
Gale, Julian D.
Gao, Jiali
Garrett, Bruce C.
Gleiter, Rolf
Gordon, Mark S.
Grein, Fritz
Grimme, Stefan
Guo, Hua
Halevi, E. Amitai
Hall, George G.
Hall, Michael B.
Hammes-Schifffer, Sharon
Harvey, Jeremy
Hess, Bernd A.
Helgaker, Trygve
Heller, Eric J.
Hirao, Kimihiko
Hirata, So
Houk, Kendall N.
Iwata, Suehiro
Jordan, Kenneth D.
Jorgensen, William L.
Jortner, Joshua
Kaldor, Uzi
Kapral, Raymond
Karplus, Martin
Kato, Shigeki
Kaupp, Martin
Ketelaar, J. A. A.
Kollman, Peter
Kolos, Włodzimierz
Kotani, Masao
Kouri, Donald J.
Koutecky, Jaroslav
Kutzelnigg, Werner
Lavery, Richard
Levine, Raphael D.
Levy, Melvin P.
Lim, Carmay
Linnett, J. W.
Lluch, José M.
Luque, F. Javier
Madden, Paul A.
Malrieu, Jean-Paul
Marcus, Rudolph A.
Martin, Jan M. L.
Martinez, Todd J.
Mennucci, Benedetta
Merz, Kenneth M. Jr.
Meyer, Hans-Dieter
Meyer, Wilfried

Miller, James A.
Miller, William H.
Morokuma, Keiji
Mukherjee, Debashis
Murcko, Mark
Nakatsuji, Hiroshi
Neckel, Adolf
Newton, Marshall D.
Nikitin, Evgueni E.
Noid, Donald W.
Oddershede, Jens
Olivucci, Massimo
Olsen, Jeppe
Ornellas, Fernando R.
Orozco, Modesto
Pacchioni, Gianfranco
Paldus, Josef
Parr, Robert G.
Parrinello, Michele
Pearson, Ralph G.
Petersson, George A.
Pettitt, B. Montgomery
Primas, Hans
Pullman, Bernard
Pyykko, Veli Pekka
Radom, Leo
Raghavachari, Krishnan
Ranby, B.
Ratner, Mark A.
Reiher, Markus
Robb, Michael A.
Roos, Björn O.
Rossky, Peter J.
Ruch, Ernst
Ruedenberg, Klaus
Salahub, Dennis R.
Sandorfy, Camille
Savin, Andreas
Schaefer III, Henry Fritz
Schatz, George C.
Schinke, Reinhard
Schlegel, H. Bernhard
Scuseria, Gustavo E.
Seideman, Tamar
Shaik, Sason
Shuai, Zhigang
Siegbahn, Per
Siepmann, J. Ilja
Simonetta, Massimo
Sinanoglu, Oktay
Skodje, Rex
Sprik, Michiel
Thiel, Walter

Thirumalai, Deverajan
Tomasi, Jacopo
Tucker, Susan C.
Tuckerman, Mark E.
Tully, John C.
van Santen, Rutger A.
van Wüllen, Christoph
Veillard, Alain
Voter, Arthur F.
Voth, Gregory A.
Warshel, Arieh
Werner, Hans-Joachim
Whangbo, Myung-Hwan
Yang, Jinlong
Yang, Weitao
Yarkony, David R.
Zahradnik, Rudolf
Zeng, Xiao Cheng
Zerner, Michael C.
Zhan, Chang-Guo
Zhang, John Z. H.

## 2 Highly cited papers

For each decade, the list contains the twelve most highly cited papers from that decade, as of the preparation of this editorial, with the most highly cited first. The lists contain many important contributors and many important contributions, and it is interesting to notice the evolution of topics represented therein. A key topic that has persisted on the lists for the full 50 years is basis sets, which have always been featured prominently in *TCA*.

### 2.1 First decade: 1962–1971 (Theoretica Chimica Acta)

- Some remarks on the Pariser-Parr-Pople method. Ohno, K. 1964. *Theoretica Chimica Acta* 2, pp. 219–227
- Possible "ferromagnetic states" of some hypothetical hydrocarbons. Mataga, N. *Theoretica Chimica Acta* 10 (4), pp. 372–376
- Recherches sur la géométrie de quelques hydrocarbures non-alternants: son influence sur les énergies de transition, une nouvelle définition de l'aromaticité. Julg, A., François, P. 1967 *Theoretica Chimica Acta* 8 (3), pp. 249–259
- Porphyrins XIV. Theory for the luminescent state in VO, Co, Cu complexes. Ake, R.L., Gouterman, M. 1969 *Theoretica Chimica Acta* 15 (1), pp. 20–42 60
- Gaussian basis set for molecular wavefunctions containing second-row atoms. Veillard, A. 1968 *Theoretica Chimica Acta* 12 (5), pp. 405–411

- SCFMO calculations of heteroatomic systems with the variable $\beta$ approximation—I. Heteroatomic molecules containing nitrogen or oxygen atoms. Nishimoto, K., Forster, L.S. 1966 *Theoretica Chimica Acta* 4 (2), pp. 155–165
- Porphyrins—VIII. Extended Hückel calculations on iron complexes. Zerner, M., Gouterman, M., Kobayashi, H. 1966 *Theoretica Chimica Acta* 6 (5), pp. 363–400
- Porphyrins—IV. Extended Hückel calculations on transition metal complexes. Zerner, M., Gouterman, M. 1966 *Theoretica Chimica Acta* 4 (1), pp. 44–63
- Valence orbital ionization potentials from atomic spectral data. Basch, H., Viste, A., Gray, H.B. 1965 *Theoretica Chimica Acta* 3 (5), pp. 458–464
- Energy partitioning with the CNDO method. Fischer, H., Kollmar, H. 1970 *Theoretica Chimica Acta* 16 (3), pp. 163–174
- The continuation of the periodic table up to $Z = 172$. The chemistry of superheavy elements. Fricke, B., Greiner, W., Waber, J.T. 1971 *Theoretica Chimica Acta* 21 (3), pp. 235–260
- Electronic wave functions for atoms—II. Some aspects of the convergence of the configuration interaction expansion for the ground states of the He isoelectronic series. Bunge, C.F. 1970 *Theoretica Chimica Acta* 16 (2), pp. 126–144.

## 2.2 Second decade: 1972–1981 (Theoretica Chimica Acta)

- The influence of polarization functions on molecular orbital hydrogenation energies. Hariharan, P.C., Pople, J.A. 1973 *Theoretica Chimica Acta* 28 (3), pp. 213–222
- An intermediate neglect of differential overlap technique for spectroscopy: Pyrrole and the azines. Ridley, J., Zerner, M. 1973 *Theoretica Chimica Acta* 32 (2), pp. 111–134 1171
- Bonded-atom fragments for describing molecular charge densities. Hirshfeld, F.L. 1977 *Theoretica Chimica Acta* 44 (2), pp. 129–138
- On the calculation of bonding energies by the Hartree–Fock Slater method—I. The transition state method. Ziegler, T., Rauk, A. 1977 *Theoretica Chimica Acta* 46 (1), pp. 1–10
- Individualized configuration selection in CI calculations with subsequent energy extrapolation. Buenker, R.J., Peyerimhoff, S.D. 1974 *Theoretica Chimica Acta* 35 (1), pp. 33–58
- An intermediate neglect of differential overlap theory for transition metal complexes: Fe, Co and Cu chlorides. Bacon, A.D., Zerner, M.C. 1979 *Theoretica Chimica Acta* 53 (1), pp. 21–54

- Triplet states via intermediate neglect of differential overlap: Benzene, pyridine and the diazines. Ridley, J.E., Zerner, M.C. 1976 *Theoretica Chimica Acta* 42 (3), pp. 223–236
- On the calculation of multiplet energies by the Hartree–Fock-Slater method. Ziegler, T., Rauk, A., Baerends, E.J. 1977 *Theoretica Chimica Acta* 43 (3), pp. 261–271
- Energy extrapolation in CI calculations. Buenker, R.J., Peyerimhoff, S.D. 1975 *Theoretica Chimica Acta* 39 (3), pp. 217–228
- A theoretical method to determine atomic pseudopotentials for electronic structure calculations of molecules and solids. Durand, P., Barthelat, J.-C. 1975 *Theoretica Chimica Acta* 38 (4), pp. 283–302
- Approximate calculation of the correlation energy for the closed shells. Colle, R., Salvetti, O. 1975 *Theoretica Chimica Acta* 37 (4), pp. 329–334
- Multiplicity of the ground state of large alternant organic molecules with conjugated bonds—(Do Organic Ferromagnetics Exist?). Ovchinnikov, A.A. 1978 *Theoretica Chimica Acta* 47 (4), pp. 297–304.

## 2.3 Third decade: 1982–1991 (Theoretica Chimica Acta)

- Energy-adjusted ab initio pseudopotentials for the second and third row transition elements. Andrae, D., Häußermann, U., Dolg, M., Stoll, H., Preuß, H. 1990 *Theoretica Chimica Acta* 77 (2), pp. 123–141
- Density matrix averaged atomic natural orbital (ANO) basis sets for correlated molecular wave functions—I. First row atoms. Widmark, P.-O., Malmqvist, P.-Å., Roos, B.O. 1990 *Theoretica Chimica Acta* 77 (5), pp. 291–306
- Medium-size polarized basis sets for high-level-correlated calculations of molecular electric properties—II. Second-row atoms: Si through Cl, Sadlej, A.J. 1991 *Theoretica Chimica Acta* 79 (2), pp. 123–140
- Energy-adjusted pseudopotentials for the rare earth elements. Dolg, M., Stoll, H., Savin, A., Preuss, H. 1989 *Theoretica Chimica Acta* 75 (3), pp. 173–194
- Density matrix averaged atomic natural orbital (ANO) basis sets for correlated molecular wave functions—II. Second row atoms. Widmark, P.-O., Persson, B.J., Roos, B.O. 1991 *Theoretica Chimica Acta* 79 (6), pp. 419–432
- $r_{12}$-Dependent terms in the wave function as closed sums of partial wave amplitudes for large l. Kutzelnigg, W. 1985 *Theoretica Chimica Acta* 68 (6), pp. 445–469
- Orbital-invariant formulation and second-order gradient evaluation in Møller-Plesset perturbation theory. Pulay, P., Saebø, S. 1986 *Theoretica Chimica Acta* 69 (5–6), pp. 357–368

- A generalized restricted open-shell Fock operator. Edwards, W.D., Zerner, M.C. 1987 *Theoretica Chimica Acta* 72 (5–6), pp. 347–361
- Theoretical investigations of molecules composed only of fluorine, oxygen and nitrogen: determination of the equilibrium structures of FOOF, (NO)2 and FNNF and the transition state structure for FNNF cis–trans isomerization. Lee, T.J., Rice, J.E., Scuseria, G.E., Schaefer III, H.F. 1989 *Theoretica Chimica Acta* 75 (2), pp. 81–98
- An overview of coupled cluster theory and its applications in physics. Bishop, R.F. 1991 *Theoretica Chimica Acta* 80 (2–3), pp. 95–148
- Recursive intermediate factorization and complete computational linearization of the coupled-cluster single, double, triple, and quadruple excitation equations. Kucharski, S.A., Bartlett, R.J. 1991 *Theoretica Chimica Acta* 80 (4–5), pp. 387–405
- A comparison of variational and non-variational internally contracted multiconfiguration-reference configuration interaction calculations. Werner, H.-J., Knowles, P.J. 1990 *Theoretica Chimica Acta* 78 (3), pp. 175–187.

## 2.4 Fourth decade: 1992–2001 (Theoretica Chimica Acta and Theoretical Chemistry Accounts)

- Towards an order-*N* DFT method. Fonseca Guerra, C., Snijders, J.G., Te Velde, G., Baerends, E.J. 1998 *Theoretical Chemistry Accounts* 99 (6), pp. 391–403 1524
- Auxiliary basis sets for main row atoms and transition metals and their use to approximate Coulomb potentials. Eichkorn, K., Weigend, F., Treutler, O., Ahlrichs, R. 1997 *Theoretical Chemistry Accounts* 97 (1–4), pp. 119–124
- RI-MP2: First derivatives and global consistency. Weigend, F., Häser, M. 1997 *Theoretical Chemistry Accounts* 97 (1–4), pp. 331-340
- An efficient data compression method for the Davidson subspace diagonalization scheme Dachsel, H., Lischka, H. 1995 *Theoretica Chimica Acta* 92 (6), pp. 339–349
- An implementation of the conductor-like screening model of solvation within the Amsterdam density functional package. Pye, C.C., Ziegler, T. 1999 *Theoretical Chemistry Accounts* 101 (6), pp. 396–408
- Reparameterization of hybrid functionals based on energy differences of states of different multiplicity. Reiher, M., Salomon, O., Hess, B.A. 2001 Theoretical Chemistry Accounts 107 (1), pp. 48–55
- A combination of quasirelativistic pseudopotential and ligand field calculations for lanthanoid compounds. Dolg, M., Stoll, H., Preuss, H. 1993 Theoretica Chimica Acta 85 (6), pp. 441–450

- Density matrix averaged atomic natural orbital (ANO) basis sets for correlated molecular wave functions—III. First row transition metal atoms. Pou-Amérigo, R., Merchán, M., Nebot-Gil, I., Widmark, P.-O., Roos, B.O. 1995 Theoretica Chimica Acta 92 (3), pp. 149–181
- The singlet and triplet states of phenyl cation. A hybrid approach for locating minimum energy crossing points between non-interacting potential energy surfaces. Harvey, J.N., Aschi, M., Schwarz, H., Koch, W. 1998 Theoretical Chemistry Accounts 99 (2), pp. 95–99
- Internally contracted multiconfiguration-reference configuration interaction calculations for excited states. Knowles, P.J., Werner, H.-J. 1992 *Theoretica Chimica Acta* 84 (1–2), pp. 95–103
- Extensions and tests of "multimode": A code to obtain accurate vibration/rotation energies of many-mode molecules. Carter, S., Bowman, J.M., Handy, N.C. 1998 *Theoretical Chemistry Accounts* 100 (1–4), pp. 191–198
- The MIDI! basis set for quantum mechanical calculations of molecular geometries and partial charges. Easton, R.E., Giesen, D.J., Welch, A., Cramer, C.J., Truhlar, D.G. 1996 *Theoretical Chemistry Accounts* 93 (5), pp. 281–301.

## 2.5 Fifth decade: 2002–2011 (Theoretical Chemistry Accounts)

- The M06 suite of density functionals for main group thermochemistry, thermochemical kinetics, noncovalent interactions, excited states, and transition elements: Two new functionals and systematic testing of four M06-class functionals and 12 other functionals. Zhao, Y., Truhlar, D.G. 2008 *Theoretical Chemistry Accounts* 120 (1–3), pp. 215–241
- Systematically convergent basis sets for transition metals. II. Pseudopotential-based correlation consistent basis sets for the group 11 (Cu, Ag, Au) and 12 (Zn, Cd, Hg) elements. Peterson, K.A., Puzzarini, C. 2005 *Theoretical Chemistry Accounts* 114 (4–5), pp. 283–296
- Quantum molecular dynamics: Propagating wavepackets and density operators using the multiconfiguration time-dependent Hartree method. Meyer, H.-D., Worth, G.A. 2003 *Theoretical Chemistry Accounts* 109 (5), pp. 251–267 188
- QM/MM: What have we learned, where are we, and where do we go from here? Lin, H., Truhlar, D.G. 2007 *Theoretical Chemistry Accounts* 117 (2), pp. 185–199
- Ab initio calculation of molecular chiroptical properties. Crawford, T.D. 2006 *Theoretical Chemistry Accounts* 115 (4), pp. 227–245

- Molecular potential-energy surfaces for chemical reaction dynamics. Collins, M.A. 2002 *Theoretical Chemistry Accounts* 108 (6), pp. 313–324
- Method of moments of coupled-cluster equations: A new formalism for designing accurate electronic structure methods for ground and excited states. Piecuch, P., Kowalski, K., Pimienta, I.S.O., Fan, P.-D., Lodriguito, M., McGuire, M.J., Kucharski, S.A., Kus, T., Musiał, M. 2004 *Theoretical Chemistry Accounts* 112 (5–6), pp. 349–393
- Investigation of the S0 → S1 excitation in bacteriorhodopsin with the ONIOM(MO:MM) hybrid method. Vreven, T., Morokuma, K. 2003 *Theoretical Chemistry Accounts* 109 (3), pp. 125–132
- Electron localization function for transition-metal compounds. Kohout, M., Wagner, F.R., Grin, Y. 2002 *Theoretical Chemistry Accounts* 108 (3), pp. 150–156
- Improper, blue-shifting hydrogen bond. Hobza, P., Havlas, Z. 2002 *Theoretical Chemistry Accounts* 108 (6), pp. 325–334
- The fundamental nature and role of the electrostatic potential in atoms and molecules. Politzer, P., Murray, J.S. 2002 *Theoretical Chemistry Accounts* 108 (3), pp. 134–142
- Similarities and differences in the structure of 3d-metal monocarbides and monoxides. Gutsev, G.L., Andrews, L., Bauschlicher Jr., C.W. 2003 *Theoretical Chemistry Accounts* 109 (6), pp. 298–308.

## 3 A glimpse of the future: the 50th anniversary issue

Each of our five guest editors has selected an important or emerging area of theoretical chemistry and has recruited experts to write about some key issues in that area. The five areas featured and their coordinators are:

- Electronic structure: present and future challenges (So Hirata)
- Relativistic quantum chemistry (Christoph van Wüllen)
- Reaction dynamics in gas phase and in solution (Hua Guo)
- Macromolecules and extended systems (Gino DiLabio)
- Spectroscopy, atmospheric chemistry, and thermochemistry (Fernando Ornellas)

We believe that the subjects of these featured articles encompass several areas that will be growing in importance over the next many years. We hope that you enjoy reading these contributions.

REGULAR ARTICLE

# Computational modeling of extended systems

Gino A. DiLabio

Received: 29 July 2011 / Accepted: 22 August 2011 / Published online: 13 January 2012
© Her Majesty the Queen in Right of Canada 2012

**Abstract** Advancements in computing architecture and in theoretical techniques allow for the modeling of complex, extended systems. This section of the 50th anniversary issue of *Theoretical Chemistry Accounts* highlights modeling work performed on nanostructured systems and underscores the enormous potential for synergy between theory and experiment in modern nanoscience.

**Keywords** Advances in computer architecture · Synergy between nanoscience theory and experiment · Molecular lines on silicon · Nanoparticle-polymer nanocomposites · Protein crystal structure modeling by density-functional theory

"What a wonderful time to be a computational chemist". One may speculate that that was the sentiment of the scientists that first published in *Theoretica Chimica Acta*, the progenitor of *Theoretical Chemistry Accounts*. In 1962, Fritz Grein (a new professor of theoretical chemistry at the University of New Brunswick in Canada)[1] published a paper in German, one of the four accepted languages for papers published in the journal, on the use of one-center wave functions for the study of symmetric $BH_4^-$, $CH_4$, and $NH_4^+$ in the 50-year-old issue of *TCA* [1]. Grein used a Royal Precision LGP-30 for his calculations, a machine that was referred to as a mobile "desk computer". These descriptors of the LGP-30 were indeed accurate: The 800-pound computer was the size of a desk and it had wheels so that it could be rolled from room to room [2]. Professor Grein recently recounted that the computer was "unbelievably slow by today's standards," but it was nevertheless a critical tool for his research. Computers have come a long way since 1962,[2] as have the theoretical methods we use for simulations. Cutting-edge modeling tools have been built to take advantage of new architectures, such as graphics processing units [3, 4][3], and are able to provide valuable information about large systems.

The central role of quantum chemical simulations in most areas of modern research has much to do with the remarkable growth in nanoscience over the last decade. There has been a convergence in the ability to perform experiments on nanosized objects (viz., atom-resolved microscopy, single-molecule experimental techniques, etc.) with advances in computing capability and in theory and simulation techniques. This convergence has resulted in unprecedented synergy between computational scientists and experimental physicists/chemists and engineers in

---

Published as part of the special collection of articles celebrating the 50th anniversary of Theoretical Chemistry Accounts/Theoretica Chimica Acta.

G. A. DiLabio (✉)
National Institute for Nanotechnology, 11421 Saskatchewan Drive, Edmonton, AB T6G 2M9, Canada
e-mail: Gino.DiLabio@nrc.ca

---

[1] Professor Grein is now a retired (but still active) professor of theoretical chemistry at the University of New Brunswick (Fredericton). He was invited to contribute to the first issue of *Theoretica Chimica Acta* by his PhD supervisor, Hermann Hartmann, who started the journal.

[2] For example, the computational infrastructure at the National Institute for Nanotechnology consists of 1140 conventional CPU-cores with 32 GPU nodes housed in four interconnected racks. Each rack has wheels (for mobility), and the entire system weighs about 11 times *more* than the LGP-30.

[3] The Theoretical and Computational Biophysics Group at the University of Illinois at Urbana-Champaign has developed a version of NAMD, a classical molecular dynamics package, to take good advantage of GPUs to accelerate certain part of the program.

efforts ranging from the creation of novel materials to the development of devices based on molecular electronics.[4]

The contributions to this section of the special issue illustrate the diversity of approaches to treating extended systems and their relevance to modern nanosystems. The work of Smeu, Wolkow and Guo [5] demonstrates the use of a non-equilibrium Green's function approach combined with density-functional theory (NEGF-DFT) to predict and visualize how current passes through molecular nanostructures that are formed in a self-directed fashion on semiconducting silicon surfaces. Nanosystems of this type may allow for the exploitation of the tunable properties of organic species with technologically ubiquitous silicon. Earlier experimental and computational work on similar nanostructures suggested they may be used as new models for molecular transistors [6] and to create self-assembled molecular "circuit" patterns [7]. The results of Smeu et al. provide insights into how substituents might be used to cause current to flow through the molecular assemblies rather than through the silicon substrate and thereby offers much needed guidance in the construction of novel, molecule-based electronic devices.

In the work of Rowen and Paci [8], DFT is used in combination with molecular dynamics simulations to predict how the optical properties of silver nanoparticles (NPs) embedded in polymer matrices depend on the size, shape and other physical properties of the particles. Nanoparticle-polymer composites hold much promise for creating novel materials that combine the tunability associated with NPs with the desired mechanical properties of polymers. Methodologies for controlling metal nanoparticle morphology are now well developed [9], allowing for control over their properties. Well-organized distributions of NPs a bulk polymers allows materials with carefully engineered and well understood properties to be assembled [10].

The contribution of Falklöf, Collyer and Reimers [11] discusses an important element in their group's effort to develop a first-principles DFT approach to protein X-ray crystal structure refinement. Recent studies have shown that there are problems with standard refinement procedures when they are applied to systems with unusual components (see the manuscript of Falklöf et al. for details). For this reason, the use of computational approaches that do not rely on classical mechanical force-field type parameterization are desirable, and the use of quantum mechanical methods in this connection has been shown to improve protein structure prediction [12].

In closing, this section of the 50th anniversary issue of *Theoretical Chemistry Accounts* celebrates the great advancements in the modeling of extended systems and showcases novel approaches to predicting the properties of nanoscale systems. The advances in theoretical approaches and in computing capability have given us the ability to engage in research never before thought possible, making this a truly wonderful time to be a computational chemist.

**Acknowledgments** I would like to thank Professor Don Truhlar for the kind invitation to take part in the occasion of *TCA*'s 50th anniversary and Manuel Smeu and Professors Jeffrey Reimers, Irina Paci, and Hong Guo for their contributions to this special event. The participation of several reviewers is also gratefully acknowledged.

# References

1. Grein F (1962) Die Anwendbarkeit yon Ein-Zentrum-Wellenfunktionen mit sphäriseher Symmetric bei $BH_4^-$, $CH_4$ und $NH_4^+$. Theor Chim Acta 1:52–65
2. http://www.computerhistory.org/revolution/early-computer-companies/5/116/1889. Accessed 22 July 2011
3. Ufimtsev IS, Martinez TJ (2008) Graphical processing units for quantum chemistry. Comput Sci Eng 10:26–34. See also later publications by the Martinez group
4. Stone JE, Phillips JC, Freddolino PL, Hardy DJ, Trabuco LG, Schulten K (2007) Accelerating molecular modeling applications with graphics processors. J Comput Chem 28:2618–2640
5. Smeu M, Wolkow RA, Guo H (2012) Conduction modulation of π-stacked ethylbenzene wires on Si(100) with substituent groups. Theor Chem Acc 131:1085
6. Piva PG, DiLabio GA, Pitters JL, Zikovsky J, Rezeq M, Dogel S, Hofer WA, Wolkow RA (2005) Field regulation of single-molecule conductivity by a charged surface atom. Nature 435: 658–661
7. Hossain MZ, Kato HS, Kawai M (2007) Selective chain reaction of acetone leading to the successive growth of mutually perpendicular molecular lines on the Si(100)-(2×1)-H surface. J Am Chem Soc 129:12304–12309
8. Rowan CK, Paci I (2012) Nanoparticle morphology and aspect ratio effects in Ag/PVDF nanocomposites. Theor Chem Acc 131:1078
9. Scaiano JC, Billone P, Gonzales CM, Maretti L, Marin ML, McGilvray KL, Yuan N (2009) Photochemical routes to silver and gold nanoparticles. Pure Appl Chem 81:635–647
10. Balazs AC, Emrick T, Russell TP (2006) Nanoparticle polymer composites: Where two small worlds meet. Science 314:1107–1110
11. Falklöf O, Collyer C, Reimers JR (2012) Towards ab initio refinement of protein X-ray crystal structures: interpreting and correlating structural fluctuations. Theor Chem Acc 131:1076
12. Ryde U, Nilsson K (2003) Quantum chemistry can locally improve protein crystal structures. J Am Chem Soc 125:14232–14233

---

[4] http://www.wtec.org/nano2/Nanotechnology_Research_Directions_to_2020/chapter01.pdf. Accessed 22 July 2011. This National Science Foundation sponsored report to the World Technology Evaluation Center (http://www.wtec.org/nano2) was published by Springer in 2010. It highlights the achievements made in nanoscience between 2000 and 2010 and describes the role that theory, modeling and simulation will play in the next decade.

REGULAR ARTICLE

# Nanoparticle morphology and aspect ratio effects in Ag/PVDF nanocomposites

Christopher K. Rowan · Irina Paci

Received: 13 June 2011 / Accepted: 1 August 2011 / Published online: 11 January 2012
© Springer-Verlag 2012

**Abstract** Optical response in silver/polyvinylidene fluoride nanocomposite materials with nonspherical inclusions was examined using direct dipolar interband transitions, from density functional theory. We discuss here the dependence of the optical response of the material on the geometry, crystallographic makeup and end-cap morphology of the metallic inclusions, as well as on their orientation relative to the polarization direction of the applied electromagnetic field. Each periodic unit cell contained a single inclusion and a polymer matrix; thus, the composite behaved as a monodisperse, perfectly oriented material. Overall, the spectral location of the composite excitation spectrum was tied to that of the metallic inclusions and correlated well to quantum confinement models for the direction of polarization: As linear size of the inclusion increased in a given direction, the excitation spectrum of light polarized in that direction was red-shifted. The effect of the polymer matrix was also examined. Coulomb repulsion from matrix energy states led to splitting of nanoparticle-based energy levels, and the matrix conduction band became involved in high-energy transitions. These effects led to extensions of the spectra of nanocomposites with less stable {100}–basal plane inclusions to very low excitation energies. Attenuation or redshifting of nanoparticle peaks with high photon energies was also observed for materials with small linear sizes along the excitation direction. Comparisons with experimental and time-dependent density functional theory results suggest that estimating the complex dielectric constant from interband transition dipole moments, in a time-independent fashion, provides reliable qualitative spectra for these systems.

**Keywords** Metal/polymer nanocomposites · Optical response · Density functional theory · Inclusion morphology · Birefrigent materials

## 1 Introduction

In metal-polymer nanocomposites (NCs), the versatility of polymers can be combined with the distinctive field-response properties of metal nanoparticles (NPs) to obtain interesting new materials with highly tunable characteristics. Even at small inclusion loadings, the absorption spectra of the material can be enhanced by many orders of magnitude when noble metal NPs are dispersed in a polymer matrix [1, 2]. This enhanced signal is due to NP electrons that collectively oscillate, known as localized surface plasmon resonances [3]. Absorption strengths and resonant frequencies are dependent on the choice of polymer and NP, particle size, shape, distribution, orientation and volume fraction (loading), and the polarization of the incident electric field. Thus, engineering materials for specific functions requires insight into the independent and correlated effects of all of these variables.

Ag NPs have electronic transitions in the visible range, and a variety of shapes and sizes have been fabricated to tune their absorption bands. Geometries have included various polyhedrals [4–6], spheres [7], disks [8], sheets [9], rods [8] and wires [10]. The distribution of NP sizes can be narrowed down to produce nearly monodisperse systems

Published as part of the special collection of articles celebrating the 50th anniversary of Theoretical Chemistry Accounts/Theoretica Chimica Acta.

C. K. Rowan · I. Paci (✉)
Department of Chemistry, University of Victoria,
PO Box 3065, Victoria, BC V8W 3V6, Canada
e-mail: ipaci@uvic.ca

through bottom-up reaction-driven approaches [11] or top-down lithographic methods [12]. NC structures are often limited to random distributions of NPs, except in 2-D lithographic arrays, but new approaches are being developed [13, 14].

Light absorption by nonspherical NPs is dependent on their orientation in the incident field. Metal NPs placed in an electrically insulating environment experience quantum confinement, and electronic absorptions occur at lower energies when polarized fields induce electronic resonances along longer NP axes [15]. When a majority of NPs are oriented in the same direction, the material response depends on the field polarization, resulting in dichroic or birefringent materials [16, 17]. The ability to control the orientation and alignment of NPs in a NC could facilitate applications as sensors [18], broadband waveguide polarizers [19], optical antennas [20] and optical fibers that employ surface enhanced Raman spectroscopy [21, 22]. Aligned and oriented Ag/polymer NCs have recently been reported with nanorods [23] and wires [14, 23].

Interfacial areas play an essential role in establishing the dielectric and optical properties of a NC. NP surfaces present various crystallographic facets, with different surface energies and atomic densities, which impact the nature of physisorption sites and generally the interactions between the inclusion and the matrix. Ag and other face-centered cubic crystals adopt primarily {111} facets, as these are the most energetically stable, followed by {100} and {110} facets [24, 25]. Experiments have shown that Ag nanodisks often have {111} basal planes [26–28]. Nanorods frequently have pentagonal cross-sections, with {100} sides and pyramidal end-caps with {111} facets [29–32]. Optical properties for these nanoparticles have been well studied experimentally [10, 33–37] and theoretically [38–40]. However, the development of NC materials built with these inclusions is still an emerging area of research.

We previously applied density functional theory (DFT) to the study of Ag/polyvinylidene fluoride (PVDF) to investigate the relationships between absorption spectra, particle size, loading and dispersion [41]. In ideal, monodisperse systems, we found that the optical properties of the NC were strongly influenced by those of the metallic inclusions. Analysis of density of states (DOS) diagrams showed that the inclusions introduced multiple orbitals and electrons in the $\approx 7$ eV polymer band gap. This led to the appearance of optical transitions in the NC where there were none in the polymer. The electronic population of the filled metallic orbitals led to a shift in the Fermi energy of the material and the involvement of the polymeric conduction band in some of the higher-energy transitions. Overall, the NP absorption peaks in the imaginary part of the dielectric constant $\epsilon_2$ absorption spectra were shifted slightly and broadened by the presence of the polymer.

Increased NP loading resulted in more occupied and available orbitals in the DOS, and thus in stronger optical response peaks. NP anisotropy led to a shift of absorption lines through quantum confinement: Polarization along the longer NP axes resulted in redshifting of the NC absorption peaks. Additionally, the absorption spectrum of a simple polydisperse system could be traced back to the individual monodisperse systems: Spectral energies and intensities of the polydisperse NC were correlated directly to interparticle spacing of NPs, rather than NP loading.

These strong relationships between NP and NC optical properties led to our interest in a systematic study of the optical behavior of NCs containing anisotropic inclusions. The intent was to perform a more in-depth examination of how material properties can be tuned by using NPs of different geometric characteristics. Thus, we describe here the optical behavior of a range of NCs, containing small Ag disks and rods of different axis ratios and crystallographic makeup, with the goal of elucidating the effect of these variables on $\epsilon_2$ in polarized electromagnetic fields. In addition to pyramidal end-caps on nanorods, nanobars and nanorice with blunt or hemispherical-like termini, respectively [42], were also considered. Comparisons with existing experimental and theoretical results are provided, where available, to probe the reliability of the time-independent method used here, for optical calculations.

## 2 Method

We considered the twenty crystalline disk- and rod-shaped Ag inclusions shown in Fig. 1. Nanodisks consisted of two atomic layers with varying numbers of atoms, and nanorods were composed of multiple stacked layers (between three and seven layers of up to seven atoms per layer were considered). For both shapes, {111} and {100} facets were investigated. Bilayer disks with larger diameters than those considered here have recently been reported experimentally [43]. The inclusions we considered were cut from the fcc crystal structure, rather than minimum-energy structures: Such structural minimizations lead to maximum-binding, near-spherical structures that would invalidate the geometrical analysis which constitutes the purpose of the present study.

To estimate the relative stability of the different geometries considered here, the binding energies of the NPs were calculated as the energy difference between bound and free Ag atoms:

$$E_b(Ag_n) = \frac{E(Ag_n)}{n} - E(Ag), \tag{1}$$

where $E_b$ is the binding energy per atom, $E(Ag_n)$ is the energy of $Ag_n$, and $n$ is the number of atoms in the cluster. Larger binding energies correspond to increased particle

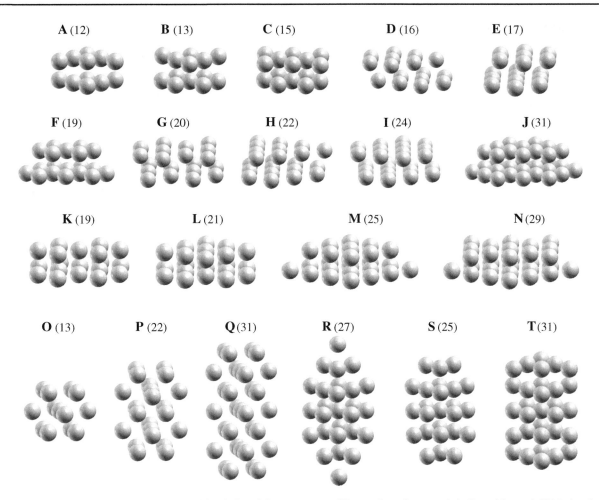

**Fig. 1** Disk- and rod-shaped Ag$_n$ inclusions. Particles A–J are bilayer disks with {111} basal planes. K–N are bilayer disks with {100} basal planes. O–Q are rods with layered {100} basal planes. R–T are rods with layered {111} basal planes and morphologically tailored end-caps. The number of atoms $n$ is indicated for each NP in brackets. NP $z$-axes run vertically and $x$-axes run horizontally, in the plane of the page. Pictures are slightly tilted to reveal the NP 3-D structure

stability. Nanoparticle dimensions, aspect ratios and binding energies are given in Table 1.

Throughout, axis or aspect ratios are given as the ratio between the NP dimension in the direction of the applied field (normal for disk-shaped NPs and longitudinal for rod-shaped NPs) and the NP dimension in the $x$ direction, which was also the alternate polarization direction considered in the study. NPs may or may not have rotational symmetry and may often have different diameters in the $y$ direction. However, polarization along this axis does not add any further information to the present discussion and was thus not considered.

From the table, and as confirmed by previous experimental observations [44], larger clusters were more stable. The presence of {111} basal planes also led to tighter binding in both disk- and rod-shaped NPs, a trend that was expected from crystal plane surface energies. Not unexpectedly, given their small sizes, all of the NPs considered here exhibited strong finite size effects: For comparison, the PBE/DZP-calculated binding energy for bulk Ag was −5.03 eV. The electronic structure of NPs (specifically, HOMO-LUMO gaps) has previously been correlated with the stability of the clusters, even away from ultrastable magic number clusters with complete electronic shells [45–49]. However, the dependence is not monotonic, and generally applicable relationships cannot be established.

Initial NC structures were compressed using classical molecular dynamics within the Gromacs [50] software package using periodic boundary conditions (PBC). Lennard–Jones parameters for structure optimizations were taken from the Universal Force Field [51]. Partial atomic charges were based on those of Chen and Shew [52]; silver atoms had zero charge. In order to provide consistency throughout, NP loading was held constant at 4.2 vol%. One implication of this was a variation of the interparticle separation between NP images in the different systems, which in turn affected absorption intensity and resonance

**Table 1** NP and NC geometric properties and binding energies

| NC | X | Y | Z | A | $E_b$ | $X_{sep}$ | $Y_{sep}$ | $Z_{sep}$ |
|---|---|---|---|---|---|---|---|---|
| A/470 | 8.7 | 9.6 | 5.3 | 1/1.65 | −3.16 | 6.3 | 8.0 | 13.4 |
| B/512 | 8.7 | 8.7 | 5.3 | 1/1.65 | −3.29 | 6.8 | 9.3 | 13.9 |
| C/590 | 8.7 | 8.7 | 5.3 | 1/1.65 | −3.34 | 7.5 | 10.2 | 14.8 |
| D/626 | 11.2 | 8.7 | 5.3 | 1/2.14 | −3.34 | 5.3 | 10.6 | 15.2 |
| E/668 | 8.7 | 11.6 | 5.3 | 1/1.66 | −3.45 | 8.2 | 8.2 | 15.7 |
| F/740 | 11.6 | 10.4 | 5.3 | 1/2.20 | −3.50 | 9.8 | 7.1 | 16.4 |
| G/782 | 11.1 | 8.5 | 5.3 | 1/2.10 | −3.46 | 6.7 | 12.2 | 16.9 |
| H/860 | 11.2 | 11.6 | 5.3 | 1/2.14 | −3.50 | 8.7 | 10.1 | 15.6 |
| I/938 | 11.2 | 11.6 | 5.3 | 1/2.14 | −3.65 | 10.0 | 11.5 | 16.4 |
| J/1214 | 14.5 | 12.9 | 5.3 | 1/2.75 | −3.68 | 7.9 | 11.3 | 18.2 |
| K/740 | 11.1 | 11.1 | 4.9 | 1/2.24 | −3.26 | 6.4 | 9.3 | 16.7 |
| L/818 | 11.0 | 11.0 | 4.9 | 1/2.24 | −3.42 | 7.0 | 10.1 | 17.5 |
| M/980 | 15.2 | 15.2 | 4.9 | 1/3.07 | −3.46 | 6.5 | 8.3 | 17.0 |
| N/1130 | 15.2 | 15.2 | 4.9 | 1/3.07 | −3.47 | 6.6 | 8.5 | 17.9 |
| O/512 | 8.7 | 8.7 | 7.0 | 1/1.24 | −3.25 | 8.7 | 8.7 | 7.0 |
| P/860 | 8.7 | 8.7 | 11.1 | 1.28 | −3.52 | 11.2 | 12.9 | 9.8 |
| Q/1214 | 8.7 | 8.7 | 15.2 | 2.17 | −3.63 | 13.7 | 15.6 | 8.3 |
| R/1052 | 8.7 | 9.6 | 17.0 | 1.97 | −3.71 | 12.6 | 14.4 | 5.3 |
| S/980 | 8.7 | 9.6 | 12.3 | 1.42 | −3.69 | 12.1 | 13.0 | 9.5 |
| T/1214 | 8.7 | 9.6 | 12.3 | 1.42 | −3.67 | 13.7 | 14.7 | 11.2 |

NP dimensions (Å), axis ratios, binding energies (eV) and edge-to-edge inclusion–inclusion distances across PBC images (Å) are included. Inclusion diameters (X, Y and Z) had an additional atomic diameter (2.9 Å) added to the largest internuclear distances. The same atomic diameter was subtracted from interparticle distances, such that, for example, $X + X_{sep} = X_{lattice}$, the simulation cell size along X. The aspect ratio (A) in the $xz$ plane is given as $z/x$

energies. Figure 2 shows examples of Gromacs-optimized unit cells for a disk- (panel a) and a rod-shaped inclusion (panel b). The construction of suitable simulation cells and computational details have been described in greater detail elsewhere [41].

The NP axes specified in Fig. 1 were oriented in line with the respective axes of the simulation cells. Specific NPs will be referred to hereafter as NP X, where X = A, B,…, T, as shown in the figure. NC mixtures will be referred to as, for example, A/470, where the letter indicates the inclusion size and geometry and the number indicates the size of the polymer chain.

DFT calculations were performed using SIESTA version 2.0.2 [53]. A PBE/DZP formalism [54] was used, with core electrons modeled using norm-conserving Troullier–Martins pseudopotentials [55] from the SIESTA website. The optical feature in SIESTA calculates $\epsilon_2$ from direct interband transitions [56, 57]. For periodic materials, allowed momentum–space transition dipole matrix elements are calculated between different eigenfunctions of the entire system Hamiltonian. The model, based on the damped, harmonic Lorentz oscillator, assumes the local field to be equal to the macroscopic field and is given by a sum of these transitions:

$$\epsilon_2(\omega) = 1 - \frac{e^2\hbar^2}{\epsilon_0 m_e^2 V} \sum_k \sum_{i,j} \frac{|\mu_{i,j}(\boldsymbol{k})|^2}{E_{ij}^2} \frac{f_0(E_j(\boldsymbol{k})) - f_0(E_i(\boldsymbol{k}))}{E_{ij}\hbar\omega - ih\Gamma}.$$

$$(2)$$

where $e$ is the electronic charge, $\hbar = h/2\pi$, $h$ is Planck's constant, $\epsilon_0$ is the permittivity of a vacuum, $m_e$ is the mass of the electron, and V is the volume of the cell. The sum ranges over all $\boldsymbol{k}$-points in reciprocal space with a double sum over all electronic energy values $i$ and $j$. $\mu_{i,j}(\boldsymbol{k})$ are transition dipole matrix elements, $f_0$ is the Fermi distribution function, $E_{ij} = E_i(\boldsymbol{k}) - E_j(\boldsymbol{k})$, $\omega$ is the angular frequency, and $\Gamma$ is a peak broadening term. All reported $\epsilon_2$ spectra use a broadening factor of 0.05 eV.

Further details of the optical SIESTA/DFT calculations, their advantages and limitations are given in Ref. [41] and references therein. Spin polarization was considered for several NCs containing NP O. The resulting spectra were virtually indistinguishable from spin-unpolarized results. The faster, spin-unpolarized calculations are reported here. DFT-based geometry optimizations, beyond the force field optimized structures, also had a minor effect on the resulting $\epsilon_2$ spectra. As most standard DFT methods, the PBE Hamiltonian underestimates band gaps [58–60]. Empirical scissor operators are often applied [61, 62] to compensate for this tendency. They were not used in this work, because of a lack of comparable experimental data. On the other hand, higher levels of theory were computationally intractable for these large systems. Therefore, spectra and relevant excitations will be discussed relative to one another, rather than as absolute values.

Polarized electric fields were directed along two symmetry axes of the NPs, as shown in Fig. 3. For nanorods,

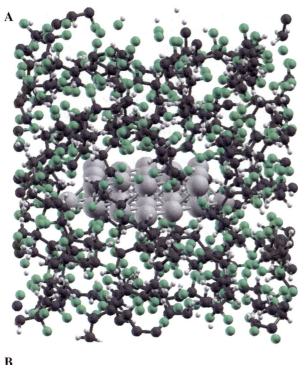

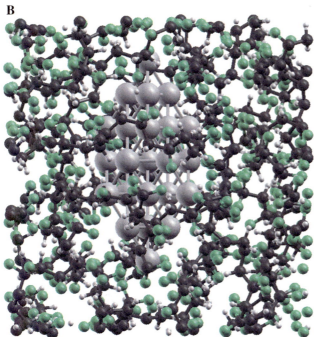

**Fig. 2** Simulation cells for NCs with disk- and rod-shaped inclusions: System N/1130 is shown in (**a**), and R/1052 in (**b**). *Large gray* atoms are Ag, C is *dark gray*, H is *white*, and F atoms are colored in *light green*

longitudinal fields were directed along the principal ($z$) axis, and transverse fields were along a perpendicular ($x$) axis. Normal mode polarization for nanodisks was along the normal vector of the disk, and the transverse mode was incident along the side, in the $x$ directions.

## 3 Results and discussion

The optical properties of the NCs were strongly determined by those of the inclusions. Due to their small size, NPs generally exhibited discrete electronic energy levels within a few eV of the Fermi energy, which led to strong, discrete peaks in their "interband" transition spectra (note that due to their small sizes, the NPs exhibit discrete energy levels rather than bands). These levels populated the large band gap of the polymer matrix in the NC. In its pure form, bulk PVDF exhibited no optical response in the UV–vis spectral region. However, in the NCs, polymer bands interacted slightly with NP orbitals, leading to some broadening of DOS peaks. Matrix electronic bands were also involved in the larger-energy transitions of the NC, as the Fermi energy of the material was repositioned due to NP electrons. In the following pages, the discussion of NC optical properties will often refer back to the respective NP spectra, due to their central role in determining the response of the material.

### 3.1 Nanodisk aspect ratio effects

Changes in the horizontal ($xy$) size of the nanodisk layers led simultaneously to changes in cluster size, axis (aspect) ratio and horizontal interparticle distance in the corresponding NCs. Spectra for normal ($z$) and transversal ($x$) excitations for NCs of three {111}-facet disks are presented in Fig. 4, along with DOS plots for the relevant NPs and NCs. As expected from confinement models, absorption for both the NPs and their NCs occurred at higher energies for excitations along the shorter normal direction (panel a), than along the transversal direction (panel b).

A pervasive feature in disk-based NC spectra was that the high-energy absorption peaks of the inclusions were shifted to lower energies and broadened in the NCs. This

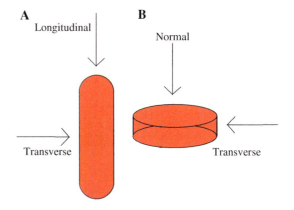

**Fig. 3** Polarization modes in nanorods and nanodisks. **a** Longitudinal and transverse modes in rods. **b** Normal and transverse modes in disks

was due to a combination of two effects: On the one hand, the conduction bands originating from the pure matrix became involved in electronic transitions at these higher photon energies, and on the other hand, high-energy NP virtual orbitals were broadened in the NC due to interaction with the matrix conduction band. Both effects can be seen, for example, by comparisons of the NC DOS plots with those of the respective NPs in Fig. 4c and d, respectively, in the 2–4 eV range.

Overall, NP and NC spectra for {111} clusters became broader as nanodisk size increased, regardless of polarization direction. NCs with 12–15 atom inclusions (NPs A–C) had few, discrete absorption peaks [see Fig. 4a(i)]. Increasing NP size caused first a coalescence of peaks [see Fig. 4a(ii)] and then for the larger inclusions considered here, an overall extension of the absorption energy range [see Fig. 4a(iii)]. These spectral changes generally arose through the broadening of wavefunction bands in the DOS plots with increasing particle sizes.

A redshift of the absorption spectrum with increasing inclusion size was observed for NCs exposed to a field polarized along the $x$ direction (Fig. 4b), in agreement with experiments [63] and discrete dipole approximation simulations [64] of NPs. Very low-energy peaks appeared in the spectra of some of the larger NPs and their NCs, where narrow energy bands formed in the vicinity of the Fermi energy and also near 1 eV. This was the case for NCs G/782 [subpanels (iii) in Fig. 4b–d], H/860, I/938 and, to a lesser extent, D/626. The spectral location of the absorption of $z$-polarized light was independent of the inclusion sizes (see, for example Fig. 4a), as disks A–J have the same thickness.

### 3.2 The effects of crystallographic facets

The crystallographic profile of NP basal planes impacted NC absorption spectra in a number of ways. Figure 5 shows $\epsilon_2$ for three {100} nanodisks (NPs K, L and M) and their NCs, from normal and transverse field polarizations,

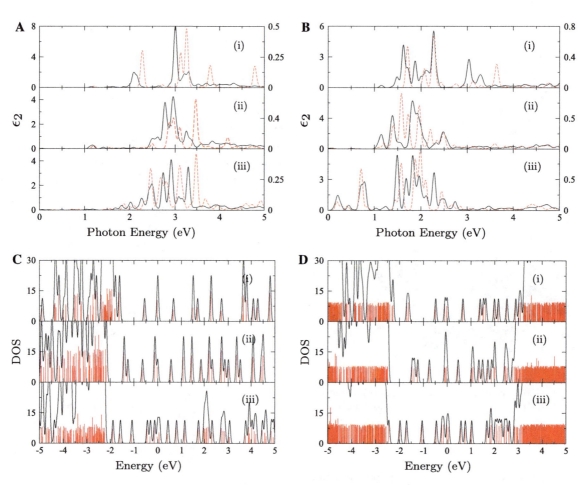

**Fig. 4** Optical response and DOS plots for NCs containing {111} basal nanodisk inclusions. **a** and **b** present optical response from normal and transverse polarization modes, respectively. *Solid black* is from NC, and *dotted red* is from NP in vacuum. $\epsilon_2$ scales are given on the *left-hand side* of the graph for the NC and on the *right-hand side* for the NP. **c** and **d** present the DOS of the NPs and NCs, respectively. Curves for $\Gamma = 0.05$ eV (*black lines*) and 0.001 eV (*red lines*) are shown. Subpanels (i)–(iii) present NP or NC systems A/470, F/740 and G/782, respectively

along with DOS plots for the NPs and NCs (panels a through d, respectively). Compared to the {111}–facet inclusions discussed above, the spectra of the less stable {100} disk NPs and their NCs were broader and also slightly blue-shifted in normal fields and red-shifted in transverse fields.

These effects are particularly visible by direct comparison of the electronic and optical properties of F/740 [Fig. 4(ii)—all panels] and K/740 [Fig. 5(i)—all panels]. The two NPs had 19 atoms each and the same axis ratios and differ in their basal plane geometries only. NPs with {100} facets exhibited less degeneracy in their electronic structure and broader bands in their DOS plots. These bands were further extended by interaction with the matrix, leading to $\epsilon_2$ spectra spread over large spectral regions, and with relatively broad features overall. Matrix-induced splitting in the region near the Fermi energy of disks K and L was stronger than in {111} disks. This led to the extension of spectral features to the low-energy (less than 1 eV) region of transversal $\epsilon_2$ spectra. The mentioned small blueshift in normal polarization was likely due entirely to quantum confinement: Disks based on the {100} facets of the fcc crystal were thinner than those based in {111} planes.

Spectral broadening in NCs based on {100} inclusions correlated well with the extent of polymer involvement in the optical response of the material: High-energy NC spectra for normal polarizations were broadened as the matrix conduction band participated in these transitions [for example, NCs K/740 and L/818 in Fig. 5a(i) and (ii)]. In fact, examination of the spectral behavior of NCs based on {111} and {100}-facet disks provided strong support to our suggested mechanism for matrix–inclusion interactions in these systems, through a contrasting case: that of NC M/980. This system exhibited, in normal polarizations, an $\epsilon_2$ spectrum very similar to that of the vacuum inclusion [see Fig. 5a(iii)]. A comparison of DOS plots for the NP and the NC, given in panels c(iii) and d(iii), respectively, reveals a peculiarity in the high-energy region. High degeneracy below the Fermi energy in NP M led to a low

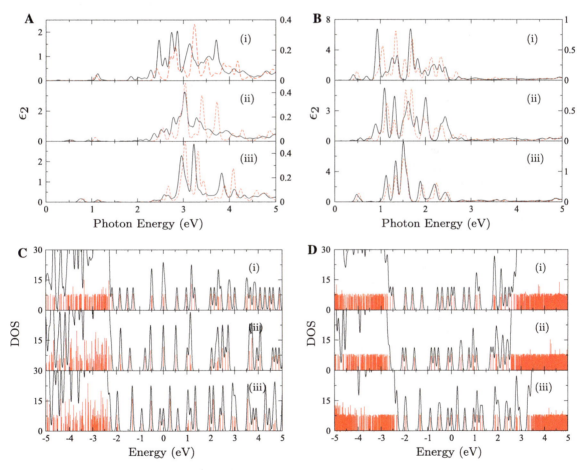

**Fig. 5** Optical response and DOS plots for NCs with disk-shaped inclusions from {100} basal planes. Optical response from normal and transverse polarization, as well as DOS plots for the NPs and the NCs, is shown in **a**–**d**, respectively. *Solid black* is from NC, and *dotted red* is from NP in vacuum. $\epsilon_2$ scales are given on the *left-hand side* of the graph for the NC and on the *right-hand side* for the NP. **c** and **d** present the DOS of the NPs and NCs, respectively. Curves for $\Gamma = 0.05$ eV (*black lines*) and 0.001 eV (*red lines*) are shown. Subpanels (i)–(iii) present results for K/740, L/818 and M/980, respectively

Fermi energy in the NC, and a relatively high energy of the matrix-based conduction band. Orbitals involved in transitions up to and over 3.5 eV in this system belonged largely to the NP, thus the similarity of the two spectra.

### 3.3 NCs with rod-like inclusions and the effect of NP elongation

Figure 6 presents $\epsilon_2$ spectra and DOS for NCs with rod-like inclusions based on several layers of {100} facets. A small set of rod-like inclusions was built by extending the stable magic number cuboctahedral NP O by adding {100} basal layers along the principal $z$ axis (resulting in NPs P and Q). These longer rods exhibited more structured DOS plots, as high degeneracy in the magic number NP was lost in the extended ones (see Fig. 6c). Further splitting of high DOS peaks occurred through interactions with the polymer matrix (Fig. 6d). These interactions were particularly evident for O/512, which exhibited strong NP–matrix interactions and significant broadening of its $\epsilon_2$ spectrum [e.g., Fig. 6a(i)].

Overall, changes in NP and NC optical response with elongation followed a quantum confinement model: A strong redshift was observed for longitudinal polarization response as the inclusion longitudinal size was extended (see Fig. 6a for example). Such redshifting and peak broadening in Ag nanorod spectra with increasing aspect ratio have been observed experimentally [10]. The extent of redshifting was consistent with recent TD-DFT studies for both cigar-shaped and pentagonal $Ag_n$ nanorods with $n$ = 12–120 [65]. Our calculated spectral shapes are comparable to those in Ref. [65], after application of a scissor operator of $\approx 1.5$ eV. Spectral broadening also occurred, as expected with increasing particle size and interaction strengths, and was particularly evident for longitudinal excitations. These excitations became more intense and more numerous with increasing nanorod length, whereas the transverse mode was largely unaffected (see Fig. 6), in agreement with experimental results for Ag nanorods [35]. The integral area under the longitudinal absorption curves from 0–5 eV (panel a) increased by $\approx 50\%$ between subsequent NPs (O, P and Q). In contrast, the number of transitions from transverse polarized fields (panel b) remained constant in the three NPs.

Matrix–inclusion interactions appeared stronger in rod-based NCs than in disk-based materials. Significant extension of the NP spectra in the low-energy region was evident in all rod-based NCs discussed here (see Figs. 6 and 7). The effect is rooted in significant matrix-induced broadening of low-energy states in the NPs (Fig. 6c–d). Only NP spectra with very low frequency response are relatively unperturbed. In those systems, the density of available states was already large, before interaction with the polymer matrix was introduced.

### 3.4 End-cap effects and {111}–facet rod-like inclusions

NP end-caps were altered to produce three different configurations: a trigonal pyramidal end-cap with {111} facets (NP R); hemispherical end-cap (NP S); and a blunt {111} plane (NP T). The central region of the rod was preserved in the three NPs. Absorption spectra and DOS plots for NPs R-T and their composites are presented in the Fig. 7. One striking feature is the strong intensity enhancement in the R/1052 NC longitudinal spectrum from the NP, and in relationship to the other NC spectra. The spectral location and width of NC response were relatively immune to changes in NP end-cap structures, though a few effects are observed in the figure. First, the single endpoint atoms in NP R were insufficient to produce redshifting in accordance with quantum confinement effects: The longitudinal spectrum of R/1052 resided in the same spectral region as the other NC spectra (see Fig. 7a). The R/1052 spectrum, however, was narrower than the others and lacked excitations over 1.5 eV.

A second observation relates to the matrix effect on the NP spectrum: The more stable (though smaller) NP S was most affected by matrix-induced splitting (see Fig.7a and b). The relatively narrow and structured NP spectra (both longitudinal and transversal) were broadened and redshifted in the NC. The effect arose through significant splitting and loss of degeneracy in the DOS plots, included in Fig. 7c and d.

Finally, examination of the longitudinal response of NC R/1052 reveals a narrow spectrum with high intensity lines. This was related to an increase in the conductivity of the material, as this NC exhibited the shortest inclusion–inclusion distance along the PBCs (about 5.3 Å). Similarly enhanced response was observed in the $x$-excitation spectrum of D/626, where interparticle distance along that axis was also 5.3 Å. However, when interparticle separation along the excitation direction increased to 6.3 Å (in NCs A/470 and K/740, for example), this strong spectral enhancement disappeared.

## 4 Conclusions

We evaluated the modulation of optical properties in NC systems, as changes in the inclusions sizes, shapes and crystallographic appearance were introduced, focusing on the participation of the two components in the complex response of the material. NCs with nanodisk and nanorod

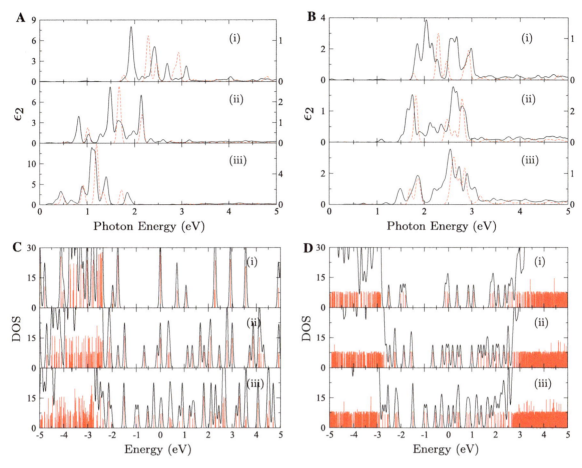

**Fig. 6** Influence of incident field polarization and nanorod aspect ratio on optical response and DOS for NCs with rod-like inclusions. **a–d** present optical response from longitudinal and transverse polarization modes, and NP and NC DOS, respectively. *Solid black* is from NC and *dotted red* is from NP in vacuum. $\epsilon_2$ scales are given on the *left-hand side* of the graph for the NC and on the *right-hand side* for the NP. **c** and **d** present the DOS of the NPs and NCs, respectively. Curves for $\Gamma = 0.05$ eV (*black lines*) and 0.001 eV (*red lines*) are shown. *Subpanels (i)–(iii)* show O/512, P/860 and Q/1214, respectively

inclusions exhibited dichroic optical response that makes these systems suitable for switching devices or sensors.

Most of the changes in absorption wavelengths observed in the $\epsilon_2$ spectra of these materials were captured in simple confinement models. These models suggest that a larger dimension of the metallic inclusion in the excitation direction corresponds to closer-spaced energy levels, lower transition energies and thus redshifts of spectra. This held true for the disks and rods examined here, regardless of crystallographic profile: Materials with the largest inclusions (about 15 Å along the polarization direction) exhibited absorption maxima between 1 and 1.5 eV, while those with narrow inclusions (around 5 Å along the polarization axis) had spectra centered around 3–3.5 eV. Note that a scissor operator would shift these values by around 1.5 eV and that approximations in the methodology and model selection, as discussed above, made these values estimative theoretical limits, rather than quantitative guides for experiments.

In view of these factors, a choice of rod-like or disk-like inclusions would be determined by the desired symmetry of the dichroic response and ease of fabrication. The inclusion's crystallographic projection affected the spacing of accessible states in the neighborhood of the Fermi energy for the NPs: The less stable {100}–facet NPs exhibited closer-spaced DOS in the Fermi region. Matrix-induced splitting of energy levels in this region had thus a higher impact on the NC excitation spectra, which were extended toward lower energies than the corresponding vacuum inclusion spectra.

The present study helped clarify the role of the matrix in establishing the optical properties of these materials. Generally, spectral location and features for the composite followed those of the NP. However, the matrix was found to impact spectra in two ways: through splitting of NP energy levels through Coulomb repulsion, and through the direct participation of the matrix conduction band in high-energy electronic transitions. Through the former effect, energy levels that were degenerate in the NP DOS were broadened, and generally, a more dense DOS was obtained for the NC. This led to broader NC spectra and the discussed {100}–facet extension into the low-energy region.

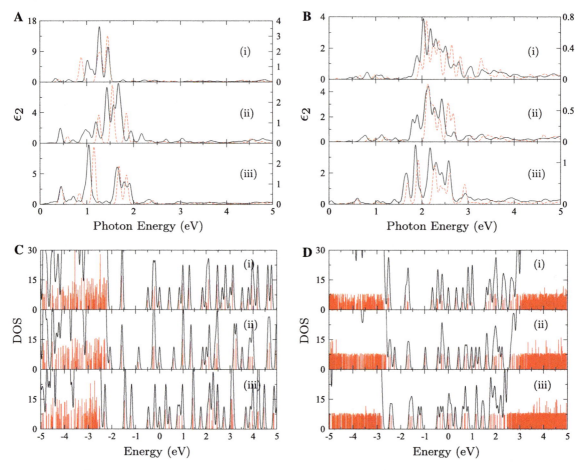

**Fig. 7** Influence of incident field polarization and nanorod end-cap structure on the optical response of Ag NPs and NCs. **a–d** present optical response from longitudinal and transverse polarization modes, and NP and NC DOS, respectively. *Solid black* is from NC and *dotted red* is from NP in vacuum. $\epsilon_2$ scales are given on the *left-hand side* of the graph for the NC and on the *right-hand side* for the NP. **c** and **d** present the DOS of the NPs and NCs, respectively. Curves for $\Gamma = 0.05$ eV (*black lines*) and 0.001 eV (*red lines*) are shown. *Subpanels (i)–(iii)* show different NC spectra as follows: R/1052, S/980 and T/1214, respectively

Repulsive splitting of energy levels was most prevalent in the high-energy region of the DOS. This effect, together with the involvement of matrix conduction bands in optical transitions, led to the obliteration of most NP peaks above 3.5 eV in addition to the redshifting and broadening of other high-energy peaks.

Sharp intensities of the imaginary dielectric constant for the systems with the smallest interparticle distances across the PBCs were related to a conductivity increase in these systems. However, conductors are not well treated with interband optical methods, and the reliability of such spectra is thus somewhat questionable. Nonetheless, for insulators, we found that the interband method implemented in SIESTA produced reasonable results, which correlated well with more computationally expensive time-dependent studies. Ultimately, however, the biggest limitation of either method is the accessible size of the simulation cell, which is too small to allow studies of experimentally feasible systems, from the points of view of inclusion size and distribution. Multiple scale approaches are being developed at present, which could also provide some flexibility in choosing methodologies and approximations according to the specific characteristics of individual systems.

**Acknowledgments** Funding was provided by the National Science and Engineering Research Council of Canada, the Canada Foundation for Innovation, the British Columbia Knowledge Development Fund and the University of Victoria. This research was performed in part using the WestGrid computing resources, which are funded in part by the Canada Foundation for Innovation, Alberta Innovation and Science, BC Advanced Education and the participating research institutions.

## References

1. Nicolais L, Carotenuto G (2005) Metal-polymer nanocomposites. Wiley, New Jersey
2. Caseri W (2000) Macromol Rapid Commun 21:705
3. Willets KA, Van Duyne RP (2007) Annu Rev Phys Chem 58:267

4. Tao A, Sinsermsuksakul P, Yang P (2006) Angew Chem 118:4713
5. Sun Y, Xia Y (2002) Science 298:2176
6. Jin R, Cao Y, Mirkin CA, Kelly KL, Schatz GC, Zheng JG (2001) Science 294:1901
7. Uznanski P, Bryszewska E (2010) J Mater Sci 45:1547
8. Sun Y, Yang B, Cai W, Guizhe Z, Zhang Q (2010) Micro Nano Lett 5:162
9. Deng Z, Mansuipur M, Muscat AJ (2007) J Phys Chem C 113:867
10. Murphy CJ, Jana NR (2002) Adv Mater 14:80
11. Park J, Joo J, Kwon SG, Jang Y, Hyeon T (2007) Angew Chem Int Ed 46:4630
12. Song Y, Elsayed-Ali HE (2010) Appl Surf Sci 256:5961
13. Zhang J, Coombs N, Kumacheva E (2002) J Am Chem Soc 124:14512
14. Xu J, Munari A, Dalton E, Matheson A, Razeeb KM (2009) J App Phys 106:124310
15. Godovsky DY (2000) Device applications of polymer-metal nanocomposites, advances in polymer science, vol 153. Springer, Berlin
16. Khlebstov NG (2010) J Nanophotonics 4:041587
17. Reyes-Esqueda JA, Torres-Torres C, Cheang-Wong JC, Crespo-Sosa A, Rodriguez-Fernandez L, Noguez C, Oliver A (2008) Opt Express 16:710
18. McFarland AD, Van Duyne RP (2003) Nano Lett 3:1057
19. Bloemer MJ, Haus JW (1996) J Lightwave Technol 14:1534
20. Smythe EJ, Dickey MD, Bao J, Whitesides GM, Capasso F (2009) Nano Lett 9:1132
21. Felidj N, Aubard J, Levi G, Krenn JR, Salerno M, Schider G, Lamprecht B, Leitner A, Aussenegg FR (2002) Phys Rev B 65:075419
22. Yang Y, Xiong L, Shi J, Nogami M (2006) Nanotechnology 17:2670
23. Bernabo M, Pucci A, Ramanitra HH, Ruggeri G (2010) Materials 3:1461
24. Yu P, Huang J, Tang J (2011) Nanoscale Res Lett 6:46
25. Muller-Plathe F (2002) Chem Phys Chem 3:754
26. Maillard M, Giorgio S, Pileni MP (2002) Adv Mater 14:1084
27. Tang B, An J, Zheng X, Xu S, Li D, Zhou J, Zhao J, Xu W (2008) J Phys Chem C 112:18361
28. Zhu YC, Geng WT (2008) J Phys Chem C 112:8545
29. Sun Y, Xia Y (2003) In: Cao G, Xia Y, Braun PV (eds.) Society of photo-optical instrumentation engineers (SPIE) conference series, vol 5224. pp 43–52
30. Kan CX, Zhu JJ, Zhu XG (2008) J Phys D Appl Phys 41:155304
31. Tang X, Tsuji M, Nishio M, Jiang P (2009) Bull Chem Soc Jpn 82:1304
32. Gao Y, Song L, Jiang P, Liu LF, Yan XQ, Zhou ZP, Liu DF, Wang JX, Yuan HJ, Zhang ZX, Zhao XW, Dou XY, Zhou WY, Wang G, Xie SS, Chen HY, Li JQ (2005) J Cryst Growth 276:606
33. Krenn JR, Schider G, Rechberger W, Lamprecht B, Leitner A, Weeber JC (2000) App Phys Lett 77:3379
34. Wilson O, Wilson GJ, Mulvaney P (2002) Adv Mater 14:1000
35. Evans PR, Hendren WR, Atkinson R, Pollard RJ (2008) Nanotechnology 19:465708
36. Zhao YP, Chaney SB, Shanmukh S, Dluhy RA (2006) J Phys Chem B 110:3153
37. Li D, Komarneni S (2010) J Nanosci Nanotechnol 10:8035
38. Jensen LL, Jensen L (2009) J Phys Chem C 113:15182
39. Cortie MB, Xu X, Ford MJ (2006) Phys Chem Chem Phys 8:3520
40. Johnson HE, Aikens CM (2009) J Phys Chem A 113:4445
41. Rowan CK, Paci I (2011) J Phys Chem C 115:8316
42. Wiley BJ, Chen Y, McLelland JM, Xiong Y, Li ZY, Ginger D, Xia Y (2007) Nano Lett 7:1032
43. Qin H, Gao Y, Teng J, Xu H, Wu K, Gao S (2010) Nano Lett 10:2961
44. Wang YL, Lai MY (2001) J Phys: Condens Matter 13:R589
45. Jackschath C, Rabin I, Schulze W (1992) Z Phys D At Mol Clusters 22:517
46. Fedrigo S, Harbich W, Belyaev J, Butter J (1993) Chem Phys Lett 211:166
47. Chiu YP, Wei CM, Chang CS (2008) Phys Rev B 78:115402
48. Kreibig U, Vollmer M (1995) Optical properties of metal clusters. Springer, Berlin
49. Wang J, Wang G, Zhao J (2003) Chem Phys Lett 380:716
50. Hess B, Kutzner C, van der Spoel D, Lindahl E (2008) J Chem Theory Comput 4:435
51. Rappe AK, Casewit CJ, Colwell KS, Goddard WA, Skiff WM (1992) J Am Chem Soc 114:10024
52. Chen Y, Shew CY (2003) J Mol Model 9:379
53. Soler JM, Artacho E, Gale JD, Garcia A, Junquera J, Ordejon P, Sanchez-Portal D (2002) J Phys Condens Matter 14:2745
54. Perdew JP, Burke K, Ernzerhof M (1996) Phys Rev Lett 77:3865
55. Troullier N, Martins JL (1991) Phys Rev B 43:1993
56. Nozieres P, Pines D (1959) Phys Rev 113:1254
57. Ehrenreich H, Cohen MH (1959) Phys Rev 115:786
58. Guzman D, Cruz M, Wang C (2008) Microelectr J 39:523
59. Martin RM (2004) Electronic structure: basic theory and practical methods. Cambridge University Press, Cambridge
60. Barone V, Hod O, Peralta JE, Scuseria GE (2011) Acc Chem Res 44:269
61. He Y, Zeng T (2010) J Phys Chem C 114:18023
62. Gonze X, Lee C (1997) Phys Rev B 55:10355
63. Germain V, Brioude A, Ingert D, Pileni MP (2005) J Chem Phys 122:124707
64. Brioude A, Pileni MP (2005) J Chem Phys B 109:23371
65. Liao MS, Bonifassi P, Leszczynski J, Ray PC, Huang MJ, Watts JD (2010) J Phys Chem A 114:12701

REGULAR ARTICLE

# Toward ab initio refinement of protein X-ray crystal structures: interpreting and correlating structural fluctuations

Olle Falklöf · Charles A. Collyer · Jeffrey R. Reimers

Received: 4 July 2011 / Accepted: 9 August 2011 / Published online: 11 January 2012
© Springer-Verlag 2012

**Abstract** The refinement of protein crystal structures currently involves the use of empirical restraints and force fields that are known to work well in many situations but nevertheless yield structural models with some features that are inconsistent with detailed chemical analysis and therefore warrant further improvement. Ab initio electronic structure computational methods have now advanced to the point at which they can deliver reliable results for macromolecules in realistic times using linear-scaling algorithms. The replacement of empirical force fields with ab initio methods in a final refinement stage could allow new structural features to be identified in complex structures, reduce errors and remove computational bias from structural models. In contrast to empirical approaches, ab initio refinements can only be performed on models that obey basic qualitative chemical rules, imposing constraints on the parameter space of existing refinements, and this in turn inhibits the inclusion of unlikely structural features. Here, we focus on methods for determining an appropriate ensemble of initial structural models for an ab initio X-ray refinement, modeling as an example the high-resolution single-crystal X-ray diffraction data reported for the structure of lysozyme (PDB entry "2VB1"). The AMBER force field is used in a Monte Carlo calculation to determine an ensemble of 8 structures that together embody all of the partial atomic occupancies noted in the original refinement, correlating these variations into a set of feasible chemical structures while simultaneously retaining consistency with the X-ray diffraction data. Subsequent analysis of these results strongly suggests that the occupancies in the empirically refined model are inconsistent with protein energetic considerations, thus depicting the 2VB1 structure as a deep-lying minimum in its optimized parameter space that actually embodies chemically unreasonable features. Indeed, density-functional theory calculations for one specific nitrate ion with an occupancy of 62% indicate that water replaces this ion 38% of the time, a result confirmed by subsequent crystallographic analysis. It is foreseeable that any subsequent ab initio refinement of the whole structure would need to locate a *globally* improved structure involving significant changes to 2VB1 which correct these identified *local* structural inconsistencies.

**Keywords** Monte Carlo · Density-functional theory · Protein refinement · Ensemble refinement · Lysozyme

Published as part of the special collection of articles celebrating the 50th anniversary of Theoretical Chemistry Accounts/Theoretica Chimica Acta.

O. Falklöf · J. R. Reimers (✉)
School of Chemistry, The University of Sydney,
Sydney, NSW 2006, Australia
e-mail: jeffrey.reimers@sydney.edu.au

O. Falklöf
Department of Chemistry, The University of Gothenburg,
Gothenburg, Sweden

*Present Address:*
O. Falklöf
Department of Physics, Chemistry and Biology,
Linköping University, 581 83 Linköping, Sweden

C. A. Collyer
School of Molecular Bioscience, The University of Sydney,
Sydney, NSW 2006, Australia

## 1 Introduction

X-ray crystallography is the most important experimental technique in protein structure determination. In order to understand biological functions and chemical mechanisms, the structures of macromolecules need to be obtained with high accuracy. In general, crystal structures of small molecules ($\sim 150$ atoms) determined at high resolution match closely with models derived from ab initio calculations due to the low complexity of their molecular structure, the overdetermination of the refinement process and the use of unbiased free atom refinement methods. In contrast, despite technological enhancements during the last 30 years, it is difficult to generate accurate atomic-resolution macromolecular structures. For macromolecular systems, the structures are often highly complex and usually the number of observations per atomic parameter is low and therefore empirical information must be included to enable a structure to be refined. Even in a handful of cases, where structures have been subjected to free atom refinement against high-resolution data, the structural models are inadequate in representing the unknown ensemble of conformers which compose the observed average structure of the molecule.

For X-ray refinement, an initial protein model structure is generated and then improved by matching it to observed reflection data while simultaneously maintaining geometrical restraints. Geometrical restraints have a large impact on the structure if the reflection data is poor but are normally weighted down in high-resolution refinements. A common approach is to use the geometrical parameters derived from accurate measurements of bond and angle parameters in small protein structures [1]. These standardized parameters are used in refinement programs such as SHELXL [2] and REFMAC5 [3] that are used widely. For a "normal" protein structure, these values are in general realistic, but for unusual structural features, such as cofactors, this methodology can yield inaccurate results. For example, we recently optimized the structure of photosystem-I by linear-scaling density-functional theory without reference to the original X-ray diffraction data [4]. This introduced only small changes to well-represented features such as main chain conformations but significant changes to the chlorophylls, cofactors, oligomerization features and group orientations. More generally, significant problems with even the determination of the backbone structure have been noted [5–12], and Eyal et al. [13] have shown that structures by the same authors in the PDB are more similar to each other than structures from independent groups.

Restrained refinement was pioneered by Konnert and Hendrickson [14–17], and the function minimized is the weighted sum of the difference of the observed and predicted intensities combined with the differences of the squared ideal and observed interatomic distances together with other types of geometrical restraints. Jack and Levitt [18] introduced a refinement approach where a sum of energy and X-ray terms was minimized. Brunger et al. [19]

further developed the methods and combined the X-ray data with a potential from molecular dynamics.

Due to the computational time and problematic scaling properties, high-level density functional calculations have typically not been applicable to protein structure refinement. During the last few years, a number of linear-scaling semi-empirical and density-functional theory (DFT) algorithms have been developed [4, 20–39] (for recent reviews, see e.g., [40–44]). The development of modern density functionals capable of describing dispersion interactions [45–61] allows such methods to properly describe all of the chemical bonding types associated with protein structure. These methods are commonly used to enhance descriptions of protein active sites but have also been used to optimize whole protein structures, but without reference to the original raw diffraction intensities used to generate them [4, 23–25, 27–29, 33]. Recently, dispersion-corrected DFT optimization led to the revision of a protein X-ray structure, reducing the $R$ factor to increase the quality of the interpretation of the diffraction data [62].

For many reasons, it is preferable to employ such methods *within* the refinement stage, however. Pioneering studies employing quantum chemical methods focused on improving low-resolution structures. Ryde et al. [63–66] have developed a method that combined quantum mechanics and molecular mechanics, where only a part of the structure used restraints from quantum mechanical calculations. Also, Merz et al. [67–73] and Stewart [23, 33, 43] made use of another approach using a semi-empirical method for the energy restraints. These approaches have yielded more chemically reasonable aspects of critical structural features than was previously obtained using traditional refinement methods, often also increasing agreement between observed and calculated diffraction intensities.

A major challenge at this time is the demonstration that quantum chemical methods can be used to enhance the refinement of *all* aspects of, in particular, high-resolution X-ray structures. These are the structures that have the most significant features resolved. The advantage that ab initio methods have to offer protein crystallography is that they know most about the highest-resolution features of the protein, precisely the area in which little information is directly available from current experiments. Complementary to this, ab initio calculations know least about the overall protein conformation and other lower-resolution features that are readily determined using crystallography. There is, however, an intermediate regime concerning torsional motions and short-range intermolecular forces in which useful information could be expected from both approaches, and indeed it is important to verify that any used computation method can accurately reproduce such features. All computational methods will suffer from some systematic failures, and perhaps now the time has come where ab initio methods can be utilized to simply add information to crystallographic refinements.

Here, we investigate one of the most basic issues facing the melding of ab initio computations of protein structure with X-ray structure refinement: the presence of multiple conformations in protein structures, a feature that often leads to poor X-ray structural models [7]. A multiple-copy refinement scheme (ensemble refinement) [74–80] is developed, which treats multiple conformations as an ensemble of fully connected structures, each of which obeys basic chemical rules. Modern X-ray structures often detect many atoms with multiple sites. Major correlations between the locations of such atoms, especially those directly related to function, are often noted in structural models, but systematic approaches to determining such correlations are rare, and PDB files are rarely constructed so as to make correlations explicit. As a result, most correlations go unnoticed, yet such features are critical to any ab initio refinement of crystal structures. Ensemble refinement can lead to significant improvements of observed and fitted diffraction intensities [75] and are currently of great interest as a means of improving structural models [11, 81–91], including refinement of NMR data [92].

The structure of lysozyme is used as a sample system to consider the issue of correlated structural fluctuations as lysozyme is one of the most studied proteins. It was discovered by Fleming [93], and its three-dimensional structure was first solved by Blake et al. [94], providing the first enzyme structure to be solved by X-ray crystallography. The number of lysozyme structures contained in the Protein Data Bank (PDB) [95] is large. Even so, the exact catalytic mechanism of lysozyme is still a subject of debate [96, 97], indicating that further structural characterization and computational analysis may be required. From an experimental point of view, lysozyme is, in comparison with other proteins, easy to purify and crystallize, and crystals diffract at high resolution. Moreover, lysozyme crystallizes in different polymorphic space groups depending on the crystallization conditions [98]. We consider only the refined "2VB1" structure from the PDB of triclinic hen egg white lysozyme (HEWL) by Wang et al. [98]. This is currently the third-highest-resolution protein structure deposited in the PDB and is at a resolution of 0.65 Å. This structure was refined in SHELXL by *removing* the restraints for the well-ordered parts of the model (those in a single conformation) and represents a close approximation to the structures resulting from free atom refinements conducted in small-molecule X-ray crystallography.

In Sect. 2, we review features of standard protein X-ray structure refinement that are critical to the development of enhanced methods. These methods focus on providing the best-possible representation of the raw X-ray diffraction data in terms of atomic coordinates, given options for the application of non-experimental features such as constraints and empirical force fields. Our aim is to replace these empirical conditions with ab initio ones, but in order to commence such an operation, many significant issues involving lack of experimental knowledge of required information must first be addressed. Section 3 describes critical data that is either present or absent from the 2VB1 X-ray structure of lysozyme [98]. Section 4 describes a Monte Carlo technique for determining an ensemble of chemically realistic structures that represent the multiple conformations detected in the original refinement. This involves adding additional constraints to the refinement that enforce basic chemical rules, but otherwise this Monte Carlo step uses the AMBER force field to distinguish between a vast number of chemically realistic structures that would each generate the *same R* factor during X-ray refinement. Section 5, however, considers basic chemical features that appear *inconsistent* with the original refinement, indicating that the construction of a chemically realistic model that is also in good agreement with the raw diffraction data is a considerable challenge.

## 2 Basic relevant features of protein X-ray diffraction analyses

The raw data collected during X-ray diffraction experiments consist of diffraction intensities $I(h, k, l)$ and associated structure-factor amplitudes $F_{obs}(h, k, l)$, with

$$I(h, k, l) \propto F_{obs}^2(h, k, l) \tag{1}$$

where $h$, $k$ and $l$ index the observed reflections. The electron density within the crystal at point $(x, y, z)$ may be generated from a Fourier transform of the structure factors,

$$\rho_{obs}(x, y, z) = \frac{1}{V} \sum_h \sum_k \sum_l F_{obs}(h, k, l)$$
$$\times \exp(-2\pi i(hx + ky + lz) + i\alpha_{obs}(h, k, l))), \tag{2}$$

where $V$ is the volume of the unit cell, if the phases $\alpha_{obs}(h, k, l)$ are known [99]. Phases and structure factors may both be readily determined from an atomic model of the crystal structure, allowing for the construction of the comparable electron density

$$\rho_{calc}(x, y, z) = \frac{1}{V} \sum_h \sum_k \sum_l F_{calc}(h, k, l)$$
$$\times \exp(-2\pi i(hx + ky + lz) + i\alpha_{calc}(h, k, l))). \tag{3}$$

However, phases are not directly determinable from the diffraction data and so estimated phases $\alpha_{\mathrm{obs}} \approx \alpha_{\mathrm{calc}}$ are utilized in Eq. 2. The aim of crystallographic refinement is to determine a reliable atomic coordinate model and therefore reliable estimates of the phases. This scenario is readily achievable for small-molecule crystallography as realistic initial phases are derived using direct methods, but this approach cannot usually be applied in the case of protein crystallography [99]. While femtosecond nano-crystallography [100] offers in the next decade a possible experimental solution to this problem, the analysis of protein X-ray diffraction data currently usually proceeds through a number of stages, with at each stage the phases deduced from the atomic coordinate model; as better phases are obtained in each cycle, the "observed" electron density $\rho_{\mathrm{obs}}\,(x, y, z)$ reveals more of the atomic structure, leading then to better phases. Use of ab initio calculations in aiding protein structure refinement should be conceived as simply adding another stage to the existing analysis sequence, striving again to improve accuracy and provide a better interpretation of the raw diffraction data.

The final steps of the structure determination of a protein involve the refinement of the structure. In a structure refinement, the model structure is refined against the experimental diffraction data to improve the $R$ factor defined as

$$R = \frac{\sum_{hkl} \left| |F_{\mathrm{obs}}| - \tilde{k}|F_{\mathrm{calc}}| \right|}{\sum_{hkl} |F_{\mathrm{obs}}|}, \tag{4}$$

where $\tilde{k}$ is a scale factor. Note that this involves comparing functions of purely observed quantities $F_{\mathrm{obs}}$ to those of purely calculated ones $F_{\mathrm{calc}}$; the precise nature of the function that is actually minimized depends on the software used, with the maximum likelihood method [3] being a common choice.

Density difference maps such as

$$
\begin{aligned}
\Delta\rho(x, y, z) &= \rho_{\mathrm{obs}}(x, y, z) - \rho_{\mathrm{calc}}(x, y, z) \\
&\approx \text{``}F_O - F_C\text{''} = \frac{1}{V} \sum_{hkl} (|F_{\mathrm{obs}}| - |F_{\mathrm{calc}}|) \\
&\quad \times \exp[-2\pi i(hx + ky + lz) + i\alpha_{\mathrm{calc}}]
\end{aligned}
\tag{5}
$$

are usually created in order to enhance refinement but because of the use of $\alpha_{\mathrm{calc}}$ in approximating $\rho_{\mathrm{obs}}$, their quality varies spatially within the structure and so they can be difficult to interpret. Of particular concern for the generation of a structural model suitable for ab initio refinement is the feature that typically $\alpha_{\mathrm{calc}}$ has a more significant influence on $F_O - F_C$ than does $|F_{\mathrm{obs}}| - |F_{\mathrm{calc}}|$, sometimes allowing the magnitude of $F_O - F_C$ to decrease while $R$ increases [1]. Note too that a commonly reported

type is also the combined map with an observed density added to the difference map [99]

$$
\begin{aligned}
\text{``}2F_O - F_C\text{''} &= \frac{1}{V} \sum_{hkl} (2|F_{\mathrm{obs}}| - |F_{\mathrm{calc}}|) \exp[-2\pi i(hx + ky \\
&\quad + lz) + i\alpha_{\mathrm{calc}}].
\end{aligned}
\tag{6}
$$

The electron density around an atomic nucleus is typically temperature independent, though large-amplitude motions of atoms with frequencies less than $kT/\hbar$ can blur the density. If all of the equivalent molecules in a protein crystal do not adopt the same conformation, then the electron density will be distributed, possibly mimicking the effects of thermal motion. The effects of large thermal motions are not usually included explicitly via ensemble representations in protein structure refinements, though the presence of multiple conformers can sometimes be detected and explicit structures modeled for each conformer [99]. Explicitly identified conformers are ascribed weights indicating the fraction of molecules adopting each particular conformer. Thermal effects and non-explicitly represented conformational effects are usually treated in protein structure refinement implicitly by smearing out the atomic electron density using

$$T_{\mathrm{iso}} = \exp\left(-B\frac{\sin^2\theta}{\lambda^2}\right) = \exp\left(-\frac{B}{4}\left(\frac{2\sin\theta}{\lambda}\right)^2\right), \tag{7}$$

where $B$ is an "atomic displacement parameter" that reflects isotropic thermal motion with mean square displacement, $\overline{u^2}$,

$$B = 8\pi^2\overline{u^2}. \tag{8}$$

If high-resolution experimental data are available, then this isotropic approximation may be inadequate so that an alternate anisotropic temperature factor

$$
\begin{aligned}
T_{\mathrm{aniso}}(h, k, l) &= \exp[-2\pi^2(U_{11}h^2a^{*2} + U_{22}k^2b^{*2} + U_{33}l^2c^{*2} \\
&\quad + 2U_{12}hka^*b^* + 2U_{13}hla^*c^* \\
&\quad + 2U_{23}klb^*c^*)]
\end{aligned}
\tag{9}
$$

is often used, where $a^*$, $b^*$ and $c^*$ are the lengths of the axes in the reciprocal space. Here, $U_{ij}$ is the element in the $i$th row and $j$th column of the displacement tensor with respect to the axes in reciprocal space.

Every observation is associated with some level of noise. Since the aim is to model real features and not noise through over-fitting, the experimental data are usually divided into a working and a test set. Only reflections from the working set are used in the refinement. The indicator of the quality of the model is the $R_{\mathrm{free}}$-factor, which is an $R$ factor calculated considering only the reflections from the test set that was not used during the refinement [101].

Hydrogen atoms give a very weak signal in the diffraction pattern. Usually, only non-hydrogen atoms are modeled in X-ray structures. If the hydrogen atoms are included, they are commonly included as riding groups and follow the movement of their neighboring heavy atom. From a chemical point of view, hydrogen atoms are crucial for the understanding of the stability and mechanisms of the structures and must be considered carefully [99]. Any ab initio calculation requires an accurate description of the locations of the hydrogen atoms.

The observed number of reflections per fitting parameter is usually low, and to enhance refinement, extra-geometrical conditions are usually included. Jack and Levitt proposed a method that takes an energy term into account,

$$Q = (1 - w_x) \cdot E + w_x \cdot \sum_{hkl} w(h, k, l)(|F_{obs}(h, k, l)| - |F_{calc}(h, k, l)|)^2 \qquad (10)$$

where the function involving an atomic energy contribution $E$ was minimized, and $w_x \in [0, 1]$ controls the relative contributions of the energy and the X-ray term [18]. It is this function $E$ that could be replaced with an ab initio estimate rather than its estimation using empirical force fields.

Large volumes in the crystal may be occupied primarily by solvent molecules, however; and in these regions, it is typically not adequate to use atomic models to describe the diffraction as the degree of disorder is typically too large. Such regions are typically modeled more realistically as a homogeneous electron gas. It is, however, critical that the best-possible description of these regions be obtained as diffraction contributions from them add to the structure factors and influence the determination of phases. A commonly used correction based on Babinet's principle [102] is

$$F_{calc} = \left[ 1 - K \exp \left[ -B_{solvent} \frac{\sin^2 \theta}{\lambda^2} \right] \right] \cdot F_{protein}. \qquad (11)$$

## 3 Hen egg white lysozyme 2VB1 structure properties

The observed [98] crystal structure of HEWL is triclinic with just one molecule in the asymmetric unit. The unit-cell parameters are $|a| = 27.07$ Å, $|b| = 31.25$ Å, $|c| = 33.76$ Å, $\alpha = 87.98°$, $\beta = 108.00°$ and $\gamma = 112.11°$. Only one protein chain, containing 129 amino acids, is present per cell. Other identified molecules at least partially included in the structure are one acetate ion, nine nitrate ions, three ethylene glycol molecules and 170 water molecules. The structure 2VB1 contains coordinates for all hydrogen atoms except those in the water molecules. Nine atoms are missing in the B conformation of TYR-20. Based on our analysis, 17.5% of the volume of the unit cell is not explicitly represented using atoms, enough volume to accommodate 146 additional water molecules. However, the identified regions of the unit cell are also positively charged and so at least some counter anions are missing, and it is impossible that some of the identified areas in hydrophobic regions contain no atoms at all.

Most significantly, 33% of the atoms are identified with multiple sites, reflecting observed conformational differences between molecules in different unit cells within the crystal. At most three configurations were identified for any one atom, named "A", "B" and "C"; conformation "A" of one atom is not necessarily correlated with conformation "A" of another atom, however. Nevertheless, the set of atoms from LEU-17 to LEU-25 have been identified as giving rise to two distinct conformations of the protein chain, depicting a correlated structural fluctuation. As such, chemical rules indicate that the occupancy of each atom in the set in each configuration must be the same, but despite this, the reported structure optimizes individual weights for *each* atom in *each* conformation. Such a feature is not allowed if ab initio computations are to be used to refine the structure. In other cases, for example, for the anions and ethylene glycols, the same occupancy is specified for all atoms in a chemical group. However, no correlations were presented between the occupancies of the vast majority of atoms with multiple identified sites. As ab initio calculations manipulate explicit knowledge of correlations between atoms, the 2VB1 structure is far from one that is suitable for such a treatment. Yet it is only because 2VB1 is obtained at such high resolution that the very presence of multiple conformations is identified, and so only for structures like this can an ab initio calculation be conceived. As an indication of the severity of the problem faced by an ab initio calculation, only 25% of the residues of the protein are surrounded by a 5-Å region in which the atomic structure is unambiguously defined in 2VB1.

The structure 2VB1 was also refined using anisotropic displacement parameters. This process is desirable in that it improves phases and allows the structure to be indentified in more regions of the crystal. However, these enhancements arise at the cost of increasing the number of parameters per heavy atom in a single conformer from 4 to 9. Just as a chemically meaningful structure suitable for ab initio refinement requires consistent values for the occupancies of each atom in a fluctuating chain conformation, so it also requires the anisotropy values of atoms need to be correlated with each other. Such correlations are *not* built into the 2VB1 analysis, however, and as a result, many parameters are introduced into the anisotropic refinement that have no physical basis but significantly improve accuracy measures such as $R$: for lysozyme, $R$ decreases from 19.48 for an isotropic analysis to 8.40 for an anisotropic one, but it is not clear what fraction of this decrease can be attributed to the inclusion of real chemical effects and how much can be attributed to over-parameterization. Indeed, 20025 free parameters were fitted to 187165 observed reflections, locating just 1814 non-hydrogen atoms, or 11 parameters per heavy atom on average. It is most likely that ab initio refinement will preclude the use of unconstrained anisotropic temperature factors, however.

## 4 A Monte Carlo scheme to determine a small set of chemically reasonable structures that represent the original 2VB1 X-ray model

In an ab initio calculation, all atoms must be explicitly represented. Variations in conformers within the protein crystal can be accounted for by performing calculations on an ensemble of possible protein structures, averaging the energies and forces. To represent truly disordered regions such as the 17.5% of the protein structure not so far explicitly represented, a large number of configurations could be necessary but the use of a small number of structures may be feasible [103, 104]. While this task is a central feature in all molecular dynamics simulations of protein function performed outside the scope of X-ray structural refinement, we focus here on other, more basic, ambiguous features of the data analysis and choose to ignore this region completely. In a subsequent improved implicit approach, this region could be represented by a dielectric material [9, 105–107] or by explicit molecules [108, 109].

We focus on the atoms in the 2VB1 structure that are attributed partial or multiple occupancies, and an experimental error bar of $\pm 10\%$ in these occupancies is stated [98]. Given this and the triclinic symmetry of the crystal, the simplest possibly realistic description of conformation variation of the represented atoms would include an ensemble of 8 chemically complete structures, allowing occupancies to be represented to an accuracy of $\pm 6.25\%$. Converting all of the stated occupancies to their nearest multiple of one-eighth increases the $R$ factor from 19.48 to 19.52% for an isotropic analysis and from 8.40 to 8.51% for an anisotropic analysis, as determined using REFMAC5 [3]; these increases are much less than what could be considered as physically meaningful. However, changing these occupancies to enforce the chemical restriction that all atoms in a particular conformation have the same weight has a larger effect, increasing the $R$ factors to 19.60 and 8.88%, respectively. The quality of some subsequent ab initio optimization should be compared to these modified $R$ factors rather than the originals as these increased values reflect mostly the chemically realistic requirement that occupancies are assigned to conserved chemical groups as whole entities.

An ensemble of 8 chemically feasible structures is then produced that maintains the generated atomic-site occupancies (i.e., this procedure essentially does not alter the $R$ factor). This ensemble is represented in two different ways. As presented to the X-ray refinement codes, 8 all-atom structures are defined in which every atom has precisely an occupancy of one-eighth. As presented to a subsequently described energy analysis program, the coordinates are obtained expanding the original unit cell into a $2 \times 2 \times 2$ superlattice. Each site of the superlattice is assigned one of the 8 structures from the ensemble, the entire structure satisfying the new boundary conditions. It is possible to transform between these two representations without difficulty, and using an explicit superlattice representation during X-ray refinement would be inconvenient as the unit-cell parameters, and hence all diffraction indices, would require modification. In either case, refinement of the expanded model using standard means would not be possible as the number of atomic coordinates required to be fitted is increased eightfold, making such an analysis overparameterized. However, if ab initio forces are included during refinement and a chemically based scheme is used to specify ab initio the thermal parameters, then a solution should be possible as our model includes 10,871 heavy atoms in the $2 \times 2 \times 2$ superlattice and 187,165 reflections are observed. It is also feasible to constrain the coordinates of some atomic fragments in the 8 cells to be equivalent, thus reducing overparameterization for chemical units dominated by only one significant structure.

Here, we focus not on such a refinement but rather on the required task of generating a starting structure for the $2 \times 2 \times 2$ superlattice. Fluctuations of conformations of nearby atoms are likely to be highly correlated (e.g., the presence or absence of a nearby nitrate ion can determine the conformation of a cationic residue), yet such correlations are not identified in the existing crystal structure [98]. A realistic configuration of the $2 \times 2 \times 2$ superlattice must take these correlations into account. We consider a vast number of possibilities, rating them using a molecular mechanics energy function. All of the structures considered in this section essentially share the same $R$ factors (there are trivial variations between them caused by differing hydrogen locations that have no quantitative effect) and hence the energy function is being used purely to add *additional* information required for a subsequent refinement process.

As previously mentioned, all atoms in the identified [98] nine-residue group LEU-17 to LEU-25 that form a loop region with two distinct conformations were given the same weights and treated as *one* chemical unit. Similarly, the residues [98] ASP-87 to ALA-90, LYS-97 to VAL-99, and ASN-103 and GLY-104 were treated each as single correlated chemical units. All other single residues were identified as chemical units and assigned the same weight in each configuration. A single chemically realistic structure was added for the nine missing atoms in the B conformation of TYR-20, with subsequent structural refinement being assumed to be sufficient to describe the required conformational variations of these atoms.

To determine all other correlations, Monte Carlo simulations were conducted using specifically developed software. If a chemical unit has a weight of say 5/8 for one conformation and 3/8 for a second, then 5 copies of the first configuration and three of the second are added in random order to occupy the 8 sections of the $2 \times 2 \times 2$ superlattice. Alternatively, if say a water molecule had an occupancy of 5/8, then water molecules were added to 5 of the 8 sections with the other 3 sections remaining vacant. The purpose of the Monte Carlo program is to determine the optimum distribution for the included chemical groups among the 8 cells of the $2 \times 2 \times 2$ superlattice, establishing correlations between the occupancies of different chemical sites. Note that the full crystallographic symmetry properties of the superlattice are used in all calculations.

The AMBER force field [110, 111] was used to drive the Monte Carlo simulations. This is made up of a local part (bond stretches, bends, torsions) and a long-range part (dispersion and electrostatic interactions). If the correlation between fluctuations of two neighboring residues needs to be determined, then bond bending and torsional motions will differ between conformations. This effect is small, however, and is also a rare occurrence, and hence we consider only the long-range part of the AMBER force field,

$$E^{\mathrm{LR}} = \sum_{i<j} \left[ \frac{A_{ij}}{R_{ij}^{12}} - \frac{B_{ij}}{R_{ij}^{6}} + \frac{q_i q_j}{\varepsilon R_{ij}} \right]. \tag{12}$$

The dielectric constant, $\varepsilon$, was set to unity and $A_{ij} = e_{ij}^*(R_{ij}^*)^{12}$ and $B_{ij} = 2e_{ij}^*(R_{ij}^*)^6$, where $R_{ij}^*$ and $e_{ij}^*$ are the equilibrium bond length and bond energy, respectively, of the Lennard–Jones potential between atoms $i$ and $j$. For interacting atoms of different types, the Lennard–Jones bond length was taken as a sum of the van der Waals radius of each atom, $R_{ij}^* = R_i^* + R_j^*$, while the well depth was taken as the geometric average of the well depths for each atoms, $e_{ij}^* = \sqrt{e_i^* e_j^*}$. The Lennard–Jones parameters and atomic charges of the residue atoms were taken from the study of Cornell et al. [111]. For acetate, nitrate and ethylene glycol, charges were calculated, after optimization, at the B3LYP/6-31G(d) level of theory [112, 113] by GAUSSIAN 09 [114] including the "prop = (fitcharge, dipole)" and "scrf = cosmo" options; the resulting charges are shown in Fig. 1. Use of a unit dielectric constant is appropriate if all atoms in the crystal are explicitly represented and properly sampled, and its use herein is based on

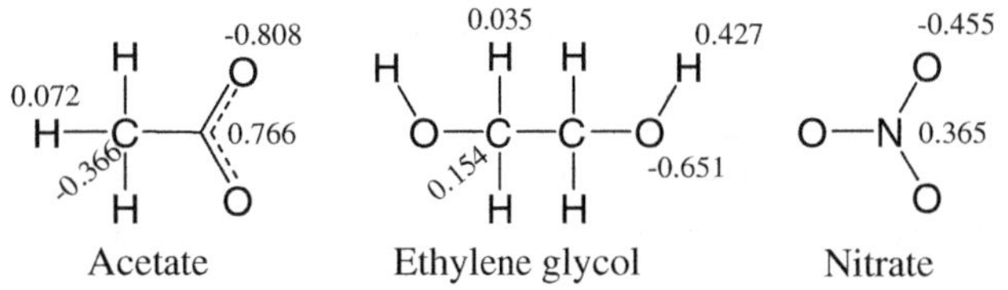

**Fig. 1** Atomic charges used for acetate, ethylene glycol and nitrate; AMBER atom types "C", "N", "H" and "O" were also applied

the somewhat crude assumption that most correlations are induced by short-range intermolecular forces rather than by long-range electrostatics. Note that the full crystallographic boundary conditions are used in all calculations.

Hydrogen atoms were added to all of the water molecules in the 2VB1 structure [98]. To describe the gross features of the possible water configurations, a grid of 72 different water orientations per water molecule was set up. These grid points were distributed on the surface of two cones pointing toward each other with an angular displacement of 30° between each point. By using such a grid, the water molecule were able to do physically unlikely movements and jump between possible alternate local minima in a single Monte Carlo step. To make the grid more flexible and address more subtle bonding features, each individual molecule was also allowed to rotate small amounts about its $x$-, $y$- and $z$-axes. The TIP3P water molecule model by Jorgensen et al. [115] was applied with both fixed bond length (0.9572 Å) and bond angle (104.45°).

The Monte Carlo scheme functioned by selecting at random one of five possible operations, making a random move for that operation, and determining the associated change in the total energy. This Monte Carlo move was accepted according to the Metropolis algorithm:

If $E_{\mathrm{new}} \leq E_{\mathrm{old}}$ always accept,
If $E_{\mathrm{new}} > E_{\mathrm{old}}$ accept if $\xi \leq \exp\left(\frac{E_{\mathrm{old}} - E_{\mathrm{new}}}{k_B T}\right),$

$$\tag{13}$$

where $\xi \in [0, 1]$ is a random number. The five possible operations and the relative weighting used in determining which type of operation to make next are described in Table 1. Four of these operations involve simply interchanging the conformations between a selected two of the eight sections of the $2 \times 2 \times 2$ superlattice containing either part of the protein chain or else interchanging present and absent sites for ethylene glycol, nitrate ions or waters, while the fifth operation involves small-angle changes to the configuration of a specific water molecule. For the water molecules, seven random numbers are generated per move: selecting the water to move, selecting a new grid point number, selecting a rotation axis, selecting the

**Table 1** Types of Monte Carlo operations

| Operation | Weight |
|---|---|
| Residue conformation (A, B or C) | 35 (1/group) |
| Ethylene glycol (present or absent) | 3 (1/group) |
| Nitrate (present or absent) | 9 (1/group) |
| Water (present or absent) | 170 (1/group) |
| Water configuration | 1,700 (10/group) |

**Table 2** Average fraction of accepted moves after each temperature step used in the Monte Carlo simulation

| Cycles | Temperature/$K$ | Acceptance rate |
|---|---|---|
| 1–20,000,000 | 4,273 | 0.522 |
| 20,000,001–40,000,000 | 4,173 | 0.518 |
| 40,000,001–60,000,000 | 4,073 | 0.514 |
| 60,000,001–80,000,000 | 3,773 | 0.506 |
| 80,000,001–100,000,000 | 0 | 0.405 |

**Table 3** Average fraction of accepted trial moves for each operation

| Group type | Acceptance rate |
|---|---|
| Residue | 0.287 |
| Ethylene glycol | 0.518 |
| Nitrate | 0.158 |
| Water | 0.348 |
| Water configuration | 0.413 |

rotation angle and optionally three specifying a small translation of the oxygen coordinates.

The Monte Carlo simulations must be run at some temperature $T$. This temperature does not reflect that used in the crystallography experiments but instead is adjusted to produce an optimized representative structure at 0 K. Hence, a run is initially performed at a temperature sufficiently high to sample all of the available parameter space. As a typical quantity associated with a configuration change is the making or breaking of a hydrogen bond, likely temperatures fall in the region of 5–10 kcal mol$^{-1}$ energy, or 2,000–4,000 K. We found a temperature of $\sim$4,300 K to be sufficient based on the generally accepted criterion that a Monte Carlo calculation works best if ca. half of the moves are accepted and half rejected. After this, the calculation was slowly quenched to 0 K; Table 2 shows the temperatures used, the total number of Monte Carlo moves made and the fraction of accepted moves. As the five classes of operations have intrinsically different associated energy scales, the fraction of accepted moves varies between each class. Acceptance ratios for each class are shown in Table 3 and indicate acceptable rates, the most difficult operation being exchange of nitrate ions with an acceptance ratio of just 0.158. In total, 10$^8$ Monte Carlo moves were made per calculation, and 10 separate calculations were performed.

Since the Monte Carlo simulations embody a constant total number for each possibility of each chemical group and hence fixed marginal distributions, Fisher's exact test [116] is a suitable independency test for the configuration of each chemical unit considered. It was calculated under the null hypothesis that the configurations of the different units are independent and 4.7% were found to have a $p$

value below 0.05 in a two-sided test. This two-sided test was performed as described by Freeman and Halton [117–119], where the $p$ value is the sum of all possibilities with a probability less or equal to the observed one. It reveals that some of the variations noted originally in the 2VB1 structure, in particular groups close to each other, are indeed strongly correlated.

In order to compare the ten individual configurations (named "A"–"J") obtained after quenching to 0 K each simulation of the $2 \times 2 \times 2$ superlattice, a similarity value $S$ was evaluated,

$$S_{pq} = \frac{\sum I_{ii}^{pq}}{N} \tag{14}$$

where

$$I_{ii}^{pq} = \begin{cases} 1 & \text{if} \quad I_i^p = I_i^q \\ 0 & \text{if} \quad I_i^p \neq I_i^q \end{cases}, \tag{15}$$

$N$ is the total number of independent chemical units correlated, and $I_i^p$ and $I_i^q$ are the states (1 if present, 0 if absent, etc.) of chemical unit number $i$ in simulation runs $p$ and $q$, respectively. As the 8 individual structures may be placed inside a $2 \times 2 \times 2$ superlattice in a number of equivalent ways, all possibilities were considered and the largest similarity value chosen. Table 4 shows the deduced similarities between the 10 final structures, and all values are close to 0.7. Since this value is larger than the lowest-possible value of about 0.5, many features are conserved throughout the 10 structure, indicating that essential correlations have indeed been identified by the Monte Carlo procedure. Furthermore, this value is less than unity, indicating that there are some significant differences between the 10 structures and hence many correlations are possibly not of great significance. Such a result is a requirement if just any one configuration of a $2 \times 2 \times 2$ superlattice is to provide a useful description of the actual inhomogeneity evidenced in the original X-ray refinement. Simulation J had the lowest total interaction energy and the relative energies of the other runs are plotted against the similarity to this structure in Fig. 2. It is found that the total energy is not dependent on the similarity, another desired feature. This result shows that the Monte Carlo simulation has sampled and obtained low energy structures in different

**Table 4** Comparison of the similarities between the 10 different Monte Carlo runs

|   | A | B | C | D | E | F | G | H | I | J |
|---|---|---|---|---|---|---|---|---|---|---|
| A | 1.00 | 0.72 | 0.72 | 0.73 | 0.73 | 0.73 | 0.72 | 0.72 | 0.72 | 0.72 |
| B | 0.72 | 1.00 | 0.73 | 0.71 | 0.71 | 0.73 | 0.74 | 0.73 | 0.72 | 0.72 |
| C | 0.72 | 0.73 | 1.00 | 0.71 | 0.71 | 0.72 | 0.72 | 0.74 | 0.72 | 0.73 |
| D | 0.73 | 0.71 | 0.71 | 1.00 | 0.71 | 0.71 | 0.72 | 0.72 | 0.72 | 0.73 |
| E | 0.73 | 0.71 | 0.71 | 0.71 | 1.00 | 0.70 | 0.72 | 0.73 | 0.71 | 0.71 |
| F | 0.73 | 0.73 | 0.72 | 0.71 | 0.70 | 1.00 | 0.71 | 0.72 | 0.71 | 0.72 |
| G | 0.72 | 0.74 | 0.72 | 0.72 | 0.72 | 0.71 | 1.00 | 0.73 | 0.72 | 0.72 |
| H | 0.72 | 0.73 | 0.74 | 0.72 | 0.73 | 0.72 | 0.73 | 1.00 | 0.71 | 0.71 |
| I | 0.72 | 0.72 | 0.72 | 0.72 | 0.71 | 0.71 | 0.72 | 0.71 | 1.00 | 0.72 |
| J | 0.72 | 0.72 | 0.73 | 0.73 | 0.71 | 0.72 | 0.72 | 0.71 | 0.72 | 1.00 |

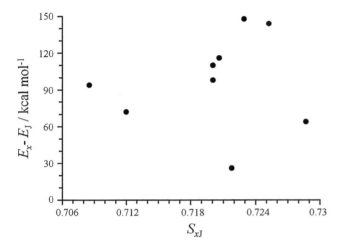

**Fig. 2** The energy of quenched optimized $2 \times 2 \times 2$ superlattice structure $x$, less that of structure J, as a function of the similarity of the two structures. This energy arises from variations in 840 $df$

parts of the configuration space. The energy variation between the 10 structures appears quite large, but there are 840 $df$ in the Monte Carlo simulations, and so the energy spread per degree of freedom is just 0.18 kcal mol$^{-1}$, less than thermal energy at 300 K (0.3 kcal mol$^{-1}$ per degree of freedom). Nevertheless, this result indicates that it is worthwhile maximizing the number of quenches made in calculations of this type.

Further, the calculated wide dispersion in the total energies indicates that the energy landscape is composed of a large number of local minima. All 10 structures have been determined to be local minima in the parameter space by considering all possible single-parameter variations, with the energy increasing in all cases. Figure 3 shows the energy increases determined for the lowest-energy structure, J, as well as for the highest-energy structure, G. These profiles are quite similar, as indeed are also those for the other 8 configurations. The left-most peaks ($\Delta E = 0 - 1$ kcal mol$^{-1}$) arise typically from ILE, TRP, PRO and VAL residues, uncharged groups that in specific instances occupy different sites without steric hindrance. Strong correlations manifest through large (say >10 kcal mol$^{-1}$) to extreme (100 kcal mol$^{-1}$) costs for a single exchange, mostly due to steric repulsion and ion–ion interactions. In summary, this energy analysis thus also indicates that while the $2 \times 2 \times 2$ superlattice model does capture many (previously unknown) essential correlations, further expansion in its size would yield an improved description. However, such an expansion would undesirably add more parameters to a subsequent X-ray refinement, so careful choice of the superlattice size must be made.

## 5 Possible changes of the solvent atoms in the X-ray structure

In this section, we consider the chemical feasibility of the partial occupancies attributed in 2VB1 to many of the nitrate ions, ethylene glycol molecules, and water molecules. In earlier models for the triclinic lysozyme, some of the molecules identified in 2VB1 as nitrate ions and ethylene glycol molecules were identified as water molecules instead [120], and it is clear that only very high-quality data and analyses can accurately discriminate between various explanations of observed electron density in these regions. In this section, we consider variations of the proposed atomic structures for the nitrates, ethylene glycols and water molecules, examining them in terms of both the associated energetics and induced changes to the $R$ factor.

Figure 4 shows the calculated AMBER energy for inserting water molecules into holes left by vacant nitrate ions, ethylene glycol molecules or water molecules in the optimized $2 \times 2 \times 2$ superlattice structure J representing the 2VB1 optimized coordinates. For a missing nitrate ion, one water molecule is used to fill the resulting cavity, whereas two water molecules are used to fill an ethylene glycol hole.

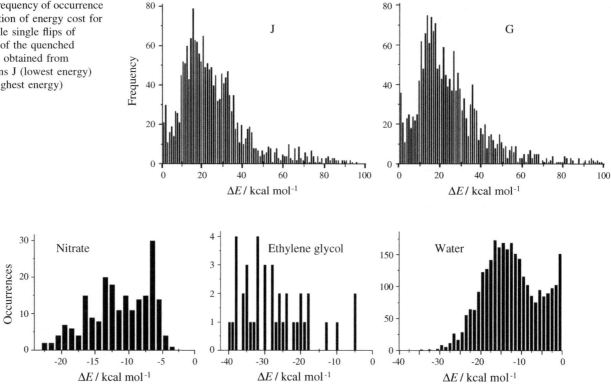

**Fig. 3** Frequency of occurrence as a function of energy cost for all possible single flips of elements of the quenched structures obtained from simulations J (lowest energy) and G (highest energy)

**Fig. 4** Distribution of the energy change when water molecules were added to the holes due to missing nitrate ions (1 water), ethylene glycol molecules (2 waters) and water molecules (1 water)

Addition of a water molecule from the gas phase to a nitrate hole was always found to be attractive, but the energy change $\Delta E$ varied from $-3$ to $-23$ kcal mol$^{-1}$. However, the calculated energy for liquid water is $-9.9$ kcal mol$^{-1}$ [115], close to the observed enthalpy change for formation of liquid water of $-10.5$ kcal mol$^{-1}$ [121], and so, ignoring entropy changes between the liquid and protein environments, only insertions that are more exothermic than this are likely to proceed. The calculations thus indicate that many of the nitrate vacancies in the 2VB1 structure will in fact be filled with water. For ethylene glycol holes, Fig. 4 shows that a similar situation arises with most substitutions resulting in exothermicities of up to 40 kcal mol$^{-1}$, well in excess of the 20 kcal mol$^{-1}$ required to extract two water molecules from the bulk liquid. However, even larger exothermicities per molecule are also depicted in Fig. 4 for the situation in which a water molecule fills a water hole, up to 35 kcal mol$^{-1}$. Strong interactions of water in these holes often arise from direct hydrogen-bonding interactions with charged residues such as LYS, ARG, GLU and ASP, and the desolvation of these residues seems highly unlikely.

A limitation of these calculations is that they ignore the 17.5 % of the volume of the unit cell for which no atomic structure has been proposed. Some of the sites considered will be near this vacant volume and so may be poorly described, missing perhaps additional local attractive forces. The long-range effect of dielectric media is to mitigate the effects of electrostatic forces, however, and so could significantly reduce some of the calculated water substitution energies. Conversely, other effects such as thermal motion of solvent could easily enhance these energies.

Despite such uncertainties, the emphatic result from these calculations, that water molecules should spontaneously fill many of the holes created by partially occupied water sites, stands in contrast to the current experimental interpretation of the occupancies of the water sites: simply increasing the water occupancies as demanded by the calculations would, by construct, automatically lead to an increase in the refined $R$ factor. This implies that the perceived electron density from Eq. 2 in the vicinity of these water sites is underestimated, an effect possibly attributable to poor calculated phases caused by an incorrect description of other either explicitly or implicitly represented regions of the unit cell.

Such a dramatic contradiction of calculated and refined properties is not automatically implied by the prediction that water molecules should fill nitrate or ethylene glycol holes, however. The changes to the assigned compositions of these sites with improved structural refinement [98] indicate that the total electron densities and its distribution in these regions are difficult to determine. Adding water to

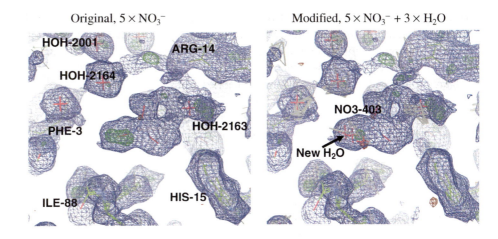

**Fig. 5** Electron density maps (*blue*: $2F_O - F_C$, *green*: $F_O - F_C$ positive, *red*: $F_O - F_C$ negative) around NO3-403 as is the 2VB1 refinement, showing significant unmodeled electron density on one side of the ion, and after the vacant, nitrate locations in the $2 \times 2 \times 2$ supercell are occupied by additional water molecules

previously vacant cells could even lead to a decrease in the R factor, indicating, by all measures, an improved structure.

In a second set of Monte Carlo simulations, all nitrate ions and ethylene glycol holes were filled with water, setting the occupancies of each species so as to conserve the total number of electrons perceived in the region. Because of the nonlinear nature of crystallographic refinement with calculated phase factors used in generating density maps (Eqs. 5–6), such conservation of net charge is not a requirement, but here it is used in a minimalist approach to modification of the original structural model. A sequence of simulations with reducing temperature was performed as before, finally quenching the structure to 0 K. This structure was verified to be a local minimum in configuration space.

When two water molecules filled an ethylene glycol hole, the optimized structure always located the waters on one side of the hole, producing an electron density map that looks quite different from the original (that coming from assuming either ethylene glycol or else nothing occupied the cavity). For the nitrate holes, the water molecules often located in regions of unaccounted electron density in the original maps, however, suggesting that an improved structure is possible. Specifically, nine nitrate ions are identified in 2VB1, named NO3-401 to NO3-409. Of these, five (NO3-402, NO3-403, NO3-405, NO3-407 and NO3-409) have $F_O - F_C$ maps showing significant observed electron density that is not accounted for by the structural model; these groups have occupancies of 73, 62, 68, 69 and 49%, respectively. Conversely, the other four nitrates (NO3-401, NO3-404, NO3-406 and NO3-408) with occupancies of 73, 52, 85 and 100%, respectively, appear to account for most of the observed electron density. Here, we focus only on the nitrates with unaccounted electron density, specifically NO3-403 whose $2F_O - F_C$ and $F_O - F_C$ maps from structure J are shown in Fig. 5. The appearance of substantial unaccounted electron density near one of the nitrate oxygen atoms is readily apparent.

Nitrate NO3-403, with 62% occupancy (thus present in only 5 of the 8 cells in the superlattice structure J) forms a hydrogen bond to ILE-88, and indeed, it is near this residue that the unaccounted electron density is centered. Other close residues include ARG-14, while LYS-1 is more distant, and many water molecules are also found nearby. Most significant, however, is ASP-87 that is located above the plane of the nitrate molecule and hence not visualized in Fig. 5. This residue has two observed conformations, one that is in van der Waals contact with NO3-403 with 82% occupancy, while the less-prevalent conformer is most distant, but in supercell J the vacant NO3-403 cells all have ASP-87 very close. As additional electron density needs to be accounted for, water molecules were added to the three cells unoccupied by NO3-403 rather than utilizing the simplistic density-conserving scheme used in the previous Monte Carlo simulations. These added molecule interact strongly with ASP-87, ILE-88 and the surrounding water molecules.

The structures of the added water molecules, as well as the surrounding water molecule orientations, were then optimized using our ONIOM-based linear-scaling DFT scheme [4] implemented in GAUSSIAN-09 [114] using the PW91 density functional [122] and the 6-31G* basis set for all atoms except those bearing anions, for which 6-31+G* was used. This scheme breaks a full optimization down into small pieces, and for the three critical units, this included 27, 29 and 60 nearby water molecules and 1213, 1290 and 1915 atoms in total; such significant variation in the size of the critical units occurs owing to unoccupied water sites in structure J as complete local hydrogen-bonded networks are included in all calculations. The energies for water molecules placed into the three holes were thus calculated to be $-18$, $-20$ and $-21$ kcal mol$^{-1}$, typical of the values found in the previous Monte Carlo simulations using the Amber force field. Single-point energy calculations using an external reaction field to model neglected paths of the structure has no effect on these energy differences, as one would expect given that the calculations involve large regions of explicitly included matter (effective cavity radii 13–15 Å). Energetic considerations thus strongly favor occupation of these three nitrate holes.

Figure 5 shows also the density maps generated using the revised structure with water molecules filling the three NO3-403 holes. It is clear from the figure that this modification results in significant reduction of the unaccounted electron density, but this is not a good measure of the quality of the modified model as the nonlinearities in the refinement procedure can easily result in the $R$ factor increasing despite the visualized apparent improvement to the model. However, using an isotropic temperature factor of 8 for the added water molecules, the $R$ factors are found to decrease by 0.01%, a value that is significant given that artificially changing the occupancy of NO3-403 from 5/8 to 1 increase the $R$ factors by 0.01% and that decreasing the occupancy to 0 increases $R$ by 0.08%. Hence, both energetic and crystallographic measures indicate that water molecules fill the site when NO3-403 is not present.

The unaccounted electron density manifest for the original structure in Fig. 5 can readily be accounted for using anisotropic B factor chosen for each atom in the nitrate ion, but such a choice would not be consistent with the actual motions of a nitrate ion. Hence, we see that while the anisotropic X-ray structure refinement resulted in a deep local minimum in its parameter space, some of the deduced structural features may not necessarily be realistic. While anisotropic $B$ factors can account for real non-spherical electron density and produce a dramatic decrease in the $R$ factor in 2VB1, they may do this simply by overparameterization, eroding the physical meaning of all deduced parameters. We thus recommend that use of anisotropic thermal displacement parameters be minimized during subsequent ab initio refinement.

# 6 Conclusions

A Monte Carlo procedure utilizing the AMBER force field was developed to find an ensemble of 8 structures represented in a $2 \times 2 \times 2$ superlattice that embodies in a properly correlated fashion all of the structural variations previously reported in the 2VB1 structure of lysozyme crystal. This calculation included explicitly all correlations between atoms identified by high-resolution X-ray refinement methods [98] and deduced many new correlations manifested through large calculated interaction energies. Statistical analyses indicate that an average structure (with only some local multiple conformations) does include many key correlations and suggest that expanded superlattices could yield improved results. Such superlattices are a critical requirement before ab initio electronic structure

computational methods can be embedded inside any X-ray refinement packages, but expansion is limited as an undesired feature of their inclusion is an increase in the number of free parameters during refinement. Further extensions of this method are required, however, in order to include the 17.5% of the crystal volume that was not explicitly represented in the original X-ray structure.

The development of such a structure by necessity introduces chemically meaningful constraints, constraints not present in the original X-ray refinement. In this case, these constraints ensured that all atoms appearing in different chain conformations have the same occupancy, an important physical property, yet their introduction by necessity *decreases* the quality of the structural refinement as perceived by increases in $R$ and $R_{\text{free}}$. In addition, use of the $2 \times 2 \times 2$ superlattice does decrease the quality of the refinement in real terms by limiting the occupancies to a discrete set of values, multiples of one-eighth for the present example. The quality of ab initio refined structures needs to be compared to the values of $R$ and $R_{\text{free}}$ determined at this stage, not the original unconstrained values.

Ab initio refinement will, in general, require some modifications to the way X-ray refinements are completed. In particular, $B$ factors should be modified to include only the effects of thermal motion and not the effects of multiple conformations as multiple conformations are now being included explicitly. Such isotropic or anisotropic displacement factors could be determined ab initio using say molecular dynamics simulations and hence not appear as free parameters during X-ray structure refinement [81, 86, 123].

The development of a single $2 \times 2 \times 2$ superlattice representing an ensemble of 8 structures to describe conformational variations by default allows for individual coordinates for all atoms in each of the 8 cell copies. While this is demanded for regions showing significant variation, it may not be justified for other regions in which one single structure dominates the X-ray diffraction. In this region, replicated copies of all coordinates could be maintained in some or all of the 8 cells, reducing considerably the number of free parameters in an X-ray refinement. Established methods for treating the $B$ factors would be appropriate for such atoms.

Initially, superlattices were developed embodying most the features of the original 2VB1 structure for lysozyme. Each considered superlattice thus generated essentially the same $R$ factor and so was equally consistent with the raw X-ray diffraction data. These structures varied considerably in terms of their perceived total energies although the maximum energy difference per degree of freedom was quite small. Many of the geometrical fluctuations apparent in the 2VB1 structure were found to be strongly correlated with each other, while many others were found to be uncorrelated. The lowest-energy structure found was considered as a starting point for future ab initio X-ray refinement. Subsequently, however, the question as to the chemical feasibility of the originally deduced nitrate, ethylene glycol and water occupancies was examined. The calculations predict that it is highly exothermic to fill many of these holes with water molecules taken from bulk solution, indicating that the original occupancies are not actually feasible. While authoritative calculations require an improved treatment of missing solvent atoms than is used herein, one specific example is considered, nitrate NO3-403, for which DFT calculations predict that water fills the nitrate holes while this change actually does result in a decrease in the $R$ factor, indicating that indeed an improved structural model is produced.

Such a simple correlation between predicted and actual model improvements will not be universal, however, and it could be that the application of ab initio computational methods to X-ray structure refinement does not easily lead to improved measures of structure quality such as simply a decrease in $R_{free}$. Because of the nonlinearities inherent in X-ray diffraction modeling, it is feasible that such a decrease could only be obtained by taking the whole optimized structure from its original local minimum in parameter space and transforming it to a quite different local minimum, affecting many of the atomic parameters of the structure, especially perhaps solvent and unassigned regions. This is a difficult task.

**Acknowledgments** We thank Aaron McGrath for technical advice, Zbigniew Dauter from the Argonne National Laboratory Biosciences Division for providing detailed information regarding the refinement of 2VB1, the Australian Research Council for funding this research, and National Computer Infrastructure (NCI) and Australian Centre for Advanced Computing and Communications (AC3) for computing resources.

# References

1. Engh RA, Huber R (1991) Acta Crystallogr A 47:392–400
2. Sheldrick G, Schneider T (1997) SHELXL: high-resolution refinement. Methods Enzymol 277:319–343
3. Murshudov GN, Vagin AA, Dobson EJ (1997) Refinement of macromolecular structures by the maximum-likelihood method. Acta Crystallogr D Biol Crystallogr 53:240–255
4. Canfield P, Dahlbom MG, Hush N, Reimers JR (2006) Density-functional geometry optimization of the 150000-atom photosystem-I trimer. J Chem Phys 124:024301
5. Kleywegt GJ (1999) Experimental assessment of differences between related protein crystal structures. Acta Crystallogr D Biol Crystallogr 55:1878–1884
6. Cruickshank DWJ (1999) Remarks about protein structure precision. Acta Crystallogr D Biol Crystallogr 55:583–601
7. DePristo MA, De Bakker PIW, Blundell TL (2004) Heterogeneity and inaccuracy in protein structures solved by X-ray crystallography. Structure 12:831–838

8. Jaskolski M, Gilski M, Dauter Z, Wlodawer A (2007) Stereochemical restraints revisited: how accurate are refinement targets and how much should protein structures be allowed to deviate from them? Acta Crystallogr D Biol Crystallogr 63:611–620
9. Chen J, Brooks CL (2007) Can molecular dynamics simulations provide high-resolution refinement of protein structure? Proteins Struct Funct Bioinform 67:922–930
10. Karplus PA, Shapovalov MV, Dunbrack RL, Berkholz DS (2008) A forward-looking suggestion for resolving the stereochemical restraints debate: ideal geometry functions. Acta Crystallogr D Biol Crystallogr 64:335–336
11. Rashin AA, Rashin AHL, Jernigan RL (2009) Protein flexibility: coordinate uncertainties and interpretation of structural differences. Acta Crystallogr D Biol Crystallogr 65:1140–1161
12. Jaskolski M (2010) From atomic resolution to molecular giants: an overview of crystallographic studies of biological macromolecules with synchrotron radiation. Acta Physica Polonica A 117:257–263
13. Eyal E, Gerzon S, Potapov V, Edelman M, Sobolev V (2005) The limit of accuracy of protein modeling: influence of crystal packing on protein structure. J Mol Biol 351:431–442
14. Konnert JH (1976) A restrained parameter structure-factor least-squares refinement procedure for large asymmetric units. Acta Crystallogr A 32:614–617
15. Hendrickson WA, Konnert JH (1979) Stereochemically restrained crystallographic least-squares refinement of macromolecule structures. In: Srinivasan R (ed) Biomolecular structure, conformation, function, and evolution, vol 1. Pergamon Press, Oxford, pp 43–57
16. Konnert JH, Hendrickson WA (1980) A restrained-parameter thermal-factor refinement procedure. Acta Crystallogr A 36:344–350
17. Hendrickson WA (1985) Stereochemically restrained refinement of macromolecular structures. Methods Enzymol 115:252–270
18. Jack A, Levitt M (1978) Refinement of large structures by simultaneous minimization of energy and R factor. Acta Crystallogr A 34:931–935
19. Brunger AT, Kuriyan J, Karplus M (1987) Crystallographic R factor refinement by molecular dynamics. Science 235:458–460
20. Ohta K, Yoshioka Y, Morokuma K, Kitaura K (1983) The effective fragment potential method. An approximate ab initio mo method for large molecules. Chem Phys Lett 101:12–17
21. Stewart JJP (1996) Application of localized molecular orbitals to the solution of semiempirical self-consistent field equations. Int J Quantum Chem 58:133–146
22. White CA, Johnson BG, Gill PMW, Head-Gordon M (1996) Linear scaling density functional calculations via the continuous fast multipole method. Chem Phys Lett 253:268–278
23. Stewart JJP (1997) Calculation of the geometry of a small protein using semiempirical methods. J Mol Struct Theochem 401:195–205
24. Lee TS, Lewis JP, Yang W (1998) Linear-scaling quantum mechanical calculations of biological molecules: the divide-and-conquer approach. Comput Mater Sci 12:259–277
25. Van Alsenoy C, Yu CH, Peeters A, Martin JML, Schäfer L (1998) Ab initio geometry determinations of proteins. 1. Crambin. J Phys Chem A 102:2246–2251
26. Artacho E, Sánchez-Portal D, Ordejón P, García A, Soler JM (1999) Linear-scaling ab initio calculations for large and complex systems. Phys Status Solidi B 215:809–817
27. Sato F, Yoshihiro T, Era M, Kashiwagi H (2001) Calculation of all-electron wavefunction of hemoprotein cytochrome c by density functional theory. Chem Phys Lett 341:645–651
28. Inaba T, Tahara S, Nisikawa N, Kashiwagi H, Sato F (2005) All-electron density functional calculation on insulin with quasi-canonical localized orbitals. J Comput Chem 26:987–993

29. Wada M, Sakurai M (2005) A quantum chemical method for rapid optimization of protein structures. J Comput Chem 26:160–168
30. Li S, Shen J, Li W, Jiang Y (2006) An efficient implementation of the "cluster-in-molecule" approach for local electron correlation calculations. J Chem Phys 125:074109
31. Sale P, Høst S, Thøgersen L, Jørgensen P, Manninen P, Olsen J, Jansik B, Reine S, Pawlowski F, Tellgren E, Helgaker T, Coriani S (2007) Linear-scaling implementation of molecular electronic self-consistent field theory. J Chem Phys 126:114110
32. Cankurtaran BO, Gale JD, Ford MJ (2008) First principles calculations using density matrix divide-and-conquer within the SIESTA methodology. J Phys Condens Matter 20:294208
33. Stewart JJP (2009) Application of the PM6 method to modeling proteins. J Mol Model 15:765–805
34. Gordon MS, Mullin JM, Pruitt SR, Roskop LB, Slipchenko LV, Boatz JA (2009) Accurate methods for large molecular systems. J Phys Chem B 113:9646–9663
35. Fedorov DG, Alexeev Y, Kitaura K (2010) Geometry optimization of the active site of a large system with the fragment molecular orbital method. J Phys Chem Lett 2:282–288
36. Kobayashi M, Kunisada T, Akama T, Sakura D, Nakai H (2010) Reconsidering an analytical gradient expression within a divide-and-conquer self-consistent field approach: exact formula and its approximate treatment. J Chem Phys 134:034105
37. Mayhall NJ, Raghavachari K (2010) Molecules-in-molecules: an extrapolated fragment-based approach for accurate calculations on large molecules and materials. J Chem Theory Comput 7:1336–1343
38. Nagata T, Brorsen K, Fedorov DG, Kitaura K, Gordon MS (2010) Fully analytic energy gradient in the fragment molecular orbital method. J Chem Phys 134:124115
39. Reine S, Krapp A, Iozzi MF, Bakken V, Helgaker T, Pawowski F, Saek P (2010) An efficient density-functional-theory force evaluation for large molecular systems. J Chem Phys 133:044102
40. Bylaska E, Tsemekhman K, Govind N, Valiev M (2011) Large-scale plane-wave-based density-functional theory: formalism, parallelization, and applications. In: Reimers JR (ed) Computational methods for large systems: electronic structure approaches for biotechnology and nanotechnology. Wiley, Hoboken, pp 77–116
41. Gale JD (2011) SIESTA: a linear-scaling method for density functional calculations. In: Reimers JR (ed) Computational methods for large systems: electronic structure approaches for biotechnology and nanotechnology. Wiley, Hoboken, pp 45–74
42. Li W, Hua W, Fang T, Li S (2011) The energy-based fragmentation approach for computing total energies, structures, and molecular properties of large systems at the ab initio levels. In: Reimers JR (ed) Computational methods for large systems: electronic structure approaches for biotechnology and nanotechnology. Wiley, Hoboken, pp 227–258
43. Clark T, Stewart JJP (2011) MNDO-like semiempirical molecular orbital theory and its application to large systems. In: Reimers JR (ed) Computational methods for large systems: electronic structure approaches for biotechnology and nanotechnology. Wiley, Hoboken, pp 259–286
44. Elstner M, Gaus M (2011) The self-consistent-charge density-functional tight-binding (SCC-DFTB) method: an efficient approximation of density functional theory. In: Reimers JR (ed) Computational methods for large systems: electronic structure approaches for biotechnology and nanotechnology. Wiley, Hoboken, pp 287–308
45. Zimmerli U, Parrinello M, Koumoutsakos P (2004) Dispersion corrections to density functionals for water aromatic interactions. J Chem Phys 120:2693–2699
46. Antony J, Grimme S (2006) Density functional theory including dispersion corrections for intermolecular interactions in a large benchmark set of biologically relevant molecules. Phys Chem Chem Phys 8:5287–5293
47. Grimme S, Antony J, Schwabe T, Mück-Lichtenfeld C (2007) Density functional theory with dispersion corrections for supramolecular structures, aggregates, and complexes of (bio)organic molecules. Org Biomol Chem 5:741–758
48. Zhao Y, Truhlar DG (2007) Density functionals for noncovalent interaction energies of biological importance. J Chem Theory Comput 3:289–300
49. Murdachaew G, De Gironcoli S, Scoles G (2008) Toward an accurate and efficient theory of physisorption. I. Development of an augmented density-functional theory model. J Phys Chem A 112:9993–10005
50. DiLabio GA (2008) Accurate treatment of van der Waals interactions using standard density functional theory methods with effective core-type potentials: application to carbon-containing dimers. Chem Phys Lett 455:348–353
51. Gräfenstein J, Cremer D (2009) An efficient algorithm for the density-functional theory treatment of dispersion interactions. J Chem Phys 130:124105
52. Liu Y, Goddard WA (2009) A universal damping function for empirical dispersion correction on density functional theory. Mater Trans 50:1664–1670
53. Sato T, Nakai H (2009) Density functional method including weak interactions: dispersion coefficients based on the local response approximation. J Chem Phys 131:224104
54. Foster ME, Sohlberg K (2010) Empirically corrected DFT and semi-empirical methods for non-bonding interactions. Phys Chem Chem Phys 12:307–322
55. Grimme S, Antony J, Ehrlich S, Krieg H (2010) A consistent and accurate ab initio parametrization of density functional dispersion correction (DFT-D) for the 94 elements H-Pu. J Chem Phys 132:154104
56. Riley KE, Pitoňčák M, Jurecčka P, Hobza P (2010) Stabilization and structure calculations for noncovalent interactions in extended molecular systems based on wave function and density functional theories. Chem Rev 110:5023–5063
57. MacKie ID, Dilabio GA (2010) Accurate dispersion interactions from standard density-functional theory methods with small basis sets. Phys Chem Chem Phys 12:6092–6098
58. Goerigk L, Grimme S (2011) A thorough benchmark of density functional methods for general main group thermochemistry, kinetics, and noncovalent interactions. Phys Chem Chem Phys 13:6670–6688
59. Grimme S, Ehrlich S, Goerigk L (2011) Effect of the damping function in dispersion corrected density functional theory. J Comput Chem 32:1456–1465
60. Steinmann SN, Corminboeuf C (2011) A density dependent dispersion correction. Chimia 65:240–244
61. Zhao Y, Truhlar DG (2011) Density functional theory for reaction energies: test of meta and hybrid meta functionals, range-separated functionals, and other high-performance functionals. J Chem Theory Comput 7:669–676
62. Brüning J, Alig E, Van De Streek J, Schmidt MU (2011) The use of dispersion-corrected DFT calculations to prevent an incorrect structure determination from powder data: The case of acetolone, C 11H11N3O3. Z Kristallogr 226:476–482
63. Ryde U, Olsen L, Nilsson K (2002) Quantum chemical geometry optimizations in proteins using crystallographic raw data. J Comput Chem 23:1058–1070
64. Ryde U, Nilsson K (2003) Quantum chemistry can locally improve protein crystal structures. J Am Chem Soc 125:14232–14233

65. Ryde U (2007) Accurate metal-site structures in proteins obtained by combining experimental data and quantum chemistry. Dalton Trans 607–625
66. Ryde U, Greco C, De Gioia L (2010) Quantum refinement of [FeFe] hydrogenase indicates a dithiomethylamine ligand. J Am Chem Soc 132:4512–4513
67. Yu N, Yennawar HP, Merz KM Jr (2005) Refinement of protein crystal structures using energy restraints derived from linear-scaling quantum mechanics. Acta Crystallogr D Biol Crystallogr 61:322–332
68. Yu N, Li X, Cui G, Hayik SA, Merz KM Jr (2006) Critical assessment of quantum mechanics based energy restraints in protein crystal structure refinement. Protein Sci 15:2773–2784
69. Yu N, Hayik SA, Wang B, Liao N, Reynolds CH, Merz KM Jr (2006) Assigning the protonation states of the key aspartates in beta-secretase using QM/MM X-ray structure refinement. J Chem Theory Comput 2:1057–1069
70. Van Der Vaart A, Suárez D, Merz KM Jr (2000) Critical assessment of the performance of the semiempirical divide and conquer method for single point calculations and geometry optimizations of large chemical systems. J Chem Phys 113:10512–10523
71. Van Der Vaart A, Gogonea V, Dixon SL, Merz KM Jr (2000) Linear scaling molecular orbital calculations of biological systems using the semiempirical divide and conquer method. J Comput Chem 21:1494–1504
72. Dixon SL, Merz KM Jr (1997) Fast, accurate semiempirical molecular orbital calculations for macromolecules. J Chem Phys 107:879–893
73. Dixon SL, Merz KM Jr (1996) Semiempirical molecular orbital calculations with linear system size scaling. J Chem Phys 104:6643–6649
74. Pellegrini M, Grønbech-Jensen N, Kelly JA, Pfluegl GMU, Yeates TO (1997) Highly constrained multiple-copy refinement of protein crystal structures. Proteins Struct Funct Bioinform 29:426–432
75. Levin EJ, Kondrashov DA, Wesenberg GE, Phillips GN Jr (2007) Ensemble refinement of protein crystal structures: validation and application. Structure 15:1040–1052
76. Terwilliger TC, Grosse-Kunstleve RW, Afonine PV, Adams PD, Moriarty NW, Zwart P, Read RJ, Turk D, Hung LW (2007) Interpretation of ensembles created by multiple iterative rebuilding of macromolecular models. Acta Crystallogr D Biol Crystallogr 63:597–610
77. Stewart KA, Robinson DA, Lapthorn AJ (2008) Type II dehydroquinase: molecular replacement with many copies. Acta Crystallogr D Biol Crystallogr 64:108–118
78. Stewart JJP (2008) Application of the PM6 method to modeling the solid state. J Mol Model 14:499–535
79. Genheden S, Ryde U (2011) A comparison of different initialization protocols to obtain statistically independent molecular dynamics simulations. J Comput Chem 32:187–195
80. Genheden S, Diehl C, Akke M, Ryde U (2010) Starting-condition dependence of order parameters derived from molecular dynamics simulations. J Chem Theory Comput 6:2176–2190
81. Delarue M (2007) Dealing with structural variability in molecular replacement and crystallographic refinement through normal-mode analysis. Acta Crystallogr D Biol Crystallogr 64:40–48
82. Knight JL, Zhou Z, Gallicchio E, Himmel DM, Friesner RA, Arnold E, Levy RM (2008) Exploring structural variability in X-ray crystallographic models using protein local optimization by torsion-angle sampling. Acta Crystallogr D Biol Crystallogr 64:383–396
83. Sellers BD, Zhu K, Zhao S, Friesner RA, Jacobson MP (2008) Toward better refinement of comparative models: predicting loops in inexact environments. Proteins Struct Funct Genet 72:959–971
84. Yao P, Dhanik A, Marz N, Propper R, Kou C, Liu G, Van Den Bedem H, Latombe JC, Halperin-Landsberg I, Altman RB (2008) Efficient algorithms to explore conformation spaces of flexible protein loops. IEEE/ACM Trans Comput Biol Bioinform 5:534–545
85. Lindorff-Larsen K, Ferkinghoff-Borg J (2009) Similarity measures for protein ensembles. PLoS One 4:e4203
86. Yang L, Song G, Jernigan RL (2009) Comparisons of experimental and computed protein anisotropic temperature factors. Proteins Struct Funct Bioinform 76:164–175
87. Dhanik A, Van Den Bedem H, Deacon A, Latombe JC (2010) Modeling structural heterogeneity in proteins from X-ray data. Springer Tracts Adv Robot 57:551–566
88. Schwander P, Fung R, Phillips GN Jr, Ourmazd A (2010) Mapping the conformations of biological assemblies. New J Phys 12:035007
89. Lang PT, Ng HL, Fraser JS, Corn JE, Echols N, Sales M, Holton JM, Alber T (2010) Automated electron-density sampling reveals widespread conformational polymorphism in proteins. Protein Sci 19:1420–1431
90. Kohn JE, Afonine PV, Ruscio JZ, Adams PD, Head-Gordon T (2010) Evidence of functional protein dynamics from X-ray crystallographic ensembles. PLoS Comput Biol 6:e1000911
91. Tyka MD, Keedy DA, André I, Dimaio F, Song Y, Richardson DC, Richardson JS, Baker D (2011) Alternate states of proteins revealed by detailed energy landscape mapping. J Mol Biol 405:607–618
92. Ramelot TA, Raman S, Kuzin AP, Xiao R, Ma L-C, Acton TB, Hunt JF, Montelione GT, Baker D, Kennedy MA (2009) Improving NMR protein structure quality by Rosetta refinement: a molecular replacement study. Proteins Struct Funct Bioinform 75:147–167
93. Fleming A (1922) On a remarkable bacteriolytic element found in tissues and secretions. Proc R Soc Ser B 93:306–317
94. Blake CCF, Fenn RH, North ACT, Phillips DC, Poljak RJ (1962) Structure of lysozyme. Nature 196:1173–1176
95. Berman HM, Henrick K, Nakamura H (2003) Announcing the world wide protein data bank. Nat Struct Biol 10:980
96. Vocadlo DJ, Davies GJ, Laine R, Withers SG (2001) Catalysis by hen egg-white lysozyme proceeds via a covalent intermediate. Nature 412:835–838
97. Bottoni A, Miscione GP, De Vivo M (2005) A theoretical DFT investigation of the lysozyme mechanism: computational evidence for a covalent intermediate pathway. Proteins Struct Funct Genet 59:118–130
98. Wang J, Dauter M, Alkire R, Joachimiak A, Dauter Z (2007) Triclinic lysozyme at 0.65 a resolution. Acta Crystallogr D Biol Crystallogr 63:1254–1268
99. Blundell TL, Johnson LN (1976) Protein crystallography. Academic Press, London
100. Chapman HN, Fromme P, Barty A, White TA, Kirian RA, Aquila A, Hunter MS, Schulz J, DePonte DP, Weierstall U, Doak RB, Maia FRNC, Martin AV, Schlichting I, Lomb L, Coppola N, Shoeman RL, Epp SW, Hartmann R, Rolles D, Rudenko A, Foucar L, Kimmel N, Weidenspointner G, Holl P, Liang M, Barthelmess M, Caleman C, Boutet S, Bogan MJ, Krzywinski J, Bostedt C, Bajt S, Gumprecht L, Rudek B, Erk B, Schmidt C, Homke A, Reich C, Pietschner D, Struder L, Hauser G, Gorke H, Ullrich J, Herrmann S, Schaller G, Schopper F, Soltau H, Kuhnel K-U, Messerschmidt M, Bozek JD, Hau-Riege SP, Frank M, Hampton CY, Sierra RG, Starodub D, Williams GJ, Hajdu J, Timneanu N, Seibert MM, Andreasson J, Rocker A, Jonsson O, Svenda M, Stern S, Nass K, Andritschke R, Schroter C-D, Krasniqi F, Bott M, Schmidt KE, Wang X, Grotjohann I,

Holton JM, Barends TRM, Neutze R, Marchesini S, Fromme R, Schorb S, Rupp D, Adolph M, Gorkhover T, Andersson I, Hirsemann H, Potdevin G, Graafsma H, Nilsson B, Spence JCH (2011) Femtosecond X-ray protein nanocrystallography. Nature 470:73–77

101. Brunger AT (1992) Free R value: a novel statistical quantity for assessing the accuracy of crystal structures. Nature 355:472–475

102. Badger J (1997) Modeling and refinement of water molecules and disordered solvent. Methods Enzymol 277:344–352

103. Podjarny AD, Howard EI, Urzhumtsev A, Grigera JR (1997) A multicopy modeling of the water distribution in macromolecular crystals. Proteins Struct Funct Bioinform 28:303–312

104. Colominas C, Luque FJ, Orozco M (1999) Monte Carlo–MST: new strategy for representation of solvent configurational space in solution. J Comput Chem 20:665–678

105. Liu Y, Beveridge DL (2002) Exploratory studies of ab initio protein structure prediction: multiple copy simulated annealing, AMBER energy functions, and a generalized born/solvent accessibility solvation model. Proteins Struct Funct Bioinform 46:128–146

106. Das B, Meirovitch H (2003) Solvation parameters for predicting the structure of surface loops in proteins: transferability and entropic effects. Proteins Struct Funct Bioinform 51:470–483

107. Hassan SA, Mehler EL, Zhang D, Weinstein H (2003) Molecular dynamics simulations of peptides and proteins with a continuum electrostatic model based on screened coulomb potentials. Proteins Struct Funct Bioinform 51:109–125

108. Dechene M, Wink G, Smith M, Swartz P, Mattos C (2009) Multiple solvent crystal structures of ribonuclease A: an assessment of the method. Proteins Struct Funct Bioinform 76:861–881

109. Kannan S, Zacharias M (2010) Application of biasing-potential replica-exchange simulations for loop modeling and refinement of proteins in explicit solvent. Proteins Struct Funct Bioinform 78:2809–2819

110. Weiner SJ, Kollman PA, Case DA, Singh UC, Ghio C, Alagona G, Profeta SJ, Weiner P (1984) A new force field for molecular mechanical simulation of nucleic acids and proteins. J Am Chem Soc 106:765–784

111. Cornell WD, Cieplak P, Bayly CI, Gould IR, Merz KM Jr, Ferguson DM, Spellmeyer DC, Fox T, Caldwell JW, Kollman PA (1995) A second generation force field for the simulation of proteins, nucleic acids, and organic molecules. J Am Chem Soc 117:5179–5197

112. Becke AD (1993) Density-functional thermochemistry. III. The role of exact exchange. J Chem Phys 98:5648–5652

113. Hehre WJ, Ditchfield R, Pople JA (1972) Self-consistent molecular orbital methods. XII. Further extensions of gaussian-type basis sets for use in molecular orbital studies of organic molecules. J Chem Phys 56:2257–2261

114. Frisch MJ, Trucks GW, Schlegel HB et al (2009) Gaussian 09, revision A.02. Gaussian, Inc., Pittsburgh

115. Jorgensen WL, Chandrasekhar J, Madura JD, Impey RW, Klein ML (1983) Comparison of simple potential functions for simulating liquid water. J Chem Phys 79:926–935

116. Fischer RA (1935) The logic of inductive inference. J R Stat Soc A 98:39–54

117. Freeman GH, Halton JH (1951) Note on an exact treatment of contingency, goodness of fit and other problems of significance. Biometrika 38:141–149

118. Agresti A (1990) Categorical data analysis. Wiley, New York

119. Bartoszyński R, Niewiadomska-Bugaj M (1996) Probability and statistical inference. Wiley, New York

120. Walsh MA, Schneider TR, Sieker LC, Dauter Z, Lamzin VS, Wilson KS (1998) Refinement of triclinic hen egg-white lysozyme at atomic resolution. Acta Crystallogr D Biol Crystallogr 54:522–546

121. Lide DR (ed) (2005) CRC handbook of chemistry and physics, 86th edn. CRC Press, Boca Raton

122. Perdew JP, Wang Y (1992) Accurate and simple analytic representation of the electron-gas correlation energy. Phys Rev B 45:13244–13249

123. Vitkup D, Ringe D, Karplus M, Petsko GA (2002) Why protein R-factors are so large: a self-consistent analysis. Proteins Struct Funct Genet 46:345–354

REGULAR ARTICLE

# Conduction modulation of π-stacked ethylbenzene wires on Si(100) with substituent groups

Manuel Smeu · Robert A. Wolkow · Hong Guo

Received: 12 July 2011 / Accepted: 9 August 2011 / Published online: 7 January 2012
© Springer-Verlag 2012

**Abstract** For the realization of molecular electronics, one essential goal is the ability to systematically fabricate molecular functional components in a well-controlled manner. Experimental techniques have been developed such that π-stacked ethylbenzene molecules can now be routinely induced to self-assemble on an H-terminated Si(100) surface at precise locations and along precise directions. Electron transport calculations predict that such molecular wires could indeed carry an electrical current, but the Si substrate may play a considerable role as a competing pathway for conducting electrons. In this work, we investigate the effect of placing substituent groups of varying electron donating or withdrawing strengths on the ethylbenzene molecules to determine how they would affect the transport properties of such molecular wires. The systems consist of a line of π-stacked ethylbenzene molecules covalently bonded to a Si substrate. The ethylbenzene line is bridging two Al electrodes to model current through the molecular stack. For our transport calculations, we employ a first-principles technique where density functional theory (DFT) is used within the non-equilibrium Green's function formalism (NEGF). The calculated density of states suggest that substituent groups are an effective way to shift molecular states relative to the electronic states associated with the Si substrate. The electron transmission spectra obtained from the NEGF–DFT calculations reveal that the transport properties could also be extensively modulated by changing substituent groups. For certain molecules, it is possible to have a transmission peak at the Fermi level of the electrodes, corresponding to high conduction through the molecular wire with essentially no leakage into the Si substrate.

**Keywords** Molecular wire · NEGF–DFT · Substituent group · Conductance modulation · Electron transport · Ethylbenzene wire

## 1 Introduction

The ability to control the properties of molecular wires is required if they are to be used in nanoelectronics. Advanced synthetic chemistry techniques can now be routinely applied to achieve very specific structures in high yields. One example is the self-directed line growth of styrene molecules on an H-terminated Si(111)−2 × 1 surface, resulting in lines of π-stacked ethylbenzene molecules covalently bonded to the Si substrate [1]. During this reaction, a styrene molecule reacts with a dangling bond on the Si surface, forming a covalent C–Si bond, and also abstracting a H atom from a Si atom on the neighboring site. This leaves a dangling bond next to the attached molecule where another styrene molecule can attach and the process is repeated, resulting in self-assembled molecular wires on the Si surface. Experimental work by Piva et al. [2] on such systems showed that the tunneling current from an STM tip through an ethylbenzene

Published as part of the special collection of articles celebrating the 50th anniversary of Theoretical Chemistry Accounts/Theoretica Chimica Acta.

M. Smeu (✉) · H. Guo
Department of Physics, Center for the Physics of Materials, McGill University, Montreal, QC, Canada
e-mail: smeum@physics.mcgill.ca

R. A. Wolkow
National Institute for Nanotechnology, National Research Council of Canada, Edmonton, AB, Canada

R. A. Wolkow
Department of Physics, University of Alberta, Edmonton, AB, Canada

molecule could be regulated by changing the charge state of a nearby dangling bond on the Si substrate, demonstrating field effect control. Basu et al. [3] have reported on the ability to control the extent of line growth using patterned TEMPO molecules on the surface. Work by Hossain et al. [4] showed the ability to grow molecular lines on H–Si(100) in the direction perpendicular to the dimer rows and then they reported line growth both perpendicular to and along dimer rows in the same sample [5], resulting in interconnected molecular lines. Subsequent work by Zikovsky et al. [6] demonstrated the ability to control the direction of molecular line growth so that contiguous molecular lines having complex shapes could be grown on a Si surface.

Because there is $\pi$–$\pi$ overlap between the ethylbenzene molecules forming the lines in such structures, it is believed that they could perform as molecular wires if connected to electrodes. Due to experimental challenges involved in studying such a hypothesis, it has been investigated theoretically by several groups [7–11]. These studies agree that a line of $\pi$-stacked ethylbenzene molecules could indeed act as a molecular wire. However, because these structures exist chemisorbed to a Si surface, it has been suggested that the transport properties may be affected by the substrate [7, 10], and this was confirmed to be the case with transport calculations that explicitly include the Si substrate and treat all atoms on an equal footing [11]. Reference [11] revealed that the nearest transmission peak to the Fermi level ($E_F$) of the electrodes was due to electron transport through the Si substrate, and this would dominate the low-bias conductance. Clearly, from the point of view of molecular conduction, it would be interesting to somehow alter the transmission spectrum for such a system to either increase transmission through the molecular wire near $E_F$ or decrease transmission through the substrate. It is conceivable that the former could be achieved with substituent groups on the ethylbenzene molecules. It has been reported that lines composed of ethylbenzene substituted in the para positions with –CH$_3$, –CF$_3$, and –OCH$_3$ have been grown experimentally [7, 12].

It is well known that substituent groups can have a profound effect on the energies of certain molecular energy levels. It has been shown that substituent groups could be used to modify the conductance properties of molecules such as benzene diamine [13] and benzenedithiol [14, 15]. Likewise, substituent groups could also alter the properties of a Si substrate. For example, Anagaw et al. [16] showed that the work function of a Si surface can be effectively tuned by chemisorbing substituted styrene molecules to it. In the present work, we investigate whether substituent groups can be used to modify the conductance properties of a molecular line composed of $\pi$-stacked ethylbenzene molecules bonded to a Si substrate. We employ density functional theory (DFT) combined with the non-equilibrium Green's function (NEGF) formalism for this study. This is an ab initio technique that treats all components of the system on an equal footing, including the molecules in the line, the Si substrate, and the electrodes connected to them.

## 2 Theoretical method

The systems studied consist of $\pi$-stacked aromatic rings bonded to a Si(100)–2 × 1 surface via ethyl chains. The substrate is represented as a Si slab of six atomic layers in thickness, with H atoms used to cap dangling bonds, as shown in Fig. 1.

The structure relaxations were carried out using the Vienna ab initio simulation package (VASP) [17, 18] on a periodic (repeating) system consisting of one ethylbenzene molecule on the Si substrate. The substrate consists of two dimer rows, containing one dimer each, as shown in Fig. 1. The ethylbenzene molecule has $\pi$-stacking interactions with its images in the direction normal to the plane of the

**Fig. 1** Periodic structure of ethylbenzene bonded to a Si slab with six atomic layers. The front views show two possible conformations: **a** the molecule over the trench between dimer rows, **b** the molecule over a dimer row. The side view **c** shows one unit and its two images to illustrate the $\pi$-stacking arrangement

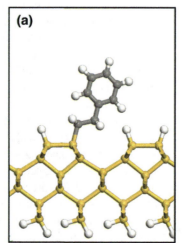
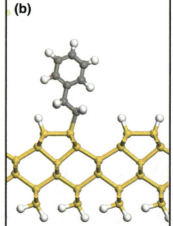
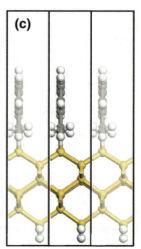

**Fig. 2** The two-probe system used for the transport calculations. *Inset* shows the alignment of the electrodes to the molecular wire from a different perspective

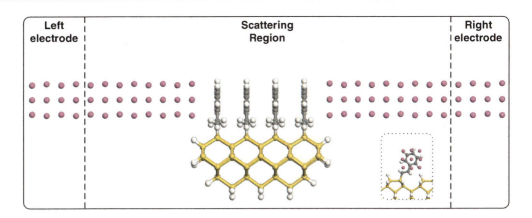

benzene ring, with a molecule–molecule separation of 3.867 Å, corresponding to the distance between two Si dimers on the same row, as shown in Fig. 1c. In the vertical direction (normal to the Si surface), at least 8 Å of vacuum space separated the top of the molecule from the lowest atom in the neighboring image so that there would be no interaction in between them.

The VASP calculations were carried out using the Perdew–Burke–Ernzerhof generalized gradient approximation (PBE–GGA) for the exchange correlation energy [19] with a semi-empirical van der Waals (vdW) correction to account for dispersion interactions of the π-stacked rings [20].[1] A projector augmented wave method was used for the ionic potentials [21, 22], with a kinetic energy cutoff for the plane wave basis of 400 eV. During the structure relaxations, the lowest two layers of Si atoms were frozen to their bulk positions while all other atoms were relaxed until the net force was less than 0.02 eV Å. In the plane of the Si slab, the Brillouin zone was sampled with sufficient $k$ points such that the energy was converged to less than 1 meV/atom.

It should be noted that there are two distinct possible conformations for the molecule bonded to the Si(100)–2 × 1 surface. In one of them, the molecule is over the trench between two dimer rows (Fig. 1a), while in the other, it is over a dimer row (Fig. 1b). From our total energy calculations, we found that the arrangement over the trench is more energetically favorable, in agreement with the experiments of Lopinski et al. [1]. Therefore, this structure was used for all further calculations.

The relaxed structure from Fig. 1a, c is then used to build the two-probe structure for the transport calculations. The system is repeated to form a line of four ethylbenzene molecules, which are then connected to 1-D Al electrodes, as shown in Fig. 2. Note that the structure is periodic in the direction normal to the page, as shown in Fig. 1a. It is also periodic in the vertical direction, but sufficient vacuum space separates the images so that there is no interaction between them. Along the direction of the electrodes, the scattering region is finite but the left and right electrode regions are periodic. The small cross-section of the electrodes was deliberately chosen to permit direct overlap with the π-stacked molecular wire, while limiting interaction with the Si substrate and the substituent groups on the molecules (see inset of Fig. 2). The electrodes used are 1-D Al(100) wires with cross-sectional area of 3 × 3 units, as shown in Fig. 2. The electrodes were relaxed in VASP in the same manner as described above. For the electrode–molecule separation, VASP calculations determined the ideal distance to be 3.40 Å for the ethylbenzene molecule. The same approach was used to relax the substituted benzene systems and to set up their two-probe geometries. For consistency, the same electrode–molecule separation was used so that similar orbital overlap could be maintained between the molecular wire and the electrodes. In each system, the electrodes were aligned with the center of the benzene ring, as shown in the inset of Fig. 2.

To calculate the electron transport through the system, we used a computational quantum transport technique that is based on the real-space Keldysh NEGF formalism combined self-consistently with DFT [23, 24]. This has been packaged in the Nanodcal electron transport code.[2]

The idea behind this approach is to calculate the Hamiltonian and electronic structure of the two-probe transport structure by DFT and the non-equilibrium statistical properties (population of electronic levels) of the scattering region is done by NEGF. The transport boundary conditions are treated by real-space numerical procedures. Interested readers are referred to Refs. [23, 24] for details of the NEGF–DFT implementation.

---

[1] The van der Waals radius cutoff was set to 15.0 Å, so that the terms corresponding to interactions over distances greater than this value are assumed to be zero.

[2] http://www.nanoacademic.ca.

This procedure is summarized as follows. The retarded Green's function at energy $E$ is obtained by inverting the Hamiltonian matrix,

$$G(E) = [(E + i\eta)S - H - \Sigma_1 - \Sigma_2]^{-1}, \quad (1)$$

where $H$ and $S$ are the Hamiltonian and overlap matrices for the central region determined with DFT [25]. $\eta$ is a positive infinitesimal and $\Sigma_{1,2}$ are self-energies that include the effect of the left and right electrodes on the scattering region. The self-energy is calculated within the NEGF–DFT formalism by an iterative technique [26]. It is a complex quantity with its real part representing a shift of the energy levels and its imaginary part representing their broadening, which can be represented as the broadening matrix, $\Gamma_{1,2} = i(\Sigma_{1,2} - \Sigma_{1,2}^{\dagger})$. The electronic density can be obtained from these quantities as,

$$\rho = (1/2\pi) \int_{-\infty}^{\infty} [f(E, \mu_1) G \Gamma_1 G^{\dagger} + f(E, \mu_2) G \Gamma_2 G^{\dagger}] dE, \quad (2)$$

where $\mu_{1,2}$ are the electrochemical potentials of the left and right electrodes and $f(E, \mu)$ is the Fermi–Dirac function that describes the population for a given energy and electrochemical potential. The density obtained is then used in a subsequent DFT iteration step, and the cycle is repeated until self-consistency is achieved. The transmission function is then calculated as

$$T(E) = \text{Tr}(\Gamma_1 G \Gamma_2 G^{\dagger}), \quad (3)$$

which represents the probability that an electron with a given energy $E$ transmits from one electrode, through the scattering region, and into the other electrode. This quantity can then be used to calculate the electric current as,

$$I = 2e/h \int_{\mu_1}^{\mu_2} T(E) dE. \quad (4)$$

For the transport calculations, norm-conserving pseudopotentials [27] were used to describe the atomic cores, and double-$\zeta$ polarized (DZP) numerical orbitals for the valence electrons. The exchange-correlation was treated using the local density approximation (LDA).

## 3 Results and discussion

This work investigates the effect of different substituent groups on the electronic and transport properties of the ethylbenzene molecule in the molecular wires. The substituents are placed at the para position relative to the ethyl group attaching to the surface and include $R = -NH_2$, $-OCH_3$, $-CH_3$, $-H$, $-CF_3$, and $-NO_2$; which are listed in order from strongest electron donating group (EDG) to strongest electron withdrawing group (EWG). To amplify the effect, di-substitution with the $-NH_2$ and $-NO_2$ groups at the 2,5-positions was also considered. The molecules are illustrated in Fig. 3.

### 3.1 Density of states

Plotting the density of states (DOS) is an effective way to compare the electronic structure for various systems. The DOS represents the proportion of electronic states as a function of energy. By projecting the DOS onto certain atoms, we can monitor which parts of a physical system contribute electronic states at certain energies. This analysis can be useful for interpreting the effect of substituent groups on the electronic structure of the different systems studied in this work. Note that the DOS calculations were

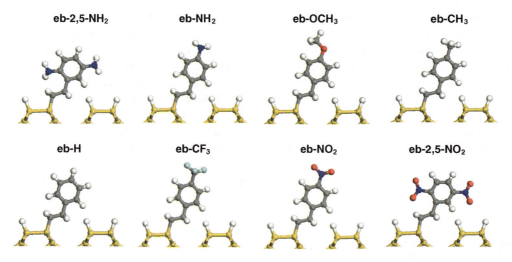

**Fig. 3** Set of substituted ethylbenzene molecules studied in this work

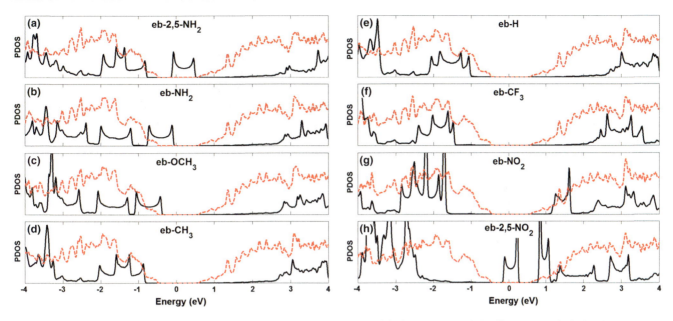

**Fig. 4** DOS projected onto atoms of substituted ethylbenzene molecules (*solid black*) and atoms of the Si substrate (*dashed red*)

carried out on the periodic systems shown in Fig. 1 with the Nanodcal code using LDA. Figure 4 shows DOS projected (PDOS) onto the atoms of the molecules (solid black plots) and the DOS projected onto the Si substrate (dashed red plots). The energy scale of the plots was shifted such that the valence and conduction bands of the Si would coincide in all systems for ease of comparison. In all cases, the Si PDOS look nearly identical showing only small differences in relative peak height. The top of the valence band is at −0.46 eV while the bottom of the conduction band is at 0.58 eV, meaning that the bandgap of the Si 1.04 eV, which is close to the known experimental value of 1.11 eV. However, it should be pointed out that this is a fortuitous agreement due to the known tendency of LDA to underestimate the bandgap, and the fact that the Si in these calculations is not truly a bulk material but a slab of finite thickness (6 atomic layers). Nevertheless, the use of LDA and a finite slab have opposite effects on the bandgap, and this results in a value close to experiment.

Beginning with the eb–H system (Fig. 4e), the HOMO state is located near −1.1 eV, HOMO-1 near −1.3 eV, and so on; while the LUMO is near 3.0 eV.[3] This gives a HOMO–LUMO gap of ca. 4.1 eV, in agreement with our previous work [11], and also the work carried out by others [7, 10]. Note that for the eb–H system, there are no molecular states in the Si bandgap, i.e. in the range [−0.5, 0.6] eV. As expected, adding EDGs to the molecule shifts the electronic states to higher energies by an amount proportional to the strength of the substituent group. In eb–CH$_3$, the HOMO level is shifted up to −0.9 eV; in eb–OCH$_3$ it is shifted to −0.4 eV; in eb–NH$_2$ it is at −0.1 eV; and with two –NH$_2$ substituents (Fig. 4a) it is at +0.5 eV. For the three strongest EDG systems (Fig. 4a–c), the molecular states actually lie inside the Si bandgap.

Conversely, the use of electron withdrawing groups has the opposite effect by shifting the molecular levels to lower energies. In this case, the LUMO level gets progressively closer to the Si bandgap for the eb–CF$_3$, and eb–NO$_2$ systems, and finally it is inside the bandgap for the eb–2,5-NO$_2$ system. Therefore, we found that substituent groups could be used to very effectively shift the molecular levels relative to those of the substrate. This approach could even be used to select the type of orbital (occupied vs. unoccupied) that is desired for electron transport to occur through. Finally, this effect can be further amplified by using multiple substituent groups on the same molecule, as shown for the disubstituted –NH$_2$ and –NO$_2$ systems (Fig. 4a, h).

### 3.2 Transmission

The transmission spectrum of a two-probe system (Fig. 2) shows the probability that an electron with a given energy originating in the left electrode will transmit through the scattering region and into the right electrode. By comparing the transmission spectra for the various substituted ethylbenzene molecules, we can determine the effect of the different substituent groups on the conductance of the ethylbenzene molecular wire.

In Fig. 5, the transmission spectra of three related systems are compared. In the middle (Fig. 5b), the

---

[3] The reason these levels have finite broadening is because of the π-stacking interaction with molecules in neighboring images of the periodic system.

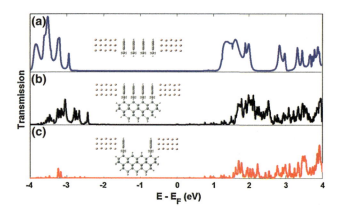

**Fig. 5** Transmission spectra for **a** four ethylbenzene molecules with no substrate, **b** four ethylbenzene molecules on top of Si, and **c** two molecules separated by a gap on Si

transmission through an ethylbenzene line on Si is plotted. At the top (Fig. 5a) is shown the transmission through an ethylbenzene stack with no underlying substrate, and at the bottom (Fig. 5c), there is a substrate but there is a gap in the molecular wire. For the system shown in Fig. 5a, clearly all of the transmission occurs through the molecules. The transmission peaks at −2.9, −3.2, −3.5, and −3.8 eV are due to transport through the HOMO, HOMO-1, and so on, of the molecular stack. The peaks in the range [1, 2] eV and higher are due to transmission through the unoccupied states. For the system at the bottom (Fig. 5c), the transmission is only possible through the substrate since there is a gap in the molecular wire. The important result is that, even with a gap in the molecular line, some transmission peaks are present, which must represent transmission through the substrate. This phenomenon was previously reported for a similar system with a smaller Si substrate [11], but it is interesting that it remains very similar when a much larger substrate is included in the calculation.

The plot in the middle (Fig. 5b) shows the transmission through the intact eb–H wire on top of Si, where transmission is possible through either the molecular wire or the substrate. One way to determine the conduction pathway for this system is by comparing its transmission spectrum to the other two systems. For example, the peaks in the range [−3.5, −2.4] eV are due to transmission through the molecules since there are similar peaks near this region in Fig. 5a but not in Fig. 5c.[4] On the other end of the spectrum, near 2 eV, the transmission seems to be through a combination of the molecular wire and through the Si substrate since there are substantial transmission peaks in both Fig. 5a and c.[5]

A subtle but important point is that there are some very small peaks at 0.35 and 0.77 eV for both systems containing the Si substrate (Fig. 5b, c), but these peaks are absent in the system without Si (Fig. 5a). It is significant because these peaks are the closest to the $E_F$ of the electrodes, and therefore will play the most important role in the low-bias conductance properties. These transmission peaks represent current leakage into the substrate.

A more direct approach for assigning transmission peaks is by analyzing the scattering states associated with each peak. Scattering states are analogous to eigenstates, but they apply to open-boundary systems such as the two-probe geometries considered in this work. They can be plotted in real space and reveal the parts of the system that the electrons are transmitting through at a given energy [23]. Figure 6 shows the transmission spectra for molecular wires composed of the molecules shown in Fig. 3. Various peaks are labeled according to which part of the system the transmission is occurring through, such as the molecular wire (mol), the substrate (sub) or through both, as determined from a scattering state analysis (vide infra).

For the unsubstituted system eb–H (Fig. 6e), we can see that the scattering state analysis supports the conclusions reached from the comparison of Fig. 5. The peaks in the range [−4.0, −2.5] eV and those near [1.5, 2.5] eV are mainly due to transmission through the molecule, while those small peaks in the range [0.0, 1.5] eV are due to transmission through the substrate. As discussed in Sect. 3.1, adding EDG substituents to the molecular wire shifts the states up in energy, therefore these groups can be used to bring the transmission peaks due to occupied states closer to the $E_F$ of the electrodes. This can be observed in Fig. 6a–d, the strongest effect being achieved with two −NH$_2$ groups on each ethylbenzene molecule (Fig. 6a), where the peak due to transmission through the HOMO is at −1.2 eV.

The use of EWGs as substituents has the opposite effect. It lowers the energy of the molecular states, bringing the peaks due to transmission through unoccupied states closer to the $E_F$, as can be seen in Fig. 6f–h. Substitution with one or two −NO$_2$ groups actually brings the transmission peaks of the LUMO states near the $E_F$ such that their tails spill over the $E_F$. In fact, were the LUMO states to be shifted below the $E_F$, these would, by definition, become occupied as electrons from the electrodes would fill them [28].

---

[4] Note that the effect of including the Si substrate is to shift the positions of the transmission peaks by ca. 0.5 eV and they become split due to hybridization with the Si states.

[5] Comparing the transmission peak positions in Fig. 5b to the PDOS positions in Fig. 4e, we see that they differ. This is because in position of the DOS peaks are relative to the Si substrate, while in the transmission spectrum they are relative to the $E_F$ of the Al electrodes. In other words, the peaks in the transmission spectra are shifted (by ca. −1.5 eV) relative to the PDOS plots.

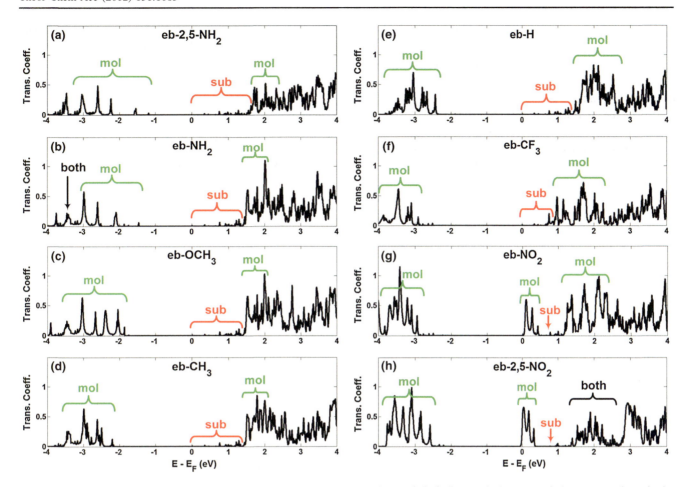

**Fig. 6** Transmission spectra for different substituted ethylbenzene molecules. *Labels* indicate whether transmission occurs through the molecular stack (mol), the substrate (sub), or through both

However, in Fig. 6g, h, only a small tail crosses the $E_F$, corresponding to a very small amount of charge transfer between the scattering region and the electrodes in these systems. The most important point is that in these cases (Fig. 6g, h), the transmission peak closest to the $E_F$ is through a molecular state. Therefore, there would be negligible leakage into the substrate at low bias, and because of the height and width of these peaks, the eb–NO$_2$ and eb–2,5–NO$_2$ molecules would make excellent molecular wires.

To further support this point, Fig. 7 shows two scattering states for the eb–NO$_2$ system (Fig. 6g). Fig. 7a is the scattering state associated with the tall molecular peak at 0.11 eV, while Fig. 7b is the scattering state associated with the small substrate peak at 0.77 eV. Note that the scattering state through the molecular stack clearly shows transmission through the π-orbitals, as indicated by the nodal planes coinciding with the phenyl rings in Fig. 7a. This sort of analysis was carried out to assign all peaks in Fig. 6.

One interesting point about Fig. 6 is that the peaks in the range [0.0, 1.5] eV which are due to transmission through the substrate are not affected by the nature of the

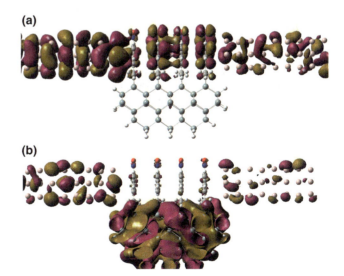

**Fig. 7** Scattering states for the eb–NO$_2$ system. **a** is associated with transmission peak at 0.11 eV, and **b** is associated with transmission peak at 0.77 eV in Fig. 6g

substituent. In fact, the small peaks at 0.35 and 0.77 eV are present in all systems (presumably including the systems containing –NO$_2$ groups, but buried underneath the tall molecular peaks). This is in contrast to the results reported by Anagaw et al. [16] who showed that the work function of a similar system is affected by the nature of the substituents. At this time, it is not clear why this sort of effect is not evident in the transport features. It may be due to the fact that the Fermi level in our two-probe system is determined by the Al electrodes, which might align the Si states. Another possibility is that the Si peaks outside the [0.0, 1.5] eV range are affected, but they are difficult to identify and compare in the different systems (i.e., in the range of [1.5, 4.0] eV). In any case, our calculations show that molecular states are clearly affected by the presence of substituent groups, even though we have not observed an effect on the Si substrate in the energy range we considered.

## 4 Summary

The electronic structure and transport properties were calculated for substituted ethylbenzene molecules that $\pi$-stack to form a wire on top of a Si substrate. The substituents varied from electron donating to electron withdrawing groups. In terms of the electronic structure, projected density of states revealed that molecular states could be tuned relative to the states of the Si substrate by judicious selection of the substituent groups. This degree of control was also reflected in the transmission spectra of such molecular wires when connected to Al electrodes. The low-bias transport mechanism can be changed from leakage into the Si substrate to high conductance through the molecular wire if the appropriate substituent groups are used.

**Acknowledgments** We thank Dr. Gino A. DiLabio for numerous helpful discussions. We are grateful to Sharcnet for access to computational resources and NSERC for financial support.

## References

1. Lopinski GP, Wayner DDM, Wolkow RA (2000) Nature 406(6791):48
2. Piva PG, DiLabio GA, Pitters JL, Zikovsky J, Rezeq M, Dogel S, Hofer WA, Wolkow RA (2005) Nature 435(7042):658
3. Basu R, Guisinger NP, Greene ME, Hersam MC (2004) Appl Phys Lett 85(13):2619
4. Hossain MZ, Kato HS, Kawai M (2005) J Am Chem Soc 127(43):15030
5. Hossain MZ, Kato HS, Kawai M (2005) J Phys Chem B 109(49):23129
6. Zikovsky J, Dogel SA, Haider MB, DiLabio GA, Wolkow RA (2007) J Phys Chem A 111(49):12257
7. Kirczenow G, Piva PG, Wolkow RA (2005) Phys Rev B 72(24):245306
8. Liu XY, Raynolds JE, Wells C, Welch J, Cale TS (2005) J Appl Phys 98(3):033712
9. Rochefort A, Boyer P, Nacer B (2007) Org Elect 8(1):1
10. Geng WT, Oda M, Nara J, Kondo H, Ohno T (2008) J Phys Chem B 112(10):2795
11. Smeu M, Wolkow RA, Guo H (2009) J Am Chem Soc 131(31):11019
12. Kirczenow G, Piva PG, Wolkow RA (2009) Phys Rev B 80(3):035309
13. Venkataraman L, Park YS, Whalley AC, Nuckolls C, Hybertsen MS, Steigerwald ML (2007) Nano Lett 7(2):502
14. Smeu M, Wolkow RA, DiLabio GA (2008) J Chem Phys 129(3):034707
15. Jalili S, Ashrafi R (2011) Physica E 43(4):960
16. Anagaw AY, Wolkow RA, DiLabio GA (2008) J Phys Chem C 112(10):3780
17. Kresse G, Hafner J (1993) Phys Rev B 47(1):558
18. Kresse G, Furthmüller J (1996) Phys Rev B 54(16):11169
19. Perdew JP, Burke K, Ernzerhof M (1996) Phys Rev Lett 77(18):3865
20. Grimme S (2006) J Comput Chem 27(15):1787
21. Blöchl PE (1994) Phys Rev B 50(24):17953
22. Kresse G, Joubert D (1999) Phys Rev B 59(3):1758
23. Taylor J, Guo H, J. Wang (2001) Phys Rev B 63(24):245407
24. Waldron D, Haney P, Larade B, MacDonald A, Guo H (2006) Phys Rev Lett 96(16):166804
25. Datta S (1995) Electronic transport in mesoscopic systems. Cambridge University Press, Cambridge
26. Sancho MPL, Sancho JML, Rubio J (1984) J Phys F: Met Phys 14(5):1205
27. Troullier N, Martins JL (1991) Phys Rev B 43(3):1993
28. Smeu M, DiLabio GA (2010) J Phys Chem C 114(41):17874

REGULAR ARTICLE

# Theoretical characterization of reaction dynamics in the gas phase and at interfaces

Hua Guo

Received: 8 August 2011 / Accepted: 22 August 2011 / Published online: 11 January 2012
© Springer-Verlag 2012

**Abstract** Reaction dynamics is a central topic in physical chemistry, and tremendous progress has been made in theoretical characterization of gas phase and surface scattering processes. Here, an overview is given on several important frontiers represented by the following articles.

**Keywords** Reaction dynamics · Scattering · Potential energy surface

Dynamics of chemical reactions in the gas phase and at interfaces is of fundamental importance in physical chemistry. Since the advent of crossed molecular beam technique, it has become possible to experimentally measure almost all quantum state resolved attributes resulting from both reactive and non-reactive scattering (see, for example, Ref. [1]). These quantities, including translational, internal, and angular distributions of the products, provide a wealth of information concerning the interaction potential between the colliding partners. The ability to accurately measure scattering attributes with quantum state resolution challenges theoreticians to develop accurate and efficient methods for characterizing reaction dynamics. On the other hand, reliable theoretical predictions can in turn provide guidance to further experimental exploration. The intimate interplay between experiment and theory has led,

Published as part of the special collection of articles celebrating the 50th anniversary of Theoretical Chemistry Accounts/Theoretica Chimica Acta.

H. Guo (✉)
Department of Chemistry and Chemical Biology,
University of New Mexico, Albuquerque, NM 87131, USA
e-mail: hguo@unm.edu

and will continue to lead, to deeper understanding of reaction dynamics in many molecular systems.

There are two major challenges in theoretical characterization of reaction dynamics. First, one has to establish an accurate global potential energy surface (PES) that covers the all necessary configuration space accessed by the scattering event. (In some systems where non-Born-Oppenheimer effects are important, one has to map out the non-adiabatic coupling as well). This is typically achieved by using a high-level electronic structure theory, followed by fitting the global PES. For atom–diatom reactions, for example, the ab initio points can be fit with the spline method. In larger systems, however, more sophisticated fitting methods are needed. The second challenge is the characterization of the reaction dynamics itself on the PES. As in other molecular processes, reaction dynamics is quantum mechanical in nature, and should be treated as such whenever possible. In many cases, nonetheless, classical mechanics with appropriate quantum modifications can also yield valuable and sometimes accurate information about reaction dynamics [2]. Here, we focus on the second challenge, namely theoretical characterization of reaction dynamics, assuming the relevant PES is known.

Since the first full-dimensional quantum scattering calculation on the $H + H_2$ reaction 35 years ago [3], tremendous progress has been made [4]. It is now routine to quantum mechanically compute nearly all observable attributes for atom–diatom reactive scattering events, even for those systems dominated by a long-lived reaction intermediate and those with no hydrogen atom [5]. While the time-independent scattering theory has made tremendous progress, it is the wave packet approach that has greatly extended our abilities to study complex reactions [6]. The basic idea is to propagate an initial state-specific wave packet from the reactant channel to the product channel, and to extract the necessary scattering information such as the S-matrix elements. Due to its time dependence, it is particularly suitable for direct reactions, which complete in a relatively short time. In addition, only one column of the S-matrix is calculated, which is less computationally demanding than that required for the entire S-matrix. Besides numerical efficiency, the wave packet approach is also physically intuitive. The crowning achievement of this approach is the recent work on the $H_2 + OH \rightarrow H + H_2O$ reaction, in which the exact quantum state-to-state differential cross sections were shown to be in excellent agreement with the latest crossed molecular beam experiment [7].

In this anniversary issue, Zhang and coworkers report a wave packet study on another tetra-atomic reactive system, namely the $OH + CO \rightarrow H + CO_2$ reaction. This is an important combustion reaction, representing the major production pathway for $CO_2$ in the flame. This exothermic reaction has a complex-forming mechanism, due to the existence of both *cis* and *trans* HOCO intermediates. Comparing with the direct $H_2 + OH$ reaction, its dynamics is much more challenging because of its involvement of three, instead of one, non-hydrogen atoms; as well as the deep HOCO potential wells. As a result, a huge number of basis functions and/or a large grid are needed to converge the scattering calculations. In addition, the reaction is essentially barrierless, necessitating a large number of partial waves to converge the cross sections. For these reasons, the work of Zhang and coworkers did not have the product state resolution. Nevertheless, this work represents the first full-dimensional quantum mechanical calculation of the total reaction cross section and rate constant for the $OH + CO$ reaction.

The high numerical costs in computing reaction cross sections for tetra-atomic systems underscores the bottleneck faced by conventional quantum mechanical approaches to reaction dynamics. In such a basis/grid-based approach, the size of the problem scales exponentially with the dimensionality of the system, and it is difficult to foresee the same kind of high-level treatment for higher dimensional systems in the near future. Consequently, it is highly desirable to explore alternative approaches. One possible solution is the multi-configuration time-dependent Hartree (MCTDH) method [8], which scales more favorably with dimensionality. Another approach resorts to the so-called Bohmian mechanics (also called quantum hydrodynamics), which is essentially a reformulation of time-dependent quantum mechanics [9]. Here, the wave packet is approximated by an ensemble of trajectories on the regular PES augmented by the so-called quantum potential, which is responsible for all quantum effects. This approach is very appealing because of the linear scaling of classical mechanics, which can easily handle larger

systems. It also offers a basis for more approximate quantum methods. However, the Bohmian approach is also fraught with difficulties, primarily due to the non-locality of its quantum potential. In this anniversary issue, progress in this direction is reported by two experts. Garashchuk discusses a Bohmian inspired imaginary time propagation method that allows the approximate calculation of the zero-point energy of large quantum systems up to 11 particles (33 dimensions). This is made possible by a low-order polynomial approximation of the quantum potential and its derivatives in Cartesian coordinates. Kendrick, on the other hand, proposes a numerical algorithm for propagating wave packet in the Bohmian formulation based on an iterative finite difference method. This method was demonstrated successfully in several model problems, including collinear reactive scattering. These advances hold promise for applications to larger systems in the future.

The final contribution by Troya deals with gas-surface collisional dynamics, which highlights the general methodology for describing extended systems. Due to the larger number of degrees of freedom, it is impractical to treat the dynamics quantum mechanically. As a result, the quasi-classical trajectory (QCT) method has been used to describe the dynamics. The appropriateness of classical mechanics in treating reaction dynamics has been extensively discussed in the literature [2], and it is widely accepted that such a treatment is reasonable, provided that quantum effects, such as tunneling and zero-point energy, are properly accounted for. QCT methods are particularly suitable for highly averaged quantities such as cross sections and rate constants. An added bonus is that the trajectories offer an intuitive description of the collisional encounters. In this particular contribution, the scattering of OH molecular from a fluorinated alkane self-assembled monolayer surface was investigated. Although no reactive channel is considered, this study addresses several important issues common in many surface processes, including reactions. Particular attention is paid to mechanistic questions, namely whether the scattering is direct or via surface trapping. In addition to the good agreement with experiment, such atomistic simulations provide valuable insights into microscopic details of the molecular encounters.

It is obviously impossible to cover the vibrant research area of reaction dynamics with only four articles. Nevertheless, these contributions, ranging from a state-of-the-art exact quantum scattering study based on the conventional basis/grid approach, the latest advances in Bohmian mechanics, to quasi-classical trajectory investigation of gas-surface scattering, offers some snapshots of the latest advances in theoretical characterization of gas phase and gas-surface reaction dynamics. Looking into the future, we expect more method development in quantum reactive scattering to mitigate the dimensionality bottleneck, as well as various semi-classical and quasi-classical treatments of collisional events with accurate inclusion of various quantum effects. An interesting future research area is concerned with the role played by dynamics in solution phase and enzymatic reactions, which is believed so far to be amenable to transition-state theory [10]. However, there is an increasing body of evidence that dynamics might play an important role in these processes [11].

## References

1. Yang X (2007) Annu Rev Phys Chem 58:433
2. Aoiz FJ, Banares L, Herrero VJ (2006) J Phys Chem A 110:12546
3. Schatz GC, Kuppermann A (1976) J Chem Phys 65:4642
4. Althorpe SC, Clary DC (2003) Annu Rev Phys Chem 54:493
5. Sun Z, Liu L, Lin SY, Schinke R, Guo H, Zhang DH (2010) Proc Natl Acad Sci USA 107:555
6. Zhang JZH (1999) Theory and application of quantum molecular dynamics. World Scientific, Singapore
7. Xiao C, Xu X, Liu S, Wang T, Dong W, Yang T, Sun Z, Dai D, Xu X, Zhang DH, Yang X (2011) Science 333:440
8. Beck MH, Jackle A, Worth GA, Meyer H-D (2000) Phys Rep 324:1
9. Wyatt RE (2005) Quantum dynamics with trajectories: introduction to quantum hydrodynamics. Springer, New York
10. Truhlar DG, Garrett BC, Klippenstein SJ (1996) J Phys Chem 100:12771
11. Hammes-Schiffer S, Benkovic SJ (2006) Annu Rev Biochem 75:519

Theor Chem Acc (2012) 131:1083
DOI 10.1007/s00214-011-1083-9

REGULAR ARTICLE

# Calculation of the zero-point energy from imaginary-time quantum trajectory dynamics in Cartesian coordinates

Sophya Garashchuk

Received: 6 July 2011 / Accepted: 19 August 2011 / Published online: 11 January 2012
© Springer-Verlag 2012

**Abstract** The imaginary-time quantum dynamics is implemented in Cartesian coordinates using the momentum-dependent quantum potential approach. A nodeless wavefunction, represented in terms of quantum trajectories, is evolved in imaginary time according to the quantum-mechanical Boltzmann operator in the Eulerian frame-of-reference. The quantum potential and its gradient are determined approximately, from the global low-order (quadratic) polynomial fit to the trajectory momenta, which makes the approach practical in high dimensions. Implementation in the Cartesian coordinates allows one to work with the Hamiltonian of the simplest form, to setup calculations in the molecular dynamics-compatible framework and to naturally mix quantum and classical description of particles. Localization of wavefunctions in the center-of-mass degrees of freedom and in the overall rotation, which makes the quadratic polynomial fitting in Cartesian coordinates accurate, is accomplished by the addition of a quadratic constraining potential, and its contribution to the zero-point energy is analytically subtracted. For illustration, the zero-point energies are computed for model clusters consisting of up to 11 atoms (33 dimensions).

**Keywords** Quantum dynamics · Quantum trajectories · Zero-point energy · Boltzmann operator · Imaginary time

Published as part of the special collection of articles celebrating the 50th anniversary of Theoretical Chemistry Accounts/Theoretica Chimica Acta.

S. Garashchuk (✉)
Department of Chemistry and Biochemistry,
University of South Carolina, Columbia, SC 29208, USA
e-mail: sgarashc@mail.chem.sc.edu

## 1 Introduction

The quantum-mechanical (QM) behavior of nuclei, manifested in the zero-point energy (ZPE) effect, tunneling and nonadiabatic transitions, is often essential for accurate description and understanding of reactions in gas phase and complex chemical environments, especially in processes involving hydrogen at low temperatures and energies. For example, ZPE stored in the vibrational modes of chemical reactants, products and transition-state species modifies reaction energy barriers, which can greatly influence the reaction rates and branching ratios [1, 2]. QM tunneling can be critical in proton transfer reactions [3–6]. Nonadiabatic dynamics involving transitions between different electronic or vibrational energy levels is always present in photochemistry [7–9]. As the system size increases, it becomes very difficult to describe molecular systems quantum-mechanically due to the exponential scaling of the standard methods of solving the time-dependent Schrödinger equation [10]. A number of multidimensional quantum approaches have been developed over the years, including those using basis contractions [11–13], Gaussian coherent state representations [14, 15] and mixed quantum/classical strategies [16–22]. Nevertheless, since 2001, up to date reaction dynamics of hydrogen and methane remains the largest reactive scattering process studied quantum-mechanically in full dimension using exact evolution method, the Multi-Configurational Time-Dependent Hartree method [12, 23].

At the same time, the classical treatment of nuclei is often appropriate, and the simulation methods based on classical trajectory dynamics [24] are applicable to molecular systems comprised of thousands of atoms. The trajectory representation of large molecular systems is appealing for several reasons. One reason is that the trajectory description of heavy particles, based on quasiclassical, semiclassical or quantum trajectory dynamics, is often more appropriate than the grid or basis representation, because close to the classical limit, $\hbar \to 0$, the wavefunctions are highly oscillatory. Another reason is that the initial conditions for the trajectories simulation can be chosen randomly, which formally circumvents the exponential scaling of the wavefunction representation with the system size. (The conventional direct product grid or basis methods scale exponentially by construction). In general, a wavefunction for an arbitrary coupled anharmonic Hamiltonian will be an exponentially complex object, but for molecular systems with most of the nuclei behaving classically one expects simpler wavefunctions due to quantum and thermal decoherence [25]. Not surprisingly, incorporation of QM effects into classical trajectory framework is a long-standing theoretical goal, which motivated the development of quasiclassical [26] and semiclassical methods (for example see [27]) traditionally based on dynamics of *independent* classical trajectories.

A number of recent trajectory methodologies [28], related to the quantum or the Madelung–de Broglie–Bohm trajectory formulation of the time-dependent Schrödinger equation [29–31], incorporate the intrinsically non-local QM effects into evolution of a trajectory *ensemble* representing a wavefunction. In this paper, we implement an approximate method of this type, namely dynamics with the momentum-dependent quantum potential (MDQP) [32] in Cartesian space: a wavefunction is evolved in imaginary time according to the QM Boltzmann operator yielding the ZPE estimates for high-dimensional systems. Implementation in the Cartesian coordinates is important because it allows one to work with the Hamiltonian of the simplest form, to setup calculations in the molecular dynamics-compatible framework and to naturally mix quantum and classical description of particles. The MDQP and its gradient are determined approximately from the global fit of the trajectory momentum, which is necessary for a practical multidimensional implementation with polynomial scaling. Section 2 describes the formalism and implementation, for simplicity in one dimension; $\nabla$ denotes the spatial derivatives throughout, including the one-dimensional case $\nabla = \partial/\partial x$. Numerical illustration for a model cluster (up to 11 particles) is given in Sect. 3. Section 4 presents discussion and summary.

## 2 Imaginary-time quantum trajectory evolution in Cartesian coordinates

### 2.1 The MDQP quantum trajectory formulation in the Eulerian frame-of-reference

The quantum trajectory approach in imaginary time is inspired by the Bohmian formulation of the Schrödinger equation,

$$\hat{H}\psi(x,t) = i\hbar \frac{\partial}{\partial t}\psi(x,t), \tag{1}$$

with the time-dependent complex wavefunction written in terms of real phase $S(x, t)$ and amplitude $A(x, t)$, $\psi(x,t) = A(x,t)\exp(iS(x,t)/\hbar)$. This yields the Hamilton–Jacobi equation

$$\frac{\partial S(x,t)}{\partial t} = -\frac{(\nabla S(x,t))^2}{2m} - V - Q \tag{2}$$

and the continuity equation on the probability density $A^2(x,t)$ [31]. Equation 2 leads to the Newtons equations of motion for the trajectories with the momenta $p(x, t) = \nabla S(x, t)$ guided by the sum of the classical potential $V$ and the quantum potential $Q$,

$$Q = -\frac{\hbar^2}{2m}\frac{\nabla^2 A(x,t)}{A(x,t)}. \tag{3}$$

Exact and approximate implementations of the real-time Bohmian methodology, including applicability and limitations, are reviewed in [28, 33]. The goal of the approximate methodology is to give estimates of the QM effects

The Boltzmann evolution of a wavefunction according to the diffusion equation with the QM Hamiltonian $\hat{H}$,

$$\hat{H}\psi(x,\tau) = -\hbar \frac{\partial}{\partial \tau}\psi(x,\tau), \quad \tau > 0 \tag{4}$$

is equivalent to Eq. 1 with the real-time variable $t$ replaced by $-\iota\tau$. This transformation, the so-called Wick rotation [34], is widely used starting with the path integral formulation of statistical mechanics [35] and including, for example, recent Gaussian-based methods [36–38]. In the semiclassical trajectory context, this transformation generates the real-time trajectory evolution on the inverted classical potential [39].

As $\tau \to \infty$, any initial wavefunction propagated in time according to Eq. 4 will evolve to the lowest energy eigenfunction (of the same symmetry if the system has a definite symmetry), since the lowest energy component is the slowest to decay. Choosing the energy scale so that eigenenergies are positive to avoid the exponential growth of the wavefunction norm $\langle\psi|\psi\rangle_\tau$, the wavefunction energy $E$ will converge to the ZPE value, $E_0$:

$$E(\tau) = \frac{\langle\psi|\hat{H}|\psi\rangle_\tau}{\langle\psi|\psi\rangle_\tau}, \quad \lim_{\tau\to\infty} E(\tau) = E_0 \tag{5}$$

This feature is central to the Diffusion Monte Carlo methods used in the largest exact QM ZPE calculations [40–44]. The imaginary-time evolution also can be viewed as "cooling" of a system to the temperature $T$, where $k_B T = 1/\beta = \hbar/\tau$, when acted upon by the Boltzmann operator $\exp(-\beta\hat{H})$ in the thermal reaction rate constant calculations [45].

In MDQP, Eq. 4 is reformulated in terms of trajectories by expressing a nodeless wavefunction as an exponent of the "action" function $S$, whose gradient is identified with the trajectory momenta [32],

$$p(x, \tau) = \nabla S(x, \tau). \tag{6}$$

We present the formalism for a particle of mass $m$ in one Cartesian dimension $x$ in the Eulerian frame-of-reference. Multidimensional generalizations and the Lagrangian frame-of-reference formulation can be found in [46, 47]; Ref. [48] contains a comparison of the Eulerian and Lagrangian formulations. Equation 4 has been implemented in terms of trajectories earlier by other groups [49, 50] in terms of "independent" quantum trajectories. These methods are based on the hierarchy of differential equations for the high-order gradients of $S$ truncated at a fixed level $q$ ($q = 4$ or 6 in practice); the number of equations scales as $(N_{\mathrm{dim}})^{q+1}$. The MDQP approach evolves only the first derivatives of $S$, $p = \nabla S$, and the approximation is implemented globally, thus making it practical for high-dimensional systems.

To obtain the classical-like equations of motion, we express a positive wavefunction via a single exponential function,

$$\psi(x, \tau) = \exp\left(-\frac{S(x, \tau)}{\hbar}\right) \tag{7}$$

which, substituted into Eq. 4, gives the equivalent of the Hamilton–Jacobi equation,

$$\frac{\partial S}{\partial \tau} = -\frac{(\nabla S)^2}{2m} + V + \frac{\hbar}{2m}\nabla^2 S. \tag{8}$$

Defining the momentum according to Eq. 6, the last term in Eq. 8 is interpreted as the momentum-dependent quantum potential (MDQP) [32, 51],

$$U(x, \tau) = \frac{\hbar \nabla p}{2m}, \tag{9}$$

responsible for *all* QM effects. It is non-local and influences the dynamics on equal footing with the external classical potential $V$. In the Lagrangian frame, Eq. 8 defines the trajectory dynamics on the inverted classical potential with MDQP of Eq. 9 added to it [32]. As a consequence, trajectories leave the region of low potential energy causing undersampling of the ground-state wavefunction at long times in high-dimensional ZPE calculations [51]. Thus, we consider the Eulerian frame-of-reference where the initial trajectory positions are stationary random grid points. In quantum trajectory dynamics, the Eulerian and Arbitrary Lagrangian-Eulerian frames were introduced by Trahan and Wyatt [52].

The trajectory momentum function at a fixed $x$ evolves according to the gradient of Eq. 8,

$$\frac{\partial p}{\partial \tau} = -\frac{p \nabla p}{m} + \nabla(V + U). \tag{10}$$

For practical multidimensional implementation, the first and second derivatives of $p$ in Eqs. 9 and 10 are computed approximately from the global Least Squares Fit [53] to $p$ in the Taylor (or monomial) basis $\vec{f}$,

$$\vec{f} = (1, x, x^2 \ldots). \tag{11}$$

The fitting coefficients $\vec{c}$ minimize the difference between the exact momenta and its fit $\tilde{p}$, $\tilde{p} = \vec{f} \cdot \vec{c}$,

$$I = \langle (p - \tilde{p})^2 \rangle. \tag{12}$$

Their optimal values are the solutions to a system of linear equations $\nabla_{\vec{c}} I = \vec{0}$,

$$\mathbf{M}\vec{c} = \vec{b}, \quad \mathbf{M} = \langle \vec{f} \otimes \vec{f} \rangle, \quad \vec{b} = \langle p\vec{f} \rangle. \tag{13}$$

The expectation values are evaluated over the trajectory ensemble,

$$\langle \hat{\Omega} \rangle = \int \Omega(x)\psi^2(x, \tau)\mathrm{d}x = \sum_i \Omega(x_\tau^{(i)})e^{-2S_\tau^{(i)}/\hbar}\delta x^{(i)}. \tag{14}$$

Superscript $i$ labels the trajectory-related quantities after discretization of the initial wavefunction; subscript defines the time. The trajectory weight, $\delta x^{(i)}$, is the contribution of the $i$th trajectory to the integrals at time $\tau = 0$; the trajectory weights are constant in time.

The linear basis is exact for Gaussian wavefunctions and gives zero quantum force in Eq. 10. The quadratic basis is the smallest one that generates evolution of $p(x, \tau)$ that is different from the classical evolution. The momentum fit $\tilde{p}$ determines the approximate MDQP in Eq. 8 and in the right-hand side of Eq. 10. The latter contains the term without $\hbar$, $p\nabla p/m$, which is an approximation we have to make in the Eulerian frame in addition to approximating the MDQP terms. (In the Lagrangian frame, this term is treated exactly as part of the full-time derivative). But the advantages of the stationary points over the evolving trajectories are considerable: classical potential and its gradient is evaluated only once and, for localized ground states, the stationary points sample the high-density region of the wavefunction at all times, while in the Lagrangian formulation, the trajectories leave this region. The MDQP formulation given by Eqs. 7, 10 and 13 has been shown to give accurate ZPE estimates for anharmonic systems, including the double well, and for the triatomic molecules using the normal mode coordinates with a reasonably small (quadratic) momentum fitting. The MDQP results converged to the QM result for larger bases of 4–6 functions [32, 51].

For a system of $N_{\mathrm{atom}}$ atoms, the Cartesian space formulation in $3N_{\mathrm{atom}}$ dimensions has the advantage of the simple equations of motion [23, 54] for an example of how complicated the Hamiltonian in internal coordinates is already for a non-rotating four-particle system. However, in imaginary time, the Cartesian description brings forward a question of how to treat the redundant degrees of freedom: the center-of-mass (CoM) motion and the overall rotation do not contribute to ZPE, but result in the wavefunction delocalization in the corresponding degrees of freedom. The practical, small-basis momentum fitting is accurate for localized wavefunctions close to Gaussians in Cartesian space. Therefore, wavefunction localization and shorter decay (to ZPE) time are highly desirable features for the imaginary-time approximate MDQP dynamics discussed in the remainder of this section.

### 2.2 The center-of-mass motion

In imaginary time, an arbitrary initial wavefunction decays to the ground state of the non-rotating system, CoM being at rest: at infinite $\tau$, the total energy is equal to the ZPE of the internal degrees of freedom. The CoM motion is, of course, decoupled from the other modes of motion, but it can affect the accuracy of the approximate implementation. Let us examine the effect of CoM motion on the convergence of the total energy to the ZPE. We take $\psi(\vec{x}, 0)$ as a product of Gaussians in each Cartesian dimension centered at the minimum $\vec{x}_0$ of $V$,

$$\psi(x_\lambda, 0) = \left(\frac{2\alpha_0^\lambda}{\pi}\right)^{1/4} \exp(-\alpha_0^\lambda(x_\lambda - x_0^\lambda)^2), \quad (15)$$

where $x_0^\lambda$ and $\alpha_0^\lambda$ correspond to the dimension $\lambda$. In terms of the atomic Cartesian positions $\vec{r}_n$, for $n = 1, 2, \ldots, N_{\text{atom}}$ and the corresponding atomic masses, the CoM position $\vec{R}$, is

$$\vec{R} = M^{-1} \sum_n m_n \vec{r}_n, \quad M = \sum_n m_n. \quad (16)$$

Here and below the latin subscripts, $i, j$ etc., are used to label atoms; the greek subscripts $\lambda$, $\mu$ etc., are used to index the elements of the vectors and matrices of dimensionality $3N_{\text{atom}}$. In particular,

$$\vec{x} = (\vec{r}_1, \vec{r}_2, \ldots, \vec{r}_{N_{\text{atom}}}) = \{x_\lambda\}, \quad \lambda = 1, 2, \ldots, 3N_{\text{atom}}. \quad (17)$$

In full dimensionality, the masses can be arranged as an array of $3N_{\text{atom}}$ elements, $(m_1, m_1, m_1, m_2, m_2, m_2, \ldots)$.

The normalized energy of a Gaussian (for one degree of freedom) evolving according to Eq. 4 in free space is

$$E^{\text{cm}} = \frac{\alpha_0}{2M(1 + 2\tau\alpha_0/M)}. \quad (18)$$

The value of $E^{\text{cm}}$ and its convergence to zero depend on $\alpha_0/M$, where $\alpha_0$ is large for the 'classical' degrees of freedom

describing heavy particles. The convergence is hyperbolic with time and implies a complete delocalization of $\psi(x, \tau)$ in the CoM degrees of freedom. Since the approximate MDQP methodology is practical and accurate in the regime of localized wavepackets, we need to subtract CoM energy, without changing the Cartesian space wavepacket setup and Hamiltonian, and to counteract the spreading. This can be achieved by constraining—in the spirit of soft constraints used in molecular mechanics methods [55]—the CoM motion with the quadratic potential in $\vec{R}$ added to $V$,

$$V^{\text{cm}} = \frac{Mw^2}{2}(\vec{R} - \langle\vec{R}\rangle)^2 = \frac{k_{\text{cm}}}{2}(\vec{R} - \langle\vec{R}\rangle)^2. \quad (19)$$

The average CoM position will be set to zero, $\langle\vec{R}\rangle = 0$, henceforth.

Examining the imaginary-time evolution of a Gaussian in a quadratic potential [51], one finds the convergence of $E^{\text{cm}}$ to a constant value to be exponential, which is also true for the wavepacket width parameter $\alpha_\tau$. Defining the coherent width $\alpha_c = Mw/2$ and $\eta = (\alpha_0 - \alpha_c)/(\alpha_0 + \alpha_c)$,

$$E^{\text{cm}}_{\tau\to\infty} = \frac{w}{2}\left(1 + \eta^2 e^{-4w\tau}\right) \quad (20)$$

$$\alpha_{\tau\to\infty} = \alpha_c\left(1 + 2\eta e^{-2w\tau}\right). \quad (21)$$

Obviously, presence of $V^{\text{cm}}$ improves convergence of ZPE with time and localizes the CoM wavepacket at the coherent value. The CoM energy, which is $3w/2$ at the end of time evolution for three dimensions of $\vec{R}$, can be analytically subtracted without knowing the explicit form of the initial CoM wavefunction. For large molecular systems, $M \to \infty$, the CoM constraint might be unnecessary if $\alpha_0/M$ is small—the CoM wavepacket remains localized during the course of evolution—and if $E^{\text{cm}}$ can be neglected with compared to the internal ZPE.

### 2.3 The overall rotation

We also need to fix the overall rotation of the molecular system and prevent delocalization of the wavefunction over the corresponding angles. This will be accomplished with the "soft constraint" as well by adding an effectively three-dimensional quadratic potential defined (after shifting CoM to zero) by three vectors $\vec{d}^{(n)}$, $n = \{1, 2, 3\}$, perpendicular to the average positions of all atoms $\vec{q}_i, i = 1 \ldots N_{\text{atom}}$, and perpendicular to the three unit vectors along the Cartesian axes $\vec{e}^{(n)}$,

$$\vec{e}^{(1)} = (1, 0, 0), \quad \vec{e}^{(2)} = (0, 1, 0), \quad \vec{e}^{(3)} = (0, 0, 1). \quad (22)$$

We have also used the three unit vectors along the principal axes of the moments of inertia [56] with the same effect as when using Eq. 22. Using the full dimensional vectors of average positions

$$\vec{q} = (\vec{q}_1, \vec{q}_2 \ldots \vec{q} N_{\text{atom}}) = \frac{\langle \psi | \vec{x} | \psi \rangle}{\langle \psi | \psi \rangle}, \tag{23}$$

and the three vectors defining the directions of the overall rotation (subscript $i$ labels atoms)

$$\vec{d}_i^{(n)} = \frac{m_i}{M} \vec{q}_i \times \vec{e}^{(n)}, \vec{d}^{(n)} = (\vec{d}_1^{(n)}, \vec{d}_2^{(n)} \ldots \vec{d}_{N_{\text{atom}}}^{(n)}) \tag{24}$$

the rotational localizing potential is

$$V^{\text{rot}} = \frac{k_{\text{rot}}}{2} \sum_{n=1}^{3} \left( \vec{d}^{(n)} \cdot (\vec{x} - \vec{q}) \right)^2. \tag{25}$$

The same is more convenient when written in using matrix and vectors of full dimensionality:

$$V^{\text{rot}} = \frac{k_{\text{rot}}}{2} (\vec{x} - \vec{q}) \cdot \mathbf{D} \cdot (\vec{x} - \vec{q}), \quad D_{\lambda\mu} = \sum_{n=1}^{3} d_\lambda^{(n)} d_\mu^{(n)}. \tag{26}$$

The vectors $\vec{d}^{(n)}$ of Eq. 24 are normalized after construction. The factor $m_i/M$ is introduced into $\vec{d}^{(n)}$ to make $V^{\text{rot}}$ independent on the particle mass, as to have the same $V^{\text{rot}}$ for $H_2$ and HD.

Since the localizing potential given by Eq. 26 is a quadratic function, the energy of the overall rotation will decay in time to the ZPE of $V^{\text{rot}}$ defined by the three (or two for a linear molecule) non-zero eigenvalues $\eta_\mu$ of the mass-weighted Hessian matrix $\mathbf{h}$,

$$h_{\lambda\mu} = \frac{k_{\text{rot}} D_{\lambda\mu}}{\sqrt{m_\lambda m_\mu}}, \quad E_{\tau \to \infty}^{\text{rot}} = \frac{1}{2} \sum_{\mu=1,3} \sqrt{\eta_\mu}. \tag{27}$$

The coordinate transformation into the internal degrees of freedom is not needed [56]; the matrix size is $3N_{\text{atom}}$. Once added to $V + V^{\text{cm}}$, $V^{\text{rot}}$ will keep the wavefunctions localized in the three directions of the overall rotation, enabling use of a cheap MDQP approximation; its analytically known contribution to the total energy should be subtracted to obtain the internal ZPE. In general, for anharmonic potentials, the rotations are not rigorously decoupled from the internal modes [57], but for the ground state, the effect of $V^{\text{rot}}$ on ZPE is small as shown in the next section. The effects of the CoM and rotational harmonic potentials are visualized in Fig. 1 for a trimer, as described detail in the next section.

## 3 ZPE calculations

The vibrational energy calculations of spectroscopic accuracy for general systems are beyond the capabilities of the approximate MDQP method. This approach is not designed to compete with the exact methods, such as the Diffusion Monte Carlo or Vibrational Self-Consistent Field [58, 59] but to give cheap estimates of various types of QM

effects. The ZPE calculations serve as a convenient test of the approximate MDQP, which is expected to give reasonable estimates for semi-rigid molecules by incorporating leading anharmonic terms of classical potentials.

As a proof-of-principle, we apply the formalism of Sect. 2 to compute ZPEs of systems consisting of up to $N_{\text{atom}} = 11$ nuclei with *pairwise nearest neighbor* interactions,

$$V = \sum_{i>j}^{\text{nearest}} V_{ij}, \quad V_{ij} = D(\exp(-z(r_{ij} - r_0)) - 1)^2, \tag{28}$$

where $r_{ij}$ are the bond distances, $r_{ij} = |\vec{x}_i - \vec{x}_j|$. The mass and parameters of the Morse potential given by Eq. 28 [60] describe $H_2$ molecule in atomic units: $D = 0.17429$ $E_h$, $r_0 = 1.4$ $a_0$ and $z = 1.0435$ $a_0^{-1}$, $m = 1836$ a.u. The classical minimum energy configuration of atoms is simply the geometry when all bonds included in the sum are equal to $r_0$, $\langle r_{ij} \rangle = r_0$. The formalism of Sect. 2 is implemented using the quadratic fit of $p$. The initial wavefunction is defined as the direct product of the Gaussians (15) centered at the minimum of the classical potential. The width parameters $\alpha_0$ were assigned the same values for all dimensions and did not correspond to the normal mode values. The sampling of the random grid points is uniform [53] within the region of the wavefunction density $\psi^2(x,0) > 10^{-\varepsilon}$. Parameter $\varepsilon$ is the sampling cutoff in a single dimension. To make the low-order polynomial fitting accurate, only the central region, $\varepsilon = 0.125$, is typically sampled. The force constants in Eqs. 19 and 26 are chosen to give energy due to the localization potentials, Eqs. 19 and 26, on the order of the internal ZPE. The time evolution of $S$ and $p$ in the Eulerian frame-of-reference according to Eqs. 8 and 10 is implemented in a straightforward manner giving linear convergence of the wavefunction norm and (unnormalized) energy with respect to the time step. (i) The polynomial fit, $\tilde{p}$, of the function $p(x)$ is used in the right-hand side (RHS) of Eqs. 8 and 10. (ii) The values of $S$ and $p$ are incremented by the corresponding RHS values multiplied by $d\tau$. This choice allows the RHS of Eq. 10 to be an analytical gradient of Eq. 8 throughout the propagation.

### 3.1 A diatomic molecule

$H_2$ molecule is described in six Cartesian coordinates with positions of protons 1 and 2 denoted as $\vec{r}_1 = (x_1, x_2, x_3)$ and $\vec{r}_2 = (x_4, x_5, x_6)$. The initial width parameter is $\alpha_0 = 12$ $a_0^{-2}$ in all dimensions; the molecule is oriented along the $z$-axis, and its CoM is at zero. The value of $\alpha_0$ is roughly defined by the normal mode frequencies, but as shown in [32] in practice, the ZPE is rather insensitive to this choice. In the Cartesian coordinates, the localizing CoM potential is

**Fig. 1** Trimer: density localization with and without constraints. The isodensity surfaces are shown after evolution up to $\tau = 0.1$ a.u. **a** without localization potentials, **b** with just $V^{cm}$ included, **c** with just $V^{rot}$ included and **d** with both CoM and overall rotation localizing potentials included

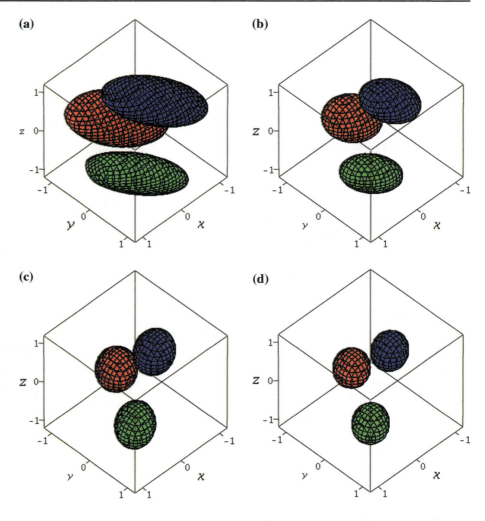

$$V^{cm} = \frac{k_{cm}}{8}\left((x_1+x_4)^2+(x_2+x_5)^2+(x_3+x_6)^2\right). \quad (29)$$

The rotational potential is

$$V^{rot} = \frac{k_{rot}}{4}\left((x_1-x_4)^2+(x_2-x_5)^2\right), \quad (30)$$

which makes physical sense—rotations from the original z-direction will increase the total energy. Figure 2 shows the total normalized energy of the wavefunction with the ZPEs of localizing potentials subtracted as appropriate. The values of the force constants were the same $k_{cm} = k_{rot} = 1.5 E_h/a_0^2$, which gave ZPEs of 2.86 $E_0$ and 3.04 $E_0$, respectively. $E_0$ is the analytical ZPE of the Morse potential, $E_0 = 1.00187 \times 10^{-2}$ $E_h$. The trajectory ensemble consisted of 2500 random points uniformly sampling the initial density within the cutoff parameter $\varepsilon = 0.125$. The momentum components were fitted with quadratic polynomials yielding the total basis size of $N_{bas} = 28$. The scaling of the quadratic basis with the dimensionality is $N_{bas} = (N_d + 1)(N_d + 2)/2$, $N_d = 3N_{atom}$. The approximate MDQP evolution is cheap: propagation of 1,000 trajectories for one thousand time steps in six dimensions takes about 2 s on a desktop workstation. The scaling of CPU is linear with respect to the number of trajectories and quadratic with respect to the basis size.

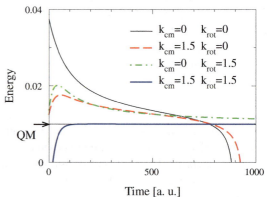

**Fig. 2** The internal energy of the $H_2$ system computed in 6D Cartesian coordinates. The force constants of the confining potentials are on the legend. *Arrow marks* the analytical internal ZPE

As seen from Fig. 2 without $V^{rot}$, the wavepacket energy $E$ sharply drops after $\tau \approx 700$ a.u., because the wavefunction spread in angles makes the quadratic momentum

**Table 1** The zero-point energy for H$_2$ (and isotope substitutions) computed from imaginary-time propagation in 6 Cartesian coordinates

| $N_{traj}$ | $t$ (a.u.) | $m$ (a.u.) | $\alpha_0$ (a$_0^{-2}$) | $E(\tau)$ (%) | $dE/d\tau$, (E$_h$/a.u.) |
|---|---|---|---|---|---|
| 100 | 250 | HH | 12 | 99.60 | $1.9 \times 10^{-8}$ |
| 500 | 250 | HH | 12 | 99.52 | $2.5 \times 10^{-8}$ |
| 2500 | 250 | HH | 12 | 99.52 | $2.3 \times 10^{-8}$ |
| 2500 | 250 | HH | 9 | 99.36 | $7.2 \times 10^{-9}$ |
| 2500 | 250 | HH | 18 | 99.70 | $6.0 \times 10^{-8}$ |
| 2500 | 250 | HD | 12 | 99.73 | $-6.6 \times 10^{-8}$ |
| 2500 | 250 | DD | 12 | 99.61 | $-3.2 \times 10^{-7}$ |
| 2500 | 250 | HT | 12 | 99.68 | $-2.1 \times 10^{-7}$ |
| 2500 | 250 | DT | 12 | 99.43 | $-5.8 \times 10^{-7}$ |
| 2500* | 250 | HH | 12 | 98.05 | $9.6 \times 10^{-7}$ |

All calculations were performed with the quadratic momentum fitting, except for the last calculation marked * performed with the linear momentum fitting

fitting inaccurate. With the rotational localizing potential in place, the energy converges to a constant, but it does so slowly without the CoM constraint. The accuracy of the internal ZPEs is given in Table 1. The initial orientation of H$_2$ was along the $z$-axis, but to verify the invariance of the formalism to rotation, the initial orientation for the isotopically substituted species was rotated. The initial Gaussian wavefunction was centered at $\vec{x}_0 = (0.73654, -0.40656, 0.40415, -0.36827, 0.20328, -0.20207)$ a$_0$. $V^{rot}$ was constructed using the principal axes of inertia rather than Cartesian directions of Eq. 22 in this case. The internal ZPE is given as percent of the analytical ZPE. The accuracy is within 0.5% in all cases. The linear fitting of the momentum gives ZPE within 2% and, more importantly, does not converge to a plateau value with time.

3.2 Clusters

First, we will examine the trimer with pairwise interactions via the Morse potential described above. The exact QM ZPEs are obtained from the Fourier-transforms of autocorrelation functions propagated in real time using the split-operator method [61]. The system was described in three dimensions in the Jacobi coordinates at zero total angular momentum using a grid of $32 \times 32$ points in the radial coordinates and 25 Discrete Variable Representation (Gauss-Legendre) points in angle [10]. The energy level resolution is $4 \times 10^{-5}$ E$_h$. For the MDQP calculation, the parameters are $\varepsilon = 0.125$, $k_{cm} = k_{rot} = 8$ E$_h$/a$_0^2$. Ensemble of 2500 trajectories was evolved up to $\tau = 200$ a.u. The initial width parameter is taken $\alpha_0 = 12$ a$_0^{-2}$ for all 9 Cartesian dimensions. The centers of the Gaussians form an equilateral triangle in $xy$-plane with the side of $r_0 = 1.4$ a$_0$. This configuration gives the lowest value of the

**Table 2** ZPE for the hydrogen (and deuterium substituted) model trimer computed from imaginary-time propagation in 9 Cartesian coordinates

| Species | $E(\tau)$ (E$_h$) | $dE/d\tau$ (E$_h$/a.u.) | $E^{QM}$ | $E(\tau)/E^{QM}$ |
|---|---|---|---|---|
| H$_3$ | 0.029496 | $3.0 \times 10^{-8}$ | 0.029368 | 1.0044 |
| H$_2$D | 0.026880 | $-1.0 \times 10^{-7}$ | 0.026777 | 1.0039 |
| HD$_2$ | 0.024026 | $-4.5 \times 10^{-7}$ | 0.023950 | 1.0039 |
| D$_3$ | 0.020929 | $-1.0 \times 10^{-6}$ | 0.020887 | 1.0020 |

The number of trajectories is 2500, final $\tau = 200$ a.u., $\varepsilon = 0.125$ and $\alpha_0 = 12$ a$_0^{-2}$ for all systems

classical potential. The internal ZPEs, given in Table 2, are computed for H$_3$ and for the deuterium substituted species and compared to the exact QM results. As seen from the table, the MDQP with quadratic fitting gives accuracy better than 0.4% and converges to the plateau value within $10^{-6}$ E$_h$. This shows that the choice of the rotational localization potential of Eq. 26 is correct. The MDQP calculation takes 6 s, whereas the exact QM propagation (using a fairly small grid described above) takes about 6 min.

There are two parameters that determine the accuracy of the MDQP calculation: the force constants of the localization potentials, $k_{cm}$ and $k_{rot}$ and the cutoff parameter $\varepsilon$. (The fitting basis size, obviously, has crucial effect on the accuracy, but due to polynomial scaling of the basis size with $N_{atom}$, we consider only the quadratic basis). The CoM motion separates from other modes of motion; the rotational motion is separable if the classical potential is quadratic. In this case, considering for simplicity $k_{cm} = k_{rot} = k$, the energy associated with the added localization potentials given by Eqs. 19 and 26 is a linear function of $\sqrt{k}$. Once this localization energy is subtracted from the total energy, the remaining internal modes ZPE should be constant. Figure 3 shows the internal modes ZPE for H$_3$ extracted from calculations with various localization force constant $k$. As seen from the plot, indeed there is a

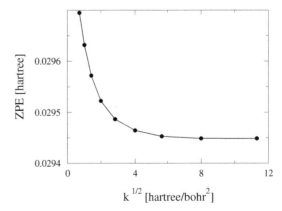

**Fig. 3** Dependence of the internal ZPE on the force constant of the localization potentials given by Eqs. 19 and 26 for the trimer; $k_{cm} = k_{rot} = k$

regime of $k$, where the internal ZPE approaches a plateau value (which corresponds to the total ZPE being linear in $\sqrt{k}$) yielding the internal ZPE estimates.

The effect of localization in CoM and rotation modes is visualized on Fig. 1. The isodensity surface plotted on the figure at $\tau = 200$ a.u. is approximated with three-dimensional Gaussian functions for each atom using single-particle fitting of $\vec{p}$. Panel (a) has zero $V^{cm}$ and $V^{rot}$; on panel (b), the CoM localization is introduced; on panel (c), just the rotational localization is included. Both, $V^{rot}$ and $V^{cm}$, were included on panel (d). The value of the force constant is $k = 2$. As we see, $V^{rot}$ and $V^{cm}$ are essential for localization of the wavepacket. The effect of $V^{cm}$ is not as dramatic as that for $V^{rot}$. As discussed in Sect. 2, $V^{cm}$ accelerates convergence of the total energy to a plateau value.

The dependence of ZPE on the sampling cutoff parameter $\varepsilon$ is illustrated on Fig. 4. The initial random points sample the region of the wavefunction density where $\rho > 10^{-\varepsilon}$ for each dimension. Sampling more compact region of space—smaller $\varepsilon$ values—improves accuracy of the fitting, but making $\varepsilon$ too small looses information on the anharmonicity of the potential. On the figure, the ZPE is shown for the linear and quadratic fitting of the momentum: the linear fit approaches the normal mode value as $\varepsilon \to 0$; the quadratic fit approaches a value about 0.3% lower than the exact QM result. The normal mode estimate is 1.9% higher. The MDQP accuracy is worse for large values of $\varepsilon$. This difference of 0.3% is the limit of the quadratic fitting basis. The parameter values $k = 8$, $\varepsilon = 0.125$ and the quadratic fitting basis will be used for larges systems below unless stated otherwise.

To test the scalability of the approximate MDQP, we extend our model system to clusters of up to 11 atoms. The cluster geometries are obtained by first adding an atom on the positive $z$-axis above the trimer in $xy$-plane to create a tetrahedron (tetramer), then adding the fifth atom on the negative $z$-axis (pentamer). More atoms are added atop of each face of the pentamer creating an hexagonal close-packed array. The initial parameters are $\alpha_0 = 12$ $a_0^{-2}$, $\varepsilon = 0.125$. Ensembles of $N_{traj} = 2500$ trajectories were evolved up to $\tau = 200$ a.u. At this time, the normalized energies reached plateau values, which are the MDQP estimates of the cluster ZPEs. These values are compared to the normal mode ZPEs in Table 3 for $N_{atom} = 3 - 8, 11$. The time derivatives, $dE(\tau)/d\tau$, indicate the convergence. The results are converged with respect to the number of trajectories within 4 digits and with respect to $\varepsilon$ within 0.5%. Figure 5a shows the ZPEs per bond in the units of the energy of the dimer bond. The difference between the MDQP estimates and normal mode estimates changes from 2% for 3 bonds to 0.5% for 27 bonds. For the trimer, for which we have accurate QM ZPE, the difference between the QM and MDQP result is 6 times smaller than between the MDQP and the normal mode estimate. Thus, MDQP calculation captures changes in the ZPE due to anharmonicity of the Morse potentials. Note that the leading, cubic, anharmonic term gives zero correction to the energy within the first-order perturbation theory. This claim is consistent with the ZPE calculation for particles that are 10 times heavier, $m = 18360$ a.u. The results are shown on Fig. 5b. For heavier particles, the eigenstates are more localized. Therefore, the normal mode approximation is more accurate, and indeed, the discrepancy between the MDQP and normal mode approximations is smaller ($\approx 3$ times smaller since the energy scales as $m^{-1/2}$). Compared to $m = 1836$ a.u. calculations, the initial width was changed to $\alpha_0 = 38$ $a_0^{-2}$ and the evolution was performed up to $\tau = 1{,}000$ a.u.

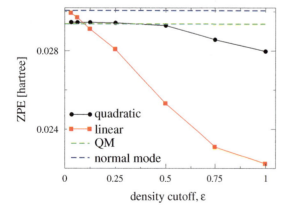

**Fig. 4** Dependence of the ZPE of the trimer on the sampling cutoff parameter $\varepsilon$, $\rho > 10^{-\varepsilon}$. The results for the linear momentum fitting converge to the normal mode ZPE; the results for the quadratic momentum fitting the QM result

**Table 3** Internal energy for the clusters, $E(\tau)$, approaching the ZPE at $\tau = 200$ a.u., computed from the imaginary-time MDQP propagation in Cartesian coordinates

| $N_{atom}$ | $N_{bond}$ | $E(\tau)$ ($E_h$) | $dE/d\tau$ ($E_h$/a.u.) | $E^{norm}$ |
|---|---|---|---|---|
| 3 | 3 | 0.02947 | $-0.81 \times 10^{-9}$ | 0.03006 |
| 4 | 6 | 0.05833 | $-0.26 \times 10^{-8}$ | 0.05926 |
| 5 | 9 | 0.08665 | $0.29 \times 10^{-9}$ | 0.08780 |
| 6 | 12 | 0.11470 | $0.66 \times 10^{-8}$ | 0.11597 |
| 7 | 15 | 0.14245 | $0.15 \times 10^{-7}$ | 0.14378 |
| 8 | 18 | 0.16977 | $0.25 \times 10^{-7}$ | 0.17123 |
| 11 | 27 | 0.24962 | $0.55 \times 10^{-7}$ | 0.25115 |

The last column contains the normal mode ZPE estimates. The parameters of calculations are described in text

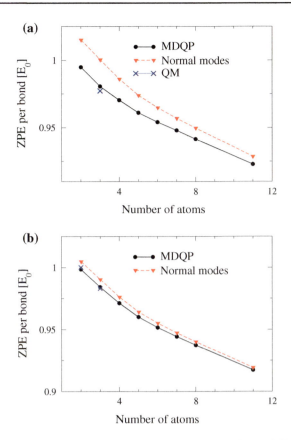

**Fig. 5** ZPE per bond obtained form the approximate MDQP evolution and in the normal mode approximation for atomic masses $m = 1836$ a.u. on panel (a) and $m = 18360$ a.u. on panel (b). The energy values are given in the units of the single bond energy of the Morse oscillator, $E_0$

## 4 Discussion and summary

We have presented an approximate approach to the wavepacket evolution in imaginary time in Cartesian coordinates and demonstrated its efficiency for high-dimensional systems. The trajectory formulation of the QM diffusion equation 4 allows one to express all the quantum effects through a single non-local potential-like term—the momentum-dependent quantum potential (MDQP). The quantum potential and the corresponding force are determined from the global Least Squares Fit of the momentum, which gives polynomial scaling with the number of dimensions to the method. For the quadratic fitting of the momenta, the number of basis functions scales quadratically with the dimensionality; the scaling of the overall numerical cost is dominated by the classical trajectory evolution, which is linear. The trajectory formulation itself allows random sampling of the initial wavefunction density, thus avoiding the exponential scaling with dimensionality of the wavefunction representation typical for the exact QM methods.

The approximate MDQP approach has been implemented in the Eulerian frame-of-reference and used for ZPE calculations. The Eulerian formulation necessitates approximation of the term, $\nabla p/m$, that does not vanish in the classical limit of $\hbar \to 0$, but it has two important practical advantages over the Lagrangian formulation with moving trajectories, at least for the potentials with localized minima: (i) the evolution equations in the Eulerian formulation are solved for stationary points—thus the representation of the ground state does not deteriorate with time; (ii) the classical potential and its gradient has to be computed only once—a tremendous saving for on-the-fly dynamics. In the ZPE calculations performed, we were able to use small ensembles of 2500 points for all cluster sizes, fairly localized around the minimum of $V$, due to the special feature of the quantum imaginary-time trajectories: the energy of each trajectory becomes equal to the ZPE as evolution unfolds. Such ensembles will be too small to represent the entire ground state, though the Gaussian approximation to it can be easily constructed. For systems with delocalized ground states or multiple minima of $V$, some combination of the Eulerian/Lagrangian formulation [48] or treatment of multiple low-energy regions as separate domains may be required. Such treatments will be more expensive requiring more trajectories and/or higher order fitting bases.

Implementation in Cartesian space is important because this is the framework of the classical mechanics methods, such as molecular dynamics, equations of motion and Hamiltonians take the simplest form, and because such formulation naturally invites mixed quantum/classical description of the nuclei. In order to include the QM correction on dynamics in Cartesian coordinates, we have introduced additional potentials that keep the wavefunction localized in the CoM and overall rotation degrees of freedom. This localization is necessary for accurate determination of MDQP in Cartesian coordinates within the minimal, quadratic basis. We have shown that there is a range of the force constants for the localizing potentials, where their effect on the internal ZPE is small and can be analytically subtracted from the total energy. The normal mode analysis can guide the choice of these parameters, as well as of the initial wavepacket widths. The concept of localizing potentials is similar to the soft constraint of molecular mechanics. For large molecular systems, they may become irrelevant, as it takes longer for a large system to move or rotate as a whole. The energy associated with the localization of the initial wavefunction in these degrees of freedom will be small compared to the internal ZPE and can be estimated from the diagonalization of the Hessian in Cartesian coordinates. Alternatively, one can consider constraining the overall motion and rotation by projection of the forces on the undesired directions, something that will be explored in the future. Application of the quadratic potential in the CoM coordinates should speed up the convergence to the ground state regardless of the propagation method.

So far, we conclude that the cost and scalability of the quadratic fitting for systems with localized ground state is promising. Cheap MDQP calculations give reasonable corrections to the normal mode ZPE estimates. The Cartesian coordinates implementation is compatible with classical simulation methods. The ultimate goal of this research is inclusion of QM corrections on dynamics of nuclei in large molecular systems. Future work will include thermal reaction rate calculations using the quantum trajectories in imaginary and real time and the mixed quantum/classical treatment of nuclei.

**Acknowledgments** This material is based on work partially supported by the South Carolina Research foundation and by the National Science Foundation under Grant No. CHE-1056188. The author is grateful to V. A. Rassolov for many stimulating discussions.

# References

1. Czako G, Bowman JM (2009) J Chem Phys 131
2. Zhang W, Kawamata H, Liu K (2009) Science 325:303
3. Dekker C, Ratner MA (2001) Phys World 14:29
4. Lear JD, Wasserman ZR, DeGrado WF (1988) Science 240:1177
5. Cha Y, Murray CJ, Klinman JP (1989) Science 243:1325
6. Knapp MJ, Klinman JP (2002) Eur J of Biochem 269:3113
7. Prezhdo OV, Rossky PJ (1997) J Chem Phys 107:5863
8. Brooksby C, Prezhdo O, Reid P (2003) J Chem Phys 119:9111
9. Prezhdo OV, Duncan WR, Prezhdo VV (2008) Acc Chem Res 41:339
10. Light JC, Carrington T Jr (2000) Adv Chem Phys 114:263
11. Meyer HD, Manthe U, Cederbaum LS (1990) Chem Phys Lett 165:73
12. Meyer HD, Worth GA (2003) Theor Chem Acc 109:251
13. Wang HB, Thoss M (2003) J Chem Phys 119:1289
14. Shalashilin DV, Child MS (2004) J Chem Phys 121:3563
15. Wu YH, Batista VS (2006) J Chem Phys 124:224305
16. Ben-Nun M, Quenneville J, Martinez TJ (2001) J Chem Phys 104:5161
17. Kim SY, Hammes-Schiffer S (2006) J Chem Phys 124:244102
18. Prezhdo O, Kisil V (1997) Phys Rev A 56:162
19. Hone TD, Izvekov S, Voth GA (2005) J Chem Phys 122:054105
20. Gao J, Truhlar DG (2002) Annu Rev Phys Chem 53:467
21. Náray-Szabó, G, Warshel, A (eds) (1997) Computational approaches to biochemical reactivity, vol. 19 of Understanding chemical reactivity. Kluwer Academic Publishers, Dordrecht
22. Gindensperger E, Meier C, Beswick JA (2000) J Chem Phys 113:9369
23. Meier C, Manthe U (2001) J Chem Phys 115:5477
24. Karplus M, Sharma RD, Porter RN (1964) J Chem Phys 40:2033
25. Rassolov VA, Garashchuk S (2008) Chem Phys Lett 464:262
26. Schatz GC, Bowman JM, Kuppermann A (1975) J Chem Phys 63:685
27. Miller WH (2001) J Phys Chem A 105:2942
28. Wyatt RE (2005) Quantum dynamics with trajectories: introduction to quantum hydrodynamics. Springer, New York
29. Madelung E (1927) Z Phys 40:322
30. de Broglie L (1930) An introduction to the study of wave mechanics. E. P. Dutton and Company Inc., New York
31. Bohm D (1952) Phys Rev 85:166
32. Garashchuk S (2010) J. Chem. Phys. 132:014112
33. Garashchuk S, Rassolov V, Prezhdo O (2011) Reviews in computational chemistry, vol 27, chap. Semiclassical Bohmian dynamics. Wiley, Hoboken, pp 111–210
34. Ramond P (1990) Field theory: a modern primer. Addison-Wesley, Reading
35. Feynman RP, Hibbs AR (1965) Quantum mechanics and path integrals. McGraw-Hill, New York
36. Frantsuzov PA and Mandelshtam VA (2008) J Chem Phys 128
37. Chen X, Wu YH, Batista VS (2005) J Chem Phys 122
38. Cartarius H and Pollak E (2011) J Chem Phys 134
39. Miller WH (1971) J Chem Phys 55:3146
40. Blume D, Lewerenz M, Niyaz P, Whaley KB (1997) Phys Rev E 55:3664
41. Ceperley DM, Mitas L (1996) Advances in chemical physics, chap. Monte Carlo methods in quantum chemistry. Wiley, London
42. Lester WA Jr, Mitas L, Hammond B (2009) Chem Phys Lett 478:1
43. Viel A, Coutinho-Neto MD, Manthe U (2007) J Chem Phys 126:024308
44. Hinkle CE, McCoy AB (2008) J Phys Chem A 112:2058
45. Miller WH, Schwartz SD, Tromp JW (1983) J Chem Phys 79:4889
46. Rassolov VA, Garashchuk S, Schatz GC (2006) J Phys Chem. A 110:5530
47. Garashchuk S, Vazhappilly T (2010) J Phys Chem C 114:20595
48. Garashchuk S, Mazzuca J, Vazhappilly T (2011b) J Chem Phys 135:034104
49. Liu J, Makri N (2005) Mol Phys 103:1083
50. Goldfarb Y, Degani I, Tannor DJ (2007) Chem Phys 338:106
51. Garashchuk S (2010b) Chem Phys Lett 491:96
52. Trahan CJ, Wyatt RE (2003) J Chem Phys 118:4784
53. Press WH, Flannery BP, Teukolsky SA, Vetterling WT (1992) Numerical recipes: the art of scientific computing. 2nd edn. Cambridge University Press, Cambridge
54. Reed SK, González-Martínez ML, Rubayo-Soneira J, and Shalashilin DV (2011) J Chem Phys 134
55. Dubbeldam D, Oxford GAE, Krishna R, Broadbelt LJ, and Snurr RQ (2010) J Chem Phys 133
56. Ochterski JW (1999) Vibrational analysis in Gaussian, http://www.gaussian.com/g_-whitepap/vib.htm
57. Meyer H (2002) Annu Rev Phys Chem 53:141
58. M. A. Ratner, Gerber RB (1986) J Phys Chem 90:20
59. Carter S, Culik SJ, Bowman JM (1997) J Chem Phys 107:10458
60. Morse PM (1929) Phys Rev 34:57
61. Feit MD, Fleck JA Jr, Steiger A (1982) J Comp Phys 47:412

REGULAR ARTICLE

# Time-dependent wave packet propagation using quantum hydrodynamics

Brian K. Kendrick

Received: 27 July 2011 / Accepted: 29 August 2011 / Published online: 11 January 2012
© Springer-Verlag 2012

**Abstract** A new approach for propagating time-dependent quantum wave packets is presented based on the direct numerical solution of the quantum hydrodynamic equations of motion associated with the de Broglie–Bohm formulation of quantum mechanics. A generalized iterative finite difference method (IFDM) is used to solve the resulting set of non-linear coupled equations. The IFDM is 2nd-order accurate in both space and time and exhibits exponential convergence with respect to the iteration count. The stability and computational efficiency of the IFDM is significantly improved by using a "smart" Eulerian grid which has the same computational advantages as a Lagrangian or Arbitrary Lagrangian Eulerian (ALE) grid. The IFDM is generalized to treat higher-dimensional problems and anharmonic potentials. The method is applied to a one-dimensional Gaussian wave packet scattering from an Eckart barrier, a one-dimensional Morse oscillator, and a two-dimensional (2D) model collinear reaction using an anharmonic potential energy surface. The 2D scattering results represent the first successful application of an accurate direct numerical solution of the quantum hydrodynamic equations to an anharmonic potential energy surface.

Published as part of the special collection of articles celebrating the 50th anniversary of Theoretical Chemistry Accounts/Theoretica Chimica Acta.

**Electronic supplementary material** The online version of this article (doi:10.1007/s00214-011-1075-9) contains supplementary material, which is available to authorized users.

B. K. Kendrick (✉)
Theoretical Division (T-1, MS-B268), Los Alamos National Laboratory, Los Alamos, New Mexico 87545, USA
e-mail: bkendric@lanl.gov

## 1 Introduction

In the de Broglie–Bohm [1–6] formulation of quantum mechanics, the polar form of the complex wave function is substituted into the time-dependent Schrödinger equation. The resulting set of non-linear coupled differential equations describe the time evolution of a flowing probability "fluid" and are often referred to as the quantum hydrodynamic equations of motion. The equations are formally exact, and solving this set of equations is identical to solving the original time-dependent Schrödinger equation. The approach is intuitively appealing since the quantum potential ($Q$) and its associated force $f_q = -\nabla Q$ appear on equal footing with the classical potential and force in the equations of motion. The quantum potential and force give rise to all quantum effects (such as tunneling and zero point energy), and the "flow lines" of the probability fluid correspond to the well-defined quantum trajectories. If $Q$ is zero, then the equations describe the motion of an uncoupled ensemble of classical trajectories. Thus, this approach provides a unique perspective for understanding quantum dynamics and a new direction for developing approximate quantum or semi-classical methods.

Despite the appealing features of this methodology, the hydrodynamic equations are notoriously difficult to solve and continue to resist a direct numerical solution to this day. These difficulties can be traced to two primary problems: (1) The non-linear coupling originating from the quantum potential $Q$ gives rise to a positive feedback loop of the numerical noise which grows quickly in magnitude and (2) the quantum potential can become singular (this problem is often referred to as the "node problem"). A variety of numerical approaches have been developed over the years based on different reference frames and approximations. The first method to successfully propagate a wave packet and obtain a transmission (tunneling) probability was called the quantum trajectory method (QTM) [7]. This method was successful in overcoming the noise feedback problem through the use of a Moving Least Squares (MLS) algorithm. The repeated MLS fitting effectively filters the solution at each time step, thereby reducing or controlling the numerical noise. Unfortunately, the QTM approach is based on a moving or Lagrangian frame which results in grid points that move apart. The increasingly sparse computational grid results in significant loss in accuracy. A fixed Eulerian frame is also not optimal. If a large Eulerian grid is used, then the method becomes unstable at the edges of the grid where the density is extremely small. On the other hand, if a small Eulerian grid is used, then the wave packet will move off the edge of the grid. A hybrid approach which combines the best features of both frames, an Arbitrary Lagrangian Eulerian (ALE) frame, proves to work much better [8, 9]. A 4th-order MLS method with an adaptive ALE frame was successful in treating several one- through four-dimensional model problems [10–12] including a quantum scattering resonance [13]. However, the MLS approach trades stability for resolution, and its numerical derivatives are not continuous. Its convergence and stability properties are also not well understood and hard to control. Furthermore, it requires significant computational resources due to the repeated MLS fitting. For these reasons, other direct numerical solution approaches are needed. A promising new approach called the iterative finite difference method (IFDM) has been recently developed and applied to several one-dimensional model problems [14–16]. In regards to the singularity issue or "node problem" [17], several different approaches have been developed to avoid this problem including: artificial viscosity [10], linearized quantum force [18, 19], bipolar decomposition [20], and covering function method [21]. Other related approaches have also been developed such as phase space methods [22, 23], complex trajectories [24–26], and the momentum-dependent quantum potential [27, 28]. For more details, the reader is referred to reviews [29, 30] and collections of recent work [15, 16].

In this work, a generalized version of the iterative finite difference method (IFDM) is presented [14]. The quantum hydrodynamic equations of motion are reviewed in Sect. 2 followed by a detailed description of the generalized IFDM. Section 2.1 derives the finite difference equations, Section 2.2 discusses the iterative solution approach, Sect. 2.3 derives the boundary conditions, Sect. 2.4 discusses the "smart" Eulerian grid, Sect. 2.5 performs the truncation error and stability analysis, and Sect. 2.6 presents the two-dimensional IFDM equations. Results of the one- and two-dimensional scattering and bound state applications are presented in Sect. 3. A notable result is the successful

application of the methodology to anharmonic potential energy surfaces. Some discussion and conclusions are presented in Sect. 4.

## 2 Quantum hydrodynamics

The derivation of the quantum hydrodynamic equations of motion is straightforward. The polar form of the wave function $\psi = \exp[C(\mathbf{x}, t) + iS(\mathbf{x}, t)/\hbar]$ (where $C$ and $S$ are real valued functions) is substituted into the time-dependent Schrödinger equation: $i\hbar\partial_t\psi = \hat{H}\psi$ where $\hat{H} = -\hbar^2\nabla^2/2m + V(\mathbf{x}, t)$, $V(\mathbf{x}, t)$ is the appropriate interaction potential, and $\partial_t$ denotes the partial derivative with respect to time. Separating the equations for the real and imaginary parts yields two equations [1–6]

$$\partial_t\rho = -\nabla \cdot (\rho\mathbf{v}), \tag{1}$$

$$\partial_t S = -(V + Q) - \frac{1}{2}m|\mathbf{v}|^2, \tag{2}$$

where the flow velocity is defined as $\mathbf{v} = \nabla S/m$, the probability density is $\rho = \exp[2\,C(\mathbf{x}, t]$, and the flux is $\mathbf{j} = \rho\mathbf{v}$. Equation 1 is the continuity equation and Eq. 2 is the *quantum* Hamilton–Jacobi equation. Both of these equations are based upon an Eulerian frame of reference (i.e., the grid points in $\mathbf{x}$ are fixed and do not move as time evolves). Taking the gradient of Eq. 2 gives an equation of motion in terms of the velocity $\mathbf{v}$

$$m\partial_t\mathbf{v} = -\nabla(V + Q) - m\mathbf{v} \cdot \nabla\mathbf{v}. \tag{3}$$

The first two terms on the right-hand side of Eq. 3 represent the classical force ($\mathbf{f}_c = -\nabla V$) and quantum force ($\mathbf{f}_q = -\nabla Q$). The last term is a convective term and can be eliminated by transforming to a moving or Lagrangian frame of reference in which the grid points $\mathbf{r}(\mathbf{x}, t)$ are chosen such that $\dot{\mathbf{r}} = \mathbf{v}$. In a Lagrangian frame of reference, the total time derivative is given by $d/dt = \partial_t + \mathbf{v} \cdot \nabla$ and Eq. 3 becomes the familiar Newton equation of motion $\mathbf{F} = m\mathbf{a}$. However, in the present work, an Eulerian frame will be used throughout. The continuity equation (1) can be expressed explicitly in terms of $C$

$$\partial_t C = -\frac{1}{2}\nabla \cdot \mathbf{v} - \mathbf{v} \cdot \nabla C. \tag{4}$$

The quantum hydrodynamic approach is intuitively appealing since the classical and quantum forces and potentials appear on equal footing in the equations of motion. The quantum potential $Q$ in Eqs. 2 and 3 gives rise to all quantum effects, such as zero point energy and tunneling. In general, $Q$ is a non-local potential and the two equations 1 and 2 describe the time evolution of a coupled ensemble of *quantum* trajectories determined from the solutions of $\dot{\mathbf{r}}(\mathbf{x},t) = \mathbf{v}(\mathbf{x},t)$. If $Q = 0$, then Eq. 2 becomes the classical Hamilton–Jacobi equation which describes the time evolution of an ensemble of uncoupled classical trajectories. The quantum potential is given by

$$Q(\mathbf{x},t) = -\frac{\hbar^2}{2m}\rho^{-1/2}\nabla^2\rho^{1/2} = -\frac{\hbar^2}{2m}(\nabla^2 C + |\nabla C|^2). \quad (5)$$

Equations 1 and 2 (or equivalently Eqs. 3 and 4) together with 5 are the quantum hydrodynamic equations of motion. These equations describe the time evolution of a "probability fluid," and the flow lines of this probability fluid are the well-defined quantum trajectories. The quantum hydrodynamic equations are exact, and a solution to these equations is identical to solving the time-dependent Schrödinger equation. However, the quantum hydrodynamic equations of motion are a coupled set of non-linear differential equations which must be solved self-consistently. The non-linear nature of these equations makes the direct numerical solution very challenging. In addition, the quantum potential can become singular whenever the amplitude $\rho^{1/2}$ approaches zero faster than $\nabla^2\rho^{1/2}$ in Eq. 5. This often occurs when there is significant interference between the incoming and reflected components of a scattered wave function and is commonly referred to as "the node problem".

## 2.1 Direct numerical solution: the iterative finite difference method

A promising new approach for the direct numerical solution of the quantum hydrodynamic equations was recently developed, which is called the iterative finite difference method (IFDM) [14–16]. This method is based on finite differencing Eqs. 3 and 4 and using Newton's method to iteratively solve them in a self-consistent way at each time step. In this section, the IFDM is reviewed and the latest improvements to the method are described. The new developments include (1) a "smart" Eulerian grid has been implemented which significantly reduces the grid size and improves the stability and accuracy of the methodology, (2) an improved (higher order) approach for computing the quantum force has been implemented, (3) new stabilizing artificial viscosity terms have been derived and implemented, (4) more general boundary conditions have been implemented at the edges of the grid relevant for treating anharmonic potentials such as the Morse potential, (5) a fast (iteration based) method for solving for the iterates ($\Delta C$ and $\Delta \mathbf{v}$) within Newton's method has been implemented (this is especially important for generalizing the method to higher dimensions), and (6) the method has been generalized and applied to a two-dimensional anharmonic scattering problem.

For simplicity, we begin by considering the one-dimensional case: $\mathbf{x} = x$ and $\mathbf{v} = v$. A "control volume" approach is implemented where the $C_i$ are defined on the grid points: $x_i = x_0 + \Delta x \cdot i\ (i = i_l, i_l + 1, \ldots, i, \ldots, i_r - 1, i_r)$ which are located at the centers of the "volume element", and the $v_{i+1/2}$ are defined on the grid points: $x_{i+1/2} = x_0 + \Delta x \cdot (i + 1/2)\ (i = i_l - 1, l_l, i_l + 1, \ldots, i, \ldots, i_r - 1, i_r)$ which are located at the boundaries between the volume elements (see Fig. 1). The left and right edges of the grid correspond respectively to $i = i_l$ and $i = i_r$ for $x$, and $i = i_l - 1$ and $i = i_r$ for $v$. Thus, the number of grid points in $x$ is given by $n_x = i_r - i_l + 1$ and for $v$ by $n_v = i_r - i_l + 2$. The use of a staggered grid significantly improves the stability of the methodology. In order to make the following equations more transparent, the velocity will be written in terms of $i' = i + 1/2$. Time is discretized between the initial $t_0$ and final $t_f$ as $t = t_0 + \Delta t \cdot n\ (n = 0, 1, 2, \ldots, n_f)$.

Central differencing the velocity (Eq. 3) and the C (Eq. 4) equations with respect to both space and time and performing a time averaging between the equations at the $n + 1$ and $n$th time steps, results in the following set of coupled equations which are second order accurate in both space and time (see Ref. [14] for more details)

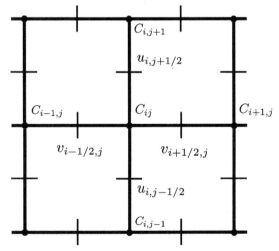

**Fig. 1** *Top panel* a one-dimensional "staggered" grid for which the $C_i$ are defined at the centers ($x_i$) of the "volume" elements, and the $v_{i\pm 1/2}$ are defined at the surfaces ($x_{i\pm 1/2}$) between the "volume" elements. *Bottom panel* a two-dimensional "staggered" grid for which the $C_{ij}$ are defined at the centers ($x_i, y_j$) of the "volume" elements, and the $x$ and $y$ velocity components $v$ and $u$ are defined at the surfaces ($x_{i\pm 1/2}, y_j$) and ($x_i, y_{j\pm 1/2}$), respectively

$$m(v_{i'}^{n+1} - v_{i'}^n)/\Delta t = (f_{i'}^{n+1} + f_{i'}^n)/2 - m\left[v_{i'}^{n+1}(v_{i'+1}^{n+1} - v_{i'-1}^{n+1})\right.$$
$$\left. + v_{i'}^n(v_{i'+1}^n - v_{i'-1}^n)\right]/(4\Delta x), \qquad (6)$$

and

$$(C_i^{n+1} - C_i^n)/\Delta t = -\left[(v_{i'+1}^{n+1} - v_{i'-1}^{n+1}) + (v_{i'}^{n+1} - v_{i'-2}^{n+1})\right.$$
$$\left. + (v_{i'+1}^n - v_{i'-1}^n) + (v_{i'}^n - v_{i'-2}^n)\right]/(16\Delta x)$$
$$- \left[(v_{i'}^{n+1} + v_{i'-1}^{n+1})(C_{i+1}^{n+1} - C_{i-1}^{n+1})\right.$$
$$\left. + (v_{i'}^n + v_{i'-1}^n)(C_{i+1}^n - C_{i-1}^n)\right]/(8\Delta x), \qquad (7)$$

where $f_{i'} = f_{c\,i'} + f_{q\,i'}$ denotes the *total* force evaluated at $x = x_{i'}$. The classical force evaluated at $i'$ is given simply by the negative gradient evaluated at $x = x_{i'}$ (i.e., $f_{ci'} = -\partial_x V(x,t)_{x=x_{i'}}$). If the interaction potential does not depend explicitly on time (i.e., $V(x, t) = V(x)$), then the $f_{ci'}$ only needs to be evaluated once at each grid point at the initial time $t = t_0$ and once at any new grid points which are added to the computational grid at later time steps. The quantum force must be continually updated, and in one dimension, it is given by

$$f_q = \frac{\hbar^2}{2m}\left(\partial_x^3 C + 2\partial_x C \partial_x^2 C\right). \qquad (8)$$

Second-order finite difference expressions for the first, second, and third derivatives of $C$ with respect to $x$ in Eq. 8 are used to compute the quantum force at each $x_i$. However, the quantum force that appears in the velocity equation (Eq. 6) is evaluated at the $x_{i'}$. In the original version of the IFDM, the quantum force at $x_{i'}$ was obtained from the average of its values at $x_i$ and $x_{i+1}$ (i.e., $f_{q\,i'} = (f_{q\,i} + f_{q\,i+1})/2$) [14]. However, better accuracy is obtained if the averaging is applied to each of the $C$ derivatives separately. That is, $f_{qi'} = \frac{\hbar^2}{2m}(\langle \partial_x^3 C \rangle_{i'} + 2\langle \partial_x C \rangle_{i'} \cdot \langle \partial_x^2 C \rangle_{i'})$ where the $\langle\,\rangle$ denotes the average $\langle A \rangle_{i'} = (A_i + A_{i+1})/2$.

### 2.2 Iterative solution: Newton's method

Equations 6 and 7 are coupled through the non-linear quantum force term. Due to this non-linear coupling, self-consistent solutions must be obtained using an iterative solution method such as Newton's method [31, 32]. In preparation for applying Newton's method, we define

$$F_1(C, v) = C_i^{n+1} - C_i^n - \Delta t \mathrm{RHS}_{\mathrm{Ceq}}, \qquad (9)$$

$$F_2(C, v) = v_{i'}^{n+1} - v_{i'}^n - \Delta t \mathrm{RHS}_{\mathrm{veq}}/m. \qquad (10)$$

where $\mathrm{RHS}_{\mathrm{Ceq}}$ and $\mathrm{RHS}_{\mathrm{veq}}$ denote the right-hand side of Eqs. 7 and 6, respectively. By construction, Eqs. 7 and 6 correspond to $F_1(C,v) = 0$ and $F_2(C, v) = 0$, respectively. In the following equations, let $C$ denote $C = C_i^{n+1}$ and $v$ denote $v = v_{i'}^{n+1}$. Newton's method in two variables consists of expanding $F_1$ and $F_2$ in a Taylor series expansion with respect to $C = C_o + \Delta C$ and $v = v_o + \Delta v$

$$F_1(C, v) = F_1(C_o, v_o) + \frac{\partial F_1}{\partial C}\Delta C + \frac{\partial F_1}{\partial v}\Delta v, \qquad (11)$$

$$F_2(C, v) = F_2(C_o, v_o) + \frac{\partial F_2}{\partial C}\Delta C + \frac{\partial F_2}{\partial v}\Delta v. \qquad (12)$$

The $C_o$ and $v_o$ are the initial values for the solution to be obtained at the $n + 1$ time step and are initially chosen to be equal to the solution from the previous $n$th time step. In one dimension, the $x_i$ and $x_{i'}$ are in one-to-one correspondence (except for the left boundary point of the velocity grid $i_l - 1$). Thus, excluding the left boundary point of the velocity grid and setting the left-hand side of Eqs. 11 and 12 to zero results in a set of two equations with two unknowns (the $\Delta C$ and $\Delta v$) at each $x_i$. The derivatives of the $F_1$ and $F_2$ with respect to $C$ and $v$ can be evaluated analytically and are given in "Appendix 1". In the original IFDM, these equations are solved analytically for the $\Delta C$ and $\Delta v$ at each $x_i$ and $x_i'$. In the new version of the IFDM, these equations are themselves solved iteratively (this procedure will be discussed in the following paragraph). Once the $\Delta C$ and $\Delta v$ are computed, the solution is updated via $C = C_o + \Delta C$ and $v = v_o + \Delta v$. The updated $C$ and $v$ at each $x_i$ and $x_{i'}$ become the new $C_o$ and $v_o$ and the process is repeated until the corrections $\Delta C$ and $\Delta v$ are below some user-specified convergence threshold (i.e., $<10^{-9}$). The residuals $F_1$ and $F_2$ evaluated at the converged $C$ and $v$ are also checked (they should be zero for an exact solution). For the model problems considered so far, numerical convergence tests have shown that the method converges exponentially with respect to the iteration count [14]. Depending upon the time step, approximately 10–20 iteration cycles are required. Analytic second-order accurate boundary conditions are applied at the edge of the grid to set the values for $C_{i_l}$, $C_{i_r}$, $v_{i'_l}$, $v_{i'_l-1}$, and $v_{i'_r}$ (the boundary conditions are discussed in Sect. 2.3). The time step is then updated $t = t + \Delta t$ and the process is repeated to obtain the solution at the next time step and so on.

For higher dimensions, the number of coupled equations increases since another velocity equation is introduced for each degree of freedom and a costly numerical linear solver is needed for a direct solution of the $\Delta C$ and $\Delta\mathbf{v}$ at each time step. Furthermore, it becomes difficult and more bookkeeping is required to ensure a one-to-one correspondence between the position and staggered velocity grids in higher dimensions. Thus, in order to extend the method to higher dimensions, a better approach for solving Eqs. 11 and 12 for $\Delta C$ and $\Delta v$ is needed. In the new version of the IFDM, Eqs. 11 and 12 are themselves solved iteratively. Equation 11 is solved for $\Delta C$ by first setting the $\Delta v = 0$ at each $x_i$. Then Eq. 12 is solved for the $\Delta v$ using the $\Delta C$ obtained in the previous step. The new estimates for the $\Delta v$ are then substituted into Eq. 11 and a new estimate for $\Delta C$ is obtained. The new $\Delta C$ is substituted into Eq. 12, and updated values for the $\Delta v$ are computed. This process is repeated until the corrections to $\Delta C$ and $\Delta v$ become smaller than a user-specified threshold (i.e., $<10^{-9}$). For the applications studied so far, the convergence is very fast and only one iteration cycle has been required. Using this approach, the one-to-one correspondence between the $x_i$ and $x_i{'}$ is no longer required, and the $v_{i_l'}$ is now included in the iterative solution so that only one boundary value $v_{i_l'-1}$ is unknown. This simplifies the treatment of the boundary conditions in $v$ at the left edge of the grid which further improves the overall accuracy and stability of the method. Most importantly, it is straightforward to extend this approach to higher dimensions (see Sect. 2.6). It is important to note that errors in the iterative solution of Eqs. 11 and 12 for $\Delta C$ and $\Delta v$ will be corrected within the main iteration cycle discussed in the preceding paragraph. Thus, a direct and highly accurate linear solver for this step of the calculation is not needed and would be waste of precious CPU time (additional discussion of this point in regards to iterative solution methods can be found in Ref. [33]).

In summary, the IFDM method consists of the following primary steps: (1) The $C$ and $v$ are initialized at $t = 0$ (i.e., $n = 0$) using known asymptotic reactant channel functions, (2) The classical and quantum forces are computed at each grid point, (3) The $n = 0$ fields are substituted into Eqs. 9 and 10 with $n = 0$, (4) The $n = 0$ fields are also used as the initial guess for the $n + 1 = 1$ fields to be determined at the future time step (i.e. they are the initial $C_o$ and $v_o$ in Eqs. 11 and 12), (5) Newton's method is used to solve Eqs. 11 and 12 iteratively for the $C^1$ and $v^1$ at the future time step: $t = (n + 1) * \Delta t = \Delta t$. The final solutions at $n + 1 = 1$ satisfy $F_1(C^1, v^1) < \epsilon$ and $F_2(C^1, v^1) < \epsilon$ at each grid point where $\epsilon$ is a user supplied convergence criteria (i.e., $\epsilon = 10^{-9}$). The quantum force at $n + 1 = 1$ is also updated during every Newton iteration cycle using the current value of $C^1$, and (6) The $C^1$ and $v^1$ are now substituted into Eqs. 9 and 10 with $n = 1$ and the steps (4)-(6) are repeated with $n$ incremented by one. This process is repeated until the desired final time step is reached. An example Fortran computer code that implements the new IFDM described above is available.[1]

## 2.3 Boundary conditions

Once all of the $\Delta C$ and $\Delta v$ have been computed at each interior $x_i$ and $x_i{'}$, the boundary conditions are applied to

obtain the values of $C$ and $v$ at the edges of the grid. For a Gaussian-like wave packet, the appropriate boundary conditions for $C$ and $v$ are given by $\partial_x^3 C = 0$ and $\partial_x^2 v = 0$, respectively. Equating the 2nd-order finite difference expression for $\partial_x^3 C$ centered at $x_{i_l+2}$ to zero and solving for $C_{i_l}$ gives

$$C_{i_l} = C_{i_l+4} - 2C_{i_l+3} + 2C_{i_l+1}. \tag{13}$$

A similar procedure on the right edge of the grid gives

$$C_{i_r} = C_{i_r-4} - 2C_{i_r-3} + 2C_{i_r-1}. \tag{14}$$

Evaluating the 2nd-order finite difference expression for $\partial_x^2 v$ centered at $x_{i'=i_l+1}$ to zero and solving for $v_{i_l'}$ gives

$$v_{i_l'} = 2v_{i_l'+1} - v_{i_l'+2}, \tag{15}$$

and similarly for the right boundary

$$v_{i_r'} = 2v_{i_r'-1} - v_{i_r'-2}. \tag{16}$$

The above boundary conditions are exact for the free Gaussian wave packet. They have also been shown to work well for the one-dimensional scattering of a Gaussian wave packet from an Eckart barrier (see Ref. [14] and Sect. 3.1). A notable feature of this methodology is that the wave packet propagates smoothly off the edge of the grid with no reflections. Thus, there is no need to introduce absorbing potentials at the edges of the grid.

For more general (anharmonic) potentials, the boundary conditions should be modified appropriately. For example, consider the one-dimensional Morse potential $V(x) = D_e [1 - \exp(-A(x - r_0))]^2$ centered at $x = r_0$ [34]. The constant $D_e$ is the dissociation energy and the $A$ and $D_e$ determine the effective force constant $k = \partial_x^2 V(x)_{x=r_0} = 2D_e A^2/\hbar^2$. The corresponding ground state wave function can be expressed as [34, 35]

$$C(x) = \ln(N) + (\lambda - 1/2)(\ln(2\lambda) - A(x - r_0)) \\ - \lambda \exp[-A(x - r_0)], \tag{17}$$

where $\lambda = \sqrt{2mD_e}/(A\hbar)$ and $N = \sqrt{A(2\lambda - 1)/\Gamma(2\lambda)}$. By taking the 1st, 2nd, and 3rd derivatives of Eq. 17 with respect to $x$, it is straightforward to show

$$\partial_x^3 C(x) = -A\partial_x^2 C(x). \tag{18}$$

Evaluating Eq. 18 using the appropriate 2nd-order finite difference expressions for $\partial_x^3 C$ and $\partial_x^2 C$ centered at $x_{i_l+2}$ and solving for $C_{i_l}$ gives

$$C_{i_l} = C_{i_l+4} - 2C_{i_l+3} + 2C_{i_l+1} + 2\Delta x A(C_{i_l+3} + C_{i_l+1} \\ - 2C_{i_l+2}), \tag{19}$$

and similarly for the right boundary

---

[1] A one-dimensional IFDM Fortran computer code (Online Resource 1) is available for download from the Theoretical Chemistry Accounts supplementary material website: www.springer.com.

$$C_{i_r} = C_{i_r-4} - 2C_{i_r-3} + 2C_{i_r-1} - 2\Delta x A(C_{i_r-3} + C_{i_r-1} - 2C_{i_r-2}).$$
$$(20)$$

In the quantum hydrodynamic approach, a stationary bound state is truly stationary (i.e., the classical and quantum forces exactly cancel) and a velocity field that is zero initially remains identically zero for all time (i.e., $v_{i'}(x, t) = 0$) [6]. Thus, the velocity boundary conditions given by Eqs. 15 and 16 are also appropriate for a stationary Morse oscillator ground state wave function. The stability of the IFDM using the generalized anharmonic Morse boundary conditions for $C$ given by Eqs. 19 and 20 is studied in Sect. 3.2. For general (ab initio) potentials, analytic boundary conditions may be determined by applying a local WKB-like expansion at the edge of the grid where the wave function tunnels into the repulsive potential. This approach is currently under development. In higher dimensions, the values of $C$ and $v$ at the edges of the grid may contain contributions from multiple degrees of freedom. In the two-dimensional application discussed in Sect. 3.3, the contributions from each boundary condition in $x$ and $y$ are averaged.

The boundary conditions in $C$ given by Eqs. 13 and 14 (or 19 and 20) are also used in the evaluation of the quantum force at the edges of the grid. In order to evaluate the quantum force at all of the internal points within the staggered velocity grid (i.e., the $x_{i'}$ with $i'_l \le i' \le i'_r - 1$), it must be evaluated at *all* of the $x_i$ grid points ($i_l \le i \le l_r$). Due to the third derivative of $C$ (see Eq. 8), the finite difference expression for the quantum force at $i = i_l$ and $i = i_r$ requires knowing the values of $C$ at two additional grid points beyond the left and right boundaries. The values of $C$ at these additional grid points are determined by simply applying the boundary conditions given by Eqs. 13 and 14 (or 19 and 20) to compute the $C_{i_l-1}$, $C_{i_l-2}$, $C_{i_r+1}$, and $C_{i_r+2}$. The appropriate averages of the 2nd-order finite difference expressions for the 1st, 2nd, and 3rd derivatives of $C$ can then be computed and used to obtain the

$$f_{qi'} = \frac{\hbar^2}{2m}(\langle \partial_x^3 C \rangle_{i'} + 2\langle \partial_x C \rangle_{i'} \cdot \langle \partial_x^2 C \rangle_{i'}).$$

## 2.4 An adaptive "smart" Eulerian grid

The analytic boundary conditions discussed in Sect. 2.3 also enable the implementation of an adaptive or "smart" Eulerian grid. In this approach, only those grid points that contain significant probability density are explicitly stored in memory and used in the IFDM solver. As the wave packet evolves in time, the density is continuously monitored at the edges of the grid. If the density falls below some user-specified cutoff (typically $10^{-6}$), then that edge point is removed from the grid. If the density increases above some user-specified cutoff (typically $10^{-5}$), then that

edge point becomes an interior grid point and a new edge point is added to the grid. The values of $C$ and $v$ at the new edge point are initialized using the boundary conditions presented in Sect. 2.3. Using this approach, the grid is always optimal at each time step and the active grid points track or follow the flow of the probability density as it evolves in time. Thus, a "smart" Eulerian grid exhibits the same computational advantages as a Lagrangian or Arbitrary Lagrangian Eulerian (ALE) grid while maintaining the desirable efficiency, stability, and accuracy associated with an Eulerian grid. This approach has also been implemented previously in our Crank–Nicholson algorithm to improve the edge stability and speed of that method [10]. Although not currently implemented, the "smart" Eulerian grid can be generalized to include a variable grid spacing. As the wave packet evolves in time and the necessary information is obtained for computing the desired scattering properties (such as the time correlation function in the product channel), the "regridding" algorithm can be turned off and the wave packet will then smoothly propagate off the edge of the grid (absorbing potentials are not needed).

## 2.5 Stability and truncation error analysis

A direct numerical solution of the non-linear quantum hydrodynamic equations of motion quickly accumulates error due to the positive feedback of numerical noise. The positive feedback produces instabilities (wiggles) that propagate inward from the edges of the grid and rapidly increase in amplitude. After only a few time steps, the entire solution can become corrupted. This effect is observed even for the propagation of a free Gaussian wave packet using the exact boundary conditions given by Eqs. 13–16. These kinds of instabilities also occur in computational fluid dynamics, and one approach for identifying the problematic terms and then controlling them is through "Truncation Error Analysis" [31]. In this approach, the $C$ and $v$ are expanded in a Taylor series with respect to the grid spacing $\Delta x$ and time interval $\Delta t$. In the following treatment, the focus will be on the velocity equations (numerical studies indicate that the instabilities originate primarily from them). The relevant expansions for $v_i^{n+1}$ and $v_{i'}^{n' \pm 1/2}$ are given by

$$v_{i'\pm1}^{n'} = v_{i'}^{n'} \pm \partial_x v_{i'}^{n'} \Delta x + \partial_x^2 v_{i'}^{n'} \Delta x^2/2, \qquad (21)$$

$$v_{i'}^{n'\pm1/2} = v_{i'}^{n'} \pm \partial_t v_{i'}^{n'} \Delta t/2 + \partial_t^2 v_{i'}^{n'} \Delta t^2/8 \pm \partial_x^3 v_{i'}^{n'} \Delta t^3/48.$$
$$(22)$$

Equations 21 and 22 are used to form the superpositions $(v_{i'}^{n'+1/2} \pm v_{i'}^{n'-1/2})$, $(v_{i'+1}^{n'} - v_{i'-1}^{n'})$, and a similar expansion for $f$ is used to form $(f_{i'}^{n'+1/2} + f_{i'}^{n'-1/2})$. All three of these terms are then substituted into Eq. 6 with $n = n' - 1/2$. The third time derivative of $v$ is then eliminated from the resulting equation by differentiating the velocity equation (i.e., $\partial_t^3 v = \partial_t^2 f / m - \partial_t^2 [v \partial_x v]$) to finally obtain the following equation

$$m \partial_t v_{i'}^{n'} = f_{i'}^{n'} - m v_{i'}^{n'} \partial_x v_{i'}^{n'} + O(\Delta x^2) + \varepsilon, \tag{23}$$

$$\varepsilon = \frac{1}{12} \Delta t^2 \left[ \partial_t^2 f_{i'}^{n'} - m \partial_t^2 (v_{i'}^{n'} \partial_x v_{i'}^{n'}) \right]. \tag{24}$$

The second-order truncation error with respect to $\Delta t$ is given explicitly by $\varepsilon$ in Eq. 24. If the classical potential does not depend explicitly on time, then the time derivative of the total force in Eq. 24 becomes equal to just the time derivative of the quantum force (i.e., $\partial_t^2 f_{i'}^{n'} = \partial_t^2 f_{qi'}^{n'}$). By repeatedly using the equations of motion for $C$ (Eq. 4) and $v$ (Eq. 3) and the definition of $f_q$ (Eq. 8), all of the time derivatives in Eq. 24 can be eliminated (see "Appendix 2"). Once this is done, Eq. 23 can be written in the form

$$\begin{aligned} m \partial_t v_{i'}^{n'} = & f_{i'}^{n'} - m v_{i'}^{n'} \partial_x v_{i'}^{n'} + O(\Delta x^2) \\ & + \Delta t^2 \left[ \gamma(C, v) \partial_x^2 v_{i'}^{n'} + G_1(C, v) \right]. \end{aligned} \tag{25}$$

Equation 25 shows that the finite difference representation of the velocity equation (Eq. 6), gives the desired equation of motion for each $v_{i'}^{n'}$ but with second order errors proportional to $\Delta x^2$ and $\Delta t^2$. This equation also shows that the error terms proportional to $\Delta t^2$ contain terms which are proportional to $\partial_x^2 v_{i'}^{n'}$. These error terms give rise to numerical diffusion and the effective diffusion coefficient is given by $\Gamma = \Delta t^2 \gamma(C, v)$ with units distance$^2$/time. The explicit functional form for $\gamma(C, v)$ is derived in "Appendix 2", and it is a function of the $C$, $v$, $f$, and various spacial derivatives of $C$ and $v$. During the course of the calculations, $\gamma$ can readily change sign (i.e., due to numerical noise). If $\gamma$ becomes negative even for a brief moment during the propagation, the effective local diffusion coefficient is negative and the solution will quickly become unstable. For a Gaussian wave packet, $\gamma$ is quadratic in $x - x_0$ where $x_0$ is the center of the wave packet (see Ref. [14] and "Appendix 2"). This explains why even for a free Gaussian wave packet the instabilities initiate at the edges of the grid and propagate inwards (the potentially negative numerical diffusion coefficients are largest there).

In order to ensure that the effective diffusion coefficient always remains positive during the IFDM calculations, an additional diffusive term of the form $\Delta t^2 \tilde{\gamma} \partial_x^2 v$ is added to the right-hand side of the velocity equation of motion (Eq. 3). This approach is well known in computational fluid dynamics and is called "artificial viscosity" [31, 32, 36–38]. Artificial viscosity has also been used previously to stabilize the MLS solutions of the quantum hydrodynamics equations [10–13]. Including the new diffusion term in the

finite difference equation for the velocity results in the $F_2(C,v)$ in Eq. 10 being replaced with $\tilde{F}_2(C, v)$ where

$$\begin{aligned} \tilde{F}_2(C, v) = & F_2(C, v) - \Delta t \left[ \Delta t^2 \tilde{\gamma}_{i'}^{n+1} (v_{i'+1}^{n+1} + v_{i'-1}^{n+1} - 2 v_{i'}^{n+1}) \right. \\ & \left. + \Delta t^2 \tilde{\gamma}_{i'}^n (v_{i'+1}^n + v_{i'-1}^n - 2 v_{i'}^n) \right] / (2 m \Delta x^2). \end{aligned} \tag{26}$$

The coefficient $\tilde{\gamma}$ is chosen such that $\tilde{\gamma} > |\gamma|$ (see Eq. 60 in "Appendix 2"). Repeating the truncation error analysis including the new diffusion term shows that the *total* effective diffusion coefficient is now given by $\Gamma = \Delta t^2 (\gamma(C, v) + \tilde{\gamma}(C, v))$ and by construction it is always positive. Thus, the numerical noise in the calculation is dissipated or suppressed, and the stability of the method is dramatically improved. In general, it is possible that some of the other error terms (i.e., those in $G_1(C,v)$ and in $O(\Delta x^2)$ in Eq. 25) could also lead to instabilities. However, a complete analysis of all these terms appears unnecessary since numerical tests indicate that (at least for the problems considered so far) these terms can be stabilized by simply increasing $\tilde{\gamma}$. Of course, care must be practiced in increasing $\tilde{\gamma}$ so that excessive diffusion is avoided. Excessive diffusion can often be identified by a significant increase in the iteration count. It is important to note that the artificial viscosity term $\Delta t^2 \tilde{\gamma} \partial_x^2 v$ is proportional to $\Delta t^2$ and it is therefore of the same order as the truncation error. As the resolution of the calculation is increased (i.e., as $\Delta x$ and $\Delta t$ are decreased) this term and the truncation error both decrease as $\Delta t^2$. Thus, the accuracy of the solution can be increased while simultaneously maintaining stability. Another finite differencing method that avoids excessive numerical diffusion while improving stability is the higher-order explicit upstream differencing QUICK method [39]. Applying this approach to the quantum hydrodynamic formulation may be investigated in future work.

### 2.6 Two-dimensional treatment

In this section, the IFDM approach is extended to two dimensions $\mathbf{x} = (x, y)$. The 2D version of the velocity equation (Eq. 3) for $\mathbf{v} = [v(x, y, t), u(x, y, t)]$ is

$$m \partial_t v = -\partial_x (V + Q) - m v \partial_x v - m u \partial_y v, \tag{27}$$

$$m \partial_t u = -\partial_y (V + Q) - m v \partial_x u - m u \partial_y u, \tag{28}$$

and the 2D version of the $C$ equation (Eq. 4) is

$$\partial_t C = -\frac{1}{2} \partial_x v - \frac{1}{2} \partial_y u - v \partial_x C - u \partial_y C. \tag{29}$$

Repeating the same steps used in the derivation of the 1D equations, the resulting 2D finite difference equation for $v$ which is 2nd-order accurate in both space and time is given by

$$m(v_{i'j}^{n+1} - v_{i'j}^n)/\Delta t$$

$$= (f_{i'j}^{n+1} + f_{i'j}^n)/2 - m\Big[v_{i'j}^{n+1}(v_{i'+1,j}^{n+1} - v_{i'-1,j}^{n+1})$$

$$+ v_{i'j}^n(v_{i'+1,j}^n - v_{i'-1,j}^n)\Big]/(4\Delta x)$$

$$- m\Big[u_{i'j}^{n+1}(v_{i'j+1}^{n+1} - v_{i'j-1}^{n+1}) + u_{i'j}^n(v_{i'j+1}^n - v_{i'j-1}^n)\Big]/(4\Delta y),$$

$$(30)$$

and similarly for $u$

$$m(u_{ij'}^{n+1} - u_{ij'}^n)/\Delta t$$

$$= (g_{ij'}^{n+1} + g_{ij'}^n)/2 - m\Big[v_{ij'}^{n+1}(u_{i+1,j'}^{n+1} - u_{i-1,j'}^{n+1})$$

$$+ v_{ij'}^n(u_{i+1,j'}^n - u_{i-1,j'}^n)\Big]/(4\Delta x)$$

$$- m\Big[u_{ij'}^{n+1}(u_{ij'+1}^{n+1} - u_{ij'-1}^{n+1}) + u_{ij'}^n(u_{ij'+1}^n - u_{ij'-1}^n)\Big]/(4\Delta y),$$

$$(31)$$

where $i' = i + 1/2$ and $j' = j + 1/2$ (see Fig. 1) and the total force in the $x$ and $y$ directions are given by $f = -\partial_x(V + Q)$ and $g = -\partial_y(V + Q)$, respectively. The evaluation of the $u_{i'j}$ in Eq. 30 is accomplished by a four-point average of the $u$ surrounding $(x_{i'}, y_j)$: $u_{i'j} = (u_{i,j'} + u_{i+1,j'} + u_{i+1,j'-1} + u_{i,j'-1})/4$. Similarly a four-point average of the $v$ surrounding $(x_i, y_{j'})$ is used in Eq. 31: $v_{ij'} = (v_{i'j} + v_{i'j+1} + v_{i'-1,j+1} + v_{i'-1,j})/4$.

The $f_q$ and $g_q$ components of the quantum force $\mathbf{f}_q = (f_q, g_q)$ are given by

$$f_q = \frac{\hbar^2}{2m}\Big(\partial_x^3 C + \partial_x\partial_y^2 C + 2\partial_x C\partial_x^2 C + 2\partial_y C\partial_{xy}^2 C\Big), \quad (32)$$

and

$$g_q = \frac{\hbar^2}{2m}\Big(\partial_y^3 C + \partial_y\partial_x^2 C + 2\partial_y C\partial_y^2 C + 2\partial_x C\partial_{xy}^2 C\Big). \quad (33)$$

The classical force components in Eqs. 30 and 31 are evaluated at the appropriate velocity grid locations by simply calling the potential energy surface with the appropriate $(x, y)$ coordinates. However, the quantum force components are native to the $C$ grid so that the appropriate averages of the various spacial derivatives of $C$ in Eqs. 32 and 33 are used to evaluate $f_q$ and $g_q$ at the appropriate locations on the velocity grids for $v$ and $u$. For example,

$$f_{qi'j} = \frac{\hbar^2}{2m}\Big(\langle\partial_x^3 C\rangle_{i'j} + \langle\partial_x\partial_y^2 C\rangle_{i'j} + 2\langle\partial_x C\rangle_{i'j}\langle\partial_x^2 C\rangle_{i'j}$$

$$+ 2\langle\partial_y C\rangle_{i'j}\langle\partial_{xy}^2 C\rangle_{i'j}\Big), \quad (34)$$

where as in the one-dimensional case (see the discussion below Eq. 8) the $\langle\,\rangle$ denote the average $\langle A\rangle_{i'} = (A_i + A_{i+1})/2$. A similar equation is used to evaluate $g_q$ at $(x_i, y_{j'})$

except that the averaging is performed in the $y$ coordinate: $\langle A\rangle_{j'} = (A_j + A_{j+1})/2$.

The 2D finite difference equation for $C$ (Eq. 29) which is 2nd-order accurate in both space and time is given by

$$(C_{ij}^{n+1} - C_{ij}^n)/\Delta t = -\Big[(v_{i'+1,j}^{n+1} - v_{i'-1,j}^{n+1}) + (v_{i'j}^{n+1} - v_{i'-2,j}^{n+1})$$

$$+ (v_{i'+1,j}^n - v_{i'-1,j}^n) + (v_{i'j}^n - v_{i'-2,j}^n)\Big]/(16\Delta x)$$

$$- \Big[(u_{ij'+1}^{n+1} - u_{ij'-1}^{n+1}) + (u_{ij'}^{n+1} - u_{ij'-2}^{n+1})$$

$$+ (u_{ij'+1}^n - u_{ij'-1}^n) + (u_{ij'}^n - v_{ij'-2}^n)\Big]/(16\Delta y)$$

$$- \Big[(v_{i'j}^{n+1} + v_{i'-1,j}^{n+1})(C_{i+1,j}^{n+1} - C_{i-1,j}^{n+1})$$

$$+ (v_{i'j}^n + v_{i'-1,j}^n)(C_{i+1,j}^n - C_{i-1,j}^n)\Big]/(8\Delta x)$$

$$- \Big[(u_{ij'}^{n+1} + u_{ij'-1}^{n+1})(C_{i,j+1}^{n+1} - C_{i,j-1}^{n+1})$$

$$+ (u_{ij'}^n + u_{ij'-1}^n)(C_{i,j+1}^n - C_{i,j-1}^n)\Big]/(8\Delta y),$$

$$(35)$$

where the appropriate averages over the velocity components $v$ and $u$ in the $x$ and $y$ directions have been done to obtain the values on the $C$ grid (see Fig. 1). The finite difference equations for $C$, $v$, and $u$ can be written in compact notation as

$$F_1(C, v, u) = C_{ij}^{n+1} - C_{ij}^n - \Delta t\,\mathrm{RHS}_{\mathrm{Ceq}}, \quad (36)$$

$$F_2(C, v, u) = v_{i'j}^{n+1} - v_{i'j}^n - \Delta t\,\mathrm{RHS}_{\mathrm{veq}}/m. \quad (37)$$

$$F_3(C, v, u) = u_{ij'}^{n+1} - u_{ij'}^n - \Delta t\,\mathrm{RHS}_{\mathrm{ueq}}/m. \quad (38)$$

where $\mathrm{RHS}_{\mathrm{Ceq}}$, $\mathrm{RHS}_{\mathrm{veq}}$, and $\mathrm{RHS}_{\mathrm{ueq}}$ denote the right-hand side of Eqs. 35, 30, and 31, respectively.

As in the one-dimensional case, an iterative solution is obtained by using Newton's method. In the following equations, let $C$ denote $C = C_{i,j}^{n+1}$, $v$ denote $v = v_{i'j}^{n+1}$ and $u$ denote $u = u_{ij'}^{n+1}$. The $F_i$ defined by Eqs. 36-38 are expanded with respect to $C = C_o + \Delta C$, $v = v_o + \Delta v$, and $u = u_o + \Delta u$

$$F_1(C, v, u) = F_1(C_o, v_o, u_o) + \frac{\partial F_1}{\partial C}\Delta C + \frac{\partial F_1}{\partial v}\Delta v + \frac{\partial F_1}{\partial u}\Delta u,$$

$$(39)$$

$$F_2(C, v, u) = F_2(C_o, v_o, u_o) + \frac{\partial F_2}{\partial C}\Delta C + \frac{\partial F_2}{\partial v}\Delta v + \frac{\partial F_2}{\partial u}\Delta u,$$

$$(40)$$

$$F_3(C, v, u) = F_3(C_o, v_o, u_o) + \frac{\partial F_3}{\partial C}\Delta C + \frac{\partial F_3}{\partial v}\Delta v + \frac{\partial F_3}{\partial u}\Delta u.$$

$$(41)$$

Recall that the $C_o$, $v_o$, and $u_o$ are the initial values for the 2D solution to be obtained at the $n + 1$ time step and are initially chosen to be equal to the 2D solution from the previous $n$th time step. Equations 39–41 are used to obtain the corrections $\Delta C$, $\Delta v$, and $\Delta u$ (the procedure for doing this will be discussed in the following paragraph). Once the $\Delta C$, $\Delta v$ and $\Delta u$ are determined at each $(x_i, y_j)$ and $(x_{i'}, y_j)$ and $(x_i, y_{j'})$, the solution is updated via $C = C_o + \Delta C$, $v = v_o + \Delta v$ and $u = u_o + \Delta u$. The updated $C$, $v$, and $u$ become the new $C_o$, $v_o$ and $u_o$ and the process is repeated until the corrections $\Delta C$, $\Delta v$, and $\Delta u$ are below some-user specified convergence threshold (i.e., $<10^{-9}$). The residuals $F_1$, $F_2$, and $F_3$ evaluated at the converged $C$, $v$, and $u$ are also checked (they should be zero for an exact solution). Depending upon the time step, approximately 10–20 iteration cycles are required. The time step is then incremented, and the iterative process is repeated at the next time step and so on.

In the generalized IFDM approach, the $\Delta C$, $\Delta v$ and $\Delta u$ in Eqs. 39–41 are themselves solved iteratively. The derivatives of the $F_i$ with respect to $C$, $v$ and $u$ can be derived analytically (see for example the one-dimensional derivation in "Appendix 1"). The left hand side of the $F_i$ are set equal to zero (the exact solution must satisfy these conditions). Equation 39 is first solved for $\Delta C$ by setting the $\Delta v$ and $\Delta u$ equal to zero. The resulting $\Delta C$ are then used in Eqs. 40 and 41. Equation 40 is then solved for $\Delta v$ with $\Delta u$ equal to zero and Eq. 41 is solved for $\Delta u$ with $\Delta v$ equal to zero. The resulting $\Delta v$ and $\Delta u$ are then substituted back into Eq. 39 and a new $\Delta C$ is obtained. This updated value of $\Delta C$ is then used in the $F_2$ and $F_3$ equations to obtain new updates for the $\Delta v$ and $\Delta u$. For example, the updated value for $\Delta v$ is obtained using the new value of $\Delta C$ and the previous value (now non-zero) of $\Delta u$. This process is repeated until the corrections are all below some user specified threshold ($10^{-9}$). In the example problems treated so far, the convergence is very fast and only one iteration cycle has been required.

The two-dimensional truncation error analysis follows the same procedure as was done for one dimension. In the following treatment, the $x$ component of the velocity ($v$) will be considered (the analysis for the $y$ component $u$ is identical). The two-dimensional Taylor series expansions for $v$ with respect to $x$, $y$, and $t$ analogous to Eqs. 21 and 22 are substituted into the finite difference equation for $v$ (Eq. 30) with $n = n' - 1/2$. The resulting equation can eventually be written as

$$m\partial_t v_{i',j}^{n'} = f_{i',j}^{n'} - m v_{i',j}^{n'} \partial_x v_{i',j}^{n'} - m u_{i',j}^{n'} \partial_y v_{i',j}^{n'} + O(\Delta x^2) + O(\Delta y^2) + \varepsilon_{2D}, \tag{42}$$

$$\varepsilon_{2D} = \frac{1}{12}\Delta t^2 \left[ \partial_t^2 f_{i',j}^{n'} - m\partial_t^2 (v_{i',j}^{n'} \partial_x v_{i',j}^{n'}) - m\partial_t^2 (u_{i',j}^{n'} \partial_y v_{i',j}^{n'}) \right]. \tag{43}$$

Using the equations of motion for $C$, $v$, and $u$, all of the time derivatives in Eq. 43 can be eliminated (see "Appendix 2"). Once this is done, Eq. 42 can be written in the form

$$m\partial_t v_{i',j}^{n'} = f_{i',j}^{n'} - m v_{i',j}^{n'} \partial_x v_{i',j}^{n'} - m u_{i',j}^{n'} \partial_y v_{i',j}^{n'} + O(\Delta x^2) + O(\Delta y^2) + \Delta t^2 \left[ {}^1\gamma(C, v, u)\partial_x^2 v_{i',j}^{n'} + {}^2\gamma(C, v, u)\partial_y^2 v_{i',j}^{n'} + \bar{G}_1(C, v, u) \right]. \tag{44}$$

An analogous equation can be derived for $u$. Equation 44 is similar to the one-dimensional result Eq. 25 except that now the diffusion occurs in two dimensions and a second diffusive term proportional to $\partial_y^2 v_{i',j}^{n'}$ appears in the truncation error. This additional term represents the numerical diffusion of $v$ in the $y$ direction. The derivation of the explicit functional forms for ${}^1\gamma(C, v, u)$ and ${}^2\gamma(C, v, u)$ is given in "Appendix 2". As was done for the one-dimensional case, the 2D finite difference equations are stabilized by introducing the appropriate artificial viscosity terms into the velocity equations for $v$ and $u$. For example, the $F_2$ defined in Eq. 37 is replaced by $F_2 \to \tilde{F}_2$ where

$$\tilde{F}_2(C, v, u) = F_2(C, v, u) - \Delta t [(\Delta t^2) {}^1\tilde{\gamma}_{i'j}^{n+1}(v_{i'+1,j}^{n+1} + v_{i'-1,j}^{n+1} - 2v_{i'j}^{n+1})$$
$$+ (\Delta t^2) {}^1\tilde{\gamma}_{i'j}^{n}(v_{i'+1,j}^{n} + v_{i'-1,j}^{n} - 2v_{i'j}^{n})]/(2m\Delta x^2)$$
$$- \Delta t [(\Delta t^2) {}^2\tilde{\gamma}_{i'j}^{n+1}(v_{i'j+1}^{n+1} + v_{i'j-1}^{n+1} - 2v_{i'j}^{n+1})$$
$$+ (\Delta t^2) {}^2\tilde{\gamma}_{i'j}^{n}(v_{i'j+1}^{n} + v_{i'j-1}^{n} - 2v_{i'j}^{n})]/(2m\Delta y^2). \tag{45}$$

The coefficients ${}^1\tilde{\gamma}$ and ${}^2\tilde{\gamma}$ are chosen such that ${}^1\tilde{\gamma} > |{}^1\gamma(C, v, u)|$ and ${}^2\tilde{\gamma} > |{}^2\gamma(C, v, u)|$, respectively. These two coefficients and the two similar ones for $u$ are chosen to be as small as possible but large enough to ensure stability (see the discussion below Eq. 26 for the one-dimensional case).

## 3 Scattering and bound state applications

In this section, the new version of the IFDM described in Sect. 2 is applied to the calculation of (1) a one-dimensional Gaussian wave packet scattering from an Eckart barrier, (2) a one-dimensional ground state (stationary) wave packet in an anharmonic Morse potential, and (3) a two-dimensional wave packet that is scattering from an Eckart barrier along the reaction path and is bound by an anharmonic Morse potential in the direction perpendicular to the reaction path. Atomic units are used throughout, so that $\hbar = 1$ and the mass is chosen to be that for the electron $m = 1$ au.

### 3.1 One-dimensional scattering from an Eckart barrier

The analytic form for the Eckart barrier is

$$V(x) = V_o \text{sech}^2[a(x - x_b)], \tag{46}$$

where $V_o = 50$ au is the barrier height, $a = 0.5a_0^{-1}$ is the width parameter, and $x_b = 0$ is the location of the barrier. The initial Gaussian wave packet is given by

$$\psi(x,0) = (2\beta/\pi)^{1/4} e^{-\beta(x-x_0)^2} e^{ik(x-x_0)}, \tag{47}$$

where $\beta = 12a_0^{-2}$ is the width parameter, $x_0 = -5a_0$ is the center of the wave packet, and $k$ determines the initial phase $S_0 = \hbar k(x - x_0)$ and flow kinetic energy $E = \hbar^2 k^2/(2m)$. The initial conditions for the $C$-amplitude and velocity $v$ are given by $C(x,0) = \ln(2\beta/\pi)^{1/4} - \beta(x - x_0)^2$ and $v(x,0) = (1/m)\partial S_0/\partial x = \hbar k/m$, respectively. As time evolves the wave packet moves from negative $x$ to positive $x$ and depending upon the initial kinetic energy, some fraction of the wave packet tunnels through the barrier located at $x = 0$ and the rest is reflected from the barrier. The grid spacing and fixed time step used in the IFDM calculations are $\Delta x = 0.025$ au and $\Delta t = 1.0 \times 10^{-5}$. At each time step, the "smart" Eulerian grid removes an edge point if the density falls below $\rho < 10^{-6}$ and it adds a new edge point if the density increases above $\rho > 10^{-5}$. A high-resolution Crank–Nicholson calculation was also performed based on a grid spacing of 0.00125 au and a time step of $5.0 \times 10^{-5}$. The Crank–Nicholson method solves the standard time-dependent Schrödinger equation [32]. Figure 2 plots the computed $C(x, t)$ amplitude at four different times for an initial kinetic energy of $E = 10$ au. The green points are from the IFDM calculation and the solid red curves are from the Crank–Nicholson calculation. Excellent agreement between the two solutions is observed at all of the time steps. Most significant is the ability of the IFDM to track the oscillations in the reflected wave packet. In particular, the "cusps" at $x = -3$ and $-4.5$ au for $t = 1.25$ au are due to the interference between the incoming and reflected components of the wave packet. This interference can lead to the formation of "nodes" (i.e., points in time and space were the density briefly becomes zero). Some small differences are observed between the IFDM and the "exact" Crank–Nicholson results in the regions where the "nodes" are forming.

Figure 3 plots the computed velocity field $v(x, t)$ at the same time steps as in Fig. 2. As in Fig. 2, the green points are from the IFDM calculation and the solid red curves are from the Crank–Nicholson calculation. The Crank–Nicholson velocities are obtained by taking the numerical derivative of the computed phase at each grid point (i.e., $v = \partial_x \phi$ where $\phi = \text{ATAN2}(\psi_{\text{real}}/\psi_{\text{imag}})$). Again excellent agreement is observed between the two methods. Some small oscillations or wiggles at the edges of the grid are observed at $t = 0.5$ au but these dissipate due to the introduction of the stabilizing artificial viscosity term (see Eq. 26 and "Appendix 2"). The optimal values for the artificial viscosity coefficients used in this calculation were $\gamma_0 = 0$ and $\gamma'_0 = 2 \times 10^6$ (see Eq. 60 in "Appendix 2") for the definitions of $\gamma_0$ and $\gamma'_0$). A significant feature of the IFDM is its ability to accurately track rapidly varying

features of the velocity field such as the sharp "spikes" near $x = -3$ and $-4.5$ au for $t = 1.25$ au. These spikes are associated with "node" formation and they can become infinitely large if a true node occurs. Fortunately, the artificial viscosity also stabilizes the IFDM against node formation. As the derivatives of the velocity increase in the nodal region, the artificial viscosity also automatically increases in that region which leads to highly localized numerical diffusion. The numerical diffusion is dissipative which causes the IFDM velocity spike to become "rounded-off," so that a singular velocity is avoided and the IFDM propagation can continue to long times. The conservation of probability during the IFDM calculations is monitored by computing the normalization error $\left|1 - \sum_i \rho_i \Delta x\right|$ which was typically less than $3 \times 10^{-5}$. The conservation of energy is also monitored by evaluating $E = \sum_i \rho_i [0.5 v_i^2 + V(x_i) + Q(x_i)]\Delta x$. The percent differences of the energy relative to its initial value at time zero were typically less than 1.0 %.

Figure 4 plots the transmission (tunneling) probability computed by integrating the density of the wave packet over the product region

$$P_t = \int_0^\infty \rho(x,t)\mathrm{d}x = \sum_{x_i > 0} \rho(x_i, t)\Delta x. \tag{48}$$

The $P_t$ for nine different initial kinetic energies are plotted as a function of time. The probabilities are well converged at all nine energies by $t = 1.5$ au. The $P_t$ computed using the IFDM (solid curves) are essentially identical to the Crank–Nicholson results (dashed curves). The IFDM results were computed but not plotted between $t = 1.4$ and 1.5 au in order to enable a better view of the Crank–Nicholson results. Table 1 lists the converged $P_t$ at $t = 1.5$ au computed using the generalized IFDM (this work), the original IFDM (Ref. [14]), and Crank–Nicholson (this work). The results from the generalized IFDM (second column) are in better agreement with the high-resolution Crank–Nicholson results (fourth column) than are the original IFDM results (third column). The improvement in accuracy is due to: (1) the use of a finer grid spacing and smaller time step and (2) the use of more accurate 2nd and 3rd numerical derivatives of $C$ with respect to $x$. The original IFDM employed a "moving window" average of the 2nd derivative of $C$ in order to improve stability and reduce the artificial viscosity coefficients [14]. This averaging was required since a full Eulerian grid was used (i.e., all of the grid points were activated and used in the calculation at all times). The new version of the IFDM described in Sect. 2 uses a truncated "smart" Eulerian grid in which only the grid points with sufficient density are included in the calculation. This greatly improves the stability (and

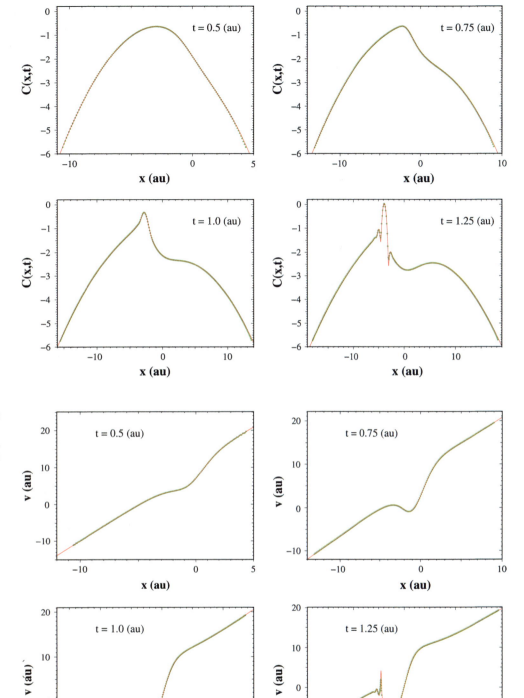

**Fig. 2** The $C(x, t)$ amplitude for a one-dimensional Gaussian wave packet scattering from an Eckart barrier with an initial kinetic energy of $E = 10$ au is plotted at four different times. The *green points* are the IFDM results and the *solid red curves* are the Crank–Nicholson results. In order to improve visualization, only every fourth IFDM data point is plotted

**Fig. 3** Same as in Fig. 2 except that the velocity $v(x, t)$ is plotted. Every fourth IFDM data point (*green*) is plotted except in the regions near the velocity spikes located at $x = -4.5$ and $x = -3$ au for $t = 1.25$ au where all of the IFDM points are plotted

computational efficiency) of the method, so that averaging the 2nd derivative of $C$ is *not* required. The new generalized IFDM uses "pure" second-order finite difference expressions for *all* spacial derivatives of $C$ and $v$ and *no* moving window average. The typical CPU times for computing a converged probability reported in Table 1 on an Intel core duo E8400 3.00 GHz processor were 1.3 min and 16.0 min for the IFDM and Crank–Nicholson,

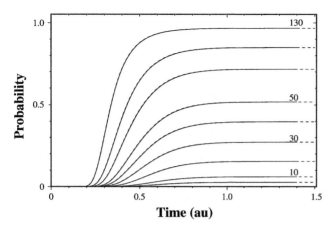

**Fig. 4** The transmission probability for a one-dimensional Gaussian wave packet scattering from an Eckart barrier is plotted as a function of time. *Each curve* corresponds to a different initial kinetic energy: $E = 5, 10, 20, 30, 40, 50, 70, 90$, and $130$ au. The *solid curves* were computed using the generalized IFDM, and the *dashed curves* were computed using the Crank–Nicholson method

**Table 1** Transmission probabilities ($P_t$) for the one-dimensional wave packet scattering from an Eckart barrier are listed for several values of initial kinetic energy ($E$)

| E (au) | $P_t$(IFDM, this work) | $P_t$(IFDM [14]) | $P_t$(CN, this work) |
|---|---|---|---|
| 0.01 | 2.4 (−3) | – | 2.5 (−3) |
| 0.1 | 3.3 (−3) | – | 3.3 (−3) |
| 1 | 7.4 (−3) | – | 7.4 (−3) |
| 5 | 0.0267 | 0.0263 | 0.0266 |
| 10 | 0.0601 | 0.0595 | 0.0600 |
| 20 | 0.1539 | 0.1531 | 0.1538 |
| 30 | 0.2709 | 0.2702 | 0.2708 |
| 40 | 0.3957 | 0.3953 | 0.3957 |
| 50 | 0.5160 | 0.5158 | 0.5159 |
| 70 | 0.7151 | 0.7154 | 0.7151 |
| 90 | 0.8474 | 0.8478 | 0.8474 |
| 130 | 0.9646 | 0.9649 | 0.9645 |

The $P_t$(IFDM) were computed using the quantum hydrodynamic approach and the IFDM. The $P_t$(CN) were computed by solving the standard time-dependent Schrödinger equation using the Crank–Nicholson method. All of these probabilities were taken from time step $t = 1.5$ au (see Fig. 4)

respectively. The total number of grid points at the final time $t = 1.5$ au was 1,700 and 56,842 for the IFDM and Crank–Nicholson calculations, respectively.

Figure 5 plots the velocity truncation error (see "Appendix 2") at $t = 1.2$ au for the IFDM solution with initial kinetic energy $E = 10$ au. The red points were computed from $\text{MAX}[|\gamma_6|, |\gamma_7|]$ and the blue points from $\text{MAX}[|\gamma_1|, |\gamma_2|, |\gamma_3|, |\gamma_4|, |\gamma_5|]$. The red points are on average greater than the blue points except in the regions near node formation ($x = -3$ and $-4$ au) and where the classical force is significant (between $x = +1$ and $+4$ au). Thus, the terms involving the 3rd derivatives of $C$ and $v$ in the velocity truncation error (i.e., the $\gamma_6$ and $\gamma_7$) are important and should be included in order to improve the stability of the method. The original IFDM method ignored the $\gamma_6$ and $\gamma_7$ and included only the $\gamma_1 - \gamma_5$ [14].

## 3.2 One-dimensional morse oscillator

In order to treat higher-dimensional scattering problems and eventually real molecules, the IFDM must be capable of treating bound anharmonic potentials. Thus, the primary goal in this section is to test the stability and accuracy of the IFDM method when it is applied to the calculation of the Morse ground state wave function. In the quantum hydrodynamic approach, a ground state wave function is truly stationary: the classical and quantum forces exactly cancel ($f_c + f_q = 0$) and a velocity field which is zero initially remains identically zero for all time ($v(x,t) = 0$) [6]. The appropriate boundary conditions for $C$ and $v$ at the edges of the grid are given by Eqs. 19 and 20, and 15 and 16, respectively. The $C$ and $v$ for the IFDM calculation were initialized at $t = 0$ using Eq. 17 for the $C(x,0)$ and using $v(x,0) = 0$ for the velocities. Since the bound state is stationary, the exact solution will also be given by Eq. 17 for all time $t > 0$. The parameters for the Morse potential were $D_e = 40$ au, $A = 0.4a_0^{-1}$, and $r_0 = 2.0a_0$. The corresponding anharmonicity constant is $\chi_e = 0.0224$ and the potential supports 21 vibrational states. For comparison, a Morse potential for $H_2$ with $D_e = 0.1743$ au, $A = 1.04a_0^{-1}$, and $r_0 = 1.4a_0$ has an anharmonicity constant of $\chi_e = 0.0292$ and supports 16 vibrational states. The IFDM time step was $\Delta t = 0.5 \times 10^{-5}$ and the artificial viscosity parameters used in the calculation were $\gamma_0 = 0$ and $\gamma'_0 = 2 \times 10^6$ (see Eq. 60 in "Appendix 2" for the definitions of $\gamma_0$ and $\gamma'_0$). The probability density at the edge of the "smart" Eulerian grid was $\rho = 1.0 \times 10^{-6}$, and since the wave packet is stationary, no points were added or subtracted at the edges. Figure 6 plots the *maximum* of the percent difference $\%C_\text{diff} = 100\%|(C_\text{IFDM} - C_\text{anal})/C_\text{anal}|$ computed at each time step for three different grid spacings $\Delta x$. The maximum at each time step was determined by computing the % $C_\text{diff}$ at all $x$ for which $\rho > 1.0 \times 10^{-4}$ and then plotting the largest value. Figure 6 shows that the IFDM stationary state calculation for the anharmonic Morse oscillator is stable for long propagation times and that the accuracy of the solution increases as $\Delta x$ is decreased over the values 0.01, 0.005, and 0.0025 au.

Figure 7 plots the velocity truncation error for the IFDM Morse bound state calculation at two different time steps denoted by A ($t = 0.01$ au) and B ($t = 1.0$ au). As in Fig. 5, the red points correspond to $\text{MAX}[|\gamma_6|, |\gamma_7|]$ and the blue points to $\text{MAX}[|\gamma_1|, |\gamma_2|, |\gamma_3|, |\gamma_4|, |\gamma_5|]$. Figure 7 shows that initially (case A: $t = 0.01$ au) the two sets (red and blue) of velocity truncation error are similar in magnitude.

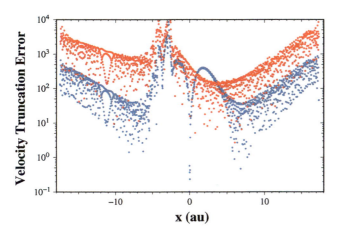

**Fig. 5** The velocity truncation error is plotted at $t = 1.2$ au for a one-dimensional Gaussian wave packet scattering from an Eckart barrier with an initial kinetic energy of $E = 10$ au. The *red points* were computed from $\text{MAX}[|\gamma_6|, |\gamma_7|]$ and the *blue points* from $\text{MAX}[|\gamma_1|, |\gamma_2|, |\gamma_3|, |\gamma_4|, |\gamma_5|]$

As time increases (case B: $t = 1.0$ au), the truncation error from $\gamma_7$ (red points) becomes several orders of magnitude larger than the error from the $\gamma_1 - \gamma_5$ (blue points). "Appendix 2" shows that $\gamma_1 - \gamma_5$ and $\gamma_6$ are all proportional to $v$, derivatives of $v$, or the total force $f$. Thus, for a stationary state, all of these are much smaller in magnitude. The $\gamma_7$ are proportional to the 3rd derivative of $v$ and are therefore most sensitive to the numerical errors. This can be seen in the red points for case A ($t = 0.01$ au) which are larger in magnitude at the left and right edges of the grid. These oscillations propagate inwards towards $x = 2$ au and increase in magnitude as time increases (see the red points for case B). The artificial viscosity (which is proportional to $\gamma_7$) stabilizes the solution and prevents further accumulation of error. Thus, the maximum value of the $\gamma_7$ remains constant at around $1 \times 10^4$ for the remainder of the calculation ($1 < t < 10$ au). These results indicate that $\gamma_7$ is critical for maintaining stability of the IFDM calculation for bound degrees of freedom (as mentioned above in Sect. 3.1 the original version [14] of the IFDM did not include this term).

### 3.3 Two-dimensional scattering

In this section, the two-dimensional (2D) version of the IFDM is applied to the scattering of a 2D wave packet propagating on an anharmonic potential energy surface consisting of an Eckart barrier along $x$ and a bound Morse potential in $y$

$$V(x,y) = V_o \text{sech}^2[a(x-x_b)] + D_e[1 - \exp(-A(y-r_0))]^2 - D_e,$$
(49)

where the Eckart parameters are $V_o = 10$ au, $a = 0.4 a_0^{-1}$, $x_b = 0$ and the Morse parameters are $D_e = 40$ au, $A = 0.25 a_0^{-1}$, $r_0 = 3 a_0$. The corresponding anharmonicity constant is $\chi_e = 0.014$ and the potential supports 35 vibrational states. The

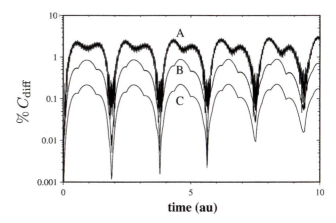

**Fig. 6** The maximum percent difference between the IFDM and analytic $C(x, t)$ (see Eq. 17) for the one-dimensional Morse ground state wave function is plotted as function of time. The IFDM results remain stable for long times and the accuracy increases as the grid spacing is decreased: (A) $\Delta x = 0.01$ au, (B) $\Delta x = 0.005$ au, and (C) $\Delta x = 0.0025$ au

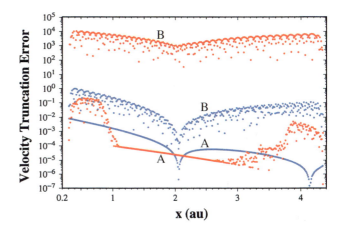

**Fig. 7** The velocity truncation error is plotted at (A) $t = 0.01$ au and (B) $t = 1.0$ au for the one-dimensional IFDM Morse ground state wave function. The *red points* were computed from $\text{MAX}[|\gamma_6|, |\gamma_7|]$ and the *blue points* from $\text{MAX}[|\gamma_1|, |\gamma_2|, |\gamma_3|, |\gamma_4|, |\gamma_5|]$. As time increases, $\gamma_7$ becomes the dominant source of the truncation error

zero of energy is chosen to be in the asymptotic limit where both $x \to \infty$ and $y \to \infty$. The initial 2D wave packet at $t = 0$ is given by

$$C(x, y, 0) = \ln(2\beta/\pi)^{1/4} - \beta(x - x_0)^2 + C_{\text{morse}}(y), \quad (50)$$

where the first two terms on the right-hand side of Eq. 50 are the contributions from the Gaussian wave packet in $x$, and the third term $C_{\text{morse}}(y)$ is the contribution from the ground state Morse oscillator wave function given by Eq. 17. The initial width of the Gaussian in $x$ is $\beta = 1.5 a_0^{-2}$ and its initial position is $x_0 = -5 a_0$. For a specified initial flow kinetic energy $E = \hbar^2 k^2/(2m)$, the value of $k$ is determined and the initial velocity component in the $x$ direction is initialized as $v(x, y, 0) = \hbar k/m$. The initial velocity component in the $y$ direction is initialized as $u(x, y, 0) = 0$ consistent with a stationary bound state.

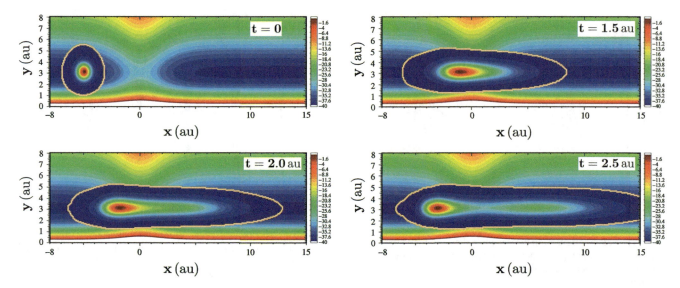

**Fig. 8** The probability density for a two-dimensional (2D) Gaussian wave packet scattering from an Eckart barrier is plotted at four different time steps. The initial kinetic energy of the wave packet is $E = 10$ au and it moves from *left* to *right*. The wave packet was computed using the 2D IFDM and is overlayed on top of a contour plot of the two-dimensional potential surface: an Eckart barrier in $x$ centered at $x = 0$ plus a Morse potential in $y$ centered at $y = 3$ au. The contours for the potential energy surface are equally spaced between: *dark blue* $= -40$ au to *red* $= 0$ au. The edge of the wave packet is plotted in *yellow*, and the contours for the wave packet are equally spaced between: *dark blue* $= 1.0 \times 10^{-5}$ to *red* $=$ maximum value

Figure 8 plots both the potential energy surface and probability density of the wave packet at four different time steps for an initial flow kinetic energy of $E = 10$ au. The colored contours for the potential energy surface are equally spaced between $V = -40$ au (dark blue) and $V = 0$ au (dark red). For large $|x|$ (i.e., in the reactant and product channels), the anharmonic Morse potential in $y$ is clearly visible with its repulsive wall near $y = 0.3a_0$ and dissociative region beyond $y > 8a_0$. The center of the Eckart barrier is clearly visible at $x = 0$, and it lies 10 au above the bottom of the asymptotic Morse potential well in the reactant and product channels. The colored contours for the probability density are equally spaced between $\rho = 1 \times 10^{-5}$ (dark blue) and $\rho_{max}$ (red). The wave packet moves from left to right and tunnels through the barrier at $x = 0$. Depending upon the value of the initial flow kinetic energy, some fraction of the wave packet tunnels through the barrier and the rest is reflected. The edge of the "smart" Eulerian grid is plotted in yellow. As time evolves and the wave packet spreads, the appropriate Eulerian grid points are activated and deactivated. By $t = 2.5$ au, the wave packet has split into two well-defined components moving to the left and right. Two MPEG movie files that show the time evolution of the wave packet plotted in Fig. 8 and the corresponding probability current density ($j_x = \rho v$) are available.[2]

A magnified view of the 2D "smart" Eulerian grid at $t = 1.0$ au is plotted in Fig. 9. The green points are internal points and the red points are the boundary (or edge) points. At each time step, the $C(x, y, t)$ are iteratively computed at each internal point. The appropriate boundary conditions are then used to set the values at the edge points. The probability density at the edge points is also monitored at each time step. If the density falls below some user-specified threshold ($\rho < 10^{-5}$ in this case), then that edge point is deactivated and the neighboring internal point is converted to an edge point. If the density increases above some user specified threshold ($\rho > 10^{-4}$ in this case), then that edge point is converted into an internal point and a new edge point or points are added around that point to ensure a valid grid structure. The $C$, $u$, and $v$ fields are then initialized at the new edge points using the appropriate boundary conditions in $x$ and $y$. The boundary conditions for $C$ are given by Eqs. 13 and 14 for a Gaussian type wave function in the $x$ direction, and by Eqs. 19 and 20 for a Morse wave function in the $y$ direction. The same boundary conditions are used for both $u$ and $v$, and they are given by Eqs. 15 and 16 for both $x$ and $y$. Depending upon the geometry of the boundary, some edge points may receive contributions from the boundary conditions in just $x$ or $y$, or they may receive contributions from both $x$ and $y$. In the later case, the contributions are averaged. The "smart" Eulerian grid is always optimal at each time step, in that only the grid points which contain significant density are active. Thus, the memory requirements and CPU time are minimized.

---

[2] Two MPEG movie files (Online Resources 2 and 3) are available for download from the Theoretical Chemistry Accounts supplementary material website: www.springer.com

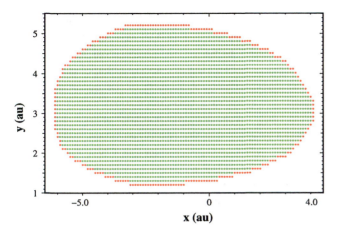

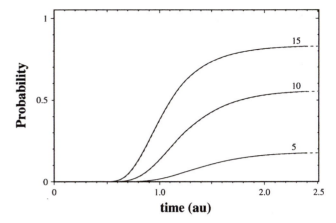

**Fig. 9** The two-dimensional "smart" Eulerian grid is plotted at $t = 1.0$ au for the wave packet propagating in Fig. 8. The *green points* are the internal points and the *red points* are the boundary points. The grid spacing is $\Delta x = \Delta y = 0.1$ au. At each time step, the $C(x, y, t)$ are iteratively computed at each internal point. The appropriate boundary conditions are then used to set the values at the edge points

**Fig. 10** The transmission (tunneling) probability for the two-dimensional (2D) wave packet scattering from an Eckart + Morse potential (see Fig. 8) is plotted as a function of time. *Each curve* corresponds to a different initial kinetic energy: $E = 5$, 10, and 15 au. The *solid curves* were computed using the 2D IFDM, and the *dashed curves* were computed using a 2D Crank–Nicholson method

Figure 10 plots the transmission (tunneling) probability computed by integrating the density of the 2D wave packet over the product region

$$P_t^{2D} = \int_0^\infty \rho(x, y, t) dx dy = \sum_{x_i > 0, y_j} \rho(x_i, y_j, t) \Delta x \Delta y. \quad (51)$$

The $P_t^{2D}$ for three different initial kinetic energies are plotted as a function of time. The probabilities are well converged by $t = 2.5$ au. The $P_i^{n+1}$ computed using the IFDM (solid curves) are essentially identical to the 2D Crank–Nicholson results (dashed curves). The IFDM results were computed but not plotted between $t = 2.4$ and 2.5 au in order to enable a better view of the Crank–Nicholson results. The final probabilities in Fig. 10 at $t = 2.5$ au are listed in Table 2. The parameters in both the IFDM and Crank–Nicholson methods were optimized to give converged probabilities to within 1%. The $P_{t=2.5}^{2D}$ listed in Table 2 show that the IFDM and Crank–Nicholson results are also in excellent agreement with each other (the differences are less than 1%). The 2D Crank–Nicholson calculations were done using a time step of $\Delta t = 1.0 \times 10^{-4}$ and a grid spacing of $\Delta x = 0.025$ au and $\Delta y = 0.05$ au. A finer grid in $x$ and $y$ is required by the Crank–Nicholson method since it must resolve the oscillatory real and imaginary parts of the wave function. In the hydrodynamic approach, a larger grid spacing can be used since the $C$ amplitude and velocity $v$ are typically smoother functions. It is important to note that both the 1D and 2D Crank–Nicholson methods also utilize a "smart" Eulerian grid. However, for stability reasons, the grids are much larger (i.e., they extend to much smaller densities $\rho \approx$

**Table 2** Transmission probabilities ($P_t^{2D}$) for the two-dimensional (2D) wave packet scattering from an Eckart + Morse potential are listed for several values of initial kinetic energy ($E$)

| E (au) | $P_t^{2D}$(IFDM) | $P_t^{2D}$(CN) |
|---|---|---|
| 5 | 0.176 | 0.175 |
| 10 | 0.553 | 0.552 |
| 15 | 0.829 | 0.828 |

The $P_t^{2D}$(IFDM) were computed using the quantum hydrodynamic approach and the 2D IFDM. The $P_t^{2D}$(CN) were computed by solving the standard time-dependent Schrödinger equation using a 2D Crank–Nicholson method. All of these probabilities were taken from time step $t = 2.5$ au (see Fig. 10)

$10^{-16}$) and the 2D grid is rectangular. An efficient sparse linear complex solver (ME57) is used to solve the resulting set of complex matrix equations [32, 40]. The typical CPU time required by the 2D Crank–Nicholson method to obtain a converged $P_{t=2.5}^{2D}$ in Table 2 was 132 h (using a Intel core duo E8400 3.00 GHz processor). In stark contrast, the typical CPU time required by the 2D IFDM method was only 1.3 h. The total number of grid points at the final time $t = 2.5$ au was 426,375 and 13,433 for the Crank–Nicholson and IFDM calculations, respectively. The optimized parameters used in the 2D IFDM calculations are the following: the grid spacings in $x$ and $y$ were $\Delta x = 0.075$ au and $\Delta y = 0.075$ au, the time step was $\Delta t = 1.0 \times 10^{-4}$, and the artificial viscosity parameters in $x$ were $^1\gamma_0 = 1 \times 10^6$, $^1\gamma'_0 = 1 \times 10^7$, $^2\gamma_0 = ^2\gamma'_0 = 5 \times 10^7$, and in $y$ they were $^1\gamma_0 = ^2\gamma_0 = 1 \times 10^8$, $^1\gamma'_0 = ^2\gamma'_0 = 2 \times 10^9$ (see "Appendix 2" Eqs. 66 and 67 for the definitions of these parameters). The conservation of probability during the 2D IFDM calculations is monitored by evaluating

$\left|1 - \sum_{ij} \rho(x_i, y_j)\Delta x\Delta y\right|$ which was less than $5 \times 10^{-4}$. The conservation of energy is also monitored by evaluating $E = \sum_{ij} \rho_{ij}[0.5v_{ij}^2 + 0.5u_{i,j}^2 + V(x_i, y_j) + Q(x_i, y_j)]\Delta x \ \Delta y$. The percent differences of the energy relative to its initial value at time zero were less than 1 %.

## 4 Conclusions

A generalized iterative finite difference method (IFDM) was presented for solving the quantum hydrodynamic equations of motion. The new methodology is based on a "smart" Eulerian grid which activates and deactivates grid points based on the time evolving density at the edges of the grid. Using this approach, the grid is always optimal in size and location which improves the overall computational efficiency and stability of the method. A more general truncation error and stability analysis was performed for both one and two dimensions. New and important stabilizing artificial viscosity terms were identified and implemented. A new approach for iteratively solving the finite difference representation of the coupled non-linear quantum hydrodynamic equations was implemented. This approach allows the IFDM to be extended to higher dimensions. Both one- and two-dimensional versions of the method were developed in this work. Generalized boundary conditions were derived for the anharmonic Morse potential. A proper treatment of anharmonic boundary conditions is important for extending the method to realistic (ab initio) potential energy surfaces.

The generalized IFDM was applied to several scattering and bound state calculations. Transmission probabilities were computed for a one-dimensional Gaussian wave packet scattering from an Eckart barrier. The probabilities are in excellent agreement (<0.4%) with those computed using the Crank–Nicholson (CN) method. The CN method solves the standard time-dependent Schrödinger equation [32]. A significant result is that the one-dimensional IFDM calculations were 12 times faster than the CN ones. The stability and accuracy of the IFDM for a one-dimensional anharmonic potential was tested by computing the stationary Morse ground state wave function. The IFDM solution for the anharmonic Morse potential was stable out to very long propagation times ($t = 10$ au). The role of a new stabilizing artificial viscosity term was also investigated. Transmission probabilities were computed for a two-dimensional Gaussian wave packet scattering on an anharmonic potential energy surface representing a model collinear reaction. The surface consisted of an Eckart barrier in $x$ and a Morse potential in $y$. The two-dimensional probabilities are in excellent agreement (<1%) with those computed using a two-dimensional CN method. These results represent the first successful application of an accurate direct numerical solution of the quantum hydrodynamic equations to a two-dimensional model collinear reaction using an anharmonic potential energy surface. A significant result is that the two-dimensional IFDM calculations are 100 times faster than the CN ones. Another advantage to using the quantum hydrodynamic approach for propagating time-dependent wave packets is that the wave packet propagates smoothly off the edges of the grid with no reflections. Thus, absorbing potentials are not needed.

Additional improvements to the IFDM approach are actively being pursued, and these include generalizing the boundary conditions for $C$ and $\mathbf{v}$ at the edges of the grid to treat ab initio potential energy surfaces, expressing the finite difference equations directly in terms of $S$, implementing a variable grid spacing, extending the method to three dimensions, and applying the approach to real molecular collisions. Both analytic and more accurate numerical approaches for treating the "node problem" are also being pursued. The method can also be generalized to compute state-to-state scattering matrix elements via the Fourier transform of the appropriate time correlation function between the scattered wave packet and a desired asymptotic product state [13, 41]. It should also be possible to apply the methodology to a variety of coordinate systems, such as bond angle, Jacobi, and hyperspherical. The intuitively appealing quantum hydrodynamic approach provides new insight as well as an alternative computational approach for studying the quantum dynamics of molecular collisions. Methods for the direct numerical solution of these equations continue to improve, and these methods together with approximate methods may provide a feasible approach for including quantum effects in large systems with many degrees of freedom.

**Acknowledgments** This work was done under the auspices of the US Department of Energy at Los Alamos National Laboratory. Los Alamos National Laboratory is operated by Los Alamos National Security, LLC, for the National Nuclear Security Administration of the US Department of Energy under contract DE-AC52-06NA25396.

## Appendix 1: Analytic derivatives of $F_1$ and $\tilde{F}_2$

The functions $F_1(C, v)$ and $\tilde{F}_2(C, v)$ are defined by Eqs. 9 and 26, respectively. The $C$ and $v$ are short-hand notation for $C = C_i^{n+1}$ and $v = v_i^{n+1}$. Thus, the analytic derivatives of $F_1(C, v)$ are obtained by taking the derivative of Eq. 9 with respect to $C_i^{n+1}$ and $v_{i'}^{n+1}$

$$\frac{\partial F_1}{\partial C} = 1, \tag{52}$$

$$\frac{\partial F_1}{\partial v} = \frac{\Delta t}{\Delta x} \left[ 1/16 + (C_{i+1}^{n+1} - C_{i-1}^{n+1})/8 \right]. \tag{53}$$

Similarly, the analytic derivatives of $\tilde{F}_2(C, v)$ are obtained by taking the derivative of Eq. 26 with respect to $C_i^{n+1}$ and $v_{i'}^{n+1}$

$$\frac{\partial \tilde{F}_2}{\partial C} = -\frac{\Delta t}{2m} \frac{\partial f_{qi'}^{n+1}}{\partial C}, \tag{54}$$

$$\frac{\partial \tilde{F}_2}{\partial v} = 1 + \frac{\Delta t}{\Delta x} \left[ (v_{i'+1}^{n+1} - v_{i'-1}^{n+1})/4 + \Delta t^2 \tilde{\gamma}_{i'}^{n+1}/(m\Delta x) \right]. \tag{55}$$

The derivative of $f_{q}^{n+1}{}_{i'}$ with respect to $C_i^{n+1}$ is given by

$$\frac{\partial f_{qi'}^{n+1}}{\partial C} = \frac{\hbar^2}{2m} \left[ 1/(2\Delta x^3) - \langle \partial_x^2 C \rangle_{i'}/(2\Delta x) - \langle \partial_x C \rangle_{i'}/\Delta x^2 \right]. \tag{56}$$

Substituting the 2nd-order finite difference expressions for the derivatives of $C$ into Eq. 56 gives

$$\frac{\partial f_{qi'}^{n+1}}{\partial C} = \frac{\hbar^2}{2m} \left[ 1 + C_i^{n+1} - C_{i+2}^{n+1} \right]/(2\Delta x^3). \tag{57}$$

In the above derivations, the derivatives of the artificial viscosity coefficient $\tilde{\gamma}$ with respect to $C$ and $v$ are ignored.


## Appendix 2: Derivation of $\gamma(C, v)$

One-dimensional case

The second-order truncation error with respect to $\Delta t$ in the finite-difference representation of the velocity equation (Eq. 6) is given by $\varepsilon$ defined in Eq. 24. If the classical potential does not depend explicitly on time, then the time derivatives of the total force become equal to the time derivatives of just the quantum force, and Eq. 24 becomes

$$\varepsilon = \frac{1}{12} \Delta t^2 \left[ \partial_t^2 f_{qi'}^{n'} - m \partial_t^2 (v_{i'}^{n'} \partial_x v_{i'}^{n'}) \right]. \tag{58}$$

By expanding the time derivatives of the two terms in brackets in Eq. 58 and using the definition of the quantum force (Eq. 8), Eq. 58 becomes a complicated expression involving many terms of mixed time and spacial derivatives of $C$ and $v$. By taking the appropriate time and spacial derivatives of the equations of motion for both $C$ (Eq. 4) and $v$ (Eq. 3), all of the various time derivatives of $C$ and $v$ can be eventually expressed in terms of derivatives of $x$ only. The coefficient of the second derivative of $v$ can then be identified as the effective

numerical diffusion coefficient (i.e., $\Gamma(C,v)\partial_x^2 v$). The algebra is straightforward but tedious, and for one dimension, it is feasible to pursue. However, it is much faster and straightforward to implement the analysis on a symbolic algebra solver such as Mathematica [42]. The resulting expression for $\Gamma(C, v) = \Delta t^2 \gamma$ is given by

$$\begin{aligned} \Gamma(C, v) = \frac{1}{12} \Delta t^2 [ & 3fv - 9mv^2 \partial_x v + 18 \frac{\hbar^2}{2m} (\partial_x C)^2 \partial_x v \\ & + 30 \frac{\hbar^2}{2m} v \partial_x C \partial_x^2 C + 36 \frac{\hbar^2}{2m} \partial_x v \partial_x^2 C \\ & + 21 \frac{\hbar^2}{2m} v \partial_x^3 C + 8 \frac{\hbar^2}{2m} \partial_x^3 v ], \end{aligned} \tag{59}$$

where for clarity, the superscripts $(n')$ and subscripts $(i')$ of the $C$ and $v$ have been dropped. The force $f$ in the first term in brackets in Eq. 59 is the *total* force ($f = f_q + f_c$). Each term in brackets will be identified as $\gamma_i$ where $i = 1\text{--}7$ so that $\gamma = \sum_{i=1}^{7} \gamma_i/12$. Equation 59 shows that $\gamma$ contains both classical and quantum contributions. The classical contributions correspond to $3 f_c v + \gamma_2$. The remaining terms are all proportional to $\hbar^2$ and are therefore quantum in nature. For a Gaussian wave packet, $C$ is quadratic and $v$ is linear with respect to $x - x_0$ where $x_0$ is the center of the packet. Thus, the first four $\gamma_{1\text{--}4}$ are quadratic (i.e., they are proportional to $(x - x_0)^2$). The fifth term $\gamma_5$ is a constant term proportional to the product of the width of the wave packet and slope of the velocity field. At the edges of the grid, this term is much smaller than the quadratic terms. The last two terms $\gamma_6$ and $\gamma_7$ involve third derivatives of $C$ and $v$ and are zero for a Gaussian wave packet. For the above reasons, the original IFDM included only the $\gamma_{1\text{--}3}$ [14]. Note that for a Gaussian wave packet, the fourth term $\gamma_4$ has the same functional form as the quantum force term in $\gamma_1$. Due to the accumulation of numerical noise, the $\gamma_6$ and especially $\gamma_7$ can become significant at the edges of the grid even for a Gaussian wave packet. As shown in Sect. 3.2, the long-term stability of the method is improved by including these terms (especially $\gamma_7$). Furthermore, for anharmonic potentials (such as the Morse potential), the third derivative of $C$ in $\gamma_6$ can be non-zero even without numerical noise.

In practice, $\tilde{\gamma}$ in Eq. 26 is computed from evaluating the following expression at each grid point

$$\begin{aligned} \tilde{\gamma} = \frac{1}{12} \{ & \gamma_0 \text{MAX}[|\gamma_1|, |\gamma_2|, |\gamma_3|, |\gamma_4|, |\gamma_5|] \\ & + \gamma_0' \text{MAX}[|\gamma_6|, |\gamma_7|] \}, \end{aligned} \tag{60}$$

where the MAX function returns the maximum value of its arguments. The first five $\gamma_i$ are grouped separately from the $\gamma_6$ and $\gamma_7$ in order to allow for separate scaling via the two independent constants $\gamma_0$ and $\gamma_0'$. The artificial viscosity constants $\gamma_0$ and $\gamma_0'$ are determined from convergence and stability tests. The goal is to choose them as small as possible but still large enough to maintain the stability of the calculation.

Two-dimensional case

The second-order truncation error with respect to $\Delta t$ in the finite-difference representation of the velocity equation (Eq. 30) is given by $\varepsilon_{2D}$ defined in Eq. 43. If the classical potential does not depend explicitly on time, then the time derivatives of the total force become equal to the time derivatives of just the quantum force, and Eq. 43 becomes

$$\varepsilon_{2D} = \frac{1}{12}\Delta t^2 \left[ \partial_t^2 f_{qi'j}^{n'} - m\partial_t^2 (v_{i'j}^{n'}\partial_x v_{i'j}^{n'}) - m\partial_t^2 (u_{i'j}^{n'}\partial_y v_{i'j}^{n'}) \right]. \tag{61}$$

By expanding the time derivatives of the two terms in brackets in Eq. 61 and using the definition of the quantum force (Eq. 32), Eq. 61 becomes a complicated expression involving many terms of mixed time and spacial derivatives of $C$, $v$, and $u$. By taking the appropriate time and spacial derivatives of the equations of motion for both $C$ (Eq. 29), $v$ (Eq. 27), and $u$ (Eq. 28), all of the various time derivatives of $C$, $v$, and $u$ can be eventually expressed in terms of derivatives of $x$ and $y$ only. The coefficient of the second derivatives of $v$ with respect to $x$ and $y$ can then be identified as the effective numerical diffusion coefficients: $^1\Gamma(C,v,u)$ and $^2\Gamma(C,v,u)$ (and similarly for $u$). The algebra is straightforward but tedious, and for two dimensions, a symbolic algebra solver such as Mathematica [42] is much preferred. The resulting expression for $^1\Gamma(C,v,u) = {}^1\gamma\Delta t^2$ is given by

$$^1\Gamma(C,v,u) = \frac{1}{12}\Delta t^2 \sum_i {}^1\gamma_i \tag{62}$$

where $^1\gamma = \sum_i {}^1\gamma_i/12$ and the individual terms are

$^1\gamma_1 = +3fv, \; ^1\gamma_2 = -9mv^2\partial_x v, \; ^1\gamma_3 = +18h(\partial_x C)^2\partial_x v,$
$^1\gamma_4 = +30hv\partial_x C\partial_x^2 C, \; ^1\gamma_5 = +36h\partial_x v\partial_x^2 C, \; ^1\gamma_6 = +21hv\partial_x^3 C,$
$^1\gamma_7 = +8h\partial_x^3 v, \; ^1\gamma_8 = -6mvu\partial_y v, \; ^1\gamma_9 = +4h\partial_x C\partial_y C\partial_y v,$
$^1\gamma_{10} = +8h\partial_x C\partial_y C\partial_x u, \; ^1\gamma_{11} = +6hv\partial_y C\partial_{xy}^2 C, \; ^1\gamma_{12} = +8h\partial_{xy}^2 C\partial_y v,$
$^1\gamma_{13} = +12hu\partial_x C\partial_{xy}^2 C, \; ^1\gamma_{14} = +16h\partial_{xy}^2 C\partial_x u, \; ^1\gamma_{15} = +4h\partial_x^2 v\partial_y C,$
$^1\gamma_{16} = +4h\partial_{xy}^2 u\partial_x C, \; ^1\gamma_{17} = +8h\partial_x^2 u\partial_y C, \; ^1\gamma_{18} = +12hu\partial_{x^2 y}^3 C,$
$^1\gamma_{19} = +4h\partial_{x^2 y}^3 u, \; ^1\gamma_{20} = +2h\partial_x^2 v\partial_x C, \; ^1\gamma_{21} = +3hv\partial_{xy^2}^3 C,$
$^1\gamma_{22} = +2h\partial_{xy^2}^3 v, \tag{63}$

where for clarity, the superscripts and subscripts on the $C$, $v$, and $u$ have been dropped and $h = \hbar^2/2m$. The first seven terms are identical to those derived for the one-dimensional case (see Eq. 59). The remaining fifteen new terms involve mixed derivatives, or $x$ derivatives of $u$ and $y$ derivatives of $v$. For the 2D application considered in this work, most of these new terms for $i > 7$ are small relative to those for $i \leq 7$.

The resulting expression for $^2\Gamma(C,v,u) = {}^2\gamma\Delta t^2$ is given by

$$^2\Gamma(C,v,u) = \frac{1}{12}\Delta t^2 \sum_i {}^2\gamma_i, \tag{64}$$

where $^2\gamma = \sum_i {}^2\gamma_i/12$ and the individual terms are

$^2\gamma_1 = 3gu, \; ^2\gamma_2 = -3mu^2\partial_x v, \; ^2\gamma_3 = -6mu^2\partial_y u,$
$^2\gamma_4 = +6hu\partial_y C\partial_x^2 C, \; ^2\gamma_5 = +5h\partial_x v\partial_x^2 C, \; ^2\gamma_6 = +3hv\partial_x^3 C,$
$^2\gamma_7 = +h\partial_x^3 v, \; ^2\gamma_8 = -6muv\partial_x u, \; ^2\gamma_9 = +2h\partial_x C\partial_y C\partial_y v,$
$^2\gamma_{10} = +4h\partial_x C\partial_y C\partial_x u, \; ^2\gamma_{11} = +4h\partial_y C\partial_x^2 C, \; ^2\gamma_{12} = +3h\partial_{xy}^2 C\partial_y v,$
$^2\gamma_{13} = +6hu\partial_x C\partial_{xy}^2 C, \; ^2\gamma_{14} = +10h\partial_{xy}^2 C\partial_x u, \; ^2\gamma_{15} = +2h\partial_x^2 v\partial_x C,$
$^2\gamma_{16} = +4h\partial_{xy}^2 u\partial_x C, \; ^2\gamma_{17} = +4h\partial_x^2 u\partial_y C, \; ^2\gamma_{18} = +9hu\partial_{x^2 y}^3 C,$
$^2\gamma_{19} = +3h\partial_{x^2 y}^3 u. \tag{65}$

The various terms in Eqs. 63 and 65 have been ordered so that they match each other in form except for the terms with $i = 3$, 11, and 15. For the 2D application considered in this work, most of the terms in Eq. 65 for $i > 7$ are small relative to those for $i \leq 7$.

In practice $^1\tilde{\gamma}$ in Eq. 45 is computed from evaluating the following expression at each grid point

$$^1\tilde{\gamma} = \frac{1}{12}\left\{ {}^1\gamma_0 \text{MAX}\left[ \sum_{i\neq 6,7} |^1\gamma_i| \right] + {}^1\gamma_0' \text{MAX}\left[ |^1\gamma_6|, |^1\gamma_7| \right] \right\}, \tag{66}$$

and similarly for $^2\tilde{\gamma}$

$$^2\tilde{\gamma} = \frac{1}{12}\left\{ {}^2\gamma_0 \text{MAX}\left[ \sum_{i\neq 6,7} |^2\gamma_i| \right] + {}^2\gamma_0' \text{MAX}\left[ |^2\gamma_6|, |^2\gamma_7| \right] \right\}, \tag{67}$$

where the MAX function returns the maximum value of its arguments. The artificial viscosity constants $^1\gamma_0$, $^1\gamma'_0$, $^2\gamma_0$, and $^2\gamma'_0$ are determined from convergence and stability tests. The goal is to choose them as small as possible but still large enough to maintain the stability of the calculation. For the 2D application considered in this work, stability was obtained using only the first seven terms in Eqs. 63 and 65 (i.e., $i \leq 7$) for the diffusion coefficients $^1\Gamma$ and $^2\Gamma$ multiplying $\partial_x^2 v$ and $\partial_y^2 v$, respectively. Similar expressions are used for the diffusion coefficients multiplying $\partial_x^2 u$ and $\partial_y^2 u$. These expressions can be obtained from those in Eqs. 63 and 65 by interchanging $x \leftrightarrow y$, $v \leftrightarrow u$, and $f \leftrightarrow g$. In evaluating the various $^1\gamma_i$ and $^2\gamma_i$, it is important to use the appropriate four-point averages so that they are properly evaluated on the staggered $v$ or $u$ grids (see Fig. 1).

## References

1. Madelung E (1926) Z Phys 40:322
2. de Broglie L (1926) CR Acad Sci Paris 183:447
3. de Broglie L (1927) CR Acad Sci Paris 184:273
4. Bohm D (1952) Phys Rev 85:166
5. Bohm D (1952) Phys Rev 85:180
6. Holland PR (1993) The quantum theory of motion. Cambridge University Press, New York
7. Lopreore C, Wyatt RE (1999) Phys Rev Lett 82:5190
8. Hughes KH, Wyatt RE (2002) Chem Phys Lett 366:336
9. Trahan CJ, Wyatt RE (2003) J Chem Phys 118:4784
10. Kendrick BK (2003) J Chem Phys 119:5805
11. Pauler DK, Kendrick BK (2004) J Chem Phys 120:603
12. Kendrick BK (2004) J Chem Phys 121:2471
13. Derrickson SW, Bittner ER, Kendrick BK (2005) J Chem Phys 123:54107-1
14. Kendrick BK (2010) J Mol Struct Theochem 943:158
15. Kendrick BK (2010) The direct numerical solution of the quantum hydrodynamic equations of motion. In: Chattaraj PK (eds) Quantum trajectories. CRC Press/Taylor & Francis Group, USA, p 325
16. Kendrick BK (2011) An iterative finite difference method for solving the quantum hydrodynamic equations of motion. In: Hughes KH, Parlant G (eds) Quantum trajectories. CCP6: Dynamics of Open Quantum Systems, Warrington, p 13
17. Wyatt RE, Bittner ER (2000) J Chem Phys 113:8898
18. Rassolov VA, Garashchuk S (2004) J Chem Phys 120:6815
19. Garashchuk S (2009) J Phys Chem A 113:4451
20. Poirier B (2004) J Chem Phys 121:4501
21. Babyuk D, Wyatt RE (2004) J Chem Phys 121:9230
22. Burghardt I, Cederbaum LS (2001) J Chem Phys 115:10303
23. Burghardt I, Moller KB, Hughes K (2007) In: Micha DA (eds) Springer series in chemical physics, p 391
24. Goldfarb Y, Degani I, Tannor DJ (2006) J Chem Phys 125:231103
25. Rowland BA, Wyatt RE (2008) Chem Phys Lett 461:155
26. Chou CC, Sanz AS, Miret-Artés S, Wyatt RE (2009) Phys Rev Lett 102:250401-1
27. Garashchuk S (2010) J Chem Phys 132:014112
28. Garashchuk S (2010) Chem Phys Lett 491:96
29. Wyatt RE (2005) Quantum dynamics with trajectories: introduction to quantum hydrodynamics. Springer, New York
30. Garashchuk S, Rassolov V, Prezhdo O (2011) Review in computational chemistry, vol 27. Wiley, London, pp 111–210
31. Scannapeico E, Harlow FH (1995) Los Alamos national laboratory report LA-12984
32. Press WH, Flannery BP, Teukolsky SA, Vetterling WT (1986) Numerical recipes; the art of scientific computing. Cambridge University Press, New York
33. Patankar SV (1980) Numerical heat transfer and fluid flow. Hemisphere Publishing Co., New York
34. Morse PM (1929) Phys Rev 34:57
35. Kais S, Levine RD (1990) Phys Rev A 41:2301
36. VonNeumann J, Richtmyer RD (1950) J Appl Phys 21:232
37. Harlow FH (1960) Los Alamos scientific laboratory report LA-2412
38. Harlow FH, Welch JE (1965) Phys Fluids 8:2182
39. Leonard BP (1979) Comput Meth Appl Mech Eng 19:59
40. HSL (2011) A collection of Fortran codes for large scale scientific computation. http://www.hsl.rl.ac.u
41. Tannor DJ, Weeks DE (1993) J Chem Phys 98:3884
42. Wolfram Research, Inc (2010) Mathematica, version 8.0. Champaign, IL

Theor Chem Acc (2012) 131:1072
DOI 10.1007/s00214-011-1072-z

REGULAR ARTICLE

# Dynamics of collisions of hydroxyl radicals with fluorinated self-assembled monolayers

**Diego Troya**

Received: 7 June 2011 / Accepted: 5 August 2011 / Published online: 11 January 2012
© Springer-Verlag 2012

**Abstract** We present a classical trajectory study of the dynamics of collisions between OH radicals and fluorinated self-assembled monolayers (F-SAMs). The gas/surface interaction potential required in the simulations has been derived from high-level ab initio calculations (focal-point-CCSD(T)/aug-cc-pVQZ) of various approaches of OH to a model fluorinated alkane. The two lowest-energy doublet potential energy surfaces considered in the electronic structure calculations have been averaged to produce a pairwise analytic potential. This analytic potential has been subsequently employed to propagate classical trajectories of collisions between OH and F-SAMs at initial conditions relevant to recent experiments on related systems. The calculated rotational distributions of the inelastically scattered OH agree well with the experiment, which serves to validate the accuracy of the simulations. Investigation of the dynamics of energy transfer for different initial rotational states of OH indicates that an increase in the initial rotation of OH results in increases in both the final average OH rotational and translational energy and in a slight decrease in the amount of energy transferred to the surface. Analysis of the dynamics as a function of the desorption angle of OH from the surface shows that while there is a correlation between the final scattering angle and OH's amount of final translational energy, the amount of rotational energy in OH is largely independent of the desorption angle. The mechanism of the collisions is found to be mostly direct; in about 90% of most trajectories, OH only

collides with the surface once before desorbing, which exemplifies the rigidity of fluorinated monolayer surfaces and their inability to efficiently accommodate gas species.

**Keywords** Gas/surface scattering · Trajectory calculations · Intermolecular potentials · Self-assembled monolayers

## 1 Introduction

There is a growing interest in determining the outcome of collisions of gases with organic surfaces, as these heterogeneous processes play an important role in a variety of fields. In the environment, our understanding of the rich and complex chemistry involving aerosols [1] is further challenged because organics present at the surface of tropospheric particles [2] might play a role in the transport of matter in and out of the particles. Additional challenges to our understanding of gas/surface chemistry in the environment emerge by the continuous processing of aerosol surface organics by gas-phase oxidizers [3], of which the hydroxyl radical is the most abundant [4].

A convenient way to perform systematic studies of the reactions of common atmospheric oxidizers with organic matter in the laboratory is to immobilize the latter in solid supports using self assembly on metal surfaces [5]. Employing self-assembled monolayer technology, D'Andrea et al. [6] have recently investigated the reactions of OH radicals with alkane- and alkene-thiol molecules absorbed on gold. Using a similar setup, Fiegland et al. [7] have studied the degradation of vinyl-terminated alkanethiol self-assembled monolayers (SAMs) by gas-phase ozone.

A technical disadvantage to the use of SAMs in the experiment to examine the degradation of organic material

Published as part of the special collection of articles celebrating the 50th anniversary of Theoretical Chemistry Accounts/Theoretica Chimica Acta.

D. Troya (✉)
Department of Chemistry, Virginia Tech,
107 Davidson Hall, Blacksburg, VA 24061-0212, USA
e-mail: troya@vt.edu

Reprinted from the journal

by common oxidizers is that the surface is constantly changing as reactions take place, and this makes it difficult to obtain atomistic information about the reaction mechanisms. Complementary experimental setups that avoid this problem are therefore rather valuable. An alternative and convenient way to perform detailed studies of the processing of organic surfaces by atmospheric oxidizers is to employ continually refreshed organic liquids of well-controlled composition. Indeed, the field of experimental gas/liquid reaction dynamics has advanced greatly in recent time, from the proton-exchange studies by Nathanson [8] and the reactions of O atoms with squalane [9] and room-temperature ionic liquids (RTILs) [10] by Minton (both using time-of-flight detection of products), to the studies of F-atom + squalane reactions by Nesbitt [11], and O-atom reactions with squalane [12] and RTILs [13] by McKendrick (both using spectroscopic detection of products). Using a gas/liquid experimental setup, the latter group has recently investigated the inelastic dynamics and reactions OH with a variety of organic liquids [14, 15]. These experiments on OH scattering from liquids have provided motivation for the simulations presented in this paper.

A common pursuit of these detailed gas/surface experiments is to unveil the factors that control whether a gas traps on a surface or recoils promptly into the gas phase after collision. Obtaining the sticking probability is important because most gas/surface reactions with appreciable barrier to reaction are thought to proceed via a Langmuir–Hinshelwood mechanism whereby gases thermally accommodate on a surface prior to reaction. Advances in detailed molecular dynamics simulations have proved useful in helping to elucidate the microscopic details of gas/surface collisions involving organics. Initial simulation efforts were devoted to rare gas scattering, with SAMs acting as the organic surface of choice because they are generally more convenient to use in the calculations than liquid surfaces [16–18]. More recent theoretical work has progressed to molecular scattering from surfaces with state-to-state resolution, with examples including $CO_2$ [19], CO [20], and HCl [21] scattering from SAMs having various functionalities at the gas/solid interface.

In this work, we begin a study of the dynamics of collisions of important atmospheric oxidizers with organic surfaces by investigating the inelastic scattering of OH from fluorinated self-assembled monolayers (F-SAMs). As mentioned before, this work is motivated recent experiments by the McKendrick group [14, 15] and by the necessity of fully understanding heterogeneous processes in the environment involving organics. A significant complication of this study compared to the prior work reviewed above is that the gas-phase collider is an open-shell species. While the surface chosen in this work is inert, and therefore, the potential energy surfaces governing the

dynamics are non-reactive, the $^2\Pi$ nature of OH's ground state implies participation of two surfaces in the collisions, which introduces new challenges in the simulations.

The reminder of this paper is as follows. In Sect. 2, we describe the electronic structure calculations and fit of the potential energy surfaces required to simulate collisions of OH with fluorinated surfaces. In Sect. 3, we present a classical trajectory study of OH collisions with F-SAMs aimed at shedding light into recent experiments. Finally, we offer some concluding remarks in Sect. 4.

# 2 Potential energy surfaces

In order to carry out reliable classical dynamics studies of gas/surface collisions, the potential energy and its gradients with respect to atomic positions need to be accurately known. The potential energy surface of a standard gas/surface system can be generally divided into three components: the potential energy surface corresponding to the organic surface, that of the gas species, and the one responsible for the interactions between the gas and the surface.

Traditionally, organic surfaces have been represented by standard force fields such as OPLS [22] or MM3 [23], as these force fields bear out the essential properties of the surfaces. MM3 has been used to describe the surface potential in this work.

Regarding the gas species, in this work, we have represented the diatomic potential energy of the hydroxyl radical by a Morse potential:

$$V_{OH} = D\left[1 - e^{-\beta(r-r_e)}\right]^2 \tag{1}$$

with $D = 101.4$ kcal/mol, $\beta = 2.353$ Å$^{-1}$, and $r_e = 0.9696$ Å. This set of parameters provides spectroscopic constants for OH in excellent agreement with experiment. For instance, the calculated vibration frequency is 3,737 cm$^{-1}$ (3,738 cm$^{-1}$ experimentally [24]) and the rotational constant is 18.91 cm$^{-1}$ (18.91 cm$^{-1}$ experimentally [24]).

Regarding the gas/surface potential, prior work has shown that the non-bonding terms of standard force fields are generally not sufficiently accurate for use in detailed scattering simulations [25]. Therefore, in this work, we have derived our own analytic potential to describe the interaction between the hydroxyl radical and a fluorinated surface. The strategy used to derive this potential is based on the assumption that the interaction between the atoms of a gas species approaching a surface and those of the surface are pairwise and additive. This strategy has been employed before in a number of simulations of gas/surface collisions with excellent results [17, 19, 20]. The pairwise potentials are obtained from a fit to ab initio calculations on relevant regions of the entire potential energy surface of the system. In the rest of this section, we describe the ab initio calculations carried out to accurately map the potential for hydroxyl radicals interacting with fluorinated species and the fit of an analytical potential to those ab initio data.

## 2.1 Ab initio calculations

To determine the intermolecular interactions of OH with fluorinated alkanes, we have computed the potential energy of various approaches of OH to the simplest perfluoroalkane, $CF_4$. The selected approaches of OH to $CF_4$ are inspired by prior work on the development of gas/surface potentials using rare gases [26] or symmetric molecules such as $CO_2$ [19]. However, there are two noteworthy differences with respect to that prior work. First, the OH molecule is not symmetric about its center, and this implies that approaches in which the H-end or O-end of the molecule face the fluoroalkane molecule both need to be explicitly considered. Operationally, this means that the sampling of potential energy surface has been significantly more costly than in prior work. Figure 1 shows the 12 approaches whose potential energy we have computed using ab initio methods denoted by the labels that will be used hereafter to identify them. (Note that in prior work with rare gases or symmetric molecules, only 2 or 3 approaches were typically considered [19, 26].)

In approaches a and b, OH nears $CF_4$ so that the OH internuclear axis is perpendicular to one of the faces of the $CF_4$ tetrahedron. Approaches c and d are similar, but OH and the closest vertex of the $CF_4$ tetrahedron are collinear. The OH internuclear axis is perpendicular to a $CF_4$ tetrahedron edge in approaches e and f. The rest of the approaches are designed to capture situations in which neither the H- nor the O-end of the hydroxyl radical approaches $CF_4$ more prominently than the other end. Thus, in these approaches, the OH bond is parallel to the face of the $CF_4$ tetrahedron (g and h), perpendicular to the closest C–F bond (i and j), and perpendicular to both the bisector of the closest F–C–F angle and a $CF_4$ tetrahedron edge (k and l).

The potential energy along each of the approaches in Fig. 1 has been computed at the MP2/aug-cc-pVQZ level of ab initio theory. To add a further level of electronic correlation to these ab initio calculations, the CCSD(T) energy with that basis set has been estimated via the focal-point approximation [27]. The focal-point approximation is based on the observation that the difference between MP2 and CCSD(T) energies for each point of a potential energy surface is constant regardless of the basis set and has been shown to perform superbly for intermolecular interactions of the type investigated in this work [26]. Thus, by computing the difference between CCSD(T) and MP2 energies for all of

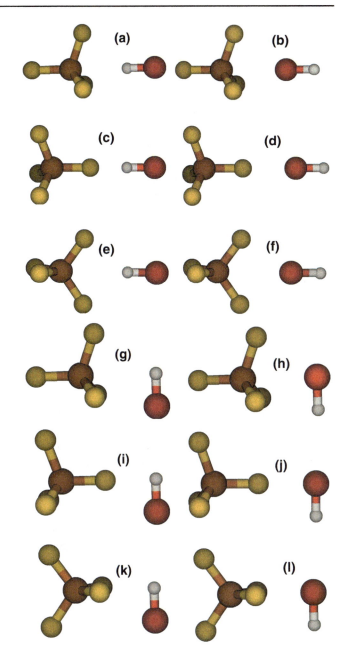

**Fig. 1** Approaches of OH to $CF_4$ explored with ab initio calculations in this work

the points of the approaches in Fig. 1 with the aug-cc-pVDZ basis set, one can conveniently estimate CCSD(T)/aug-cc-pVQZ energies from MP2/aug-cc-pVQZ calculations. Hereafter, we will refer to the estimated CCSD(T) energy using the focal-point approximation as fp-CCSD(T). All electronic structure calculations in this work have been computed using the Gaussian 09 package of programs.

A second major difference between the ab initio calculations aimed at mapping the gas/surface potential in this work and those in prior efforts is in the open-shell nature of the OH radical. The two components of the OH($^2\Pi$) state give rise to two different potential energy surfaces that can be accessed during the interaction with a closed-shell colliding partner. While some of the approaches in Fig. 1 have higher point-group symmetries than $C_s$ (approaches a–d are nominally $C_{3v}$), all of the calculations have been carried out in $C_s$ point-group symmetry, and the two potential energy surfaces emerging from the interaction of OH with $CF_4$ are thus of A′ and A″ symmetry. To accurately consider the interaction of OH with fluorinated alkanes, we have therefore needed to carry out calculations on both A′ and A″ potential energy surfaces for each of the approaches in Fig. 1, which significantly increases the computational expense.

Each potential energy surface has been scanned for each approach from the asymptote until repulsive energies of about 30 kcal/mol. The geometries of the OH and $CF_4$ molecules have been held frozen at their respective minimum energy values, and the coordinate that connects the C atom of $CF_4$ and the closest of the atoms in OH has been scanned at 0.05 Å steps. Approximately, 50 points have been computed per approach and potential energy surface. Overall, 1,224 ab initio points have been calculated in this work at the fp-CCSD(T)/aug-cc-pVQZ level to map the A′ and A″ potential energy surfaces responsible for the interaction between OH and $CF_4$.

Figure 2 presents the potential energy profiles of all of the scans that we have calculated in this work for both the A′ and A″ surfaces distributed among the same approaches as depicted in Fig. 1. Only energies of up to 5 kcal/mol are shown in the figure so that the differences at the minima can be appreciated. The energy curves have the expected profile of a weakly attractive potential well at long distances and a repulsive wall at shorter intermolecular separations. The A′ and A″ surfaces are essentially degenerate for the approaches of nominal $C_{3v}$ geometry (a–d). In the rest of approaches, the A″ surface is always more repulsive than the A′ surface except for the g and h approaches, in which the well is slightly deeper for the A″ surfaces, even if the wall is more repulsive for this surface too. This rather nuanced behavior of the A′ and A″ surfaces draws attention to the fact that one cannot estimate one surface from the other in a straightforward manner; instead, both surfaces must be computed explicitly.

Table 1 lists the location and energy of the attractive well for each approach and surface, with which the trends emerging from Fig. 2 can be corroborated: except when degeneracy exists, the well in the A″ surface always appears at longer separation distances than in the A′ surface. In addition, the A″ wells are shallower than in the A′ surface, except for approaches g and h. An immediate conclusion of the data in Fig. 2 and Table 1 is that the attraction between OH and fluoralkanes is rather weak. Therefore, one can anticipate that if OH approaches a fluorinated surface at superthermal energies, such as in the experiments of McKendrick and co-workers [14, 15], the

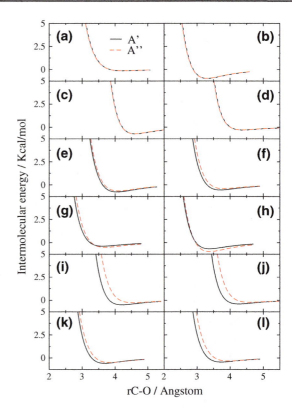

**Fig. 2** Intermolecular potential energy of approaches of OH to the $CF_4$ molecule as a function of the distance between the O and C atoms calculated at the fp-CCSD(T)/aug-cc-pVQZ. **a–l** Correspond to the approaches in Fig. 1

probability for physisorption of OH on the surface leading to long residence times should not be very large. The mechanistic studies that we present in Sect. 3 will quantify the extent to which OH traps on the F-SAM surfaces or recoils after a brief interaction.

In the following section, we describe how we have used these ab initio data to obtain an analytic potential for use in chemical dynamics simulations.

### 2.2 Fit of analytic potentials

To enlist the high-level electronic structure calculations described before in classical trajectory simulations of OH + F-SAM collisions, we have derived analytical potentials directly from the ab initio data. In this work, the intermolecular potential of an atom in a fluoroalkane molecule interacting with one of the OH atoms is represented by a Buckingham function:

$$V_{ij} = A_{ij} e^{-B_{ij} r_{ij}} + \frac{C_{ij}}{r_{ij}^6} \quad (2)$$

in which the $A$, $B$, and $C$ parameters are adjustable and specific to each pair of atoms, and $r_{ij}$ is the distance between the atoms. In general, the exponential term is responsible for representing the repulsive part of the

**Table 1** Energy and location and of the attractive well in approaches of OH to $CF_4$ in the A', A'', and fitted surfaces

| Approach | A' | A'' | $V_{sum}$ (fit)[a] |
|---|---|---|---|
| a | 0.15 (4.23) | 0.15 (4.23) | 0.20 (4.28) |
| b | 0.99 (3.30) | 0.99 (3.30) | 0.84 (3.40) |
| c | 0.71 (4.61) | 0.71 (4.61) | 0.70 (4.56) |
| d | 0.30 (4.38) | 0.30 (4.38) | 0.29 (4.58) |
| e | 0.76 (4.03) | 0.66 (4.13) | 0.46 (4.13) |
| f | 0.55 (3.70) | 0.42 (3.80) | 0.58 (3.80) |
| g | 0.39 (3.63) | 0.53 (3.73) | 0.60 (3.63) |
| h | 0.68 (3.35) | 0.99 (3.40) | 0.78 (3.45) |
| i | 0.50 (4.19) | 0.28 (4.58) | 0.42 (4.39) |
| j | 0.40 (4.28) | 0.24 (4.63) | 0.34 (4.48) |
| k | 0.59 (3.68) | 0.52 (3.83) | 0.63 (3.78) |
| l | 0.45 (3.75) | 0.36 (3.95) | 0.59 (3.80) |

Energy is in kcal/mol below the asymptote. The values in parentheses correspond to the minimum energy C–O distance in Angstrom. For the A' and A'' surfaces, the data are from fp-CCSD(T)/aug-cc-pVQZ calculations

[a] Data corresponding to the global analytic function. See text

**Table 2** Parameters of the pairwise Buckingham potential fitted to $V_{sum}$ fp-CCSD(T)/aug-cc-pVQZ ab initio points

| Pair | $A_{ij}$ (kcal/mol) | $B_{ij}$ (Å$^{-1}$) | $C_{ij}$ (kcal Å$^6$/mol) |
|---|---|---|---|
| O–F | 34,616 | 3.6736 | −409.55 |
| O–C | 36,328 | 3.8348 | −876.52 |
| H–F | 19,827 | 4.5130 | −209.58 |
| H–C | −66,879 | 4.6535 | 1,127.3 |

potential, while the second term is expected to capture the attractive region (the $C$ coefficient is thus generally negative). While an inverse sixth-power dependence of the attractive term on the separation distance is the traditional form of the potential because it has the same dependence on distance as van der Waals interactions, in some earlier work, the power of the distance has also been optimized for added flexibility [21]. In this work, we attempted to fit the ab initio data with both Eq. 2 and letting the exponent in the second term vary. The latter approach did not offer a dramatic improvement in the fit, and we thus chose to use the traditional Buckingham potential in Eq. 2 for simplicity.

Since two potential energy surfaces intervene in the dynamics of collisions of OH with an F-SAM, an important question is how to combine these surfaces to be able to compute dynamics properties that can be compared directly to experiment. The formalism of the combination was developed by Alexander in the 1980s for an open-shell diatomic molecule colliding with a structureless closed-shell species [28]. For a $^2\Pi$ molecule captured by Hund's case (a), state-to-state calculations that aim at reproducing experimental transitions in a given spin–orbit manifold require use of the $V_{sum}$ potential, which is the straight average of the A' and A'' surfaces. The $V_{dif}$ potential, which is the half the difference between the two surfaces, is responsible for transitions between different spin–orbit manifolds. This formalism for Hund's case (a) has commonly been applied to quantum dynamics calculations of inelastic scattering of OH($^2\Pi$) from rare gases [29]. Since in this work the nuclear dynamics is purely classical, the fine structure of the rotating OH radical (both the Λ

doubling and spin–orbit states) is entirely neglected, and all calculations are carried out with the $V_{sum}$ potential. Therefore, our study will primarily capture the dynamics of collisions that conserve the spin–orbit state of OH in the experiment, as has been the norm in classical trajectory simulations of energy transfer involving open-shell species in the gas phase [30].

To derive an analytical expression for the $V_{sum}$ potential, we have averaged the energies of each calculated point on the A' and A'' surface at the fp-CCSD(T)/aug-cc-pVQZ level and fitted them using the Buckingham potential of Eq. 2. A non-linear least-squares parameter optimization is utilized to obtain the parameters $A$, $B$, $C$ of each pair of atoms (O–F, O–C, H–F, and H–C) that best fit the ab initio data. These pair-specific parameters are listed in Table 2. In the fit, the parameters were not restrained so that the exponential term is repulsive and the polynomial term is attractive, and for the H–C pair, the signs of the $A$ and $C$ parameters are negative and positive, respectively, which is not the norm, but provides the best fit to the electronic structure points. In fact, the analytic function fits the ab initio data well, with a global root-mean-square deviation of 0.11 kcal/mol in the attractive region of the potential and of 0.47 kcal/mol for all points up to repulsive energies of 30 kcal/mol. More quantitative information of the fitted surface can be obtained from Table 1, where we show the location and energy of the minima along all of the approaches in the fitted $V_{sum}$ potential. In the next section, we describe how we have employed this analytic potential to propagate trajectories of collisions of OH with a fluorinated surface and compare our results with recent experimental measurements.

## 3 Classical trajectory calculations

To simulate recent experiments on the inelastic scattering of OH from fluorinated alkane surfaces, we have computed the dynamics of collisions of OH from a semifluorinated alkanethiol self-assembled monolayer via classical trajectories. While the experiments considered the scattering of OH from perfluoropolyether (PFPE, a liquid at ambient conditions) [14, 15], extensive theoretical and experimental work on the scattering of $CO_2$ from that liquid and F-SAMs has nicely shown that both surfaces are comparable [19, 31].

The F-SAM surface is composed of 25 S–$(CH_2)_2$–$(CF_2)_5$–$CF_3$ chains arranged in the pattern they adopt on a Au(111) surface [32]. Periodic boundary conditions are used to replicate the unit cell formed by the 25 chains and simulate an extended surface. Our choice of a semifluorinated surface instead of a fully fluorinated surface is motivated by the commercial availability of the former. However, we note that the presence of non-fluorinated methylene groups in the unexposed region of the SAM should be irrelevant to the work presented here, because not a single trajectory exhibits deep penetration of OH into the surface. The separation between the semifluorinated chains in the SAM is either 4.98 or 5.77 Å. 4.98 Å corresponds to the chain separation in non-fluorinated SAMs [32], while 5.77 Å has been measured for semifluorinated SAMs [33].The bulk densities of these surfaces are 2.5 and 1.8 g/cm$^3$, respectively, with the density of only the fluorinated part being approximately 15% higher. The bulk density of the liquid used in the experiments is 1.9 g/cm$^3$. As we show below, density seems to play a minor role in the dynamics of OH collisions with fluorinated surfaces at low energy.

To generate initial coordinates and momenta for the surface atoms, we have performed a 2 ns canonical (constant temperature and volume) simulation of only the SAM at 300 K. During the last 0.5 ns of this simulation, intermediate coordinates and momenta are saved every 500 fs for use as initial conditions for the surface in collisions with OH.

Regarding the initial conditions for the gas species, the OH used in the experiment emerges from photolysis of HONO at 355 nm [14]. The resulting kinetic-energy distribution of the OH radical is relatively narrow and peaks at 12.8 kcal/mol. All of the calculations in this work therefore correspond to that collision energy ($E_{coll}$). The photoinduced OH radical is vibrationally cold, but has substantial rotational population up to $N = 7$ states. In the calculations of this work, the OH is initially in the ground vibrational state ($v = 0$), and various initial rotational states are probed. While the experiments measure the total angular momentum $N$ of OH (exclusive of electron spin), in our classical dynamics calculations, we only access the nuclear rotational angular momentum $J$. To establish a correspondence between both, we use $N = J + 1$, which is a correct mapping for low levels of rotational excitation [34].

A final consideration on the initial conditions of the classical trajectories pertains to the angle of incidence formed by the initial velocity vector of OH and the surface normal. In the photolysis experiments [14], OH is generated in all directions of space. However, the laser-induced fluorescence detection is most sensitive to OH that collides normal to the surface. Therefore, all of our simulations consider normal incidence.

The OH molecule is placed initially at least 10 Å from the closest atom in the surface, and the trajectories are propagated until the recoiling OH reaches a distance of at least 10 Å from the closest surface atom after collision. The impact point on the surface is randomly sampled from a plane that represents the unit cell of the SAM. Batches of 3,000 trajectories per initial set of OH rovibrational states have been computed. To calculate the rotational state of OH after collision, the components of the rotational angular momentum ($J_x$, $J_y$, and $J_z$) are obtained from the cross product of the final diatomic position and momentum vectors, and the rotational quantum number is the nearest integer of the quantity $J = \sqrt{J_x^2 + J_y^2 + J_z^2} - 0.5$. To compare with the experiment, we assign $N = J + 1$, as explained before.

### 3.1 Influence of surface density

As mentioned before, we have considered two SAM surfaces with interchain separations of either 4.98 or 5.77 Å. Prior experimental [35] and theoretical [17] work has showed that in the case of high-energy (19.1 kcal/mol) Ar scattering from regular hydrogenated alkanethiol SAMs, the surface density plays a notable role on the amount of gas/surface energy transfer. More dense surfaces represent SAMs in which the chains are packed more tightly and thus have more difficulty in dissipating energy transferred upon gas impingement.

To verify the effect of surface density in collisions of OH with F-SAMs at conditions relevant to the experiment, we have analyzed 3,000 trajectories with OH in its ground rovibrational state impinging on the surfaces with normal incidence at $E_{coll} = 12.8$ kcal/mol. Remarkably, the properties of OH recoiling from both F-SAMs are nearly indistinguishable. The average final OH translational energy ($\langle E'_T \rangle$) in both simulations is 4.5 kcal/mol, which represents a translational energy transfer efficiency of $(E_{coll} - \langle E'_T \rangle)/E_{coll} = 0.65$. OH does not gain essentially any vibrational excitation upon collision with either surface (the change in vibrational energy upon collision is less than 2% of the zero-point energy under all conditions examined), and its average rotational energy is 0.4 kcal/mol in the less dense surface, and 0.3 kcal/mol in the surface with an interchain spacing of 4.98 Å. Figure 3 shows the full rotational distributions obtained in the simulations with both surfaces, where the slight excess of rotational energy in the low-density surface anticipated in the average rotational energies can be appreciated.

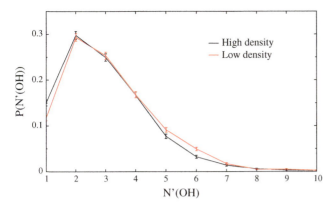

**Fig. 3** OH rotational distributions obtained in collisions of OH with fluorinated SAM surfaces at $E_{coll} = 12.8$ kcal/mol and normal incidence. The low-density SAM surface corresponds to an interchain separation of 5.77 Å and the high-density surface to 4.98 Å

The fact that the amount of energy absorbed by both F-SAM surfaces is essentially the same in the simulations is in contrast with the work on Ar scattering from regular hydrogenated surfaces of different densities mentioned above. Leaving dissimilitudes in the potential energy surfaces aside, there are two main differences that can be responsible for the OH + F-SAM collisions dynamics in this work not following the Ar + H-SAM results. First, the rare gas scattering studies were performed at collision energies slightly higher than those in this work (19.1 kcal/mol vs. 12.8 kcal/mol). Second, there is an obvious mismatch in the masses of the gas (40 amu vs. 19 amu) and surface atoms (CH$_3$- vs. CF$_3$-termini). Studies of the influence of collision kinematics on energy exchange in gas/SAM collisions indicate that surfaces with rather different masses behave similarly if the momentum of the impinging gas is not large enough to induce significant lateral motion of the SAM chains [18]. Since OH is approximately twice lighter than Ar, and its collision energy in this work is lower than that of Ar in the prior work, its momentum might not be sufficient to induce significant lateral motion of a heavy fluorinated SAM chain, regardless of the SAM packing density. This inability of OH to transfer enough momentum to the surface so that the inertial barrier for chain motion can be overcome explains the insensitivity of translational energy transfer to surface density.

In search for an explanation to the slightly enhanced OH rotation when colliding with a lower density surface, we note that the larger separation between chains in the lower density SAM makes it slightly more corrugated than the tighter-packed higher density SAM. This microscopic corrugation might elicit larger torques to an impinging OH molecule, which could result in enhanced rotational excitation. Since the scattering results at the initial conditions of this work are largely insensitive to the surface density,

all of the subsequent simulations have been performed with only the F-SAM that has an interchain separation of 4.98 Å.

As mentioned before, a complication of the current work is in the involvement of two potential energy surfaces in the scattering process. To characterize the effect of using only one of these surfaces or the $V_{sum}$ potential, we have computed trajectories for the surface of lower density with a potential obtained by fitting the data coming only from the $^2A'$ state ($1^2A$ in C$_1$ symmetry). The trajectories have used the same initial conditions as for the surface density studies just presented ($E_{coll} = 12.8$ kcal/mol, OH in ground rovibrational state, normal incidence). The average final translational energy of OH obtained with the $1^2A$ surface (5.70 kcal/mol) is significantly larger than with the $V_{sum}$ potential (4.48 kcal/mol). The average final rotational energy is not so largely affected by the choice of surface (0.29 kcal/mol vs. 0.32 kcal/mol, respectively for the $1^2A$ and $V_{sum}$ surfaces), but other scattering properties are also noticeably influenced by the surface. For instance, the average desorption angle with respect to the surface normal (36.4 and 41.6° with the $1^2A$ and $V_{sum}$ surfaces, respectively) shows deviations between the two potentials that, as we shall see below, are larger than between different initial rotational states of OH using the $V_{sum}$ surface. The conclusion of this investigation is that for the system under study, calculating the dynamics with only one surface provides results that differ significantly from the average of the two surfaces involved. Using the average potential is the common strategy in gas-phase studies, and thus, all of the results presented hereafter refer to the $V_{sum}$ potential.

### 3.2 Energy transfer

We now turn our attention to characterizing how the properties of the recoiling OH depend on its initial rotational state. The importance of this study resides on the fact that in the experiments of the McKendrick group, OH strikes a fluorinated surface having a distribution of initial rotational energies [14]. Thus, the product energy distributions emerge from the contributions of trajectories started with various OH initial rotational states. Figure 4 displays final OH rotational distributions for collisions at $E_{coll} = 12.8$ kcal/mol, normal incidence, and initial OH rotational states $N = 1, 3, 5$, and 7. Clearly, there is a correlation between the level of initial rotational excitation and product rotation. For ground-state OH, $N = 1$, there is a substantial gain of rotational excitation during the collision, which indicates that some of the collision energy is channeled into product OH rotation (T → R' energy transfer). The rest of the initial collision energy is retained by OH as final translational energy or is absorbed by the surface (as mentioned above, the collisions are

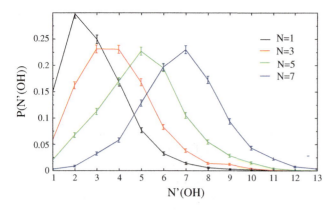

**Fig. 4** OH rotational distributions obtained in collisions of OH with fluorinated SAM surfaces at $E_{coll} = 12.8$ kcal/mol and normal incidence for various initial rotational states of OH

**Table 3** Average final translational and rotational energy of OH after collisions with a fluorinated surface

| Initial OH state | $E_R$ | $\langle E'_R \rangle$ | $\langle E'_T \rangle$ | $\langle \Delta E_{surf} \rangle$ |
|---|---|---|---|---|
| $N = 1$ | 0 | 0.32 | 4.48 | 8.00 |
| $N = 3$ | 0.20 | 0.58 | 4.55 | 7.87 |
| $N = 5$ | 0.83 | 1.09 | 4.77 | 7.77 |
| $N = 7$ | 1.89 | 1.98 | 4.99 | 7.72 |

All energies in kcal/mol, error bars are approximately 1%

vibrationally adiabatic). For OH initially rotationally excited, collisions with the fluorinated surface result in either a loss or a gain of initial rotational energy, as OH rotational states lower and higher than the initial one become populated. Average final rotational energies ($\langle E'_R \rangle$) are presented in Table 3 as a function of the initial rotational state. Those data show that for the rotational levels explored in this work, there is always a net gain of rotational energy upon collision. However, the net gain is rather small and decreases with larger initial rotational excitation.

Figure 5 presents a comparison between the final OH rotational distributions calculated in this work and the ones measured in the experiment [14]. In order to produce simulation results that can be compared with experiment, the rotational distributions generated in calculations in which OH initially possesses rotational states $N = 1-7$ are weighted by the rotational populations measured immediately after photolysis in the experiment. The laser-induced fluorescence experiments determined populations using the $R_1$, $R_2$, and $Q_1$ branches, and here, we present our calculated results in comparison with both the $R_1$ branch and the average of all measurements (obtained as $R_1 + R_2 + Q_1(R_1/R_2)$ as recommended by the experimentalists). The simulations reproduce experiment quite well, lending support to the techniques used in the calculations. There are several caveats

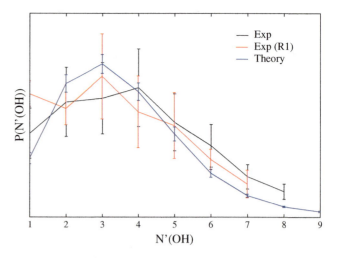

**Fig. 5** OH rotational distributions obtained in collisions of OH with fluorinated SAM surfaces at $E_{coll} = 12.8$ kcal/mol and normal incidence. The experiments are from Ref. [14]. The calculated results have been obtained by averaging rotational distributions from specific initial rotational states of OH weighted by their photolytic populations (see text)

in the theory–experiment comparison exhibited in Fig. 5. First, the experiments correspond to scattering of OH from liquid PFPE, while the simulations are for an F-SAM. However, as discussed before, there is evidence supporting the resemblance of these two surfaces toward scattering gases [19]. In addition, the comparison between surfaces of different densities introduced earlier in this paper further supports the notion that the details of the surface are perhaps not critical in this work. Second, the calculations have been performed with the $V_{sum}$ potential, which is expected to capture rotational transitions within the same spin–orbit manifold. On the other hand, the experiments indicate that non-adiabatic dynamics is not negligible [14, 15], so not all of the flux measured in the experiment should in principle be reproduced by the calculations in this work. Finally, the initial conditions in the experiment are not as well defined as in the calculations. While the experiment is most sensitive to normal-incidence OH, contributions from other incident angles are possible but difficult to quantify experimentally. Notwithstanding the difficulties in establishing a quantitative comparison between theory and experiment, the emerging conclusion from Fig. 5 is that the potential energy surfaces derived in this work provide an at least reasonable description of experiment.

To complete our understanding of energy transfer in OH + F-SAM collisions, we now examine the amount of initial energy carried by OH as final translation and the amount of initial energy transferred to the surface. Table 3 lists the average final rotational and translational of OH, and the average amount of energy transferred to the surface as a function of the initial OH rotational state. Comparison of the initial and final translational and rotational energies

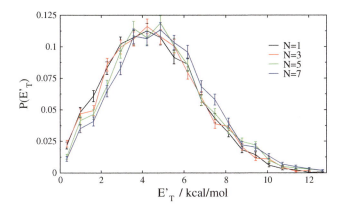

**Fig. 6** OH product translational energy distributions obtained in collisions of OH with fluorinated SAM surfaces at $E_{coll} = 12.8$ kcal/mol and normal incidence for various initial rotational states of OH

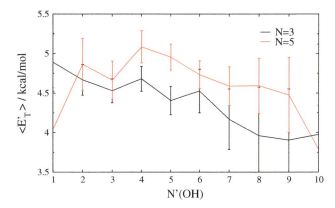

**Fig. 7** OH final average translational energy as a function of the final OH rotational state in collisions of OH ($N = 3,5$) with fluorinated SAM surfaces at $E_{coll} = 12.8$ kcal/mol and normal incidence

clearly shows that most of the initial OH energy is lost to the surface, with a slight amount going to OH rotation. Interestingly, there is also a direct correlation between initial OH rotation and final OH translation, even if the increase of final translational energy with initial rotational energy is modest. Figure 6 displays the OH final translational energy distributions for various initial rotational states of OH, where the influence of initial OH rotation on the amount of energy retained by OH as translation after collision can be fully appreciated. This increase in final OH translational energy with increasing initial rotational excitation in OH, tied to the net gain in OH rotation upon collision, implies that there is less and less energy being transferred to the surface with increasing initial rotational excitation of OH. This trend can be substantiated by the average energy gain by the surface in Table 3 ($\Delta E_{surf}$), which clearly shows how initial OH rotation acts to slightly inhibit energy transfer to the surface.

The examination of OH final rotational and translational energies in Figs. 4 and 6 reveals rather broad distributions. An interesting question then is to analyze whether trajectories in which OH is translationally hot also cause rotational excitation or whether rotational and translational modes compete for the energy available to the OH molecule during collision. The importance of this question is exacerbated by recent experimental [11] and theoretical [36] work on the *reactive* scattering of low-energy F atoms ($E_{coll} < 1$ kcal/mol) from alkane surfaces, where it was seen that HF products that result translationally excited are also rotationally excited. Figure 7 shows OH rotational-state-specific average final translational energies for two initial rotational states of OH. For OH initially in $N = 3$, in spite of the sizeable error bars, we see the absence of a positive correlation between final translation and rotation. On the contrary, there seems to be a somewhat faint decrease in the amount of final translational energy when OH becomes rotationally excited. This trend would seem to point toward the competition of rotational and translational degrees of freedom for available energy. For $N = 5$, an interesting phenomenon happens. We distinguish among three regimes of final OH rotation to facilitate discussion. First, OH that is rotationally cold after collision is also relatively translationally cold. This results implies when the surface efficiently absorbs OH rotational energy and quenches it significantly ($N = 5 \rightarrow N' = 1$), it also efficiently absorbs energy from OH initial translation. Second, collisions that are rotationally adiabatic or either gain or lose a modest amount of OH rotation ($N = 5 \rightarrow N' = 4\text{--}6$) show the largest amount of OH final translation. Finally, if OH gains substantial rotational energy during the collision ($N = 5 \rightarrow N' > 8$), it does not retain as much of its initial translational energy as when the collisions are rotationally adiabatic. These trends are significantly different from those seen in the highly exothermic HF-forming reactions of F atoms with condensed-phase alkanes, which highlights the wealth of dynamic routes that can come into play when a gas strikes an energy-absorbing organic surface, with which it might exchange energy, react, or both.

### 3.3 Angular distributions

While the experiments currently available for OH collisions with fluorinated surfaces are not exquisitely specific to the angle at which OH desorbs with respect to the surface normal, future use of molecular beam techniques with time-of-flight detection might furnish angular information about the recoiling OH. Figure 8 shows the distributions of the angle of desorption of OH (with respect to the surface normal, $\theta_f$), for various initial rotational states of OH. The distributions are very broad and peaked in the 30°–45° range. A first conclusion of the study of desorption angles is that the scattering of OH from fluorinated surfaces is that instead or recoiling in the direction of impingement, a wide range of angles are accessed in the desorption process of

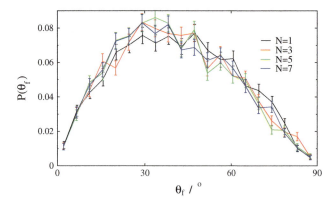

**Fig. 8** OH final polar scattering angle distributions in collisions of OH with fluorinated SAM surfaces at $E_{\text{coll}} = 12.8$ kcal/mol and normal incidence for various initial rotational states of OH

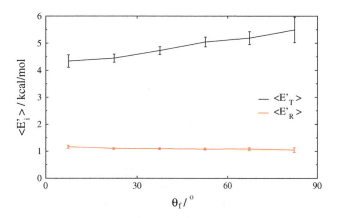

**Fig. 9** OH final average translational and rotational energy as a function of the polar scattering angle distributions in collisions of OH ($N = 5$) with fluorinated SAM surfaces at $E_{\text{coll}} = 12.8$ kcal/mol and normal incidence

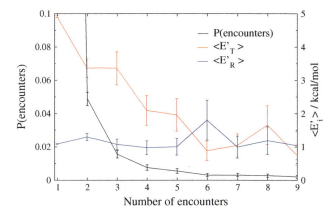

**Fig. 10** Probability distribution of the number of encounters of OH with the F-SAM (*left axis*) and final average translational and rotational energy as a function of the number of gas/surface encounters in collisions of OH ($N = 5$) with fluorinated SAM surfaces at $E_{\text{coll}} = 12.8$ kcal/mol and normal incidence. The probability of 1 encounter is 0.9

OH. A second conclusion is that the initial rotational state of OH does not influence the scattering angle, as there is substantial overlap in the distributions shown in Fig. 8.

To examine whether there is a correlation between the final OH scattering angle and the extent of gas/surface energy transfer, we have calculated the final translational and rotational energy of OH as a function of its desorption angle. Figure 9 shows how the average final translational and rotational energies depend on the polar scattering angle of the desorbing OH. The figure indicates that OH desorbing near parallel to the surface has a larger translational energy than OH desorbing near normal to the surface. An increase in the translational energy with deflection angle is expected and has been seen before in various gas/organic surface detailed calculations, including Ar scattering from F-SAMs at hyperthermal energies [37], and CO scattering from CF$_3$-terminated SAMs [20]. Such trend is also predicted by a binary collision model in which both the projectile and surface are structureless particles with a given mass, which in the case of the surface is not well defined [38]. In the next section, we present a mechanistic analysis of the collisions to further establish whether simple models could capture the dynamics of OH + F-SAM collisions.

Counter to the trend just analyzed for final translational energy, there is little to no correlation between final rotational energy and scattering angle. The differing trends in final translational and rotational energies suggest that the mechanisms controlling whether OH retains a larger fraction of its initial collision energy are not necessarily connected to an increased rotation in OH, and this is consistent with the correlation between final OH rotation and translation seen in Fig. 7.

### 3.4 Mechanism

An attractive feature of atomistic molecular dynamics simulations is that they enable one to probe the microscopic mechanism of collisions. Via examination of the coordinates and momenta throughout a trajectory, the behavior a molecule during collision can be deciphered, which adds a layer of understanding to gas/surface dynamics phenomenology.

In this work, we analyze the number of encounters of the hydroxyl radical with the surface by monitoring the number of times during a trajectory that OH's center-of-mass coordinate in the surface normal direction experiences a minimum. Trajectories in which only one minimum is found correspond to collisions in which OH encounters the surface only once. Analysis of trajectories started with OH in various rotational states reveals that approximately 90% of collisions only experience one encounter with the surface, regardless of the initial rotational state. These results clearly indicate that the mechanism of the collisions is mostly direct, i.e., the hydroxyl radical does not have a large tendency to become trapped on the surface. In the rest of trajectories, OH collides with the surface more than once, but the probability of long trapping is vanishingly small. On average, in less than 1 in 200 trajectories does OH remain trapped on the surface for longer that our integration time cutoff ($10^5$ atomic time units, or 24.2 ps). The lack of substantial long trapping was anticipated while mapping the potential energy surface in Sect. 2, as relatively shallow minima afford only weak attraction between OH and the surface. In addition, the rigid character of fluorinated alkane chains inhibits efficient accommodation of OH.

Even if most of the OH recoils from the surface after just one collision, it is still interesting to analyze how the final energy of OH depends on the residence time on the surface for the few trajectories in which it collides more than once with the surface. To this end, we present in Fig. 10 average final translational and rotational energies of OH as a function of the number of encounters with the F-SAM. The final translational energy clearly decreases with the number of encounters with the surface. This trend parallels recent simulations of the post-reaction dynamics of HF on alkanethiol SAM surfaces formed by hydrogen abstraction by incoming F atoms [36]. It also is consistent with the results of rare gas scattering from SAMs [18]. On the other hand, even if the statistics are poor due to the low number of trajectories that collide more than once with the surface, one cannot conclude that the number of encounters with the surface has a large effect on the amount of final rotational energy of OH. This seems to be a new paradigm, as it does not follow the results of the dynamics of HF on alkanethiol SAMs, where it was reported that the larger the number of encounters of the gas with the surface, the smaller the amount of rotational energy present in the gas after desorption [36]. A difference between the OH/F-SAM system studied in this work and the HF/H-SAM system of Ref. [36] is that in this work, there is a significant mass mismatch between the gas and the surface groups it interacts with that is not as remarkable in the case of HF interacting with $CH_3$-terminated SAMs. H-SAM surfaces can move laterally with more ease than F-SAMs to accommodate a gas species and efficiently absorb its rotational energy. In addition, the residence time of HF on a H-SAM surface at the conditions studied is substantially longer than that of OH on the F-SAM of this work, and this likely enables enhanced quenching of the gas rotational angular momentum by the surface. Given the intriguing trend of gas rotation with residence time on a surface unveiled in this work, in the future, it will be interesting to examine how rotational energy is transferred by a trapped species to a surface as a function of parameters such as gas

and surface masses and the strength of gas/surface interactions.

# 4 Concluding remarks

Recent experiments on the dynamics of OH scattering from organic surfaces have motivated the detailed theoretical study presented in this work. In order to model the interfacial scattering process with confidence via atomistic molecular dynamics simulations, we have developed gas/surface potential energy surfaces based on extensive and high-accuracy ab initio calculations. These electronic structure calculations have various complications, including the fact that OH is a non-symmetrical molecule and thus mapping the entire potential energy surface is more laborious than otherwise and that OH is an open-shell species, which in this case requires one to consider two potential energy surfaces.

Derivation of pairwise analytic potentials to model the OH/fluorinated surface interaction has enabled us to propagate a few thousand trajectories of OH collisions with F-SAMs at initial conditions pertinent to the experiment. After learning that the surface density seems to play a minor role on the dynamics of the collisions, we have investigated how the product properties depend on the initial rotational state of OH. Such studies have provided the possibility to compare calculated post-collision OH rotational distributions with experiment, with which we have verified the suitability of our model. Investigation of the extent of gas/surface energy transfer indicates that some modest amount of the initial collision energy is transferred to OH rotational energy and that if OH is initially rotationally excited, there is a decrease in the amount of energy transferred to the surface.

OH desorbs from the surface in a wide range of angles, and while larger deflection angles are correlated with OH that recoils faster from the surface, the amount of rotational angular momentum is essentially independent of the angle of desorption. In addition, detailed studies of the microscopic mechanism of the collisions reveal that in most of the trajectories, there is only one collision between the gas and the surface. In trajectories where OH has the opportunity to collide a few times with the surface, a decided decrease in OH's final translational energy is appreciated. This trend contrasts sharply with the apparent resilience of OH's rotational energy to be transferred to the surface.

**Acknowledgments** This work has been supported by NSF Grant No. CHE-0547543 and AFOSR Grant No. FA9550-06-1-0165. D.T. is a Cottrell Scholar of Research Corporation. Computational resources have been provided by NSF Grant No. CHE-0741927. The authors would also like to thank Ken McKendrick (Heriot-Watt University) for supplying raw experimental data and valuable advice in their interpretation.

# References

1. Finlayson-Pitts BJ, Pitts JN Jr (2000) Chemistry of the upper and lower atmosphere: theory, experiments, and applications. Academic Press, San Diego
2. Duce RA, Mohnen VA, Zimmerman PR, Grosjean D, Cautreels W, Chatfield R, Jaenicke R, Ogren JA, Pellizzari ED, Wallace GT (1983) Organic material in the global troposphere. Rev Geophys 21(4):921–952
3. Ellison GB, Tuck AF, Vaida V (1999) Atmospheric processing of organic aerosols. J Geophys Res Atmos 104(D9):11633–11641
4. Montzka SA, Krol M, Dlugokencky E, Hall B, Jockel P, Lelieveld J (2011) Small interannual variability of global atmospheric hydroxyl. Science 331(6013)
5. Love JC, Estroff LA, Kriebel JK, Nuzzo RG, Whitesides GM (2005) Self-assembled monolayers of thiolates on metals as a form of nanotechnology. Chem Rev 105(4):1103–1169
6. D'Andrea TM, Zhang X, Jochnowitz EB, Lindeman TG, Simpson C, David DE, Curtiss TJ, Morris JR, Ellison GB (2008) Oxidation of organic films by beams of hydroxyl radicals. J Phys Chem B 112(2):535–544
7. Fiegland LR, Saint Fleur MM, Morris JR (2005) Reactions of C=C-terminated self-assembled monolayers with gas-phase ozone. Langmuir 21(7):2660–2661
8. Ringeisen BR, Muenter AH, Nathanson GM (2002) Collisions of HCl, DCl, and HBr with liquid glycerol: Gas uptake, D → H exchange, and solution thermodynamics. J Phys Chem B 106(19):4988–4998
9. Zhang JM, Garton DJ, Minton TK (2002) Reactive and inelastic scattering dynamics of hyperthermal oxygen atoms on a saturated hydrocarbon surface. J Chem Phys 117(13):6239–6251
10. Wu BH, Zhang JM, Minton TK, McKendrick KG, Slattery JM, Yockel S, Schatz GC (2010) Scattering dynamics of hyperthermal oxygen atoms on ionic liquid surfaces: emim NTf2 and C(12)mim NTf2. J Phys Chem C 114(9):4015–4027
11. Zolot AM, Dagdigian PJ, Nesbitt DJ (2008) Quantum-state resolved reactive scattering at the gas-liquid interface: F plus squalane (C30H62) dynamics via high-resolution infrared absorption of nascent HF (v,J). J Chem Phys 129(19):194705
12. Kohler SPK, Allan M, Costen ML, McKendrick KG (2006) Direct gas-liquid interfacial dynamics: the reaction between O(P-3) and a liquid hydrocarbon. J Phys Chem B 110(6):2771–2776
13. Waring C, Bagot PAJ, Slattery JM, Costen ML, McKendrick KG (2010) O(P-3) atoms as a probe of surface ordering in 1-alkyl-3-methylimidazolium-based ionic liquids. J Phys Chem Lett 1(1):429–433
14. Bagot PAJ, Waring C, Costen ML, McKendrick KG (2008) Dynamics of inelastic scattering of OH radicals from reactive and inert liquid surfaces. J Phys Chem C 112(29):10868–10877
15. Waring C, King KL, Bagot PAJ, Costen ML, McKendrick KG (2011) Collision dynamics and reactive uptake of OH radicals at liquid surfaces of atmospheric interest. Phys Chem Chem Phys 13(18):8457–8469
16. Bosio SBM, Hase WL (1997) Energy transfer in rare gas collisions with self-assembled monolayers. J Chem Phys 107(22):9677–9686
17. Day BS, Morris JR, Alexander WA, Troya D (2006) Theoretical study of the effect of surface density on the dynamics of Ar plus alkanethiolate self-assembled monolayer collisions. J Phys Chem A 110(4):1319–1326
18. Alexander WA, Day BS, Moore HJ, Lee TR, Morris JR, Troya D (2008) Experimental and theoretical studies of the effect of mass on the dynamics of gas/organic-surface energy transfer. J Chem Phys 128(1):014713:1–014713:11
19. Nogueira JJ, Vazquez SA, Mazyar OA, Hase WL, Perkins BG, Nesbitt DJ, Martinez-Nunez E (2009) Dynamics of $CO_2$ scattering off a perfluorinated self-assembled monolayer. Influence of the incident collision energy, mass effects, and use of different surface models. J Phys Chem A 113(16):3850–3865
20. Alexander WA, Morris JR, Troya D (2009) Experimental and theoretical study of CO collisions with CH3- and CF3-terminated self-assembled monolayers. J Chem Phys 130(8):084702
21. Alexander WA, Troya D (2011) Theoretical study of the dynamics of collisions between HCl and omega-hydroxylated alkanethiol self-assembled monolayers. J Phys Chem C 115(5):2273–2283
22. Jorgensen WL, Maxwell DS, Tirado-Rives J (1996) Development and testing of the OPLS all-atom force field on conformational energetics and properties of organic liquids. J Am Chem Soc 117:11225–11236
23. Allinger NL, Yuh YH, Lii J-H (1989) Molecular mechanics. The MM3 force field for hydrocarbons. 1″. J Am Chem Soc 111:8551–8556
24. http://webbook.nist.gov
25. Day SB, Morris JR, Troya D (2005) Classical trajectory study of collisions of Ar with alkanethiolate self-assembled monolayers: potential-energy surface effects on dynamics. J Chem Phys 122(21):214712
26. Alexander WA, Troya D (2006) Theoretical study of the Ar–, Kr–, and Xe–CH4, –CF4 intermolecular potential-energy surfaces. J Phys Chem A 110(37):10834–10843
27. East ALL, Allen WD (1993) The heat of formation of NCO. J Chem Phys 99(6):4638–4650
28. Alexander MH (1982) Rotationally inelastic collisions between a diatomic molecule in a2Π electronic state and a structureless target J. Chem Phys 76:5974–5988
29. Scharfenberg L, Klos J, Dagdigian PJ, Alexander MH, Meijer G, van de Meerakker SYT (2010) State-to-state inelastic scattering of Stark-decelerated OH radicals with Ar atoms. Phys Chem Chem Phys 12(36):10660–10670
30. Aoiz FJ, Verdasco JE, Herrero VJ, Rabanos VS, Alexander MA (2003) Attractive and repulsive interactions in the inelastic scattering of NO by Ar: a comparison between classical trajectory and close-coupling quantum mechanical results. J Chem Phys 119(12):5860–5866
31. Perkins BG, Nesbitt DJ (2008) Stereodynamics in state-resolved scattering at the gas–liquid interface. Proc Natl Acad Sci USA 105(35):12684–12689
32. Camillone N, Chidsey CED, Eisenberger P, Fenter P, Li J, Liang KS, Liu GY, Scoles G (1993) Structural defects in self-assembled organic monolayers via combined atomic-beam and X-ray-diffraction. J Chem Phys 99(1):744–747
33. Liu GY, Fenter P, Chidsey CED, Ogletree DF, Eisenberger P, Salmeron M (1994) An unexpected packing of fluorinated n-alkane thiols on Au(111): a combined atomic force microscopy and x-ray diffraction study. J Chem Phys 101:4301–4306
34. Klos J, Aoiz FJ, Cireasa R, ter Meulen JJ (2004) Rotationally inelastic scattering of OH((2)Pi) by HCl((1)Sigma). Comparison of experiment and theory. Phys Chem Chem Phys 6(21):4968–4974
35. Day BS, Morris JR (2005) Packing density and structure effects on energy-transfer dynamics in argon collisions with organic monolayers. J Chem Phys 122(23):234714
36. Layfield JP, Troya D (2010) Theoretical study of the dynamics of F plus alkanethiol self-assembled monolayer hydrogen-abstraction reactions. J Chem Phys 132(13):134307
37. Tasic U, Troya D (2008) Theoretical study of the dynamics of hyperthermal collisions of Ar with a fluorinated alkanethiolate self-assembled monolayer. Phys Chem Chem Phys 10(37):5776–5786
38. Lahaye R, Kang H (2001) Energy exchange in structure scattering: a molecular dynamics study for Cs+ from Pt(111). Surf Sci 490(3):327–335

Theor Chem Acc (2012) 131:1068
DOI 10.1007/s00214-011-1068-8

REGULAR ARTICLE

# A full-dimensional time-dependent wave packet study of the OH + CO → H + CO₂ reaction

**Shu Liu · Xin Xu · Dong H. Zhang**

Received: 1 July 2011 / Accepted: 3 August 2011 / Published online: 11 January 2012
© Springer-Verlag 2012

**Abstract** Full-dimensional time-dependent wave packet calculations were made to study the $OH + CO \rightarrow H + CO_2$ reaction on the Lakin–Troya–Schatz–Harding potential energy surface. Because of the presence of deep wells supporting long-lived collision complex, one needs to propagate the wave packet up to 450,000 a.u. of time to fully converge the total reaction probabilities. Our calculation revealed that the CO bond was substantially excited vibrationally in the complex wells, making it necessary to include sufficient CO vibration basis functions to yield quantitatively accurate results for the reaction. We calculated the total reaction probabilities from the ground initial state and two vibrationally excited states for the total angular momentum $J = 0$. The total reaction probability for the ground initial state is quite small in magnitude with many narrow and overlapping resonances due to the small complex-formation reaction probability and small probability for complex decaying into product channel. Initial OH vibrational excitation considerably enhances the reactivity because it enhances the probability for complex decaying into product channel, while initial CO excitation has little effects on the reactivity. We also calculated the reaction probabilities for a number of $J > 0$ states by using the centrifugal sudden approximation. By doing some

calculations with multiple $K$-blocks included, we found that the centrifugal sudden approximation can be employed to calculate the rate constant for the reaction rather accurately. The calculated rate constants only agree with experimental measurements qualitatively, suggesting more theoretical studies be carried out for this prototypical complex-formation four-atom reaction.

**Keywords** Complex-forming reactions · Quantum scattering · Reaction resonance · Time-dependent wave packet method

## 1 Introduction

Because of its crucial role in the conversion of CO to $CO_2$, the $OH + CO \rightarrow H + CO_2$ reaction is important to both atmospheric [1] and combustion chemistry [2]. Due to the presence of two deep wells along the reaction path which support long-lived collision complex HOCO in both *trans* and *cis* configurations, the reaction dominated by pronounced resonances has become a prototype recently for complex-forming four-atom reactions, just as $H_2 + OH \rightarrow H + H_2O$ is for direct four-atom reactions.

The OH + CO reaction has been the subject of many experimental studies [3–12]. The measured thermal rate constants show a strong non-Arrhenius dependence on temperature [13–15]. They are nearly independent on temperature between 80 and 500 K. On the contrary, at temperatures higher than 500 K, they sharply increase with temperature. This behavior has been attributed to the presence of an intermediate HOCO complex with non-zero barriers in the entrance (OH + CO) and exit (H + CO₂) channels. The molecular beam experiment carried out by Alagia et al. [16] showed strong peaks in both the forward

---

Published as part of the special collection of articles celebrating the 50th anniversary of Theoretical Chemistry Accounts/Theoretica Chimica Acta.

S. Liu · X. Xu · D. H. Zhang (✉)
State Key Laboratory of Molecular Reaction Dynamics and Center for Theoretical and Computational Chemistry, Dalian Institute of Chemical Physics, Chinese Academy of Sciences, Dalian 116023, People's Republic of China
e-mail: zhangdh@dicp.ac.cn

Reprinted from the journal

and backward directions, indicating the existence of intermediate species. Another unusual aspect of the OH + CO reactant system is that it can form two hydrogen-bonded, van der Waals complexes, OH–CO and OH–OC, prior to the entrance channel transition state. Lester and co-workers have experimentally measured the existence of the linear OH–CO complex using infrared action spectroscopy [17]. Complementary electronic structure calculations show that there is a reaction pathway connecting the OH–CO complex to the HO–CO transition state.

Extensive theoretical studies have been carried out for this reaction and its reverse. In 1987, the first global analytic potential energy surface (PES) was constructed by Schatz, Fitzcharles, and Harding (denoted as SFH) based on the many-body expansion approach [18]. Following that, a few more PESs, such as KSW [19] PES, BS [20] PES, and YMS [21] PES, were constructed to study the reaction more accurately. In 2003, Lakin, Troya, Schatz, and Harding modified the existing many-body expansion PES further based on their new ab initio calculations to give a more accurate description of the reactant channel complexes [22]. Schatz and co-workers performed substantial quasi-classical trajectory (QCT) calculations on the new PES, known as the LTSH PES, to assess the influence of the surface changes on the reaction dynamics. It was found that the presence of the reactant channel wells enhanced the reactivity at intermediate range of energies. While the thermal rate constants obtained from the QCT calculation were smaller than experiment results, likely due to an excessively high and/or broad exit channel barrier on the surface [22, 23].

The OH + CO reaction presents a huge challenge to quantum dynamics. The combination of a relatively long-lived collision complex and three heavy atoms in this reaction makes the rigorous quantum scattering calculations difficult. In 1995, Zhang and Zhang [24] carried out a potential-averaged five-dimensional (PA5D) quantum dynamical study on SFH PES. A few years later, Kroes and co-workers obtained the PA5D total reaction probabilities for total angular momentum $J = 0$ on YMS and LTSH PES and performed 6D calculations on BS PES [25]. Medvedev et al. carried out a six-dimensional quantum wave packet study on LTSH PES, but only with 2 vibrational basis functions used for non-reactive CO bond.

In the present work, we used the time-dependent wave packet (TDWP) method to study the title reaction on LTSH PES in full dimensions. We calculated the initial-state selected total reaction probabilities from the ground state as well as some vibrational excited states for the total angular momentum $J = 0$. The total reaction probabilities for the ground initial state for $J > 0$ were also computed under the centrifugal sudden (CS) approximation. To assess the accuracy of the CS approximation, we also calculated the total reaction probabilities for a few $J > 0$ initial states with multiple $K$ components included. The rest of this paper is organized as follows. In Sect. 2, we present the theory of the full-dimensional TDWP treatment to diatom–diatom reactions and some computational details. Section 3 contains our results, including total reaction probabilities, studies of reagent vibrational excitation, and the accuracy of the CS approximation for this reaction. In Sect. 4, we summarize our conclusions.

## 2 Theory

We outline the theory of the TDWP method for calculating the initial state selected total reaction probability for a diatom–diatom reaction AB + CD → A + BCD in full dimensions. For details, please refer Refs. [26, 27]. The Hamiltonian expressed in the reactant Jacobi coordinates shown in Fig. 1 for a given total angular momentum $J$ can be written as

$$\hat{H} = -\frac{\hbar^2}{2\mu}\frac{\partial^2}{\partial R^2} + h_1(r_1) + h_2(r_2) + \frac{\mathbf{j_1}}{2\mu_1 r_1^2} + \frac{\mathbf{j_2}}{2\mu_2 r_2^2} + \frac{(\mathbf{J}-\mathbf{j_{12}})^2}{2\mu R^2} + V(r_1, r_2, R, \theta_1, \theta_2, \varphi) \quad (1)$$

where $\mu$ is the reduced mass between the center of mass of OH and CO, $\mathbf{J}$ is the total angular momentum operator, and $\mathbf{j_1}$ and $\mathbf{j_2}$ are the rotational angular momentum operators of OH and CO, which are coupled to form $\mathbf{j}_{12}$. The diatomic reference Hamiltonian $h_i(r_i)$ is defined as

$$h_i(r_i) = -\frac{\hbar^2}{2\mu_i}\frac{\partial^2}{\partial r_i^2} + V_i(r_i). \quad (2)$$

The time-dependent wave function can be expanded in terms of the translational basis of $R$, the vibrational basis $\phi_{v_i}(r_i)$, and the BF rovibrational eigenfunction as

$$\Psi^{JM\epsilon}_{v_0 j_0 K_0}(\mathbf{R},\mathbf{r_1},\mathbf{r_2},t) = \sum_{n,v,j,K} F^{JM\epsilon}_{nv j K, v_0 j_0 K_0}(t) u^{v_1}_n(R)\phi_{v_1}(r_1)\phi_{v_2}(r_2) Y^{JM\epsilon}_{jK}(\hat{R},\hat{r}_1,\hat{r}_2). \quad (3)$$

The BF total angular momentum eigenfunction can be written as

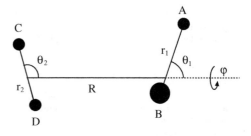

**Fig. 1** Jacobi coordinates for the reaction AB + CD → A + BCD. The angle $\varphi$ is the out-of-plane torsional angle

$$Y_{jK}^{JM\epsilon} = (1+\delta_{K0})^{-1/2}\sqrt{\frac{2J+1}{8\pi}}$$
$$\times \left[D_{K,M}^{J} Y_{j_1 j_2}^{j_{12}K} + \epsilon(-1)^{j_1+j_2+j_{12}+J} D_{-K,M}^{J} Y_{j_1 j_2}^{j_{12}-K}\right], \quad (4)$$

where $D_{K,M}^{J}$ is the Wigner rotation matrix [28], $\epsilon$ is the total parity of the system and $K$ is the projection of total angular momentum on the BF axis, and $Y_{j_1 j_2}^{j_{12}K}$ is the angular momentum eigenfunction of $j_{12}$,

$$Y_{j_1 j_2}^{j_{12}K} = \sum_{m_1} \langle j_1 m_1 j_2 K - m_1 | j_{12} K \rangle$$
$$\times y_{j_1 m_1}(\theta_1, 0) y_{j_2 K-m_1}(\theta_2, \phi) \quad (5)$$

where $y_{jm}$ are spherical harmonics. From Eq. 4, one can see that the $K = 0$ block can only appear when $\epsilon(-1)^{j_1+j_2+j_{12}+J} = 1$.

We construct an initial wave packet $\psi_i(0)$ and propagate it using the split-operator method. The total reaction probability for that specific initial state $i$ for a whole range of energies can be obtained by evaluating the reactive flux at a dividing surface $s = s_0$

$$P_i^R(E) = \frac{\hbar}{m_s} \text{Im}[\langle \psi_{iE}^{+} | \delta(s-s_0) \frac{\partial}{\partial s} | \psi_{iE}^{+} \rangle]. \quad (6)$$

$\psi_{iE}^{+}$ denotes the time-independent (TI) wavefunction, which can be obtained by performing a Fourier transform of the time-dependent wave function as

$$\langle \psi_{iE}^{+} \rangle = \frac{1}{a_i(E)} \int_{-\infty}^{\infty} e^{\frac{i}{\hbar}(E-H)t} |\psi_i(0)\rangle \, dt. \quad (7)$$

The coefficient $a_i(E)$ is the overlap between the initial wave packet and the energy-normalized asymptotic scattering function, $a_i(E) = \langle \phi_{iE} | \psi_i(0) \rangle$.

The numerical parameters for the wave packet propagation were as follows: a total of 228 sine functions (among them 36 for the interaction region) were employed for the translational coordinate $R$ in a range of [3.0, 16.0] $a_0$. Five OH vibrational basis functions were used in the asymptotic region, while 32 were used in the interaction region to expand the wave function for $r_1$ going from 1.0 $a_0$ to 5.5 $a_0$. For rotational degrees of freedom, we used 36 OH rotational states and 91 CO rotational states. A total number of 7 vibrational basis functions were used for non-reactive CO bond to get converged reaction probabilities for collision energy up to 0.4 eV. The initial Gaussian wave packet was centered at 12.5 $a_0$. A dividing surface at 3.2 $a_0$ was used for flux analysis. The wave packet propagation was carried out using a time increment of 10 a.u. A typical calculation for the total angular momentum $J = 0$ takes about 300 h on 8 workstations each with 8 CPU cores.

## 3 Results

Figure 2 shows the convergence of total reaction probabilities for the ground initial state with respect to the propagation time for $J = 0$. At short propagation time, the probabilities are small and the resonance structures are not significant. As propagation time extends, the overall magnitude gradually increases and the resonance structures become more pronounced. The probabilities are well converged after the wave packet is propagated for around 450,000 a.u. As can be seen, the converged total reaction probability shows many narrow but overlapping resonances in the low collision energy region, apparently due to the long-lived HOCO complex. It gradually becomes relatively smooth for collision energy above 0.3 eV.

To give some clues on the role of CO vibration in the reaction, we depict in Fig. 3a the vibrational potentials for CO in asymptotic region and in the wells. It can be seen that the CO equilibrium distance in the wells is longer than that for the free CO by about 0.06 $a_0$. As a result, the CO bond initially in the ground state has some population on the excited states once moving into the wells as shown in Fig. 3b. As the time propagation goes on, the CO bond for the complex gets more excited as monitored from the average bond energy shown in Fig. 3c. At $T \sim 35,000$ a.u., it more or less reaches a steady state with a population shown in Fig. 3b. Thus, it is clear that the CO bond is highly excited in the complex, and it absorbs about 0.35 eV of energy resulting from the complex formation.

The substantial excitation of CO bond in the complex wells indicates that the non-reactive CO bond does not act as a spectator, as it is described by the PA5D model, in the reaction. Kroes and co-workers have obtained the PA5D

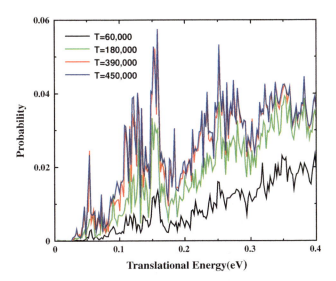

**Fig. 2** Total reaction probabilities for the ground initial state of the OH + CO reaction on the LTSH PES at wave packet propagation time of $T = 60,000, 180,000, 390,000,$ and $450,000$ a.u.

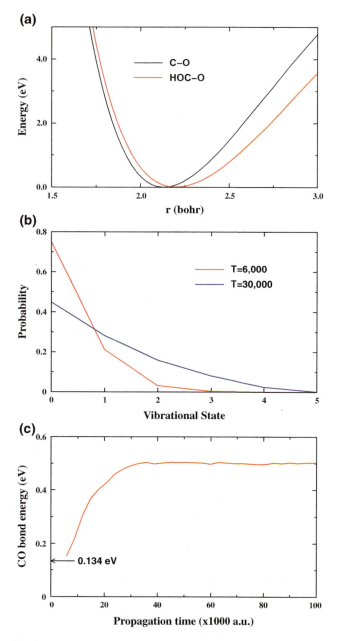

**Fig. 3** **a** Vibrational potentials for CO in asymptotic region and in the wells; **b** population probability of CO in the wells at propagation time of 6,000 and 30,000 a.u.; **c** average bond energy of the CO bond in the wells as a function of time. We start to calculate it from $T =$ 6,000 when a non-negligible portion of wave packet has moved in the wells. The *horizontal arrow* indicates the ground state energy of free CO

reaction probability on LTSH PES [25], which is quite different from our full-dimensional reaction probability in detail. So the PA5D treatment is incapable of yielding quantitative results for the title reaction, although it works very well for the $H_2 + OH$ reaction.

Although the $OH + CO \rightarrow H + CO_2$ reaction on the LTSH PES is exothermic by about 0.967 eV and has only a low barrier in the entrance channel along the minimum energy path, the reaction probability is generally small. It turned out that the majority of wave packet was reflected back to the entrance valley at $R = 4.6\ a_0$ prior to entering the HOCO well, in accord with the molecular beam experimental results of Liu and coworkers [8]. Figure 4a shows the complex-formation probability, $P_c$, as a function of collision energy obtained by calculating the flux into the wells [29]. As seen, the complex-formation probability rises quickly at the threshold energy of 0.02 eV, reaches 8% at Ec = 0.05 eV, and then increases slowly with further increase in collision energy to 21% at Ec = 0.4 eV, indicating that the entrance cone into the HOCO complex well is rather small. The ratio between the reaction probability and complex-formation probability shown in Fig. 4c gives the probability for the complex decaying into product channel, $P_d$. In low collision energy region, $P_d$ fluctuates substantially from a few percent to up to 50% around Ec = 0.15 eV. For higher collision energies, it oscillates about 20%. This states that the complex decays primarily back to reactant channel, because the barrier separating complexes from products is higher and broader than the barrier separating complexes from reactants.

The small entrance cone to the complex well and relatively high and broad barrier separating complexes from products on the LTSH PES not only make the reaction probability small, but also make the decay of complex very slow. Consequently, to converge the reaction probability, one needs to propagate the wave packet for a very long time, much longer than that for the SFH PES of about 90,000 a.u. [24].

Figure 4a and b also show the total reaction probabilities and complex-formation probabilities for the OH or CO vibrationally excited initial states as a function of translational energy. As expected, the complex-formation probabilities are very close to each other for these three initial states, in particular in low collision energy region, while the reaction probabilities behave rather differently. Although the OH vibrational excitation essentially does not reduce the threshold energy for the reaction because the reaction threshold is determined by the entrance barrier as found for the complex-formation probabilities, it enhances the reactivity considerably. The reaction probability for that initial state rises quickly in the collision energy range from threshold to about 0.06 eV. It then increases slower with the further increase in the collision energy, essentially with the same slope as that for the ground initial state. At the collision energy of 0.4 eV, the probability for the OH ($v = 1$) initial state is larger than that for the ground state by a factor of 1.8. It also can be seen that the OH ($v = 1$) reaction probability is much less oscillatory compared to the ground initial state in the low collision energy region. Consequently, it converges at a propagation time of 270,000 a.u., much shorter than that for the ground initial

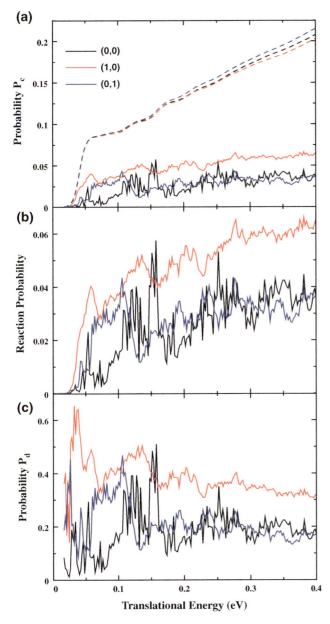

**Fig. 4** **a** Complex-formation probabilities (in *dashed lines*), $P_c$, and corresponding total reaction probabilities (in *solid lines*) for the ground and two vibrationally excited states as a function of collision energy on the LTSH PES; **b** total reaction probabilities as shown in **a** except in a smaller *y*-axis range; **c** probability for the complex decaying into product channel, $P_d$, as a function of collision energy

Fig. 4a, different reaction probabilities mean that the probabilities for the complex decaying into product channel, $P_d$, for the three states are different. As shown in Fig. 4c, in almost entire energy region, $P_d$ for OH ($v = 1$) is considerably larger than the other two states. In high-energy region $P_d$ for CO ($v = 1$) is very close to that for the ground initial state. In low-energy region, each of them has an energy window with substantially larger $P_d$, and somehow the energy window for CO ($v = 1$) is lower in energy by about 0.045 eV than that for the ground initial state. Therefore, it is clear that increase in complex energy on the coordinates other than on the OH coordinate does not increase the probability for complex decaying into product channel except near energy threshold, while increase in energy on OH bond in complex does increase the probability, and the excitation energy initially deposited on OH bond remains to some extent on the bond in the complex.

The total reaction probabilities from the ground initial state for a number of total angular momentum $J$ were calculated under the CS approximation. Figure 5 presents the results for $J = 30$ and 50. As seen, the $J > 0$ CS reaction probability curves resemble the $J = 0$ curve very well, except for a threshold energy shift caused by the centrifugal potential and a small decrease in overall magnitude with the increase of $J$.

To check the accuracy of the CS approximation for the reaction with only $K = 0$ block included, we made some $J > 0$ calculations with more than one $K$-block. This kind of calculations for $J > 0$ is extremely time-consuming as the computational effort increases by a factor of $2 \times NK + 1$ (where $NK$ is the number of $K$-blocks) compared with the CS calculation. Figure 6 shows the $J = 20$ total reaction probabilities calculated with $NK = 1$ (CS),

state of 450,000 a.u. On the other hand, no substantial enhancement on the reaction probability is observed for the vibrational excitation of CO to the first excited state as shown in Fig. 4b, except in a small energy range around 0.1 eV. We also note that Kroes and co-workers also calculated the 6D QCT reaction probabilities for CO ($v = 1$) on the LTSH PES [30], which are in reasonable agreement with our QM result.

Since the complex-formation probabilities, $P_c$, for these three initial states are essentially the same as shown in

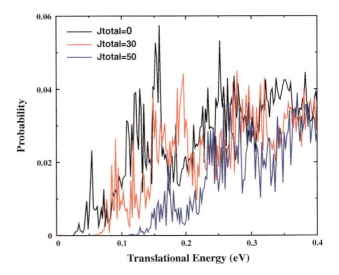

**Fig. 5** Total reaction probabilities for $J = 30$ and 50 with CS approximation for the OH + CO reaction as a function of collision energy

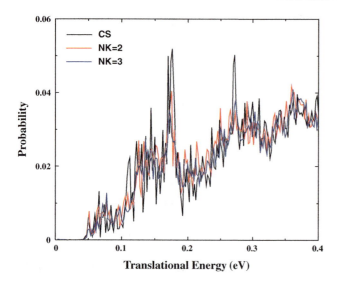

**Fig. 6** Comparison of the CS reaction probability for $J = 20$ with the $NK = 2, 3$ results

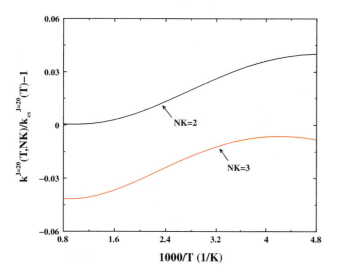

**Fig. 7** Differences between CS and $NK = 2, 3$ rate constants for partial wave $J = 20$

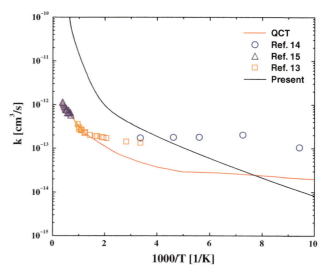

**Fig. 8** CS rate constant for the ground initial state for the title reaction compared with the previous QCT result on the same PES and experimental data of Frost et al. ([14], *circles*), Golden et al. ([15], *triangles*), and Ravishankara and Thompson ([13], *squares*)

$NK = 2$, and $NK = 3$. Overall, the CS probability agrees with the $NK > 1$ results rather well, but the probability curve becomes increasingly smoother as $NK$ increases. Although the difference between the $NK = 2$ and 3 probabilities is rather small, it seems that 3 $K$-blocks are not sufficient to yield a full convergence on the total reaction probability even for $J = 20$.

However, if one is only interested in the rate constant, the CS approximation actually can be employed without any significant deterioration in accuracy for this reaction. In Fig. 7, we show the difference between CS and $NK = 2$, 3 rate constants for $J = 20$ measured in term of $k^{J=20}(T, NK)/k_{cs}^{J=20}(T) - 1$, where $k^J(T, NK)$ is rate constant for partial wave $J$ given by

$$k^J(T, NK) = \int_0^\infty dE e^{-E/k_b T} P^J(E, NK). \quad (8)$$

In the entire temperature region shown, $k^{J=20}(T, NK = 2)$ is only a few percents larger than $k_{cs}^{J=20}(T)$, and $k^{J=20}(T, NK = 3)$ is only by a few percents smaller than $k_{cs}^{J=20}(T)$, despite the fact that the details of the corresponding reaction probabilities differ considerably as shown in Fig. 6. Our test calculation also showed that the difference between the CS rate constant for $J = 50$ and the corresponding $NK = 3$ rate constant is <10% in the valid temperature region. Therefore in the temperature region of Fig. 7, where the thermal rate constant is dominated by contributions from partial waves with $J$ up to 50, the CS approximation can be used to calculate the thermal rate constant for this reaction with a reasonable accuracy.

With long wave packet propagation time in this reaction, it is a formidable task at present to calculate the total reaction probabilities for many partial waves to obtain integral cross sections for the reaction. Instead, we calculated the rate constant for the ground initial state by using the J-shifting scheme [31]. Figure 8 shows the CS rate constant in a temperature range from 100 to 1,700 K, in comparison with the QCT result on the same PES [23] and some experimental results [13–15]. At $T = 100$ K, the QM rate is smaller than both QCT and experimental results. However, as temperature increases, the QM rate increases much faster than QCT and experimental results, which becomes the largest at $T \sim 250$ K. One should note that our calculation is for initial ground rovibrational state, while the experimental measurements are for the thermal averaging over initial rotations of the reagents. Thus, the discrepancy between the experiment and present calculation may come from the defects of the PES, or the neglect of influence of reagent rotational excitations, or both. More theoretical studies should be carried out to shed light on the discrepancy, to eventually lead a quantitative agreement between theory and experiment on this prototypical complex-formation four-atom reaction.

## 4 Conclusions

We carried out extensive TDWP calculations on the $OH + CO \rightarrow H + CO_2$ reaction on the LTSH PES. Because of the presence of deep wells supporting long-lived collision complex HOCO, one needs to propagate wave packets up to 450,000 a.u. of time to fully converge the pronounced resonance structures in the total reaction probabilities. We calculated the total reaction probabilities for the ground initial state as well as two vibrational excited states for total angular momentum $J = 0$. The total reaction probability for the ground initial state is quite small in magnitude with many narrow and overlapping resonances, because (1) the complex-formation reaction probability is quite small due to the small entrance cone into the HOCO complex wells; (2) the reactive complex decays primarily back to reactant channel as the barrier separating complexes from products is higher and broader than the barrier separating complexes from reactants. Our calculation also revealed that the CO bond was substantially excited vibrationally in the complex wells, making the PA5D method incapable of yielding quantitative results for the reaction.

We also investigated the effects of initial vibrational excitation on the reaction. It is found that initial vibrational excitation of OH considerably enhances the reactivity of the system, while initial vibrational excitation of CO has no significant enhancement to reaction probability. Because the complex-formation probabilities for these three initial states are found to be very close to each other, reactivity enhancement with the OH excitation is clearly due to the fact that initial OH excitation enhances the probability for the complex decaying into product channel.

The total reaction probabilities were calculated for a number of $J > 0$ states by using the CS approximation. To assess the validity of the CS approximation for the reaction, we also calculated some $J > 0$ reaction probabilities for the ground initial state with multiple $K$-blocks included. Overall, the agreement between the CS and $NK = 2, 3$ calculations is reasonably good, but the reaction probabilities become increasingly smoother as the number of $K$-blocks increases. While one is only interested in the thermal rate constant, the CS approximation can be employed without any significant deterioration in accuracy for this reaction. We found that the calculated CS rate constants only agree with experimental measurements qualitatively, suggesting more theoretical studies be carried out for this prototypical complex-formation four-atom reaction.

**Acknowledgments** This work was supported by the National Natural Science Foundation of China (Grant Nos. 20833007 and 90921014), the Chinese Academy of Sciences, and Ministry of Science and Technology of China.

## References

1. Finlayson-Pitts BJ, Pitts JN (2000) Chemistry of the upper and lower atmosphere. Academic press, San Diego
2. Miller JA, Kee RJ, Westbrook CK (1990) Annu Rev Phys Chem 41:345
3. Smith IWM, Zellner R (1973) J Chem Soc Faraday Trans 2(69):1617
4. Jonah CD, Mulac WA, Zeglinski P (1984) J Phys Chem (US) 88
5. Gardiner WC Jr (1977) Acc Chem Res 10:326
6. Brunning J, Derbyshire DW, Smith IWM, Williams MD (1988) J Chem Soc Faraday Trans 2 84:105
7. Fulle D, Hamann HF, Hippler H, Troe J (1996) J Chem Phys 105:983
8. Sonnenfroh DM, Macdonald RG, Liu K (1991) J Chem Phys 94:6508
9. van Beek MC, Ter Meulen JJ (2001) J Chem Phys 115:1843
10. Scherer NF, Khundkar LR, Bernstein RB, Zewail AH (1987) J Chem Phys 87:1451
11. Rice JK, Baronavski AP (1991) J Chem Phys 94:1006
12. Ionov SI, Brucker GA, Jaques C, Valachovic L, Wittig C (1993) J Chem Phys 99:6553
13. Ravishankara AR, Thompson RL (1983) Chem Phys Lett 99:377
14. Frost MJ, Sharkey P, Smith IWM (1991) Faraday Discuss Chem Soc 91:305
15. Golden DM, Smith GP, McEwen AB, Yu CL, Eiteneer B, Frenklach M, Vaghjiani GL, Ravishankara AR, Tully FP (1998) J Phys Chem A 102:8598
16. Alagia M, Balucani N, Casavecchia P, Stranges D, Volpi GG (1993) J Chem Phys 98:8341
17. Lester MI, Pond BV, Anderson DT, Harding LB, Wagner AF (2000) J Chem Phys 113:9889
18. Schatz GC, Fitzcharles MS, Harding LB (1987) Faraday Discuss Chem Soc 84:359
19. Kudla K, Schatz GC, Wagner AF (1991) J Chem Phys 95:1635
20. Bradley KS, Schatz GC (1997) J Chem Phys 106:8464
21. Yu HG, Muckerman JT, Sears TJ (2001) Chem Phys Lett 349:547
22. Lakin MJ, Troya D, Schatz GC, Harding LB (2003) J Chem Phys 119:5848
23. Medvedev DM, Gray SK, Goldfield EM, Lakin MJ, Troya D, Schatz GC (2004) J Chem Phys 120:1231
24. Zhang DH, Zhang JZH (1995) J Chem Phys 103:6512
25. Valero R, McCormack DA, Kroes GJ (2004) J Chem Phys 120:4263
26. Zhang DH, Zhang JZH (1994) J Chem Phys 101:1146
27. Zhang DH, Lee SY (1999) J Chem Phys 110:4435
28. Rose M (1957) Elementary theory of angular momentum. Wiley, New York
29. Lin S, Guo H (2004) J Chem Phys 120:9907
30. Valero R, Kroes GJ (2004) Phys Rev A 70:040701
31. Zhang DH, Zhang JZH (1999) J Chem Phys 110:7622

Theor Chem Acc (2012) 131:1071
DOI 10.1007/s00214-011-1071-0

REGULAR ARTICLE

# Electronic structure theory: present and future challenges

**So Hirata**

Received: 8 August 2011 / Accepted: 14 August 2011 / Published online: 11 January 2012
© Springer-Verlag 2012

**Abstract** The aim of electronic structure theory is to solve the electronic Schrödinger equation, which is a coupled high-dimensional partial differential equation with numerous singularities in the operator and complex boundary conditions arising from the fermion symmetry. This article briefly summarizes how electronic structure theorists have overcome the immense difficulties of solving this with quantitative accuracy. This has been achieved by elucidating the structure of wave functions and exploiting this knowledge to drastically expedite the numerical solutions, enabling predictive simulations for a broad range of chemical properties and transformations. It also lists some of the outstanding challenges that are to be or being addressed.

**Keywords** Electronic structure theory · Explicitly correlated methods · Condensed matter · Complex systems · Density-functional theory · Strong correlation

The aim of electronic structure theory is to computationally solve the electronic Schrödinger equation (with modification by the special theory of relativity, if necessary) to understand and even predict the properties and transformations of chemical systems. Since the equation is the exact equation of motion of constituent particles, electronic structure theory can in principle provide complete,

---

Published as part of the special collection of articles celebrating the 50th anniversary of Theoretical Chemistry Accounts/Theoretica Chimica Acta.

---

S. Hirata (✉)
Department of Chemistry,
University of Illinois at Urbana-Champaign,
600 South Mathews Avenue, Urbana, IL 61801, USA
e-mail: sohirata@illinois.edu

quantitative details of these properties and transformations [1], turning an increasingly wider area of chemistry into a computational science. This may appear to be a hyperbole, but for small gas-phase molecules made of light nuclei, electronic structure theory has indeed begun to predict a variety of their properties (e.g., shape and color) with such high accuracy [2] that it can compete with experimental techniques and change the ways in which research is conducted in areas like combustion and interstellar chemistry. Electronic structure theorists are determined to bring such changes in all areas of chemistry by increasing the accuracy, efficiency, and applicability of their computational methods.

However, the challenges faced by electronic structure theorists to achieve this goal can be immediately seen by inspecting the equation of motion itself (in atomic units):

$$\hat{H}\Psi(\boldsymbol{x}_1,\ldots,\boldsymbol{x}_n) = E\Psi(\boldsymbol{x}_1,\ldots,\boldsymbol{x}_n), \tag{1}$$

where $\Psi$ is the wave function, $E$ is the energy, and $\hat{H}$ is the electronic Hamiltonian given by

$$\hat{H} = -\frac{1}{2}\sum_{i=1}^{n}\nabla^2 - \sum_{i=1}^{n}\sum_{I=1}^{N}\frac{Z_I}{|\boldsymbol{x}_i - \boldsymbol{x}_I|} + \sum_{i=1}^{n-1}\sum_{j=i+1}^{n}\frac{1}{|\boldsymbol{x}_i - \boldsymbol{x}_j|}$$
$$+ \sum_{I=1}^{N-1}\sum_{J=I+1}^{N}\frac{Z_I Z_J}{|\boldsymbol{x}_I - \boldsymbol{x}_J|}, \tag{2}$$

where $n$ is the number of electrons, $N$ is the number of nuclei, $\boldsymbol{x}_i$ is the spin-orbital coordinates of the $i$th electron, and $\boldsymbol{x}_I$ and $Z_I$ are the spatial coordinates and charge of the $I$th nucleus. The wave function, $\Psi(\boldsymbol{x}_1,\ldots,\boldsymbol{x}_n)$, must also satisfy the anti-symmetry requirement:

$$\Psi(\boldsymbol{x}_1,\boldsymbol{x}_2,\boldsymbol{x}_3,\ldots,\boldsymbol{x}_n) = -\Psi(\boldsymbol{x}_2,\boldsymbol{x}_1,\boldsymbol{x}_3,\ldots,\boldsymbol{x}_n)$$
$$= -\Psi(\boldsymbol{x}_3,\boldsymbol{x}_2,\boldsymbol{x}_1,\ldots,\boldsymbol{x}_n) = \ldots \tag{3}$$

Reprinted from the journal

For benzene, Eq. 1 is a coupled 126-dimensional partial differential equation. The corresponding Hamiltonian has 1,365 singularities at electron–nucleus and electron–electron coalescence. Equation 3 is a single equation with $42! \approx 1.4 \times 10^{51}$ terms. Worse still, to achieve predictive accuracy in chemistry, one must solve Eq. 1 with errors in relative energies on the order of one millihartree or less, which is less than 0.001% of $|E|$.

Hence, when viewed naively as a general mathematical or computational problem, it is utterly unthinkable to solve Eq. 1 with such high accuracy for all but the tiniest molecules. One must instead view this as a chemistry or physics problem and eliminate many of these difficulties (the high dimensionality, singularities, and complex dependencies due to anti-symmetry) by elucidating and exploiting the structures of electronic wave functions.

Let us consider the singularities in the Hamiltonian due to electron-nucleus coalescence in this regard. The presence of these singularities mean that the electrons sufficiently near a nucleus in a molecule experience only the nucleus' strong attractive forces and thus largely behave like atomic electrons. This suggests the use of atomic-orbital (AO) basis functions and linear-combination-of-atomic-orbital (LCAO) expansions of molecular spin-orbitals (MO) as a rational description of molecular wave functions. The electron–electron interactions also cause many singularities in the Hamiltonian, but they are repulsive. This means that the electrons try to be as far away as possible from one another and their motions are, therefore, only weakly correlated. This suggests an approximation of a wave function by an anti-symmetrized product (Slater determinant) of LCAO MOs, each of which accommodates one electron. This product form of the wave function introduces an approximate separation of variables, drastically reducing the effective dimension of the equation of motion. The Hartree–Fock (HF) method is defined as such and consistently recovers *ca.* 99% of $E$ for a broad range of chemical systems.

Though the HF method achieves remarkable accuracy at a small computational cost, the accuracy is far from satisfactory for quantitative chemistry. One must recover the so-called electron-correlation energy, which is defined as the difference between the HF and exact energies. One mathematically straightforward, but physically unappealing way (see below for an explanation) to do this is to add excited determinants to the many-electron basis in the wave function expansion:

$$\Psi = c_0\Phi_0 + \sum_a \sum_i c_i^a \Phi_i^a + \sum_{a<b} \sum_{i<j} c_{ij}^{ab} \Phi_{ij}^{ab}$$
$$+ \sum_{a<b<c} \sum_{i<j<k} c_{ijk}^{abc} \Phi_{ijk}^{abc} + \ldots, \qquad (4)$$

where $i$, $j$, and $k$ ($a$, $b$, and $c$) are occupied (unoccupied) MOs in the HF determinant ($\Phi_0$) and $\Phi_i^a$, $\Phi_{ij}^{ab}$, and $\Phi_{ijk}^{abc}$ are one-, two-, and three-electron excited determinants. This mathematical description of the wave function raises the difficult question as to how one should determine the expansion coefficients, $c_i^a$, $c_{ij}^{ab}$, etc. and the correlation energy. The answer comes from a rather unexpected direction: scaling properties of energy with respect to system size [3]. The time-tested methods to determine these coefficients in a physically sound fashion are based on diagrammatic many-body theories that are *size extensive* in the sense that they maintain uniform accuracy across various system sizes and are applicable equally to problems in quantum chemistry as well as in nuclear, atomic, solid-state, and condensed-matter physics. The many-body perturbation (MBPT) and coupled-cluster (CC) methods [4] are perhaps the most successful examples in this category and are used predominantly in quantitative chemical simulations today. This is in contrast with the configuration-interaction (CI) method, which is neither a diagrammatic many-body method nor size extensive and thus slowly exiting from the essential electron-correlation arsenals.

Despite the substantial progress in the field, challenges remain. In the following are listed some of the most significant research problems today, some nearly resolved and others still being in an earlier stage of investigation. The three invited articles [5–7] in the section of electronic structure theory describe the cutting edges of research addressing different aspects of these.

# 1 Explicitly correlated methods

The weak electron correlation is captured by CC and MBPT that describe mutually avoiding motion of electrons as a superposition of excited determinants. This is an awkward physical picture of correlation, which has a grave practical drawback: the convergence of correlation energies thus obtained is extremely slow with respect to the size of the AO basis set. A more physically appealing and efficient approach is to introduce a basis function that depends explicitly on interelectronic distances ($r_{12}$), which can describe cusps in wave functions that occur because of the electron–electron singularities in the Hamiltonian [8]. The last decade has seen remarkable progress in this class of methods (the R12 method [9], in particular), virtually eradicating the problem of slow basis-set convergence. Two of the breakthroughs behind this progress are the discovery [10] of a nearly optimal form of the $r_{12}$-dependent basis function (the so-called Slater-type correlation factor, introducing the F12 method) and the proposal [11] to determine the expansion coefficient multiplying the $r_{12}$-dependent basis function analytically, both of which have been made by Ten-no [5].

## 2 Condensed-phase and complex systems

Systematic, size-extensive electron-correlation approximations (CC, MBPT, and their combinations) converging toward the exact basis-set solutions of the Schrödinger equation have been developed (increasingly with the aid of computer algebra [12]). Systematic basis sets converging toward completeness are also available. They have been instrumental in making electronic structure theory, a predictive science at least for small molecules in the gas phase. Today, these methods also raise the prospect of having condensed-phase systems and large complex systems (such as biological systems, in which the strengths of the pertinent interactions, including dispersion and stacking interactions [13], span three orders of magnitude) in the applicable domain of predictive simulations, potentially transforming quantitative aspects of these fields. Considerable progress has already been made to the use of spatially local basis functions that exploits the essentially local nature of correlation and speeds up these calculations dramatically [14]. In addition to the speedup procedures that benefit all these methods, a diagrammatic many-body method that strikes superior balance between accuracy and computational cost and is subject to an extension to metallic solids or to an integration with statistical thermodynamics (e.g., a finite-temperature extension) continues to be warranted. The random-phase approximation (RPA), introduced to chemistry by Furche [6, 15], is one such method which is only marginally more expensive than HF yet is capable of capturing much of electron correlation. RPA, like CC [16], includes the sum of perturbation corrections of a certain type to an infinite order and can thus resist a breakdown typical of MBPT when quasi-degenerate states are present near the ground state such as occur in metals. It promises to be a practical, systematic, and size-extensive electron-correlation method useful for small and large molecules as well as solids alike.

## 3 Density-functional theory

Density-functional theory (DFT) does not conform to the framework of wave-function theory (WFT), namely, the one based on the expansion of an exact wave function by determinants. It instead aims at recovering the exact energy as a functional of electron density. Abolishing at least partly the high-dimensional wave function as a quantity to be determined, DFT tends to achieve superior cost-

accuracy performance and greater practical utility than the existing counterparts in WFT. While neither systematic nor predictive in the sense of CC, MBPT, or RPA, approximations in DFT have seen an explosive advance in the past two decades, eliminating such shortcomings in earlier approximations as their poor descriptions of ionization energies [17], Rydberg excitation energies [17], charge-transfer states [18], (hyper)polarizabilities [19], and dispersion [20] and stacking interactions [21]. The use of electron density in lieu of wave functions as a basic variable must, therefore, somehow exploit a hidden structure of wave functions, the detail of which has not been understood. Further theoretical studies of fundamental nature are thus still warranted to elucidate why some approximations of DFT perform so well for a range of problems. Ultimately, like WFT, DFT needs to be characterized by systematic series of approximations converging toward the exactness, in which electron density may be a low-rank member of a new hierarchy of basic physical variables such as density matrices [22], intracules, and two-electron density parameters, the last two of which have been advocated by Gill and coworkers [7, 23].

## 4 Strong correlation

Strongly correlated systems are ones in which there are two or more (quasi-)degenerate states that are interacting with one another. They include such systems as spin lattices, transition-metal complexes with many metal centers, and breaking and formation of chemical bonds. The HF method, in which electron correlation is assumed implicitly to be weak (see above), becomes a poor approximation, rendering MBPT and occasionally even CC based on a single reference determinant also inadequate. There has been an impressive array of original methods recently proposed to resolve this problem. Some [24], including various multi-reference and active-space electron-correlation methods, selectively sum critical determinant contributions within the framework of WFT. Others [25] do so by allowing the number of electrons to be varied! Perhaps the most successful ones thus far essentially resort to the full CI method but with vastly improved algorithms based on quantum Monte Carlo [26], density matrix renormalization group [27], graphically contracted functions [28], etc. These methods and algorithms imply that certain regular structures exist in strongly correlated wave functions, which are yet to be fully understood.

The field of electronic structure theory has undergone steep, linear growth since 1990, capitalizing upon the foundational work in the preceding decades, into a new level of maturity. It has transformed neighboring experimental fields of chemistry, which, as a result, rely increasingly heavily on black-box computational methods and software electronic structure theorists have developed. It also continues to serve as an essential infrastructure of higher quality and greater convenience for the sister fields of theoretical chemistry that is quantum dynamics and statistical thermodynamics. There is hardly any question that this trend will continue, and the prime of electronic structure theory is still ahead.

## References

1. Schaefer HF III (1986) Science 231:1100
2. Bytautas L, Ruedenberg K (2006) J Chem Phys 124:174304
3. Hirata S (2011) Theor Chem Acc 129:727
4. Bartlett RJ, Musiał M (2007) Rev Mod Phys 79:291
5. Ten-no S (2012) Theor Chem Acc 131:1070
6. Eshuis H, Bates JE, Furche F (2012) Theor Chem Acc 131:1084
7. Gill PMW, Loos P-F (2012) Theor Chem Acc 131:1069
8. Klopper W, Manby FR, Ten-no S, Valeev EF (2006) Int Rev Phys Chem 25:427
9. Kutzelnigg W (1985) Theor Chim Acta 68:445
10. Ten-no S (2004) Chem Phys Lett 398:56
11. Ten-no S (2004) J Chem Phys 121:117
12. Hirata S (2006) Theor Chem Acc 116:2
13. Sinnokrot MO, Sherrill CD (2006) J Phys Chem A 110:10656
14. Saebø S, Pulay P (1985) Chem Phys Lett 113:13
15. Furche F (2001) Phys Rev B 64:195120
16. Scuseria GE, Henderson TM, Sorensen DC (2008) J Chem Phys 129:231101
17. Tozer DJ, Handy NC (1998) J Chem Phys 109:10180
18. Iikura H, Tsuneda T, Yanai T, Hirao K (2001) J Chem Phys 115:3540
19. Kamiya M, Sekino H, Tsuneda T, Hirao K (2005) J Chem Phys 122:234111
20. Kamiya M, Tsuneda T, Hirao K (2002) J Chem Phys 117:6010
21. Zhao Y, Truhlar DG (2005) Phys Chem Phys 7:2701
22. Nakatsuji H, Yasuda K (1996) Phys Rev Lett 76:1039
23. Gill PMW, O'Neill DP, Besley NA (2003) Theor Chem Acc 109:241
24. Parkhill JA, Head-Gordon M (2010) J Chem Phys 133:024103
25. Tsuchimochi T, Scuseria GE (2009) J Chem Phys 131:121102
26. Booth GH, Alavi A, Thom AJW (2009) J Chem Phys 131:054106
27. Chan GK-L, Head-Gordon M (2002) J Chem Phys 116:4462
28. Shepard R (2005) J Phys Chem A 109:11629

Theor Chem Acc (2012) 131:1084
DOI 10.1007/s00214-011-1084-8

REGULAR ARTICLE

# Electron correlation methods based on the random phase approximation

Henk Eshuis · Jefferson E. Bates · Filipp Furche

Received: 15 June 2011 / Accepted: 16 September 2011 / Published online: 14 January 2012
© Springer-Verlag 2012

**Abstract** In the past decade, the random phase approximation (RPA) has emerged as a promising post-Kohn–Sham method to treat electron correlation in molecules, surfaces, and solids. In this review, we explain how RPA arises naturally as a zero-order approximation from the adiabatic connection and the fluctuation-dissipation theorem in a density functional context. This is contrasted to RPA with exchange (RPAX) in a post-Hartree–Fock context. In both methods, RPA and RPAX, the correlation energy may be expressed as a sum over zero-point energies of harmonic oscillators representing collective electronic excitations, consistent with the physical picture originally proposed by Bohm and Pines. The extra factor 1/2 in the RPAX case is rigorously derived. Approaches beyond RPA are briefly summarized. We also review computational strategies implementing RPA. The combination of auxiliary expansions and imaginary frequency integration methods has lead to recent progress in this field, making RPA calculations affordable for systems with over 100 atoms. Finally, we summarize benchmark applications of RPA to various molecular and solid-state properties, including relative energies of conformers, reaction energies involving weak and covalent interactions, diatomic potential energy curves, ionization potentials and electron affinities, surface adsorption energies, bulk cohesive energies and lattice constants. RPA barrier heights for an extended benchmark set are presented. RPA is an order of magnitude more accurate than semi-local functionals such as B3LYP for non-covalent interactions rivaling the best empirically parametrized methods. Larger but systematic errors are observed for processes that do not conserve the number of electron pairs, such as atomization and ionization.

**Keywords** Electronic structure theory · Density functional theory · Random phase approximation · Resolution-of-the-identity (RI) approximation · Van-der-Waals forces · Thermochemistry

## 1 Introduction

The random phase approximation (RPA) is one of the oldest non-perturbative methods for computing the ground-state correlation energy of many-electron systems. In 1962, when the first issue of *Theoretica Chimica Acta* was published, RPA had already been in existence for 11 years and was reaching its first bloom in solid-state physics. The term "random phase approximation" was apparently first used in the groundbreaking series of three papers entitled "A Collective Description of Electron Interactions," which appeared between 1951 and 1953 [1–3]. In this work, Bohm and Pines attempted to solve the many-electron problem for the uniform electron gas by a transformation to a much simpler coupled harmonic oscillator problem describing long-range plasma oscillations plus a short-range correction.

Published as part of the special collection of articles celebrating the 50th anniversary of Theoretical Chemistry Accounts/Theoretica Chimica Acta.

**Electronic supplementary material** The online version of this article (doi:10.1007/s00214-011-1084-8) contains supplementary material, which is available to authorized users.

H. Eshuis · J. E. Bates · F. Furche (✉)
Department of Chemistry, University of California,
1102 Natural Sciences II, Irvine, CA 92697-2025, USA
e-mail: filipp.furche@uci.edu

H. Eshuis
e-mail: henk.eshuis@uci.edu

J. E. Bates
e-mail: batesj@uci.edu

Bohm and Pines showed that this is possible if cross-terms arising from density oscillations with different phases can be neglected, hence the name RPA. Within the RPA of Bohm and Pines, the ground-state correlation energy is given by the zero-point vibrational energy (ZPVE) of the oscillators plus a short-range correction that can be treated perturbatively. This physically appealing picture probably had a strong influence on the development of diagrammatic many-body perturbation theory (MBPT), which treats the ground-state correlation energy in analogy to the vacuum self-energy of quantum electrodynamics. Indeed, in 1957, Gell-Mann and Brueckner obtained the RPA result for the uniform gas correlation energy by summation of all ring diagrams in the MBPT expansion [4], see Fig. 1.

The random phase approximation made its first appearance in chemistry in 1964, when McLachlan and Ball pointed out that time-dependent Hartree–Fock (HF) theory is equivalent to RPA with exchange (RPAX) [5]. In the following decades, RPAX was widely used to compute molecular electronic excitation energies and transition moments and triggered the development of polarization propagator methods in the 1970s [6]. RPAX was rarely applied to molecular correlation energies, partly due to its comparatively high computational cost and partly because applications to small systems showed little promise [7]. Perhaps the most serious limitation of RPAX is its sensitivity to instabilities of the HF reference state. This and the emergence of coupled cluster theory effectively halted further development of RPAX-based correlation methods in the late 1970s. Attempts to make RPAX more stable by iteration of excitation operators ("higher-order RPA") [8, 9] were later shown to be related to the antisymmetrized geminal power method [10].

On the side of density functional theory, Langreth and Perdew [11, 12] and Gunnarsson and Lundqvist [13] established the adiabatic connection (AC) formalism in the mid 1970s. The AC underlies most modern post-Kohn–Sham (KS) correlation treatments that attempt to compute the ground-state correlation energy using the KS determinant as a reference. A crucial difference between the AC formalism and post-HF methods that the ground-state density is *independent* of the coupling strength in the former but not in the latter. Langreth and Perdew showed that RPA arises as a natural zero-order approximation if the AC

$$E^{\text{C RPA}} = \text{diagram} + \text{diagram} + \cdots$$
$$= -\frac{1}{2} \sum_{iajb} \frac{(ia|jb)(jb|ia)}{\epsilon_a + \epsilon_b - \epsilon_i - \epsilon_j} + \sum_{iajbkc} \frac{(ia|jb)(jb|kc)(kc|ia)}{(\epsilon_a + \epsilon_b - \epsilon_i - \epsilon_j)(\epsilon_a + \epsilon_c - \epsilon_i - \epsilon_k)} - \cdots$$

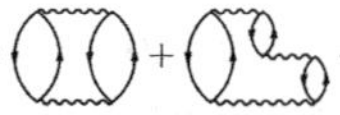

**Fig. 1** MBPT expansion of the RPA correlation energy using Goldstone diagrams. $(ia|jb)$ denotes an electron repulsion integral in Mulliken notation, and $\epsilon_p$ is a canonical Kohn–Sham orbital eigenvalue. Indices $i, j, \ldots$ denote occupied, $a, b, \ldots$ virtual, and $p, q, \ldots$ general orbitals

framework is combined with the fluctuation-dissipation theorem [14]. RPA was the basis for the development of the first van der-Waals density functionals of Langreth–Lundqvist type in the 1990s [15]. Dobson pioneered the use of RPA for the seamless treatment of long-range dispersion interactions [16, 17]. Only in 2001, RPA using a KS reference was first applied to molecules [18].

In the past decade, RPA has seen a remarkable revival. This review aims to introduce RPA in its modern form to a wider audience and explain why RPA-based electron correlation methods seem more attractive than ever. The present work reflects our personal views on the subject, and we do not claim to be exhaustive in view of a rapidly growing literature on RPA. In Sect. 2, we outline the derivation of RPA starting from the AC formalism. We also include a section on RPAX using a HF reference, highlighting the common aspects and differences between the two approaches (see also the recent review by Heßelmann and Görling [134]). An important feature of RPA is that it can be systematically improved; some beyond-RPA approaches are briefly addressed in Sect. 2.3. Section 3 contains an overview of recent RPA implementations. Recent progress in this area has helped transform RPA from a theory of mostly formal interest to a viable computational tool. In Sect. 4, we summarize recent results on the performance of RPA for a variety of molecular properties. Section 5 concludes our review.

## 2 Theory

### 2.1 Kohn–Sham reference

#### 2.1.1 Adiabatic connection

The starting point for post-KS density functional theory is the adiabatic connection Hamiltonian [11, 13]

$$\hat{H}^\alpha = \hat{T} + \hat{V}^\alpha[\rho] + \alpha\hat{V}_{ee}. \tag{1}$$

$\hat{T}$ denotes the kinetic energy operator of the electrons. The dimensionless coupling strength parameter $\alpha$ scales the electron–electron interaction $\hat{V}_{ee}$. The local multiplicative one-particle potential $\hat{V}^\alpha[\rho]$ constrains [19] the one-particle density $\rho^\alpha$ of the ground state $|\Psi_0^\alpha\rangle$ to equal the interacting ground-state density $\rho$ for all $\alpha$,

$$\rho^\alpha(x) = \rho^\alpha(x)|_{\alpha=1} = \rho(x). \tag{2}$$

$x = (\mathbf{r}, \sigma)$ denotes space–spin coordinates. Thus, for $\alpha = 1$, the physical system of $N$ interacting electrons is recovered, while $\alpha = 0$ corresponds to the KS system whose ground state is the KS determinant $|\Phi_0\rangle = |\Psi_0^\alpha\rangle|_{\alpha=0}$.

The interacting ground-state energy functional may be expressed as the energy expectation value of the Kohn–Sham determinant plus a correction for correlation,

$$E_0[\rho] = \langle \Phi_0[\rho]|\hat{H}|\Phi_0[\rho]\rangle + E_C[\rho], \tag{3}$$

where $\hat{H} = \hat{H}^\alpha|_{\alpha=1}$ is the physical Hamiltonian. While the explicit form of $E_C[\rho]$ as a functional of the density is unknown, it can be expressed by a coupling strength integral [11, 13],

$$E_C[\rho] = \int_0^1 d\alpha W_C^\alpha[\rho]. \tag{4}$$

The coupling strength integrand is the difference potential energy of the electron interaction

$$W_C^\alpha[\rho] = \langle \Psi_0^\alpha[\rho]|\hat{V}_{ee}|\Psi_0^\alpha[\rho]\rangle - \langle \Phi_0[\rho]|\hat{V}_{ee}|\Phi_0[\rho]\rangle. \tag{5}$$

#### 2.1.2 Density fluctuations

An important hierarchy of approximations to $W_C^\alpha[\rho]$ is based on the idea to reexpress the potential energy of the electron–electron interaction in terms of density fluctuations. To this end, it is convenient to express $\hat{V}_{ee}$ in second quantized form,

$$\hat{V}_{ee} = \frac{1}{2}\sum_{pqrs}\langle pq|rs\rangle \hat{c}_p^\dagger \hat{c}_q^\dagger \hat{c}_s \hat{c}_r, \tag{6}$$

where $\hat{c}_p^\dagger$ and their adjoints denote electron creation and annihilation operators, and $\langle pq|rs\rangle$ is a two-electron repulsion integral in Dirac notation. To take advantage of the locality of the electron–electron interaction, we introduce electron field operators,

$$\hat{\psi}^\dagger(x) = \sum_p \phi_p^*(x)\hat{c}_p^\dagger; \tag{7}$$

$\phi_p(x)$ denotes a Kohn–Sham spin orbital. Defining the two-particle density operator,

$$\hat{P}(x_1,x_2) = \frac{1}{2}\hat{\psi}^\dagger(x_1)\hat{\psi}^\dagger(x_2)\hat{\psi}(x_2)\hat{\psi}(x_1), \tag{8}$$

the electron interaction operator may be written as

$$\hat{V}_{ee} = \int dx_1 dx_2 \frac{\hat{P}(x_1,x_2)}{|\mathbf{r}_1 - \mathbf{r}_2|}. \tag{9}$$

Our goal is to factorize $\hat{P}(x_1,x_2)$ into products of one-particle operators. Using the fermion anticommutation relations, we obtain

$$\hat{P}(x_1,x_2) = \frac{1}{2}(\hat{\rho}(x_1)\hat{\rho}(x_2) - \delta(x_1 - x_2)\hat{\rho}(x_1)), \tag{10}$$

where $\hat{\rho}$ is the one-particle density operator

$$\hat{\rho}(x) = \hat{\psi}^\dagger(x)\hat{\psi}(x). \tag{11}$$

This may be further rewritten using the density fluctuation operator $\Delta\hat{\rho}(x) = \hat{\rho}(x) - \rho(x)$,

$$\hat{P}(x_1,x_2) = \frac{1}{2}(\Delta\hat{\rho}(x_1)\Delta\hat{\rho}(x_2) + \hat{\rho}(x_1)\rho(x_2) + \rho(x_1)\hat{\rho}(x_2) - \rho(x_1)\rho(x_2). -\delta(x_1 - x_2)\hat{\rho}(x_1)). \tag{12}$$

The last identity and Eq. 9 may be combined to evaluate the coupling strength integrand, Eq. 5. Because the density is independent of $\alpha$, all one-electron terms cancel, yielding the simple result

$$W_C^\alpha = \frac{1}{2}\int dx_1 dx_2$$
$$\times \frac{\langle \Psi_0^\alpha|\Delta\hat{\rho}(x_1)\Delta\hat{\rho}(x_2)|\Psi_0^\alpha\rangle - \langle \Phi_0|\Delta\hat{\rho}(x_1)\Delta\hat{\rho}(x_2)|\Phi_0\rangle}{|\mathbf{r}_1 - \mathbf{r}_2|}. \tag{13}$$

#### 2.1.3 Fluctuation-dissipation theorem

Equation 13 expresses the coupling strength integrand $W_C^\alpha$ as an expectation value of *products* of the density fluctuation operator $\Delta\hat{\rho}$. Using the completeness of the electronic states $|\Psi_n^\alpha\rangle$ at any $\alpha$,

$$\sum_n |\Psi_n^\alpha\rangle\langle\Psi_n{}^\alpha| = \mathbf{1}, \tag{14}$$

the expectation value of $\Delta\hat{\rho}(x_1)\Delta\hat{\rho}(x_2)$, a *two-particle* operator, may be factorized into products of *one-particle* transition densities,

$$\langle\Psi_0^\alpha|\Delta\hat{\rho}(x_1)\Delta\hat{\rho}(x_2)|\Psi_0^\alpha\rangle = \sum_{n\neq 0}\rho_{0n}^\alpha(x_1)\rho_{0n}^\alpha(x_2). \tag{15}$$

Here, we used that the ground-state expectation value of $\Delta\hat{\rho}$ is zero, and

$$\rho_{0n}^\alpha(x) = \langle\Psi_0^\alpha|\hat{\rho}(x)|\Psi_n^\alpha\rangle = \langle\Psi_0^\alpha|\Delta\hat{\rho}(x)|\Psi_n^\alpha\rangle. \tag{16}$$

Combining Eqs. 4, 13, and 15, we obtain an exact expression for the ground-state correlation energy,

$$E_C = \int_0^1 d\alpha \sum_{n\neq 0}\left(E_H[\rho_{0n}^\alpha] - E_H[\rho_{0n}^{(0)}]\right). \tag{17}$$

$E_H[\rho]$ denotes the Hartree or classical Coulomb energy functional,

$$E_H[\rho] = \frac{1}{2}\int dx_1 dx_2 \frac{\rho(x_1)\rho(x_2)}{|\mathbf{r}_1 - \mathbf{r}_2|}. \tag{18}$$

This result is remarkable, because it expresses the correlation energy entirely in terms of one-particle quantities, the transition densities. Transition densities are accessible from response theory. This becomes obvious if the zero-temperature fluctuation-dissipation theorem [14] is used to express the sum in Eq. 15 by a frequency integral of the density–density response function at coupling strength $\alpha$,

$$\langle \Psi_0^\alpha | \Delta \hat{\rho}(x_1) \Delta \hat{\rho}(x_2) | \Psi_0^\alpha \rangle = -\int_0^\infty \frac{d\omega}{\pi} \text{Im} \chi^\alpha(\omega, x_1, x_2). \quad (19)$$

The latter follows, e.g., from the Lehmann representation [20] of $\chi^\alpha$,

$$\chi^\alpha(\omega, x_1, x_2) = -\sum_{n \neq 0} \left( \frac{\rho_{0n}^\alpha(x_1) \rho_{0n}^\alpha(x_2)}{\Omega_{0n}^\alpha - \omega - i\eta} + \frac{\rho_{0n}^\alpha(x_1) \rho_{0n}^\alpha(x_2)}{\Omega_{0n}^\alpha + \omega + i\eta} \right), \quad (20)$$

where $\Omega_{0n}^\alpha$ denote excitation energies, and $i\eta$ is a small contour distortion making $\chi^\alpha$ analytical in the upper complex plane. By Eqs. 19, 15, and 4, the ground-state correlation energy may be entirely expressed in terms of $\chi^\alpha$ [11, 12],

$$E_C[\rho] = -\frac{1}{2} \int_0^1 d\alpha \int_0^\infty \frac{d\omega}{\pi} \text{Im}$$
$$\times \int dx_1 dx_2 \frac{\chi^\alpha(\omega, x_1, x_2) - \chi^{(0)}(\omega, x_1, x_2)}{|\mathbf{r}_1 - \mathbf{r}_2|}. \quad (21)$$

#### 2.1.4 Connection to time-dependent density functional theory

The density–density response function $\chi^\alpha$ or, alternatively, the transition densities $\rho_{0n}^\alpha$, are accessible from time-dependent density functional theory (TDDFT) [21]. $\chi^\alpha$ satisfies the Dyson-type equation [22]

$$\chi^\alpha = \chi^{(0)} + \chi^{(0)} f_{\text{HXC}}^\alpha \chi^\alpha. \quad (22)$$

The frequency-dependent Hartree, exchange, and correlation kernel can be decomposed into the bare Coulomb interaction and an exchange and correlation (XC) piece,

$$f_{\text{HXC}}^\alpha(\omega, x_1, x_2) = \alpha \frac{1}{|\mathbf{r}_1 - \mathbf{r}_2|} + f_{\text{XC}}^\alpha(\omega, x_1, x_2). \quad (23)$$

Alternatively, the transition densities can be computed from the response of the time-dependent KS density matrix, leading to the symplectic eigenvalue problem [23, 24]

$$(\Lambda^\alpha - \Omega_{0n}^\alpha \Delta) | X_{0n}^\alpha, Y_{0n}^\alpha \rangle = 0. \quad (24)$$

The super-vectors $X_{0n}^\alpha$ and $Y_{0n}^\alpha$ are defined on the product space $L_{\text{occ}} \times L_{\text{virt}}$ and $L_{\text{occ}} \times L_{\text{virt}}$, respectively, where $L_{\text{occ}}$ and $L_{\text{virt}}$ denote the one-particle Hilbert spaces spanned by occupied and virtual static KS molecular orbitals (MOs). We use indices $i, j, \ldots$ for occupied, $a, b, \ldots$ for virtual, and $p, q, \ldots$ for general MOs and assume that all MOs are real. The super-operator

$$\Lambda^\alpha = \begin{pmatrix} \mathbf{A}^\alpha & \mathbf{B}^\alpha \\ \mathbf{B}^\alpha & \mathbf{A}^\alpha \end{pmatrix} \quad (25)$$

contains the so-called orbital rotation Hessians,

$$(A + B)_{iajb}^\alpha = (\epsilon_a - \epsilon_i)\delta_{ij}\delta_{ab} + 2\alpha\langle ij|ab \rangle + 2f_{\text{XC}iajb}^\alpha, \quad (26)$$

$$(A - B)_{iajb}^\alpha = (\epsilon_a - \epsilon_i)\delta_{ij}\delta_{ab}. \quad (27)$$

$\epsilon_i$ and $\epsilon_a$ denote the energy eigenvalues of canonical occupied and virtual KS MOs. The eigenvectors $|X_{0n}^\alpha, Y_{0n}^\alpha \rangle$ satisfy the symplectic normalization constraint

$$\langle X_{0n}^\alpha, Y_{0n}^\alpha | \Delta | X_{0n}^\alpha, Y_{0n}^\alpha \rangle = 1, \quad (28)$$

where

$$\Delta = \begin{pmatrix} \mathbf{1} & \mathbf{0} \\ \mathbf{0} & -\mathbf{1} \end{pmatrix}. \quad (29)$$

The real-space transition density can be extracted from the eigenvectors $|X_{0n}^\alpha, Y_{0n}^\alpha \rangle$ according to

$$\rho_{0n}^\alpha(x) = \sum_{ia} (X + Y)_{ia}^\alpha \phi_i(x) \phi_a(x). \quad (30)$$

Using Eq. 17, the correlation energy in terms of $X_{0n}^\alpha$ and $Y_{0n}^\alpha$ is thus [18]

$$E_C = \frac{1}{2} \int_0^1 d\alpha \sum_{iajb} \langle ij|ab \rangle \left( \sum_{n \neq 0} (X + Y)_{0nia}^\alpha (X + Y)_{0njb}^\alpha - \delta_{ij}\delta_{ab} \right). \quad (31)$$

#### 2.1.5 Random phase approximation

The XC kernel as a functional of the ground-state density is not explicitly known. Within RPA, the XC kernel is set to zero,

$$f_{\text{XC}}^{\alpha \text{RPA}}(\omega, x_1, x_2) = 0. \quad (32)$$

Thus, RPA within a density functional context is identical to the time-dependent Hartree approximation.

Within RPA, the correlation energy takes a particularly simple form. Applying the Hellmann–Feynman theorem to the eigenvalue problem (Eq. 24) within RPA, it follows that

$$\frac{d\Omega_{0n}^{\alpha \text{RPA}}}{d\alpha} = \sum_{iajb} (X + Y)_{ia}^{\alpha \text{RPA}} \langle ij|ab \rangle (X + Y)_{jb}^{\alpha \text{RPA}}$$
$$= 2E_H[\rho_{0n}^{\alpha \text{RPA}}]. \quad (33)$$

Thus, the RPA correlation energy may be written as

$$E_C^{\text{RPA}} = \frac{1}{2} \int_0^1 d\alpha \sum_n \left( \frac{d\Omega_{0n}^{\alpha \text{RPA}}}{d\alpha} - \frac{d\Omega_{0n}^{\alpha \text{RPA}}}{d\alpha} \bigg|_{\alpha=0} \right). \quad (34)$$

The coupling strength integration can thus be carried out analytically, yielding

$$E_{\mathrm{C}}^{\mathrm{RPA}} = \frac{1}{2} \sum_n \left( \Omega_{0n}^{\mathrm{RPA}} - \Omega_{0n}^{\mathrm{D}} \right). \tag{35}$$

$\Omega_{0n}^{\mathrm{RPA}}$ is an RPA excitation energy at full coupling, and $\Omega_{0n}^{\mathrm{D}}$ is an RPA excitation energy to first order in $\alpha$. Using the unitary invariance of the trace in Eq. 35, the sum over the $\Omega_{0n}^{\mathrm{D}}$ may be replaced by the sum over the RPA excitation energies within the Tamm–Dancoff approximation (TDA) [25],

$$E_{\mathrm{C}}^{\mathrm{RPA}} = \frac{1}{2} \sum_n \left( \Omega_{0n}^{\mathrm{RPA}} - \Omega_{0n}^{\mathrm{TDARPA}} \right). \tag{36}$$

Equation 36 expresses the RPA correlation energy as a ZPVE difference of harmonic oscillators, where each oscillator corresponds to an electronic excitation, in agreement with Bohm's and Pines's original idea.

### 2.1.6 Connection to Green's function methods

RPA may be derived from the GW approximation to the one-electron self energy [26, 27]. This self-energy is the derivative of the functional

$$\Phi^{\mathrm{GW}} = \frac{1}{2} \mathrm{tr} \{ \ln(1 + iVG_{\mathrm{S}}G_{\mathrm{S}}) \}, \tag{37}$$

with respect to the non-interacting Green's function $G_{\mathrm{S}}$ [28]. $V$ denotes the bare Coulomb interaction. $-i\Phi^{\mathrm{GW}}$ plays the role of the exchange-correlation energy in the Klein functional of the total energy. The GW method is widely used to compute band structures of solids [29, 30] because it incorporates basic screening physics beyond the single-particle picture.

## 2.2 Hartree–Fock reference

### 2.2.1 Correlation energy from coupling strength integration

Post-HF methods are usually based on the Møller–Plesset partitioning of the Hamiltonian [31],

$$\hat{H}_{\mathrm{HF}}^\alpha = \hat{H}_{\mathrm{HF}}^{(0)} + \alpha \hat{H}_{\mathrm{HF}}^{(1)}. \tag{38}$$

The zero-order Hamiltonian is the Fock operator,

$$\hat{H}_{\mathrm{HF}}^{(0)} = \hat{T} + \hat{V}_{ne} + \hat{J} + \hat{K}, \tag{39}$$

where $\hat{V}_{ne}$ denotes the operator of the nucleus–electron attraction; $\hat{J}$ and $\hat{K}$ are the Coulomb and non-local exchange operators, respectively.

$$\hat{H}_{\mathrm{HF}}^{(1)} = \hat{H} - \hat{H}_{\mathrm{HF}}^{(0)} = \hat{V}_{ee} - \hat{J} - \hat{K} \tag{40}$$

is the so-called fluctuation potential. At zero coupling strength, the ground state $|\Psi_{\mathrm{HF0}}^\alpha\rangle$ equals the HF

determinant $|\Phi_{\mathrm{HF0}}\rangle$. Except for the physical system at $\alpha = 1$, $\hat{H}_{\mathrm{HF}}^\alpha$ and the ground state $|\Psi_{\mathrm{HF0}}^\alpha\rangle$ are different from the adiabatic connection Hamiltonian $\hat{H}^\alpha$ and its ground state $|\Psi_0^\alpha\rangle$ defined by Eq. 1: $|\Psi_0^\alpha\rangle$ is constrained to yield the interacting ground-state density for any $\alpha$, while the density of $|\Phi_{\mathrm{HF0}}\rangle$ changes with $\alpha$; on the other hand, $|\Phi_{\mathrm{HF0}}\rangle$ minimizes the energy expectation value

$$\langle \Phi_{\mathrm{HF0}} | \hat{H} | \Phi_{\mathrm{HF0}} \rangle = E_{\mathrm{HF}} = E_{\mathrm{HF}}^{(0)} + E_{\mathrm{HF}}^{(1)}, \tag{41}$$

while the KS determinant $|\Phi_0\rangle$ does not.

For a HF reference, the interacting ground-state energy $E_0$ equals the HF energy plus the HF correlation energy,

$$E_0 = E_{\mathrm{HF}} + E_{\mathrm{HFC}}. \tag{42}$$

The latter thus differs from the KS correlation energy defined in Eq. 3. The HF correlation energy may be expressed as an integral over coupling strength,

$$E_{\mathrm{HFC}} = \int_0^1 d\alpha \frac{dE_{\mathrm{HF0}}^\alpha}{d\alpha} - E_{\mathrm{HF}}^{(1)} = \int_0^1 d\alpha W_{\mathrm{HFC}}^\alpha \tag{43}$$

By virtue of the Hellmann–Feynman theorem, the coupling strength integrand is given by an expectation value difference of the fluctuation potential,

$$W_{\mathrm{HFC}}^\alpha = \langle \Psi_{\mathrm{HF0}}^\alpha | \hat{H}_{\mathrm{HF}}^{(1)} | \Psi_{\mathrm{HF0}}^\alpha \rangle - \langle \Phi_{\mathrm{HF0}} | \hat{H}_{\mathrm{HF}}^{(1)} | \Phi_{\mathrm{HF0}} \rangle. \tag{44}$$

This may be contrasted with the adiabatic connection integrand, Eq. 5, which does not contain any one-particle terms.

### 2.2.2 Factorization of the fluctuation potential

To apply factorization techniques along the lines of Sect. 2.1.2, we rewrite the fluctuation potential operator as

$$\hat{H}_{\mathrm{HF}}^{(1)} = \sum_{pqrs} \langle pq | rs \rangle \hat{U}_{pqrs}, \tag{45}$$

where

$$\hat{U}_{pqrs} = \frac{1}{2} \hat{c}_p^\dagger \hat{c}_q^\dagger \hat{c}_s \hat{c}_r - \hat{\gamma}_{pr} \gamma_{\mathrm{HF}qs} + \hat{\gamma}_{ps} \gamma_{\mathrm{HF}qr}. \tag{46}$$

$\hat{\gamma}_{pq} = \hat{c}_p^\dagger \hat{c}_q$ is the density matrix operator in second quantization, and

$$\gamma_{\mathrm{HF}pq} = \langle \Phi_{\mathrm{HF0}} | \hat{\gamma}_{pq} | \Phi_{\mathrm{HF0}} \rangle \tag{47}$$

is the density matrix of the HF ground-state determinant. Since the ground-state density matrix

$$\gamma_{pq}^\alpha = \langle \Psi_{\mathrm{HF0}}^\alpha | \hat{\gamma}_{pq} | \Psi_{\mathrm{HF0}}^\alpha \rangle \tag{48}$$

(and the density) varies with $\alpha$, fluctuation operators do not lead to simplifications here. However, using the Fermion anticommutation relations, $\hat{U}$ can still be factorized into one-particle operators,

$$\hat{U}_{pqrs} = \frac{1}{2}\hat{\gamma}_{pr}\hat{\gamma}_{qs} - \hat{\gamma}_{pr}\gamma_{\mathrm{HF}qs} + \hat{\gamma}_{ps}\left(\gamma_{\mathrm{HF}qr} - \frac{1}{2}\delta_{qr}\right). \quad (49)$$

While $\hat{U}_{pqrs}$ as a whole is antisymmetric under the exchange of $p$ and $q$ or $r$ and $s$, the individual terms in the last equation are not. To preserve the antisymmetry of $\hat{U}$ and thus of the two-particle density matrix when approximations are introduced, it is convenient to use the explicitly antisymmetrized expression

$$\hat{U}_{pqrs} = \frac{1}{4}(\hat{\gamma}_{pr}\hat{\gamma}_{qs} - \hat{\gamma}_{ps}\hat{\gamma}_{qr}) - \hat{\gamma}_{pr}\left(\gamma_{\mathrm{HF}qs} - \frac{1}{4}\delta_{qs}\right)$$
$$+ \hat{\gamma}_{ps}\left(\gamma_{\mathrm{HF}qr} - \frac{1}{4}\delta_{qr}\right). \quad (50)$$

### 2.2.3 Fluctuation-dissipation theorem

In analogy to Sect. 2.1.3, we use the completeness of the states $|\Psi_{\mathrm{HF}n}^{\alpha}\rangle$ to factorize expectation values of products of two density matrix operators,

$$\langle\Psi_{\mathrm{HF}0}^{\alpha}|\hat{\gamma}_{pr}\hat{\gamma}_{qs}|\Psi_{\mathrm{HF}0}^{\alpha}\rangle = \sum_{n\neq 0}\gamma_{0npq}^{\alpha}\gamma_{0nqs}^{\alpha\dagger} + \gamma_{pq}^{\alpha}\gamma_{qs}^{\alpha}. \quad (51)$$

Instead of transition densities $\rho_{0n}^{\alpha}(x)$, we now encounter the more general one-particle transition density matrices at coupling strength $\alpha$,

$$\gamma_{0npq}^{\alpha} = \langle\Psi_{\mathrm{HF}0}^{\alpha}|\hat{\gamma}_{pq}|\Psi_{\mathrm{HF}n}^{\alpha}\rangle. \quad (52)$$

The additional ground-state term arises because fluctuation operators are not used here, as explained previously.

Defining the Hartree plus exchange functional of the one-particle density matrix

$$E^{(1)}[\gamma] = \frac{1}{2}\sum_{pqrs}\langle pq|rs\rangle(\gamma_{pr}\gamma_{qs}^{\dagger} - \gamma_{ps}\gamma_{qr}^{\dagger}), \quad (53)$$

we obtain the following exact expression for the HF correlation energy:

$$E_{\mathrm{HFC}} = \frac{1}{2}\int_0^1 d\alpha \sum_{n\neq 0}(E^{(1)}[\gamma_{0n}^{\alpha}] - E^{(1)}[\gamma_{0n}^{(0)}]) + \Delta E_{\mathrm{HFC}} \quad (54)$$

It is instructive to compare this result to its density functional equivalent, Eq. 17. The additional factor of 1/2 in Eq. 54 arises from enforcing the antisymmetry of $U_{pqrs}$ in Eq. 50. The correction term $\Delta E_{\mathrm{HFC}}$, which is not present in the density functional case, is due to the change of the ground-state density matrix with $\alpha$,

$$\Delta E_{\mathrm{HFC}} = \int_0^1 d\alpha \sum_{pqrs}\langle pq|rs\rangle\left\{\frac{1}{4}\left(\gamma_{pr}^{\alpha}\gamma_{qs}^{\alpha} - \gamma_{ps}^{\alpha}\gamma_{qr}^{\alpha} - \gamma_{\mathrm{HF}pr}\gamma_{\mathrm{HF}qs}\right.\right.$$
$$+ \gamma_{\mathrm{HF}ps}\gamma_{\mathrm{HF}qr}) - \left(\gamma_{pr}^{\alpha} - \gamma_{\mathrm{HF}pr}\right)\left(\gamma_{\mathrm{HF}qs} + \frac{1}{4}\delta_{qs}\right)$$
$$+ \left(\gamma_{ps}^{\alpha} - \gamma_{\mathrm{HF}ps}\right)\left(\gamma_{\mathrm{HF}qr} + \frac{1}{4}\delta_{qr}\right)\bigg\}. \quad (55)$$

Using the zero-temperature fluctuation-dissipation theorem, the sum over transition density matrices may be expressed by a frequency integral over the density matrix—density matrix response function at coupling strength $\alpha$,

$$\sum_{n\neq 0}\gamma_{0npr}^{\alpha}\gamma_{0nqs}^{\alpha\dagger} = -\int_0^{\infty}\frac{d\omega}{\pi}\mathrm{Im}\Pi_{prqs}^{\alpha}(\omega). \quad (56)$$

This follows, e.g., from the Lehmann representation of $\Pi^{\alpha}$ [20],

$$\Pi_{prqs}^{\alpha}(\omega) = -\sum_{n\neq 0}\left(\frac{\gamma_{0npr}^{\alpha}\gamma_{0nqs}^{\alpha\dagger}}{\Omega_{0n}^{\alpha} - \omega - i\eta} + \frac{\gamma_{0npr}^{\alpha\dagger}\gamma_{0nqs}^{\alpha}}{\Omega_{0n}^{\alpha} + \omega + i\eta}\right). \quad (57)$$

The ground-state density matrix $\gamma_{pr}^{\alpha}$ occurring in $\Delta E_{\mathrm{c\,HF}}$ is accessible from $\Pi^{\alpha}$ via a partial trace of the two-particle density matrix. Thus, knowledge of $\Pi^{\alpha}$ is sufficient to compute the HF correlation energy.

### 2.2.4 Connection to polarization propagator theory

The density matrix–density matrix response function $\Pi^{\alpha}(\omega)$ is identical to the causal polarization propagator. $\Pi^{\alpha}(\omega)$ satisfies the Bethe–Salpeter equation [20]

$$\Pi^{\alpha}(\omega) = \Pi^{(0)}(\omega) + \Pi^{(0)}(\omega)K^{\alpha}(\omega)\Pi^{\alpha}(\omega). \quad (58)$$

The frequency-dependent kernel $K^{\alpha}(\omega)$ may be computed perturbatively. The lowest non-vanishing order in $\alpha$ is

$$K_{prqs}^{(1)}(\omega) = \langle pr|qs\rangle - \langle ps|qr\rangle. \quad (59)$$

### 2.2.5 Random phase approximation with exchange

The random phase approximation with exchange consists in replacing the kernel $K^{\alpha}(\omega)$ in Eq. 58 with its first-order approximation $\alpha K^{(1)}$, which is frequency independent. In addition, the ground-state density matrices $\gamma_{pq}^{\alpha}$ are replaced by $\gamma_{\mathrm{HF}pq}$, making $\Delta E_{\mathrm{HFC}}$ vanish. Thus, the RPAX coupling strength integrand is correct to first order in $\alpha$, and the RPAX correlation energy is correct to $O(\alpha^2)$, that is, it reduces to second-order Møller–Plesset perturbation theory

(MP2) for small $\alpha$. Within RPAX, the transition density matrices may be written as

$$\gamma_{0npr}^{\text{RPAX}\alpha} = X_{npr}^{\text{RPAX}\alpha} n_p(1 - n_r) + Y_{nrp}^{\text{RPAX}\alpha} n_r(1 - n_p), \quad (60)$$

where $n_p$ and $n_r$ denote orbital occupation numbers. The super-vectors $|X_n^{\text{RPAX}\alpha}, Y_n^{\text{RPAX}\alpha}\rangle$ satisfy a symplectic eigenvalue problem of the same form as Eq. 24. The RPAX orbital rotation Hessians are [32]

$$(A + B)_{iajb}^{\text{RPAX}\alpha} = (\epsilon_a - \epsilon_i)\delta_{ij}\delta_{ab} + \alpha[2\langle ij|ab\rangle - \langle ij|ba\rangle - \langle ia|jb\rangle], \quad (61)$$

$$(A - B)_{iajb}^{\text{RPAX}\alpha} = (\epsilon_a - \epsilon_i)\delta_{ij}\delta_{ab} + \alpha[\langle ij|ba\rangle - \langle ia|jb\rangle]. \quad (62)$$

Using the Hellmann–Feynman theorem once more, one finds that the excitation energies $\Omega_{0n}^{\text{RPAX}\alpha}$ satisfy

$$\frac{d\Omega_{0n}^{\text{RPAX}\alpha}}{d\alpha} = \frac{1}{2} \sum_{iajb} \left\{ (X + Y)_{ia}^{\text{RPAX}\alpha} [2(ij|ab) - (ij|ba) - (ia|jb)](X + Y)_{jb}^{\text{RPAX}\alpha} + (X - Y)_{ia}^{\text{RPAX}\alpha} [(ij|ba) - (ia|jb)](X - Y)_{jb}^{\text{RPAX}\alpha} \right\} = 2E^{(1)}[\gamma_{0n}^{\text{RPAX}\alpha}] \quad (63)$$

Again, the coupling strength integration may be carried out analytically, yielding

$$E_C^{\text{RPAX}} = \frac{1}{4} \sum_n (\Omega_{0n}^{\text{RPAX}} - \Omega_{0n}^{\text{TDARPAX}}). \quad (64)$$

Compared to Eq. 36, Eq. 64 contains an extra factor 1/2 resulting from the antisymmetrization applied in Eq. 50. If the antisymmetrization is not applied, and the exchange part of $K^{(1)}$ is neglected, Eq. 36 is recovered, now using a HF reference. Other combinations such as the RPAx method of Toulouse and coworkers [33, 135] generally do not lead to an analytically integrable coupling strength integral, because the Hellmann–Feynman theorem cannot be used. This also applies to "hybrid" schemes that incorporate only a fraction of exchange [34].

A drawback of RPAX is that it includes spin-flip excitations. For example, for a spin-restricted closed-shell HF ground state, Eq. 64 becomes

$$E_C^{\text{RPAX}}(\text{RHF}) = \frac{1}{4} \sum_n (\Omega_{0nS}^{\text{RPAX}} - \Omega_{0nS}^{\text{TDARPAX}}) + \frac{3}{4} \sum_n (\Omega_{0nT}^{\text{RPAX}} - \Omega_{0nT}^{\text{TDARPAX}}), \quad (65)$$

where subscripts S and T denote singlet and triplet excitation energies, respectively. RPAX spin-flip excitation energies are very sensitive to the quality of the HF reference. Triplet instabilities are common, leading to a breakdown of RPAX and imaginary correlation energies

[35]. In contrast, spin-flip excitations cancel out of Eq. 36, because they are the same with and without TDA.

RPAX contains all third-order particle-hole diagrams, albeit with incorrect pre-factors [6, 7, 36, 37]. Direct RPA, Eq. 36, contains ring diagrams only, but the pre-factors are correct.

### 2.2.6 Connection to coupled cluster theory

From a coupled cluster perspective, RPA is a simplified coupled cluster doubles (CCD) method [38, 39]. The RPAX correlation energy may be written as

$$E_C^{\text{RPAX}} = \frac{1}{4} \text{tr}(\mathbf{B}^{\text{RPAX}} \mathbf{T}), \quad (66)$$

where $B_{iajb}^{\text{RPAX}} = \langle ij|ab\rangle - \langle ij|ba\rangle$. The doubles amplitudes $\mathbf{T}$ are related to the vectors $|X_n^{\text{RPAX}}, Y_n^{\text{RPAX}}\rangle$ at full coupling according to [40]

$$\mathbf{T} = \mathbf{Y}^{\text{RPAX}} \mathbf{X}^{\text{RPAX}-1} \quad (67)$$

and satisfy the ring-CCD equation [9, 41]

$$\mathbf{B}^{\text{RPAX}} + \mathbf{A}^{\text{RPAX}} \mathbf{T} + \mathbf{T} \mathbf{A}^{\text{RPAX}} + \mathbf{T} \mathbf{B}^{\text{RPAX}} \mathbf{T} = \mathbf{0}. \quad (68)$$

Similar relations hold in the KS case for direct RPA [136].

### 2.3 Beyond RPA

Beyond-RPA methods aim to approximate the beyond-RPA correlation energy

$$E_C^{\text{bRPA}}[\rho] = E_C[\rho] - E_C^{\text{RPA}}[\rho]. \quad (69)$$

Due to the deficiencies of RPAX, most work so far uses direct RPA and a KS reference. Beyond-RPA methods are a rapidly growing field, and we give a brief overview of some important directions only.

### 2.3.1 Density functional corrections

In the uniform electron gas, corrections to RPA arise from short-range electron interactions. If the same holds for molecules and solids, a generalized gradient approximation (GGA) to $E_C^{\text{bRPA}}[\rho]$ should be accurate, since GGAs are generally believed to work best for short-range interactions [42]. The so-called RPA+ [43] thus approximates beyond-RPA correlation by the beyond-RPA piece of the PBE GGA,

$$E_C^{\text{bRPA}}(\text{RPA}+)[\rho] = E_C^{\text{PBE}}[\rho] - E_C^{\text{PBE RPA}}[\rho]. \quad (70)$$

While RPA+ improves total molecular correlation energies considerably, energy differences such as atomization energies are essentially unchanged compared to RPA [18]. Recently, Ruzsinszky et al. explained this puzzling result by non-local multi-center correlation that is present in molecules but not in the uniform electron gas [44]. This hypothesis is corroborated by a considerable improvement upon RPA atomization energies obtained from a non-local correction to RPA [45].

Range-separation methods go one step further, removing all short-range correlation from RPA using a screened interaction [33, 46, 137, 47, 138]. Range separation improves the basis set convergence of RPA correlation energies by largely eliminating the electron coalescence cusp in the pair density. The results of range-separated RPA depend on the adjustable range-separation parameter.

### 2.3.2 Perturbative corrections

The lowest order non-vanishing correction to RPA is second-order exchange,

$$E_C^{bRPA}(X2) = \frac{1}{2} \sum_{iajb} \frac{\langle ij|ab \rangle \langle ij|ba \rangle}{\epsilon_a + \epsilon_b - \epsilon_i - \epsilon_j}. \tag{71}$$

However, $E_C^{bRPA}(X2)$ does not improve upon RPA systematically because it suffers from the shortcomings of second-order Görling–Levy perturbation theory [48]. The latter is very sensitive to the KS HOMO-LUMO gap and breaks down already for moderately small-gap cases [49].

Based on early work by Freeman [38], Kresse et al. proposed a second-order screened exchange (SOSEX) correction to RPA [50],

$$E_C^{bRPA}(SOSEX) = -\frac{1}{2} \sum_{iajb} \langle ij|ba \rangle T_{iajb}, \tag{72}$$

where $\mathbf{T}$ is computed by solving the ring-CCD equation (without exchange). SOSEX is free of one-electron self-correlation, yields much improved total correlation energies, and appears to improve upon RPA atomization energies [50–52]. SOSEX is more stable than bare second-order exchange and finite for small-gap systems and metals. The trade-off for eliminating self-interaction error is the reintroduction of an incorrect dissociation behavior for covalent bonds in a spin-restricted closed-shell treatment [53], see Sect. 4.2.

Recently, Ren and coworkers showed that a perturbative single-excitation correction, which arises in second-order Görling–Levy perturbation theory, leads to an improvement of RPA for weakly interacting systems [54]. Lotrich and Bartlett also obtained improved results for such systems using a correction based on external coupled cluster perturbation theory [139].

### 2.3.3 Local field corrections

Local field corrections to RPA were first explored for the uniform electron gas by Singwi, Tosi, Land, and Sjölander (STLS) [55], who introduced an XC kernel depending on the pair distribution function to correct RPA at short range. In a broader sense, local field corrections use non-zero XC kernels to improve upon RPA.

A straightforward choice motivated by TDDFT is to use XC kernels derived from semi-local functionals within the adiabatic approximation [56], which replaces $f_{XC}^\alpha$ by its static limit,

$$f_{XC}^\alpha(0, x_1, x_2) = \frac{\delta^2 E_{XC}[\rho]}{\delta\rho(x_1)\delta\rho(x_2)}. \tag{73}$$

However, the spacial locality of these kernels leads to an unphysical divergence in the RPA pair density [34, 57]. Even for the uniform electron gas correlation energy, the adiabatic local-spin density approximation (ALDA) kernel yields poor results [58].

Dobson et al. [59] extended the original STLS model to inhomogeneous systems. The resulting ISTLS method is equivalent to a non-local tensor exchange-correlation kernel in the context of time-dependent current density functional theory [60]. ISTLS yields highly accurate jellium surface energies [61] but apparently has not been applied to molecular systems so far.

### 2.3.4 Optimized effective potential methods

Kotani and Akai [62] were the first to combine the exact Kohn–Sham exchange potential and RPA correlation [RPA(EXX)]. The Kohn–Sham exchange potential is obtained from the exact exchange using the framework of the optimized effective potential (OEP) method. The exact exchange potential is both non-local and frequency-dependent and is, therefore, expected to show no divergence of the pair density at short interelectronic distances. Hellgren and von Barth [63] applied RPA(EXX) to atoms, reporting good agreement with accurate configuration interaction results.

A disadvantage of OEP methods is the need to compute the inverse of the Kohn–Sham response matrix, which leads to numerical instabilities in conjunction with finite basis set methods [64, 65]. Heßelmann, Ipatov, and Görling developed an approximate scheme that avoids inverses of the Kohn–Sham response matrix [66]. Heßelmann and Görling reported RPA(EXX) correlation energies for small organic molecules that improve considerably upon RPAX correlation energies [67]. Reaction energies involving these molecules are less affected.

## 3 Implementation

### 3.1 Diagonalization

In quantum chemistry, the molecular orbitals are usually approximated by a finite linear combination of atom-centered basis functions (LCAO-MO approach). The expansion coefficients are determined variationally. The finite number of basis functions used results in a finite number of virtual orbitals, which in turn leads to a finite dimension of the RPA eigenvalue problem, Eq. 36. Thus, in a finite basis, the total number of excitation energies is given by $L_{\mathrm{occ}} \times L_{\mathrm{virt}}$.

The RPA eigenvalue problem may be transformed to a Hermitian eigenvalue problem [18]

$$\mathbf{MZ} = \mathbf{Z}\mathbf{\Omega}^2, \qquad \mathbf{ZZ}^{\mathrm{T}} = \mathbf{1}, \tag{74}$$

where

$$\mathbf{M} = (\mathbf{A} - \mathbf{B})^{1/2}(\mathbf{A} + \mathbf{B})(\mathbf{A} - \mathbf{B})^{1/2}, \tag{75}$$

is easily computed since $(\mathbf{A} - \mathbf{B})^{1/2}$ is diagonal within RPA, see Eq. 27. Thus, Eq. 36 may be rewritten in terms of the square root of $\mathbf{M}$,

$$E^{\mathrm{C\,RPA}} = \frac{1}{2}\mathrm{tr}(\mathbf{M}^{1/2} - \mathbf{A}). \tag{76}$$

Straightforward evaluation of all excitation energies in Eq. 36 leads to a $\mathcal{O}(N^6)$ scaling with system size $N$. This follows from the dimension of the eigenvalue problem ($L_{\mathrm{occ}} \times L_{\mathrm{virt}}$) and from the $\mathcal{O}(N^3)$ scaling of the diagonalization procedure.

The first RPA correlation energies on molecules were obtained by full diagonalization of Eq. 75 [18, 34] combined with a grid-based coupling constant integration. The steep increase in computational cost with system size limited these applications to small molecules (approximately 10 atoms).

## 3.2 Resolution-of-the-identity approximation

A more efficient method can be obtained by taking advantage of the rank deficiency of the two-electron Coulomb integrals present in Eq. 36. The key idea is to represent the rank deficient orbital product densities in a small number of auxiliary basis functions. Several such methods exist [39, 68–70], but the resolution-of-the-identity (RI) approximation [71] with the Coulomb metric [72] represents an optimum choice for RPA, as explained later. The RI approximation is used extensively and successfully in the context of ground-state density functional theory (DFT) [73] and TDDFT excitation energy calculations [74–76] as well as MP2 [77, 78]. In the RI approximation, the two-electron integrals can be factorized, using Mulliken notation, as

$$(ia|jb)_{\mathrm{RI}} = \sum_{PQ}(ia|P)(P|Q)^{-1}(Q|jb) = \sum_{P} S_{iaP}S_{jbP}, \tag{77}$$

where $S_{iaP} = \sum_{R}(ia|R)L_{RP}^{-1}$, $P$, $Q$, $R$ denote atom-centered Gaussian auxiliary basis functions, and $L_{RP}$ is determined by Cholesky decomposition of the two-electron integrals $(P|Q)$

$$(P|Q) = \sum_{R} L_{PR}L_{QR}. \tag{78}$$

Since the number of auxiliary basis functions $N_{\mathrm{aux}}$ increases only linearly with $N$, it is much more efficient to work with matrix $\mathbf{S}$ instead of the full two-electron matrix $(ia|jb)$. $E^{\mathrm{C\,RIRPA}}$ is evaluated by replacing the two-electron integrals in Eq. 76 with $(ia|jb)_{\mathrm{RI}}$. Importantly, the error due to RI is quadratic in each RPA excitation energy [79]. Therefore, high accuracy can be achieved with moderately sized auxiliary basis sets. The variational stability of the excitation energies carries over to the RPA correlation energy. Thus, the RIRPA correlation energy is a variational upper bound to the exact RPA correlation energy [79]. This variational property also allows for systematic optimization of the auxiliary basis sets.

Four-index quantities are entirely avoided by evaluating $\mathbf{M}^{1/2}$ as an integral over imaginary frequency [80]

$$\mathrm{tr}\left(\mathbf{M}^{1/2}\right) = 2\int_{-\infty}^{\infty}\frac{d\omega}{2\pi}\mathrm{tr}\left(\mathbf{1} - \omega^2(\mathbf{M} + \omega^2\mathbf{1})^{-1}\right). \tag{79}$$

Within RI, $\mathbf{M} = \mathbf{D}^2 + 2\,\mathbf{D}^{1/2}\,\mathbf{S}\,\mathbf{S}^{\mathrm{T}}\,\mathbf{D}^{1/2}$, where $D_{iajb} = D_{ia}\delta_{ij}\delta_{ab}$, and $D_{ia} = \epsilon_a - \epsilon_i$, is the diagonal matrix of bare orbital energy differences. The inverse of $\mathbf{M} + \omega^2\,\mathbf{1}$ may be written as

$$\begin{aligned}(\mathbf{M} + \omega^2\mathbf{1})^{-1} = \ &\mathbf{D}^{-1}\mathbf{G}(\omega) - 2\mathbf{D}^{-1/2}\mathbf{G}(\omega) \\ &\mathbf{S}(\mathbf{1}_{\mathrm{aux}} + \mathbf{Q}(\omega))^{-1}\mathbf{S}^{\mathrm{T}}\mathbf{G}(\omega)\mathbf{D}^{-1/2},\end{aligned} \tag{80}$$

where

$$\mathbf{G}(\omega) = \mathbf{D}(\mathbf{D}^2 + \omega^2\mathbf{1})^{-1} \tag{81}$$

is diagonal in the canonical Kohn–Sham orbital basis, and

$$\mathbf{Q}(\omega) = 2\mathbf{S}^{\mathrm{T}}\mathbf{G}(\omega)\mathbf{S} \tag{82}$$

is $N_{\mathrm{aux}} \times N_{\mathrm{aux}}$. Using an analogous expression for $\mathrm{tr}(\mathbf{A})$, the RPA correlation energy is expressed as

$$E^{\mathrm{C\,RIRPA}} = \int_{-\infty}^{\infty}\frac{d\omega}{2\pi}F^{\mathrm{C}}(\omega), \tag{83}$$

where the integrand contains $N_{\mathrm{aux}} \times N_{\mathrm{aux}}$ quantities only,

$$F^{\mathrm{C}}(\omega) = \frac{1}{2}\mathrm{tr}(\ln(\mathbf{1}_{\mathrm{aux}} + \mathbf{Q}(\omega)) - \mathbf{Q}(\omega)). \tag{84}$$

Taylor expansion of $\ln(\mathbf{1}_{\mathrm{aux}} + \mathbf{Q})$ in Eq. 83 leads to the diagrammatic RPA, where $\mathbf{Q}$ is the equivalent of a single ring diagram [4], see Fig. 1. The variational upper bound property of RIRPA is maintained in Eq. 83.

The efficient evaluation of Eq. 83 hinges on an efficient quadrature scheme to compute the frequency integral. Eshuis, Yarkony, and Furche used an exponentially converging Clenshaw–Curtis quadrature [81] with a single scaling parameter, which is determined from the orbital energies and the diagonal elements of **M**. RIRPA scales as $\mathcal{O}(N^4 \log N)$. It is straightforward to achieve an additional speed-up by parallelization over the frequency integration grid.

The RI approximation leads to sub-millihartree errors when MP2-optimized auxiliary basis sets are used [83], as shown in Fig. 2 for the HEAT test set [84]. The RI errors are much smaller than the intrinsic errors of RPA. The dependence of the RIRPA correlation energy for the HEAT test set on the number of grid points is shown in Fig. 3. Submillihartree accuracy is achieved for the numerical integration with typically 40–50 grid points using a basis of quadruple-zeta quality. Larger grids are required when core electrons are included. Exceptions arise when contributions from different eigenvalues to the integrand have very different scaling, for example, in small-gap systems. A measure for the spread of the excitation energies is the condition number $\kappa = \Omega_{\max}/\Omega_{\min}$. A large number of grid points are required for large $\kappa$.

To address this difficulty, a hybrid scheme was developed, which computes a few small eigenvalues explicitly and the remainder on the grid. Starting from the exact spectral representation of $F^C(\Omega)$ for the $n_{\text{lowest}}$ eigenvalues

$$\tilde{F}^C = \frac{1}{2}\sum_{ia}^{n_{\text{lowest}}}\left[\ln\left(1 + \frac{\Omega_{ia}^2 - D_{ia}^2}{\omega^2 + D_{ia}^2}\right) - \frac{M_{iaia} - D_{ia}^2}{\omega^2 + D_{ia}^2}\right], \quad (85)$$

we can compute

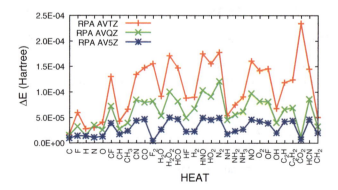

**Fig. 2** RI errors ($\Delta E = E^{C\,\text{RIRPA}} - E^{C\,\text{RPA}}$) in all electron RPA correlation energies for the HEAT test set using three of Dunning's correlation consistent basis sets (AVTZ = aug-cc-pVTZ, etc. [82]). (Reprinted with permission from Ref. [79]. Copyright 2010 American Institute of Physics)

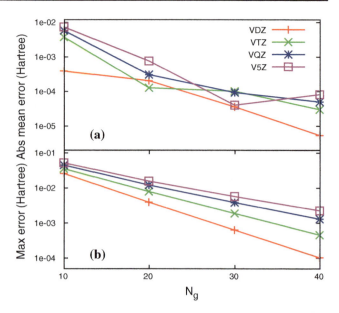

**Fig. 3** Absolute mean errors (*a*) and maximum errors (*b*) due to numerical integration for RI-RPA HEAT atomization energies with varying number of grid points $N_g$ and increasing basis set size (V*X*Z = cc-pV*X*Z). (Reprinted with permission from Ref. [79]. Copyright 2010 American Institute of Physics)

$$\tilde{E}^{C\,\text{RIRPA}} = \frac{1}{2}\int_{-\infty}^{\infty}\frac{d\omega}{2\pi}\tilde{F}^C(\omega)$$
$$= \frac{1}{2}\sum_{ia}^{n_{\text{lowest}}}\left(\Omega_{ia} - (D_{ia} + (ia|ia)_{\text{RI}})\right) \quad (86)$$

efficiently if $n_{\text{lowest}} \ll L_{\text{occ}} \times L_{\text{virt}}$. The hybrid RIRPA correlation energy is now given by

$$E^{C\,\text{RIRPA}} = \frac{1}{2}\int_{-\infty}^{\infty}\frac{d\omega}{2\pi}\left(F^C(\omega) - \tilde{F}^C(\omega)\right)$$
$$+ \frac{1}{2}\sum_{ia}^{n_{\text{lowest}}}\left(\Omega_{ia} - (D_{ia} + (ia|ia)_{\text{RI}})\right). \quad (87)$$

The new integrand, $F^C - \tilde{F}^C$, is associated with a smaller condition number $\tilde{\kappa} = \Omega_{\max}/\Omega_{n_{\text{lowest}}+1}$. Preliminary results show microhartree accuracy can be reached if eigenvalues less than approximately 0.05 Hartree are treated explicitly.

The RIRPA scheme leads to considerable speed-ups, as demonstrated for a set of polyacenes in Fig. 4. Also, the memory requirements are much smaller, because 4-index objects are avoided altogether. For the Grubbs II Ru catalyst, which consists of 117 atoms, using def2-TZVP basis sets [85] on C,H, and N and def2-QZVPP [85] on the other atoms, see Sect. 4.1, the RIRPA correlation energy can be computed in approximately 8 CPU hours on 16 processors of a single 2.2 GHz AMD Opteron 6174 node using multithreaded BLAS routines and 10 Gb of memory [86]. The number of excitations involved is 287036, which is clearly out of reach for full diagonalization.

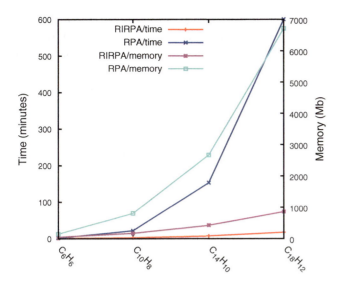

**Fig. 4** RIRPA timings and memory usage for polyacenes of increasing size compared to RPA results obtained through direct diagonalization. Def2-TZVPP basis sets were used, and the calculations were performed on a single CPU of a Xeon X5560 2.80 GHz workstation

## 3.3 Other implementations

### 3.3.1 Ring coupled cluster doubles

Random phase approximation correlation energies can also be computed from the ring approximation to coupled cluster doubles theory (rCCD) [39] by solving for the amplitudes in Eq. 68. This can be easily done starting from an existing CCD code and removing all but ring contractions. rCCD correlation energies can, therefore, be obtained from minor modifications. However, the computational cost scales as $\mathcal{O}(N^6)$ and I/O as $\mathcal{O}(N^5)$ because of the need to store all doubles amplitudes on disk.

Scuseria et al. proposed the use of Cholesky decompositions to reduce the scaling by writing

$$(ia|jb) = \sum_A u_{ia}^A u_{jb}^A, \qquad (88)$$

and

$$-T_{ij}^{ab} = \sum_A \theta_{ia}^A \theta_{jb}^A. \qquad (89)$$

This is possible in direct RPA, because both $(ia|jb)$ and $-T_{ij}^{ab}$ are positive definite. The rCCD expression can then be rewritten as

$$-T_{ij}^{ab} = -\frac{1}{\epsilon_i + \epsilon_j - \epsilon_a - \epsilon_b} \left( \sum_A u_{ia}^A u_{jb}^A - \sum_{AB} u_{ia}^A N^{AB} \theta_{jb}^B \right.$$
$$\left. - \sum_{AB} \theta_{ia}^A M^{AB} \theta_{jb}^B - \sum_{ABC} \theta_{ia}^A N^{AB} M^{BC} \theta_{jb}^C \right), \qquad (90)$$

where $M^{AB} = \sum_{kc} \theta_{kc}^A u_{kc}^B$ and $N^{AB} = \sum_{kc} u_{kc}^A \theta_{kc}^B$. Iterative methods can be used to solve Eq. 90 for the amplitudes, although no implementations seem to be available so far.

### 3.3.2 Plane wave implementations

In plane wave implementations of RPA, the use of auxiliary basis expansions is fairly straightforward. Core electrons must be treated by pseudopotentials or projector augmented wave methods [87]. The starting point for plane wave codes [87–90] is the expression by Langreth and Perdew [12]

$$E^{\mathrm{CRPA}} = \frac{1}{2} \int_{-\infty}^{\infty} \frac{d\omega}{2\pi} \mathrm{tr}(\ln(\mathbf{1} + \chi(i\omega)W) - \chi(i\omega)W), \qquad (91)$$

where $\chi$ is the Kohn–Sham density–density response function, and $W$ is the bare Coulomb interaction. The number of virtual orbitals depends on the size of the plane wave basis, which is determined by an energy cutoff $E_{\mathrm{cut}}$. In addition, the correlation energy depends on the length of the maximum reciprocal lattice vector $G_{\mathrm{cut}}^\chi$ and its energy $E_{\mathrm{cut}}^\chi$. These quantities determine the size of the plane wave expansion of the response function. Convergence with respect to $E_{\mathrm{cut}}^\chi$ is slow, but good results can be obtained by using an extrapolation scheme [87]. The frequency integration in Eq. 91 is done using Gauss–Legendre quadrature. Alternatively, the rCCD expression can be used as a starting point [50].

## 4 Applications

### 4.1 Non-covalent interactions

Non-covalent interactions play an important role in chemistry, physics, and biology, e.g., in DNA, enzymes, and graphene sheets [91–93]. A striking example are the relative energies of alkane conformers, which crucially depend on mid-range non-covalent electronic interactions between two bonds separated by another bond (1–3 interactions) [94–97]. These systems exemplify the failure of semi-local DFT to describe weak interactions [98, 99]. This well-known problem has resulted in many corrective schemes to account for weak interactions. The semi-classical corrections designed by Grimme are particularly popular, because they can be included at almost no extra cost and give much improved results for weak interactions [97, 100, 101]. A promising alternative are the van der Waals (vdW) density functionals of Langreth–Lundqvist type [102–104], but they depend strongly on the exchange functionals used [105]. From the wavefunction perspective, highly accurate interaction energies for weakly bound systems can be obtained using CCSD(T), but the steep $O(N^7)$ scaling of this method prohibits application to larger systems. A less costly alternative is MP2, but here long-range interactions are only included at the uncoupled monomer level, leading to mixed results [98]. In contrast, dispersion is naturally included in RPA in a seamless fashion as RPA describes interaction energies at the coupled monomer level [7]. In addition, RPA is the only method apart from CCSD(T) to properly capture the non-pairwise-additive nature of long-range interactions [106, 107].

The random phase approximation was applied extensively to weakly bound dimers, where non-covalent interactions play a crucial role [47, 108]. Here, RPA describes the long-range part of the PES correctly, but underbinds around the equilibrium distance and gives somewhat too large equilibrium bond distances.

Figure 5 shows errors in the relative energies of *n*-butane, *n*-hexane, and *n*-pentane conformers for several methods [86]. Quadruple-zeta basis sets (def2-QZVP, Ref. [85]) were used throughout. The average isomer energy difference is only 1.8 kcal/mol [109]. The branched conformers are more stable than the linear ones, due to the medium-range electronic interactions between bonds. Semi-local DFT fails to include these medium-range interactions and, therefore, leads to incorrect relative energetics for these systems, with errors that are of equal magnitude to the average isomer energy difference. RPA yields a spectacular improvement over semi-local DFT; the errors are now similar to the errors in the reference method. Grimme's double-hybrid B2PLYP [94], which adds a correction based on MP2 theory, is not accurate for this set, but better results are obtained when adding a semi-empirical dispersion correction, particularly the recent -D3 correction [100, 110]. The Minnesota functional M06-2X [111, 112] yields smaller errors than the semi-local functionals, but does not reach the accuracy of RPA. MP2 performs well; somewhat surprisingly, the spin-component-scaled version of MP2 [113] leads to slightly inferior results. Overall, RPA performs very well for these systems where intramolecular medium-range interactions play a crucial role. Its accuracy is comparable to the best available density functionals, such as B2PLYP-D3 and M06-2X, but without any empirical parameters.

The S22 set, designed by Jurecka et al. [115], is a widely used benchmark for weak interactions in biologically relevant systems. The set consists of 22 small- to medium-size dimers, of which 7 systems are primarily hydrogen bound and 8 dispersion bound; the remaining systems contain both interactions. Accurate theoretical reference results are available for this test set [114].

Figure 6 compares RPA results for the S22 set with other commonly used functionals and wave function-based methods. The MAE for RPA is 0.41 kcal/mol, a substantially smaller error than obtained from semi-local DFT,

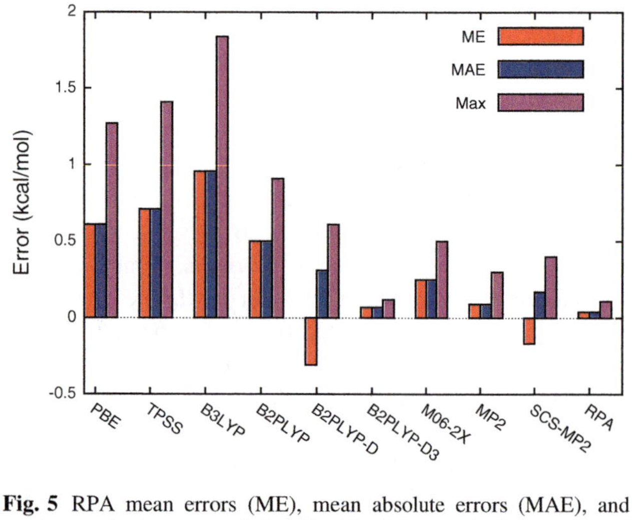

**Fig. 5** RPA mean errors (ME), mean absolute errors (MAE), and maximum absolute errors (Max) (kcal/mol) in the relative energies of *n*-alkane conformers compared to other methods. Geometries and reference energies, and non-RPA results were taken from Ref. [110], except for the MP2 and SCS-MP2 results, which were taken from Ref. [109]. def2-QZVP basis sets were used, and no counterpoise correction was applied; the MP2 and SCS-MP2 results were obtained using cc-pVQZ basis sets [82]; RPA results were taken from Ref. [86] and were obtained using self-consistent TPSS orbitals

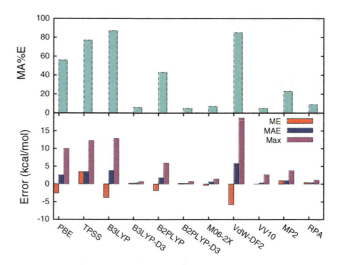

**Fig. 6** RPA mean absolute percentage errors (MA%E), mean errors (ME), mean absolute errors (MAE), and maximum absolute errors (Max) (kcal/mol) for the S22 test set compared to other methods. Geometries and reference energies were taken from Ref. [114]. No counterpoise correction was applied, and def2-QZVP basis sets were used, except for the VdW-DF2 and VV10 results [103], which were counterpoise corrected, and were obtained using aug-cc-pVTZ basis sets [82]; RPA results were taken from Ref. [86] and other results from Ref. [110]. RPA results were obtained from self-consistent TPSS orbitals

which seriously fails. The addition of a dispersion correction improves results for semi-local DFT considerably, reducing the MAE to 0.36 kcal/mol for B3LYP-D. The M06-2X functional yields errors of the same order of magnitude [111]. MP2 errors are somewhat larger [110, 115], which is not surprising given the well-known mixed performance of MP2 for dispersion interactions [98]. The recent non-local van der-Waals functional by Vydrov and van Voorhis yields results comparable to RPA, though with a larger maximum error [103]. The vdW-DF2 functional has mean errors comparable to MP2, but with a very large maximum error [103]. In comparison, RPA yields interaction energies of high accuracy (MAE = 0.41 kcal/mol), though it appears that results for weakly bound systems may not be fully converged with respect to basis set size. Further work to assess the basis set convergence of RPA shows an increase of the RPA MAE to 0.82 kcal/mol in the basis set limit. Changing input orbitals from TPSS to self-consistent PBE orbitals hardly effects the errors (MAE = 0.37 kcal/mol). This is in line with previous results [116, 117], showing little variation of the results with the choice of semi-local functional used to compute the orbitals.

Eshuis and Furche [86] applied RPA to the predissociation of a second-generation Grubbs catalyst used for olefin metathesis [118, 119]. The dissociation energy of this complex depends strongly on the medium-range interactions between the bulky ligands [120–122]. Semi-local DFT fails badly with errors of more than half of the experimental dissociation enthalpy of 36.8 kcal/mol [120]. Addition of dispersion corrections reduces the error. Of the Minnesota family of functionals, the M06 and M06-2X functionals achieve high accuracy. RPA improves much upon semi-local DFT, but does not achieve the accuracy of M06-2X. The error in RPA is possibly due to the poor description of short-range correlation within RPA, which is important here because of bond-breaking.

Though results of comparable accuracy can be achieved with other methods, RPA is the only one that is simultaneously parameter-free, non-dependent on a partitioning of the system and computationally efficient. In addition, RPA does not break down for zero-gap systems.

### 4.2 Self-interaction error and static correlation

Figure 7 shows the PES for $H_2$ obtained from several methods [53]. Stretched $H_2$ is a prototype for static correlation; at infinite separation, the orbital energies for the two electrons are exactly degenerate. RPA based on a closed-shell reference determinant describes this limit correctly [53, 123]. On the other hand, the RPA correlation energy in this limit is non-zero, due to the inherent one-electron self-interaction of direct RPA. Semi-local density functionals lead to large errors in the dissociation limit. Spin symmetry breaking produces qualitatively correct potential energy curves, but also causes unphysical spin-polarization. RPA, in contrast, leads to the correct dissociation limit based on a single-determinant *singlet* reference state [123]. RPA (EXX) also leads to the correct limit [140].

SOSEX rigorously removes all one-electron self-interaction, as demonstrated by Henderson and Scuseria [53]. But it simultaneously destroys the correct asymptotic behavior, due to non-vanishing ionic terms in the pair density. Self-interaction error causes RPA to fail completely for (effective) single electron cases such as $H_2^+$ and $He_2^+$ (Fig. 8), because of many-electron self-interaction error. Here, SOSEX improves the dissociation limit greatly, by largely eliminating the self-interaction error. In short, SOSEX removes the one-electron self-interaction error, reduces the many-electron self-interaction error, but also removes some desirable left–right correlation.

At medium range, the RPA curve has an unphysical maximum, which is also observed in $N_2$ [18] and in the challenging case of $Be_2$ dimer [47]. It has been speculated that the origin of this bump is due to self-interaction error or due to the non-self-consistent nature of the RPA scheme [53]. Though RPA underbinds at the equilibrium distance, it gives good equilibrium bond lengths compared to experiment.

### 4.3 Atomization energies

The random phase approximation systematically underbinds in molecular atomization energy benchmarks [18].

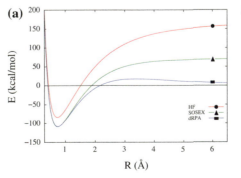
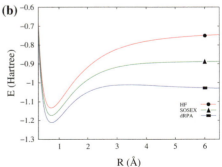

**Fig. 7** $H_2$ dissociation curves for RPA and SOSEX evaluated with self-consistent PBE orbitals using aug-cc-pVQZ basis sets: **a** relative to two H atoms, **b** on an absolute energy scale. (Reprinted with permission from Ref. [53]. Copyright 2010 Taylor & Francis.) The *circle*, *rectangle*, and *triangle* symbols were added to help distinguish the *curves*

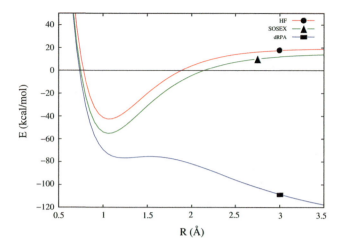

**Fig. 8** Dissociation of $He_2^+$ for RPA and SOSEX evaluated with self-consistent PBE orbitals using aug-cc-pVQZ basis sets. (Reprinted with permission from Ref [53]. Copyright 2010 Taylor & Francis.) The *circle*, *rectangle*, and *triangle* symbols were added to help distinguish the *curves*

**Table 1** Calculated atomization energies (kcal/mol) compared to experiment

| System | PBE | x-only | RPA | RPA+ | Expt.[a] |
|---|---|---|---|---|---|
| $H_2$ | 105 | 84 | 109 | 110 | 109 |
| $N_2$ | 244 | 111 | 223 | 223 | 228 |
| $O_2$ | 144 | 25 | 113 | 111 | 121 |
| $F_2$ | 53 | −43 | 30 | 29 | 38 |
| $Ne_2$[b] | 0.11 | −0.15 | 0.01 | −0.08 | 0.08[c] |
| $Si_2$ | 81 | 38 | 70 | 70 | 75 |
| HF | 142 | 96 | 133 | 132 | 141 |
| CO | 269 | 170 | 244 | 242 | 259 |
| $CO_2$ | 416 | 234 | 364 | 360 | 389[d] |
| $C_2H_2$ | 415 | 291 | 381 | 378 | 405[d] |
| $H_2O$ | 234 | 155 | 223 | 222 | 232[d] |
| $C_6H_5$–H[e] | 115 | 100 | 112 | 112 | 120 ± 1[f] |

Dunning's cc-pVQZ and cc-pV5Z basis sets [82] were used. Results were obtained using 4–5 extrapolation and basis set superposition correction. (Reprinted with permission from Ref. [18]. Copyright 2001 APS)

[a] Ref. [124], unless otherwise stated
[b] All electron results
[c] Ref. [125]
[d] Ref. [126]
[e] TZVPP basis, no counterpoise
[f] Ref. [127]

Table 1 lists RPA results for small molecules and compares them to other methods [18]. RPA errors are of the same order of magnitude as semi-local DFT. Similar results were obtained by Harl et al. [117] for 24 solids. The relatively poor performance of RPA for atomization energies

**Table 2** Mean errors (ME), mean absolute errors (MAE), and maximum errors (Max) (kcal/mol) for the G21IP ionization potential test set for various density functionals using def2-QZVP basis sets

|  | PBE | TPSS | B3LYP | B2PLYP | RPA[a] |
|---|---|---|---|---|---|
| ME | −0.12 | −0.87 | −0.12 | −0.73 | −5.01 |
| MAE | 3.85 | 3.95 | 3.55 | 2.31 | 5.11 |
| Max | 10.17 | 11.74 | 9.59 | 5.89 | 13.96 |

Core electrons were kept frozen; RPA results were obtained from self-consistent TPSS orbitals. Geometries, reference energies, and non-RPA results were taken from Ref. [129]. See supporting information for complete results

[a] Computed for this review

**Table 3** Mean errors (ME), mean absolute errors (MAE), and maximum errors (Max) (kcal/mol) for the G21EA electron affinities test set for various density functionals using def2-QZVP basis sets

|  | PBE | TPSS | B3LYP | B2PLYP | RPA[a] |
|---|---|---|---|---|---|
| ME | 2.96 | 0.42 | 0.30 | −0.79 | −0.70 |
| MAE | 3.43 | 2.21 | 1.81 | 1.37 | 3.02 |
| Max | 7.72 | 5.83 | 2.33 | 3.84 | 21.60 |

Core electrons were frozen for the RPA correlation energy calculations. RPA results were obtained from self-consistent TPSS orbitals. Geometries, reference energies, and non-RPA results were taken from Refs. [129] and [110]. See supporting information for complete results

[a] Computed for this review

might be linked to its deficient description of short-range correlation: in the process of atomization, all covalent bonds are broken, leading to large rearrangements of the electronic structure at short-range. RPA+ [43] does not improve upon RPA for atomization energies [18, 34, 117], though it corrects total correlation energies to a large extent.

The good performance for atomization energies of many semi-empirical density functionals is linked to the inclusion of atomization energies in the fitting reference set. The behavior of RPA is similar to non-empirical wavefunction-based methods; as shown for *n*-homodesmotic reactions, RPA errors become systematically smaller with decreasing change in the electronic environment [86, 128], see Sect. 4.5.

### 4.4 Ionization potentials and electron affinities

Jiang and Engel [116] first published RPA ionization potentials for atoms. They found that the semi-local functional BLYP performs better than RPA, but also report a significant improvement for RPA+. Here, we present a more extended set of ionization potentials and electron affinities that include molecular systems. Tables 2 and 3 compare RPA errors in ionization potentials and electron

**Table 4** Mean errors (ME), mean absolute errors (MAE), and maximum errors (Max) (kcal/mol) for the BH76RC reaction energy set for various density functionals. def2-QZVP basis sets were used and core electrons were kept frozen

|     | PBE | TPSS | B3LYP | B2PLYP | RPA/PBE[a] | RPA/TPSS[a] |
|-----|------|------|-------|--------|-----------|------------|
| ME  | 0.96 | 0.59 | −0.25 | −0.11 | −0.05 | 0.39 |
| MAE | 4.33 | 3.78 | 2.34 | 1.17 | 2.61 | 1.89 |
| Max | 22.69 | 12.90 | 7.24 | 5.16 | 8.75 | 4.55 |

RPA results were obtained from self-consistent TPSS or self-consistent PBE orbitals. Geometries, reference energies, and non-RPA results taken from Ref. [129]. See supporting information for complete results

[a] Computed for this review

**Table 5** Mean Error (ME), mean absolute error (MAE), and maximum error (Max) (kcal/mol) for the G2RC test set for various density functionals using def2-QZVP basis sets

|     | PBE | TPSS | B3LYP | B2PLYP | RPA[a] |
|-----|------|------|-------|--------|-------|
| ME  | 0.32 | 3.12 | 0.53 | −0.22 | −1.03 |
| MAE | 6.20 | 6.42 | 2.60 | 1.71 | 2.72 |
| Max | 18.96 | 19.97 | 7.00 | 5.63 | 14.47 |

Core electrons were kept frozen. RPA results were obtained from self-consistent TPSS orbitals. Geometries, reference energies and non-RPA results taken from Ref. [129]. See supporting information for complete results

[a] Computed for this review

**Table 6** Mean errors (ME), mean absolute errors (MAE), and maximum errors (Max) (kcal/mol) for the BH76 barrier heights test set for several density functionals using def2-QZVP basis sets

|     | PBE | TPSS | B3LYP | B2PLYP | RPA/PBE[a] | RPA/TPSS[a] |
|-----|------|------|-------|--------|-----------|------------|
| ME  | −9.18 | −8.55 | −4.56 | −2.08 | −1.65 | −1.79 |
| MAE | 9.23 | 8.57 | 4.66 | 2.24 | 3.10 | 2.77 |
| Max | −30.68 | −23.60 | −10.98 | −6.47 | −11.4 | −11.19 |

Core electrons were kept frozen. RPA results were obtained from either self-consistent TPSS or self-consistent PBE orbitals. Geometries, reference energies, and non-RPA results taken from Ref. [129]. See supporting information for complete results

[a] Computed for this review

affinities for the G21 set to other methods [129]. In contrast to other results presented here, these reactions are not isoelectronic, that is, they do not conserve the number of electrons. The errors for RPA are larger than those of semi-local functionals. However, the nearly equal magnitude of the ME and MAE for the ionization potentials in RPA suggests that RPA lends itself well to systematic improvement. Due caution is necessary for electron affinities, which do not have a well-defined basis set limit if a semi-local density functional is used [130].

## 4.5 Reaction energies

The accurate calculation and prediction of reaction energies are of great importance for thermochemistry. Most chemical processes involve covalent bondbreaking and making. In recent work, Eshuis and Furche [86] assessed the quality of RPA for the $n$-homodesmotic reaction hierarchy presented by Wheeler et al. [128]. They showed that for isogyric and isodesmic reactions, which involve bondbreaking and making typical for chemical reactions, RPA leads to smaller errors than B3LYP or M06-2X.

Here, we present new results for two test sets. Table 4 contains results for a subset of the 76 reactions studied for barrier heights (*cf.* Sect. 4.6) [129], and Table 5 shows results for a subset of the G2/97 test set. See the supporting information for complete results.

The G2/97 subset consists of 25 reactions involving closed-shell molecules containing first and second row elements [129]. The reaction energies vary from just 1 kcal/mol to over 200 kcal/mol. RPA achieves an accuracy comparable to B3LYP, but with a larger maximum error. The double-hybrid B2PLYP performs somewhat better. The MAE for RPA is approximately two times smaller than the MAE for the PBE or TPSS functional. Similar results are observed for the BH76 subset, which consists of 30 reactions involving atoms and small molecules containing first and second row elements. Here, open-shell free atoms

and molecules as well as anions are present, making this a challenging set for semi-local DFT. The reaction energies vary from 1 to 100 kcal/mol. Two sets of RPA results are presented to show that for these energy differences, a relatively large dependence on the input orbitals is observed. It is likely that this is caused by the open-shell atoms and anions, which pose a particular challenge for semi-local DFT. In summary, RPA is accurate even for non-isodesmic reactions, rivaling B3LYP as a general purpose functional for reaction energies.

## 4.6 Barrier heights

Table 6 shows errors in RPA results compared to other functionals for 76 barrier heights of hydrogen and heavy-atom transfers, nucleophilic substitution, unimolecular and association reactions [129]. The set consists of 38 reactions, and results for forward and backward barriers are presented. The average barrier height is 18.5 kcal/mol, and the reference energies are obtained from W1 and theoretical estimates. GGA functionals fail with MAEs of approximately 40% of the average barrier height. B3LYP improves upon that considerably, while B2PLYP performs best, though chemical accuracy is not reached. RPA performs at the same level as B2PLYP. Grimme's dispersion

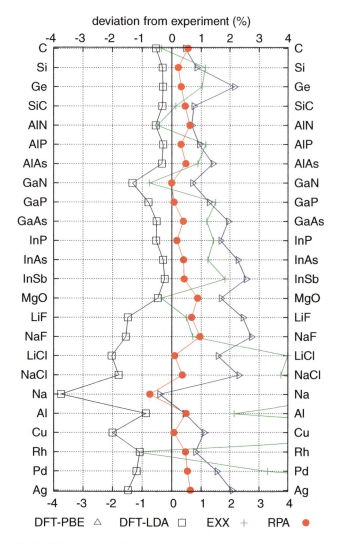

**Fig. 9** Relative error (%) of theoretical lattice constants with respect to experiment. (Reprinted with permission from Ref [117]. Copyright 2010 The American Physical Society)

correction does not seem to improve barrier heights. These data show, that RPA is also promising for chemical kinetics.

4.7 Adsorption on surfaces and solid-state properties

Ren et al. [90] applied RPA to the adsorption of CO on a Cu(111) surface. The results were compared to two GGA functionals, PBE and AM05, and two hybrid functionals, PBE0 and HSE03. RPA adsorption energies were found to be most accurate. Importantly, RPA correctly predicts the atop site to be preferred for adsorption over the hollow site. The latter is preferred by the GGA functionals, whereas the hybrid functionals yield a slight preference for the atop site.

Harl et al. [117] evaluated lattice constants for a set of insulators, semiconductors, and metals. RPA using PBE orbitals improves considerably upon semi-local DFT (Fig. 9). In contrast to semi-local DFT, RPA errors do not increase with system size. Similar results were found for bulk moduli and heats of formation. The influence of the input orbitals was found to be small, when starting from LDA orbitals instead of PBE orbitals.

In recent work, Lebègue et al. [131] studied the cohesive properties of graphite using RPA. Graphite is the parent compound of carbon-based materials, which have drawn much interest, both experimentally and theoretically. Of special interest is the interaction between graphene sheets. Here, van der Waals interactions play a crucial role. LSDA binding energies (24 meV/atom) are significantly too small, compared to quantum Monte Carlo results (56 ± 5 meV/atom).

Using RPA, the interlayer equilibrium distance, the elastic constant, and the net layer binding energy were found to be in excellent agreement with experiment, though the computed binding energy of 48 meV/atom is 15% smaller than the binding energy obtained from quantum Monte Carlo methods. In comparison with vdW-type functionals, RPA is more accurate for all three computed properties, though computationally significantly more expensive. It was also found that the correlation energy behaves as $1/d^3$ for large interlayer distances $d$, in agreement with previous analytical work [132, 133]. This illustrates that RPA captures the non-pairwise additive character of the long-range interactions.

## 5 Conclusions

RPA has many attractive features: (1) It dramatically improves upon semi-local density functionals for non-covalent interactions; (2) it is non-perturbative and finite for small-gap systems and thus more widely applicable than perturbation theory; (3) it is compatible with 100% exact exchange and captures significant static correlation; (4) it is based on sound physics and does not contain empirical parameters; (5) it is computationally affordable for molecules with well over 100 atoms. For chemical processes that conserve the number of electron pairs, RPA rivals the best available methods in the size range >20 atoms.

Nevertheless, there is room for improvement, especially for processes that change the number of electron pairs. In many ways, RPA is to the correlation energy what the Hartree energy is to the total ground-state energy: An important part that captures essential physics and chemistry, yet is not sufficient for many purposes. But this analogy also suggests that the development of beyond-RPA exchange and correlation methods is a worthwhile endeavor.

**Acknowledgments** The authors would like to acknowledge Asjbörn Burow for helpful comments. Two of us (H.E. and JE.B.) thank TURBOMOLE GmbH for support. This work was supported by the National Science Foundation, grant No. CHE-0911266.

# References

1. Bohm D, Pines D (1951) Phys Rev 82:625
2. Pines D, Bohm D (1952) Phys Rev 85:338
3. Bohm D, Pines D (1953) Phys Rev 92:609
4. Gell-Mann M, Brueckner KA (1957) Phys Rev 106:364
5. McLachlan AD, Ball MA (1964) Rev Mod Phys 36:844
6. Oddershede J (1978) Adv Quant Chem 11:275
7. Szabo A, Ostlund NS (1977) J Chem Phys 67:4351
8. Shibuya T-I, McKoy V (1970) Phys Rev A 2:2208
9. Ostlund N, Karplus M (1971) Chem Phys Lett 11:450
10. Öhrn Y, Linderberg J (1979) Int J Quant Chem 15:343
11. Langreth DC, Perdew JP (1975) Solid State Commun 17:1425
12. Langreth DC, Perdew JP (1977) Phys Rev B 15:2884
13. Gunnarsson O, Lundqvist BI (1976) Phys Rev B 13:4274
14. Callen HB, Welton TA (1951) Phys Rev 83:34
15. Andersson Y, Langreth DC, Lundqvist BI (1996) Phys Rev Lett 76:102
16. Dobson JF, Wang J (1999) Phys Rev Lett 82:2123
17. Dobson J (2006) In: Time-dependent density functional theory, vol. 706. Springer, Berlin, p 443
18. Furche F (2001) Phys Rev B 64:195120
19. Levy M (1979) Proc Natl Acad Sci USA 76:6062
20. Fetter AL, Walecka JD (1971) Quantum theory of many-particle systems, international series in pure and applied physics. MacGraw-Hill, New York
21. Runge E, Gross EKU (1984) Phys Rev Lett 52:997
22. Petersilka M, Gossmann UJ, Gross EKU (1996) Phys Rev Lett 76:1212
23. Casida ME (1995) In: Chong DP (ed) Recent advances in density functional methods, vol. 1 of Recent advances in computational chemistry. World Scientific, Singapore, p 155
24. Furche F (2001) J Chem Phys 114:5982
25. Furche F (2008) J Chem Phys 129:114105
26. Hedin L (1965) Phys Rev 139:A796
27. Toulouse J, Zhu W, Ángyán JG, Savin A (2010) Phys Rev A 82:032502
28. Hellgren M, von Barth U (2007) Phys Rev B 76:075107
29. Onida G, Reining L, Rubio A (2002) Rev Mod Phys 74:601
30. Bechstedt F, Fuchs F, Kresse G (2009) Phys Status Solidi (B) 246:1877
31. Møller C, Plesset MS (1934) Phys Rev 46:618
32. Ball MA, McLachlan AD (1964) Mol Phys 7:501
33. Toulouse J, Gerber IC, Jansen G, Savin A, Ángyán JG (2009) Phys Rev Lett 102:096404
34. Furche T, Van Voorhis T (2005) Chem Phys 122:164106
35. Klopper W, Teale AM, Coriani S, Pedersen TB, Helgaker T (2011) Chem Phys Lett 510:147
36. Jansen G, Liu RF, Ángyán JG (2010) J Chem Phys 133:154106
37. Heßelmann A (2011) J Chem Phys 134:204107
38. Freeman DL (1977) Phys Rev B 15:5512
39. Scuseria GE, Henderson TM, Sorensen DC (2008) J Chem Phys 129:231101
40. Hansen AE, Bouman TD (1979) Mol Phys 37:1713
41. Sanderson EA (1965) Phys Lett 19:141
42. Burke K, Perdew JP, Langreth DC (1994) Phys Rev Lett 73:1283
43. Yan Z, Perdew JP, Kurth S (2000) Phys Rev B 61:16430
44. Ruzsinszky A, Perdew JP, Csonka GI (2010) J Chem Theory Comput 6:127
45. Ruzsinszky A, Perdew JP, Csonka GI (2011) J Chem Phys 134:114110
46. Janesko BG, Henderson TM, Scuseria GE (2009) J Chem Phys 130:081105
47. Zhu W, Toulouse J, Savin A, Ángyán JG (2010) J Chem Phys 132:244108
48. Görling A, Levy M (1993) Phys Rev B 47:13105
49. Ernzerhof M (1996) Chem Phys Lett 263:499
50. Grüneis A, Marsman M, Harl J, Schimka L, Kresse G (2009) J Chem Phys 131:154115
51. Paier J, Janesko BG, Henderson TM, Scuseria GE, Grüneis A, Kresse G (2010) J Chem Phys 132:094103
52. Paier J, Janesko BG, Henderson TM, Scuseria GE, Grüneis A, Kresse G (2010) J Chem Phys 133:179902
53. Henderson TM, Scuseria GE (2010) Mol Phys 108:2511
54. Ren XG, Tkatchenko A, Rinke P, Scheffler M (2011) Phys Rev Lett 106:153003
55. Singwi KS, Tosi MP, Land RH, Sjölander A (1968) Phys Rev 176:589
56. Gross EKU, Kohn W (1990) Adv Quant Chem 21:255
57. Dobson JF, Wang J (2000) Phys Rev B 62:10038
58. Lein M, Gross EKU, Perdew JP (2000) Phys Rev B 61:13431
59. Dobson JF, Wang J, Gould T (2002) Phys Rev B 66:081108
60. Dobson JF (2009) Phys Chem Chem Phys 11:4528
61. Constantin LA, Pitarke JM, Dobson JF, Garcia-Lekue A, Perdew JP (2008) Phys Rev Lett 100:036401
62. Kotani T, Akai H (1998) J Magn Magn Mater 177(181):569
63. Hellgren M, von Barth U (2008) Phys Rev B 78:115107
64. Hirata S, Ivanov S, Grabowski I, Bartlett RJ (2002) J Chem Phys 116:6468
65. Shigeta Y, Hirao K, Hirata S (2006) Phys Rev A 73:010502
66. Heßelmann A, Ipatov A, Görling A (2009) Phys Rev A 80:012507
67. Heßelmann A, Görling A (2010) Mol Phys 108:359
68. Beebe N, Linderberg J (1977) Int J Quant Chem 12:683
69. Friesner RA (1991) Ann Rev Phys Chem 42:341
70. Ko C, Malick DK, Braden DA, Friesner RA, Martínez TJ (2008) J Chem Phys 128:104103
71. Baerends EJ, Ellis DE, Ros P (1973) Chem Phys 2:41
72. Dunlap BI, Connolly JWD, Sabin JRJ (1979) Chem Phys 71:3396
73. Eichkorn K, Treutler O, Öhm H, Häser M, Ahlrichs R (1995) Chem Phys Lett 240:283
74. Bauernschmitt R, Hser M, Treutler O, Ahlrichs R (1997) Chem Phys Lett 264:573
75. Neese F, Olbrich G (2002) Chem Phys Lett 362:170
76. Rappoport D, Furche F (2005) J Chem Phys 122:064105
77. Weigend F, Häser M, Patzelt H, Ahlrichs R (1998) Chem Phys Lett 294:143
78. Weigend F, Häser M (1997) Theor Chim Acta 97:331
79. Eshuis H, Yarkony J, Furche F (2010) J Chem Phys 132:234114
80. Hale N, Higham NJ, Trefethen LN (2008) SIAM J Num Anal 46:2505
81. Boyd JP (1987) J Sci Comput 2:99
82. Dunning J (1989) J Chem Phys 90:1007
83. Weigend F, Köhn A, Hättig C (2002) J Chem Phys 116:3175
84. Tajti A, Szalay PG, Császár AG, Kállay M, Gauss J, Valeev EF, Flowers BA, Vázquez J, Stanton JF (2004) J Chem Phys 121:11599
85. Weigend F, Ahlrichs R (2005) Phys Chem Chem Phys 7:3297
86. Eshuis H, Furche F (2011) J Phys Chem Lett 2:983
87. Harl J, Kresse G (2008) Phys Rev B 77:045136
88. Fuchs M, Gonze X (2002) Phys Rev B 65:235109
89. Niquet YM, Fuchs M, Gonze X (2003) Phys Rev A 68:032507

90. Ren X, Rinke P, Scheffler M (2009) Phys Rev B 80:045402
91. Bashford D, Chothia C, Lesk AM (1987) J Mol Biol 196:199
92. Dabkowska I, Gonzalez HV, Jurecka P, Hobza P (2005) J Phys Chem A 109:1131
93. Meyer EA, Castellano RK, Diederich F (2003) Angew Chem Int Ed 42:1210
94. Grimme S (2006) J Chem Phys 124:034108
95. Kemnitz CR, Mackey JL, Loewen MJ, Hargrove JL, Lewis JL, Hawkins WE, Nielsen AF (2010) Chem Eur J 16:6942
96. Wodrich MD, Jana DF, von Rague Schleyer P, Corminboeuf C (2008) J Phys Chem A 112:11495
97. Grimme S, Djukic J (2010) Inorg Chem 49:2911
98. Sherrill CD (2009) In: Rev Comp Chem, Wiley, New York, pp 1–38
99. Černý J, Hobza P (2007) Phys Chem Chem Phys 9:5291
100. Schwabe T, Grimme S (2007) Phys Chem Chem Phys 9:3397
101. Grimme S (2004) J Comp Chem 25:1463
102. Dion M, Rydberg H, Schröder E, Langreth DC, Lundqvist BI (2004) Phys Rev Lett 92:246401
103. Vydrov OA, Van Voorhis T (2010) Phys Rev A 81:062708
104. Vydrov OA, Van Voorhis T (2009) J Chem Phys 130:104105
105. Langreth D, Lundqvist B, Chakarova-Käck S, Cooper V, Dion M, Hyldgaard P, Kelkkanen A, Kleis J, Kong L, Li S et al (2009) J Phys Cond Matter 21:084203
106. Axilrod BM, Teller E (1943) J Chem Phys 11:299
107. Lu D, Nguyen H, Galli G (2010) J Chem Phys 133:154110
108. Janesko BG, Scuseria GE (2009) J Chem Phys 131:154106
109. Gruzman D, Karton A, Martin JML (2009) J Phys Chem A 113:11974
110. Goerigk L, Grimme S (2011) Phys Chem Chem Phys 13:6670
111. Zhao Y, Truhlar DG (2008) Acc Chem Res 41:157167
112. Zhao Y, Truhlar DG (2011) Chem Phys Lett 502:1
113. Grimme S (2005) J Phys Chem A 109:3067
114. Takatani T, Hohenstein EG, Malagoli M, Marshall MS, Sherrill CD (2010) J Chem Phys 132:144104
115. Jurecka P, Sponer J, Cerny J, Hobza P (2006) Phys Chem Chem Phys 8:1985
116. Jiang H, Engel E (2007) J Chem Phys 127:184108
117. Harl J, Schimka L, Kresse G (2010) Phys Rev B 81:115126
118. Vougioukalakis GC, Grubbs RH (2010) Chem Rev 110:1746
119. Sanford MS, Love JA, Grubbs RH (2001) J Am Chem Soc 123:6543
120. Śliwa P, Handzlik J (2010) Chem Phys Lett 493:273
121. Zhao Y, Truhlar DG (2009) J Chem Theory Comput 5:324
122. Benitez D, Tkatchouk E, Goddard WA (2009) Organometallics 28:2643
123. Fuchs M, Niquet Y-M, Gonze X, Burke K (2005) J Chem Phys 122:094116
124. Huber KP, Herzberg G (1979) Constants of diatomic molecules, vol. IV of Molecular spectra and molecular structure. Van Nostrand Reinhold, New York
125. Oglivie JF, Wang FYH (1992) J Mol Struct 273:277
126. Adamo C, Ernzerhof M, Scuseria GE (2000) J Chem Phys 112:2643
127. May K, Dapprich S, Furche F, Unterreiner BV, Ahlrichs R (2000) Phys Chem Chem Phys 2:5084
128. Wheeler SE, Houk KN, vR Schleyer P, Allen WD (2009) J Am Chem Soc 131:2547
129. Goerigk L, Grimme S (2010) J Chem Theory Comput 6:107
130. Lee D, Furche F, Burke K (2010) J Phys Chem Lett 1:2124
131. Lebègue S, Harl J, Gould T, Ángyán JG, Kresse G, Dobson JF (2010) Phys Rev Lett 105:196401
132. Dobson JF, White A, Rubio A (2006) Phys Rev Lett 96:073201
133. Gould T, Simpkins K, Dobson JF (2008) Phys Rev B 77:165134
134. Heßelmann A, Görling A (2011) Mol Phys 109:2473
135. Ángyán JG, Liu R-F, Toulouse J, Jansen G (2011) J Chem Theory Comput 7:3116
136. Toulouse J, Zhu W, Savin A, Jansen G, Ángyán JG (2011) J Chem Phys 135:084119
137. Toulouse J, Zhu W, Ángyán JG, Savin A (2010) Phys Rev A82:032502
138. Irelan RM, Henderson TM, Scuseria GE (2011) J Chem Phys 135:094105
139. Lotrich V, Bartlett RJ (2011) J Chem Phys 134:184108
140. Heßelmann A, Görling A (2011) Phys Rev Lett 106:093001

# REGULAR ARTICLE

# Uniform electron gases

**Peter M. W. Gill · Pierre-François Loos**

Received: 15 June 2011 / Accepted: 15 July 2011 / Published online: 13 December 2011
© Springer-Verlag 2012

**Abstract** We show that the traditional concept of the uniform electron gas (UEG)—a homogeneous system of finite density, consisting of an infinite number of electrons in an infinite volume—is inadequate to model the UEGs that arise in finite systems. We argue that, in general, a UEG is characterized by at least two parameters, viz. the usual one-electron density parameter $\rho$ and a new two-electron parameter $\eta$. We outline a systematic strategy to determine a new density functional $E(\rho, \eta)$ across the spectrum of possible $\rho$ and $\eta$ values.

**Keywords** Uniform electron gas · Homogeneous electron gas · Jellium · Density functional theory

## 1 Introduction

The year 2012 is notable for both the journal and one of us for, in the months ahead, both TCA and PMWG will achieve their half-centuries. It is inevitable and desirable that such occasions lead to retrospection, for it is often by looking backwards that we can perceive most clearly the way ahead. Thus, as we ruminate on the things that we ought not to have done, we also dream of the things that we ought now to do.

Published as part of the special collection of articles celebrating the 50th anniversary of Theoretical Chemistry Accounts/Theoretica Chimica Acta.

P. M. W. Gill (✉) · P.-F. Loos
Research School of Chemistry, Australian National University, Canberra, ACT 0200, Australia
e-mail: peter.gill@anu.edu.au

P.-F. Loos
e-mail: loos@rsc.anu.edu.au

The final decade of the 20th century witnessed a major revolution in quantum chemistry, as the subject progressed from an esoteric instrument of an erudite cognoscenti to a commonplace tool of the chemical proletariat. The fuel for this revolution was the advent of density functional theory (DFT) [1] models and software that were sufficiently accurate and user-friendly to save the experimental chemist some time. These days, DFT so dominates the popular perception of molecular orbital calculations that many non-specialists now regard the two as synonymous.

In principle, DFT is founded in the Hohenberg–Kohn theorem [2] but, in practice, much of its success can be traced to the similarity between the electron density in a molecule and the electron density in a hypothetical substance known as the uniform electron gas (UEG) or jellium [3–18]. The idea—the local density approximation (LDA)—is attractively simple: if we know the properties of jellium, we can understand the electron cloud in a molecule by dividing it into tiny chunks of density and treating each as a piece of jellium.

The good news is that the properties of jellium are known from near-exact Quantum Monte Carlo calculations [19–29]. Such calculations are possible because jellium is characterized by just a *single* parameter $\rho$, the electron density.

The bad news is that jellium has an infinite number of electrons in an infinite volume and this unboundedness renders it, in some respects, a poor model for the electrons in molecules. Indeed, the simple LDA described above predicts bond energies that are much too large and this led many chemists in the 1970s to dismiss DFT as a quantitatively worthless theory.

Most of the progress since those dark days has resulted from concocting ingenious corrections for jellium's deficiencies. For example, significant improvements in accuracy can be achieved by using both the density $\rho(\mathbf{r})$ and its gradient $\nabla \rho(\mathbf{r})$, an approach called gradient-corrected DFT

[30]. Even better results can be achieved by including a fraction of Hartree-Fock exchange (yielding hybrid methods [31, 32]) or higher derivatives of $\rho(\mathbf{r})$ (leading to meta-GGAs [33]).

However, notwithstanding the impressive progress since the 1970s, modern DFT approximations still exhibit fundamental deficiencies in large systems [34], conjugated molecules [35], charge-transfer excited states [36], dispersion-stabilized systems [37], systems with fractional spin or charge [38], isodesmic reactions [39] and elsewhere. Because DFT is in principle an exact theory, many of these problems can be traced ultimately to the use of jellium as a reference system and the ad hoc corrections that its use subsequently necessitates. It is not a good idea to build one's house on sand!

In an attempt to avoid some of the weaknesses of jellium-based DFT, we have invented and explored an alternative paradigm called intracule functional theory (IFT) [40–42]. In this approach, the one-electron density $\rho(\mathbf{r})$ is abandoned in favor of two-electron variables (such as the interelectronic distance $r_{12}$) and we have discovered that the latter offer an efficient and accurate route to the calculation of molecular energies [43–48]. Nonetheless, IFT is not perfect and has shortcomings that are complementary to those of DFT. As a result, one should seek to combine the best features of each, to obtain an approach superior to both. That is the goal of the present work and we will use atomic units throughout this article.

## 2 Electrons on spheres

In recent research, we were led to consider the behavior of electrons that are confined to the surface of a ball. This work yielded a number of unexpected discoveries [49–57] but the one of relevance here is that such systems provide a beautiful new family of uniform electron gases (see also [58]).

### 2.1 Spherium atoms

The surface of a three-dimensional ball is called a 2-sphere (for it is two-dimensional) and its free-particle orbitals (Table 1) are the spherical harmonics $Y_{lm}(\theta, \phi)$. It is known that

$$\sum_{m=-l}^{l} |Y_{lm}(\theta, \phi)|^2 = \frac{2l+1}{4\pi} \tag{1}$$

and doubly occupying all the orbitals with $0 \leq l \leq L$ thus yields a uniform electron gas. We call this system $L$-spherium and will compare it to two-dimensional jellium [18].

**Table 1** The lowest free-particle orbitals on a 2-sphere

| Name | $l$ | $m$ | $\sqrt{4\pi}Y_{lm}(\theta, \phi)$ |
| --- | --- | --- | --- |
| $s$ | 0 | 0 | 1 |
| $p_0$ | 1 | 0 | $3^{1/2}\cos\theta$ |
| $p_{+1}$ | 1 | +1 | $(3/2)^{1/2}\sin\theta\exp(+i\phi)$ |
| $p_{-1}$ | 1 | −1 | $(3/2)^{1/2}\sin\theta\exp(-i\phi)$ |
| $d_0$ | 2 | 0 | $(5/4)^{1/2}(3\cos^2\theta - 1)$ |
| $d_{+1}$ | 2 | +1 | $(15/2)^{1/2}\sin\theta\cos\theta\exp(+i\phi)$ |
| $d_{-1}$ | 2 | −1 | $(15/2)^{1/2}\sin\theta\cos\theta\exp(-i\phi)$ |
| $d_{+2}$ | 2 | +2 | $(15/8)^{1/2}\sin^2\theta\exp(+2i\phi)$ |
| $d_{-2}$ | 2 | −2 | $(15/8)^{1/2}\sin^2\theta\exp(-2i\phi)$ |

**Table 2** Number of electrons in $L$-spherium and $L$-glomium atoms

| $L$ | 0 | 1 | 2 | 3 | 4 | 5 | 6 | 7 |
| --- | --- | --- | --- | --- | --- | --- | --- | --- |
| $L$-spherium | 2 | 8 | 18 | 32 | 50 | 72 | 98 | 128 |
| $L$-glomium | 2 | 10 | 28 | 60 | 110 | 182 | 280 | 408 |

The number of electrons (Table 2) in $L$-spherium is

$$n = 2(L+1)^2 \tag{2}$$

the volume of a 2-sphere is $V = 4\pi R^2$ and, therefore,

$$\rho = \frac{(L+1)^2}{2\pi R^2}. \tag{3}$$

### 2.2 Glomium atoms

The surface of a four-dimensional ball is a 3-sphere (or "glome" [59]) and its free-particle orbitals (Table 3) are the hyperspherical harmonics $Y_{lmn}(\chi, \theta, \phi)$. It is known [60] that

$$\sum_{m=0}^{l}\sum_{n=-m}^{m} |Y_{lmn}(\chi, \theta, \phi)|^2 = \frac{(l+1)^2}{2\pi^2} \tag{4}$$

and doubly occupying all the orbitals with $0 \leq l \leq L$ thus yields a uniform electron gas. We call this system $L$-glomium and will compare it to three-dimensional jellium [18].

The number of electrons (Table 2) in $L$-glomium is

$$n = (L+1)(L+2)(2L+3)/3 \tag{5}$$

the volume of a 3-sphere is $V = 2\pi^2 R^3$ and, therefore,

$$\rho = \frac{(L+1)(L+2)(2L+3)}{6\pi^2 R^3}. \tag{6}$$

### 2.3 Exactly solvable systems

One of the most exciting features of the two-electron atoms 0-spherium and 0-glomium is that, for certain values of the

**Table 3** The lowest free-particle orbitals on a glome (i.e. a 3-sphere)

| Name | $l$ | $m$ | $n$ | $\pi\, Y_{lmn}(\chi, \theta, \phi)$ |
|---|---|---|---|---|
| $1s$ | 0 | 0 | 0 | $2^{-1/2}$ |
| $2s$ | 1 | 0 | 0 | $2^{1/2}\cos\chi$ |
| $2p_0$ | 1 | 1 | 0 | $2^{1/2}\sin\chi\cos\theta$ |
| $2p_{+1}$ | 1 | 1 | $+1$ | $\sin\chi\sin\theta\exp(+i\phi)$ |
| $2p_{-1}$ | 1 | 1 | $-1$ | $\sin\chi\sin\theta\exp(-i\phi)$ |
| $3s$ | 2 | 0 | 0 | $2^{-1/2}(4\cos^2\chi - 1)$ |
| $3p_0$ | 2 | 1 | 0 | $12^{1/2}\sin\chi\cos\chi\cos\theta$ |
| $3p_{+1}$ | 2 | 1 | $+1$ | $6^{1/2}\sin\chi\cos\chi\sin\theta\exp(+i\phi)$ |
| $3p_{-1}$ | 2 | 1 | $-1$ | $6^{1/2}\sin\chi\cos\chi\sin\theta\exp(-i\phi)$ |
| $3d_0$ | 2 | 2 | 0 | $\sin^2\chi(3\cos^2\theta - 1)$ |
| $3d_{+1}$ | 2 | 2 | $+1$ | $6^{1/2}\sin^2\chi\sin\theta\cos\theta\exp(+i\phi)$ |
| $3d_{-1}$ | 2 | 2 | $-1$ | $6^{1/2}\sin^2\chi\sin\theta\cos\theta\exp(-i\phi)$ |
| $3d_{+2}$ | 2 | 2 | $+2$ | $(3/2)^{1/2}\sin^2\chi\sin^2\theta\exp(+2i\phi)$ |
| $3d_{-2}$ | 2 | 2 | $-2$ | $(3/2)^{1/2}\sin^2\chi\sin^2\theta\exp(-2i\phi)$ |

radius $R$, their Schrödinger equations are exactly solvable [51]. The basic theory is as follows.

The Hamiltonian for two electrons on a sphere is

$$\mathbf{H} = -\frac{\nabla_1^2}{2} - \frac{\nabla_2^2}{2} + \frac{1}{u} \tag{7}$$

where $u$ is the interelectronic distance $r_{12} \equiv |\mathbf{r}_1 - \mathbf{r}_2|$. If we assume that the Hamiltonian possesses eigenfunctions that depend only on $u$, it becomes

$$\mathbf{H} = \left[\frac{u^2}{4R^2} - 1\right]\frac{d^2}{du^2} + \left[\frac{(2\mathscr{D}-1)u}{4R^2} - \frac{\mathscr{D}-1}{u}\right]\frac{d}{du} + \frac{1}{u} \tag{8}$$

where $\mathscr{D}$ is the dimensionality of the sphere. Three years ago, we discovered that $\mathbf{H}$ has polynomial eigenfunctions, but only for particular values of $R$. (This is analogous to the discovery that hookium[1] has closed-form wavefunctions, but only for particular harmonic force constants [62, 63].) We showed that there exist $\lfloor (n+1)/2 \rfloor$ $n$th-degree polynomials of this type and that the associated energies and radii satisfy

$$4R_{n,m}^2 E_{n,m} = n(n + 2\mathscr{D} - 2) \tag{9}$$

where the index $m = 1, \ldots, \lfloor (n+1)/2 \rfloor$.

For 0-spherium (i.e. $\mathscr{D} = 2$), by introducing $x = u/(2R)$ and using Eq. (9), we obtain the Sturm-Liouville equation

$$\frac{d}{dx}\left[\frac{x(1-x^2)}{2}\frac{d\Psi}{dx}\right] + \frac{n(n+2)x}{2}\Psi = R\Psi \tag{10}$$

The eigenradii $R$ can then be found by diagonalization in a polynomial basis which is orthogonal on [0,1]. The shifted Legendre polynomials are ideal for this [64].

For 0-glomium (i.e. $\mathscr{D} = 3$), proceeding similarly yields

$$\frac{d}{dx}\left[\frac{x(1-x^2)}{2}w(x)\frac{d\Psi}{dx}\right] + \frac{n(n+4)x}{2}w(x)\Psi = Rw(x)\Psi \tag{11}$$

where the weight function $w(x) = x\sqrt{1-x^2}$. The eigenradii are found by diagonalization in a basis which is orthogonal with respect to $w(x)$ on [0,1].

An exact energy can be partitioned into its kinetic part

$$E_T = (-1/4)\langle\Psi|\nabla_1^2 + \nabla_2^2|\Psi\rangle \tag{12}$$

and its two-electron part

$$E_{ee} = (1/2)\langle\Psi|u^{-1}|\Psi\rangle \tag{13}$$

and the resulting reduced energies (i.e. the energy per electron) of the ground states of 0-spherium and 0-glomium, for the first two eigenradii, are shown in the left half of Table 4.

# 3 Single-determinant methods

## 3.1 Hartree–Fock theory [65, 66]

In the Hartree-Fock (HF) partition, the reduced energy[2] of an $n$-electron system is

$$E = T_S + E_V + E_J + E_K + E_C \tag{14}$$

where the five contributions are the non-interacting-kinetic, external, Hartree, exchange and correlation energies, respectively. The first four of these are defined by

$$T_S = -\frac{1}{2n}\sum_i^n \int \psi_i^*(\mathbf{r})\nabla^2\psi_i(\mathbf{r})d\mathbf{r} \tag{15}$$

$$E_V = +\frac{1}{n}\int \rho(\mathbf{r})v(\mathbf{r})d\mathbf{r} \tag{16}$$

$$E_J = +\frac{1}{2n}\iint \rho(\mathbf{r}_1)r_{12}^{-1}\rho(\mathbf{r}_2)d\mathbf{r}_1 d\mathbf{r}_2 \tag{17}$$

$$E_K = -\frac{1}{2n}\sum_{i,j}^n \iint \psi_i^*(\mathbf{r}_1)\psi_j(\mathbf{r}_1)r_{12}^{-1}\psi_j^*(\mathbf{r}_2)\psi_i(\mathbf{r}_2)d\mathbf{r}_1 d\mathbf{r}_2 \tag{18}$$

where $\psi_i(\mathbf{r})$ is an occupied orbital, $\rho(\mathbf{r})$ is the electron density, and $v(\mathbf{r})$ is the external potential. The correlation energy $E_C$ is defined so that Eq. (14) is exact.

---

[1] The hookium atom consists of two electrons that repel Coulombically but are bound to the origin by a harmonic potential [61].

[2] Henceforth, all energies will be reduced energies.

**Table 4** Exact and Kohn–Sham reduced energies of the ground states of 0-spherium and 0-glomium for various eigenradii $R$

| | $2R$ | Exact | | | Jellium-based Kohn–Sham DFT | | | | | | Error |
|---|---|---|---|---|---|---|---|---|---|---|---|
| | | $E_T$ | $E_{ee}$ | $E$ | $T_S$ | $E_V$ | $E_J$ | $-E_X$ | $-E_C^{jell}$ | $E_{KS}$ | $E_{KS} - E$ |
| 0-spherium | $\sqrt{3}$ | 0.051982 | 0.448018 | 1/2 | 0 | 0 | 1.154701 | 0.490070 | 0.1028 | 0.562 | 0.062 |
| | $\sqrt{28}$ | 0.018594 | 0.124263 | 1/7 | 0 | 0 | 0.377964 | 0.160413 | 0.0593 | 0.158 | 0.015 |
| 0-glomium | $\sqrt{10}$ | 0.014213 | 0.235787 | 1/4 | 0 | 0 | 0.536845 | 0.217762 | 0.0437 | 0.275 | 0.025 |
| | $\sqrt{66}$ | 0.007772 | 0.083137 | 1/11 | 0 | 0 | 0.208967 | 0.084764 | 0.0270 | 0.097 | 0.006 |

## 3.2 Kohn–Sham density functional theory [67]

In the Kohn–Sham (KS) partition, the energy is

$$E_{KS} = T_S + E_V + E_J + E_X + E_C^{KS} \tag{19}$$

where the last two terms, which are sometimes combined, are the KS exchange and correlation energies. The correlation energy $E_C^{KS}$ is defined so that Eq. (19) is exact.

Many formulae have been proposed for $E_X$ and $E_C^{KS}$, but the most famous are those explicitly designed to be exact for $\mathscr{D}$-jellium. In the case of exchange, one finds

$$E_X = X_{\mathscr{D}} \int \rho^{1/\mathscr{D}} d\mathbf{r} \tag{20}$$

where Dirac [5] determined the coefficient $X_3$ in 1930 and Glasser [68] found the general formula for $X_{\mathscr{D}}$ in 1983. The correlation functional is not known exactly, but accurate Quantum Monte Carlo calculations on jellium in 2D [20–23, 25, 26, 28, 29] and 3D [19, 24, 27] have been fitted [26, 69] to functions of the form

$$E_C^{jell} = \int C_D(\rho) d\mathbf{r} \tag{21}$$

By construction, Eq. (21) yields the correct energy when applied to the uniform electron gas in jellium. But what happens when we apply it to a uniform electron gas on a sphere?

## 4 The non-uniqueness problem

The deeply disturbing aspect of jellium-based DFT models—and the launching-pad for the remainder of this paper—is the countercultural claim, that

*The uniform electron gas with density $\rho$ is not unique.*

Though it may seem heretical to someone who has worked with jellium for many years, or to someone who suspects that the claim violates the Hohenberg–Kohn theorem, we claim that two $\mathscr{D}$-dimensional uniform electron gases with the same density parameter $\rho$ may have different energies. To illustrate this, we now show that density functionals [5, 26, 69] which are exact for jellium are wrong for 0-spherium and 0-glomium.

## 4.1 Illustrations from exactly solvable systems

The energy contributions for 0-spherium and 0-glomium are easy to find. There is no external potential, so $E_V = 0$. The density $\rho(\mathbf{r})$ is constant, so the Kohn–Sham orbital $\psi(\mathbf{r}) = \sqrt{\rho(\mathbf{r})}$ is constant, and $T_S = 0$. The Hartree energy is the self-repulsion of a uniform spherical shell of charge of radius $R$ and one finds [50]

$$E_J = \frac{\Gamma(\mathscr{D} - 1)}{\Gamma(\mathscr{D} - 1/2)} \frac{\Gamma(\mathscr{D}/2 + 1/2)}{\Gamma(\mathscr{D}/2)} \frac{1}{R} \tag{22}$$

The exchange energy is predicted [57] by Eq. (20) to be

$$E_X = -\frac{2\mathscr{D}}{(\mathscr{D}^2 - 1)\pi R} \left( \frac{\mathscr{D}!}{2} \right)^{1/\mathscr{D}} \tag{23}$$

and the correlation energy predicted by Eq. (21) is simply

$$E_C^{jell} = C_{\mathscr{D}}(2/V) \tag{24}$$

Applying these formulae to the exactly solvable states of 0-spherium and 0-glomium considered in Sect. 2.3 yields the results in the right half of Table 4. In all cases, the KS-DFT energies are too high by 10 – 20%, indicating that the correlation functional that is exact for the uniform electron gas in jellium grossly underestimates the correlation energy of the uniform electron gases in 0-spherium and 0-glomium.

## 4.2 Limitations of the one-electron density parameter $\rho$

The results in Table 4 demonstrate conclusively that not all uniform electron gases with the density $\rho$ are equivalent. The simplest example of this is the ground state of two electrons on a 2-sphere with $R = \sqrt{3}/2$. The exact wavefunction, reduced energy and density of this system are

$$\Psi = 1 + u \tag{25}$$

$$E = 1/2 \tag{26}$$

$$\rho(\mathbf{r}) = 2/(3\pi) \tag{27}$$

but, when fed this uniform density, the exchange-correlation functional that is exact for two-dimensional jellium grossly overestimates the energy, yielding $E_{KS} = 0.562$.

This discovery has worrying chemical implications. Contrary to the widespread belief that the LDA (e.g. the S-VWN functional) is accurate when applied to regions of a molecule where $\rho(\mathbf{r})$ is almost uniform (such as near bond midpoints), our results reveal that it actually performs rather poorly.

The discovery also has counterintuitive implications at a theoretical level. It implies that the years of effort that have been expended in calculating the properties of jellium do *not* provide us with a complete picture of homogeneous electron gases. On the contrary, although they inform us in detail about the *infinite* uniform electron gas, they tell us very little about the properties of finite electron gases.

In a nutshell, the results in Table 4 reveal that a UEG is not completely characterized by its one-electron density parameter $\rho$. Evidently, something else is required.

### 4.3 Virtues of two-electron density parameters

We know that it is possible for two uniform electron gases to have the same density $\rho$ but different reduced energies $E$. But how can this be, given that the probability of finding an electron in a given volume is identical in the two systems? The key insight is that the probability of finding *two* electrons in that volume is different.

This is illustrated in Figs. 1 and 2, which compare the probability distributions of the interelectronic distance [52, 70, 71] in various two-dimensional uniform electron gases. These reveal that, although similar for $u \approx 0$ (because of the Kato cusp condition [72]), the specific Coulomb holes (i.e. the holes per unit volume [40]) in two gases with the same one-electron density $\rho$ can be strikingly different. In each case, the jellium hole is both deeper and wider than the corresponding spherium hole, indicating that the jellium electrons exclude one another more strongly, and one is much less likely to find two electrons in a given small volume of jellium than in the same volume of spherium.

We conclude from these comparisons that (at least) two parameters are required to characterize a uniform electron gas. Although the parameter choice is not unique, we believe that the first should be a one-electron quantity, such as the density $\rho$ (or, equivalently, the Seitz radius $r_s$) and the second should be a two-electron quantity such as $\eta = h(\mathbf{r}, \mathbf{r})$, where $h$ is the pair correlation function defined by

$$\rho_2(\mathbf{r}_1, \mathbf{r}_2) = \frac{1}{2}\rho(\mathbf{r}_1)\rho(\mathbf{r}_2)[1 + h(\mathbf{r}_1, \mathbf{r}_2)] \tag{28}$$

and $\rho_2$ is the diagonal part of the second-order spinless density matrix [1].

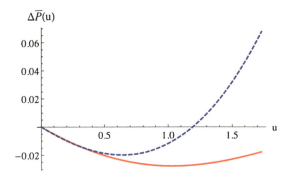

**Fig. 1** Specific Coulomb holes for 0-spherium (*dashed*) with $R = \sqrt{3}/2$ and 2D jellium (*solid*). Both are uniform gases with $\rho = 2/(3\pi)$

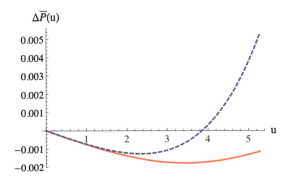

**Fig. 2** Specific Coulomb holes for 0-spherium (*dashed*) with $R = \sqrt{7}$ and 2D jellium (*solid*). Both are uniform gases with $\rho = 1/(14\pi)$

## 5 Lessons from spherium and glomium

### 5.1 A modest proposal

The discovery that uniform electron gases are characterized by *two* parameters ($\rho$ and $\eta$) has many ramifications but one of the most obvious is that the foundations of the venerable Local Density Approximation need to be rebuilt.

The traditional LDA writes the correlation energy of a molecular system as

$$E_C^{KS} \approx \int C(\rho) d\mathbf{r} \tag{29}$$

thereby assuming that the contribution from each point $\mathbf{r}$ depends only on the one-electron density $\rho(\mathbf{r})$ at that point. However, now that we know that the energy of a uniform electron gas depends on $\rho$ and $\eta$, it is natural to replace Eq. (29) by the generalized expression

$$E_C^{KS} \approx \int C(\rho, \eta) d\mathbf{r} \tag{30}$$

In a sense, this two-parameter LDA represents a convergence in the evolution of Density functional theory (which stresses the one-electron density) and intracule functional theory (which focuses on the two-electron density).

How can we find the new density functional $C(\rho, \eta)$? One could take an empirical approach but that is an overused option within the DFT community [73] and we feel that it is more satisfactory to derive it from the uniform electron gas. As we show in Sect. 5.3, it turns out that it is easy to compute the exact values of $T_S$, $E_V$, $E_J$ and $E_X$ for any $L$-spherium or $L$-glomium atom and, therefore, if one knew the exact wavefunctions and energies of $L$-spherium and $L$-glomium for a wide range of $L$ and $R$, one could extract the exact Kohn–Sham correlation energies

$$E_C^{KS} = E - T_S - E_V - E_J - E_X \qquad (31)$$

and determine the exact dependence of these on $\rho$ and $\eta$.

Accordingly, we propose to embark on a comprehensive study of $L$-spherium and $L$-glomium atoms, in order eventually to liberate the LDA from jellium's yoke through a process of radical generalization. The results of these spherium and glomium calculations will generalize the known properties of jellium, because we have shown that the energies of $L$-spherium and $L$-glomium approach those of 2D jellium and 3D jellium, as $L$ becomes large.

In the remaining sections, we will confine our attention to the (two-dimensional) $L$-spherium atoms. However, exactly the same approach can and will be used to address the (three-dimensional) $L$-glomium atoms in the future.

### 5.2 Basis sets and integrals

The Hamiltonian for $L$-spherium is

$$\mathbf{H} = -\frac{1}{2}\sum_{i=1}^{n}\nabla_i^2 + \sum_{i<j}^{n}\frac{1}{r_{ij}} \qquad (32)$$

and the natural basis functions for HF and correlated calculations on this are the spherical harmonics $Y_{lm}(\theta, \phi)$ introduced in Sect. 2. These functions are orthonormal [64]

$$\langle Y_{lm} | Y_{l'm'} \rangle = \delta_{l,l'}\,\delta_{m,m'} \qquad (33)$$

and are eigenfunctions of the Laplacian, so that

$$\langle Y_{lm} | \nabla^2 | Y_{l'm'} \rangle = -l(l+1)\delta_{l,l'}\,\delta_{m,m'} \qquad (34)$$

The required two-electron repulsion integrals can be found using the standard methods of two-electron integral theory [74]. For example, the spherical harmonic resolution of the Coulomb operator [75–79]

$$r_{12}^{-1} = R^{-1}\sum_{lm}\frac{4\pi}{2l+1}Y_{lm}^*(\mathbf{r}_1)Y_{lm}(\mathbf{r}_2) \qquad (35)$$

yields the general formula

$$\langle Y_{l_1 m_1} Y_{a_1 b_1} | r_{12}^{-1} | Y_{l_2 m_2} Y_{a_2 b_2} \rangle$$
$$= R^{-1}\sum_{lm}\frac{4\pi}{2l+1}\langle Y_{l_1 m_1} Y_{l_2 m_2} Y_{lm} \rangle \langle Y_{a_1 b_1} Y_{a_2 b_2} Y_{lm} \rangle \qquad (36)$$

where the one-electron integrals over three spherical harmonics involve Wigner $3j$ symbols [64] and the sum over $l$ and $m$ is limited by the Clebsch–Gordan selection rules.

### 5.3 Hartree–Fock calculations

Unlike our calculations for electrons in a cube [80], the HF calculations for our present systems are trivial. Because the shells in $L$-spherium are filled, the restricted[3] HF and KS orbitals are identical and are simply the spherical harmonics. These orbitals yield the reduced energy contributions

$$T_S = +\frac{L(L+2)}{4R^2} \qquad (37)$$

$$E_V = +0 \qquad (38)$$

$$E_J = +\frac{(L+1)^2}{R} \qquad (39)$$

$$E_K = -\frac{L+1}{2R}\,_4F_3\!\left[\begin{matrix}-L, & -1/2, & 1/2, & L+2 \\ -L-1/2, & 2, & L+3/2\end{matrix};1\right] \qquad (40)$$

$$E_X = -\frac{4(L+1)}{3\pi R} \qquad (41)$$

where $_4F_3$ is a balanced hypergeometric function [64] of unit argument.[4] It is encouraging to discover that Eq. (40) approaches Eq. (41) in the large-$L$ limit [57].

We have used Eqs. (37)–(40) to compute the exact HF energies of $L$-spherium for a number of the eigenradii discussed in Sect. 2.3. These, together with the $T_S$, $E_J$ and $E_X$ values, and $\rho$ values from Eq. (3), are shown in Table 5.

### 5.4 Orbital-based correlation methods

Although it is easy to find the HF energy of $L$-spherium, the calculation of its *exact* energy is not a trivial matter. How can this best be achieved? The fact that the occupied and virtual orbitals are simple functions (spherical harmonics), so the AO → MO integral transformation is unnecessary, suggests that orbital-based correlation methods may be particularly effective. We now consider some of these.

---

[3] We note that, in low-density cases, the RHF solutions are unstable with respect to lower-energy, symmetry-broken UHF wavefunctions [49]. However, we will not consider the latter in the present study.

[4] This hypergeometric may be related to a Wigner $6j$ symbol [64].

### 5.4.1 Configuration interaction [81]

This was the original scheme for proceeding beyond the HF approximation. It has fallen out of favor with many quantum chemists, because its size-inconsistency and size-inextensivity seriously hamper its efficacy for computing the energetics of chemical reactions. Nevertheless, for calculations of the energy of $L$-spherium, its variational character, systematic improvability and lack of convergence issues make it an attractive option.

### 5.4.2 Møller-Plesset perturbation theory [82]

In an earlier paper [49], we showed that the MP2, MP3, MP4 and MP5 energies of 0-spherium can be found in closed form, for any value of $R$. However, although we observed that the MP series seems to converge rapidly for small $R$, its convergence was much less satisfactory for $R \gtrsim 1$, where $\rho \lesssim 0.15$. Unfortunately, this is useless for our present purposes, because many of the $L$-spherium systems in Table 5 have much larger radii and lower densities than these values.

### 5.4.3 Coupled-cluster theory [83]

The coupled-cluster hierarchy (viz. CCSD, CCSDT, CCSDTQ, ...) probably converges much better than the MP$n$ series, and this should certainly be explored in the

future, but we suspect that it will nonetheless perform poorly in the large-$R$, small-$\rho$ systems where static correlation dominates dynamic correlation [49, 84, 85] and the single-configuration HF wavefunction is an inadequate starting point.

### 5.4.4 Explicitly correlated methods [86]

The CI, MP and CC approaches expand the exact wavefunction as a linear combination of determinants and it has been known since the early days of quantum mechanics that this ansatz struggles to describe the interelectronic cusps [72]. The R12 methods overcome this deficiency by explicitly including terms that are linear in $r_{ij}$ in the wavefunction and converge much more rapidly as the basis set is enlarged but, unfortunately, they also require a number of non-standard integrals. If these integrals are not computed exactly, they are approximated by resolutions in an auxiliary basis set. In either case, the computer implementation is complicated.

### 5.5 Other correlation methods

It is possible that the unusually high symmetry of $L$-spherium makes it well suited to correlation methods that are not based on the HF orbitals. We now consider two of these.

**Table 5** Reduced energies, densities and $\eta$ values of $L$-spherium with the four smallest eigenradii $R$

| $L$ | $R$ | Wavefunction-based energies | | | Kohn–Sham energies | | | Ingredients of the new model | | |
|---|---|---|---|---|---|---|---|---|---|---|
| | | $E_{HF}$ | $-E_C$ | $E$ | $T_S$ | $E_J$ | $-E_X$ | $-E_C^{KS}$ | $\rho$ | $\eta$ |
| 0 | $R_1$ | 0.577350 | 0.077350 | 0.500000 | 0.000000 | 1.154701 | 0.490070 | 0.164630 | 0.212207 | −0.896037 |
| | $R_2$ | 0.188982 | 0.046125 | 0.142857 | 0.000000 | 0.377964 | 0.160413 | 0.074694 | 0.022736 | −0.991159 |
| | $R_3$ | 0.092061 | 0.028497 | 0.063564 | 0.000000 | 0.184122 | 0.078144 | 0.042414 | 0.005396 | −0.999496 |
| | $R_4$ | 0.054224 | 0.018941 | 0.035282 | 0.000000 | 0.108447 | 0.046026 | 0.027138 | 0.001872 | −0.999976 |
| 1 | $R_1$ | 4.579572 | | | 1.000000 | 4.618802 | 0.980140 | | 0.848826 | |
| | $R_2$ | 1.278833 | | | 0.107143 | 1.511858 | 0.320826 | | 0.090946 | |
| | $R_3$ | 0.596204 | | | 0.025426 | 0.736488 | 0.156288 | | 0.021582 | |
| | $R_4$ | 0.345007 | | | 0.008821 | 0.433789 | 0.092053 | | 0.007487 | |
| 2 | $R_1$ | 11.543198 | | | 2.666667 | 10.392305 | 1.470210 | | 1.909859 | |
| | $R_2$ | 3.191241 | | | 0.285714 | 3.401680 | 0.481239 | | 0.204628 | |
| | $R_3$ | 1.483203 | | | 0.067802 | 1.657098 | 0.234431 | | 0.048560 | |
| | $R_4$ | 0.857188 | | | 0.023522 | 0.976025 | 0.138079 | | 0.016846 | |
| 3 | $R_1$ | 21.477457 | | | 5.000000 | 18.475209 | 1.960281 | | 3.395305 | |
| | $R_2$ | 5.929228 | | | 0.535714 | 6.047432 | 0.641652 | | 0.363783 | |
| | $R_3$ | 2.754531 | | | 0.127128 | 2.945952 | 0.312575 | | 0.086328 | |
| | $R_4$ | 1.591634 | | | 0.044103 | 1.735156 | 0.184106 | | 0.029949 | |

$R_1 = \frac{1}{2}\sqrt{3}$; $R_2 = \frac{1}{2}\sqrt{28}$; $R_3 = \frac{1}{2}\sqrt{63 + 12\sqrt{21}}$; $R_4 = \frac{1}{2}\sqrt{198 + 6\sqrt{561}}$

### 5.5.1 Iterative Complement Interaction (ICI) method [87, 88]

In this approach, the Schrödinger equation itself is used to generate a large set of $n$-electron functions that are then linearly combined to approximate the true wavefunction. It has been spectacularly successful in systems with a small number of electrons but has not yet been applied to systems as large as 2- or 3-spherium (with 18 and 32 electrons, respectively).

### 5.5.2 Quantum Monte Carlo methods [89, 90]

Of the various methods in this family, Diffusion Monte Carlo (DMC) usually yields the greatest accuracy. In this approach, the Schrödinger equation is transformed into a diffusion equation in imaginary time $\tau$ and, in the limit as $\tau \to \infty$, the ground-state energy is approached. Unfortunately, to be practically feasible, the method normally requires that the wavefunction's nodes be known and, despite some recent progress [91], the node problem remains unsolved. The quality of the trial wavefunction and finite-size errors are other potential restrictions [92, 93].

## 5.6 Results and holes

The discovery [51] that the Schrödinger equation for 0-spherium (and 0-glomium) is exactly solvable for each of its eigenradii is extremely helpful, for it allows us to generate the exact energies $E$ and resulting Kohn–Sham correlation energies $E_C^{KS}$ for the first four atoms in Table 5, without needing to perform any of the correlated calculations described above. However, these are the easiest cases, the "low-hanging fruit" so to speak, and there remain large gaps in the Table. We could have filled these gaps with rough estimates of the exact energies but we prefer to leave them empty, to emphasize that there is much to do in this field and to challenge the correlation experts in the wavefunction and density functional communities to address these beautifully simple systems.

Once this is done, the data in the final three columns of Table 5 will provide the ingredients for the construction of a new correlation density functional $E_C^{KS}(\rho, \eta)$ that will be exact for all $L$-spherium atoms and for 2D jellium.

Finally, of course, an analogous strategy will yield a new functional that is exact for all $L$-glomium atoms and for 3D jellium. We will recommend that these functionals replace the LDA correlation functionals that are now being used.

## 6 Concluding remark

All uniform electron gases are equal, but some are more equal than others.

**Acknowledgments** We would like to thank the Australian Research Council for funding (Grants DP0771978, DP0984806 and DP1094170) and the National Computational Infrastructure (NCI) for generous supercomputer grants.

## References

1. Parr RG, Yang W (1989) Density-functional theory of atoms and molecules. Clarendon Press, Oxford
2. Hohenberg P, Kohn W (1964) Phys Rev 136:B864
3. Fermi E (1926) Z Phys 36:902
4. Thomas LH (1927) Proc Cam Phil Soc 23:542
5. Dirac PAM (1930) Proc Cam Phil Soc 26:376
6. Wigner E (1934) Phys Rev 46:1002
7. Macke W (1950) Z Naturforsch A 5a:192
8. Gell-Mann M, Brueckner KA (1957) Phys Rev 106:364
9. Onsager L, Mittag L, Stephen MJ (1966) Ann Phys 18:71
10. Stern F (1973) Phys Rev Lett 30:278
11. Rajagopal AK, Kimball JC (1977) Phys Rev B 15:2819
12. Isihara A, Ioriatti L (1980) Phys Rev B 22:214
13. Hoffman GG (1992) Phys Rev B 45:8730
14. Seidl M (2004) Phys Rev B 70:073101
15. Sun J, Perdew JP, Seidl M (2010) Phys Rev B 81:085123
16. Loos PF, Gill PMW (2011) Phys Rev B 83:233102
17. Loos PF, Gill PMW (in press) Phys Rev B
18. Giuliani GF, Vignale G (2005) Quantum theory of electron liquid. Cambridge University Press, Cambridge
19. Ceperley DM, Alder BJ (1980) Phys Rev Lett 45:566
20. Tanatar B, Ceperley DM (1989) Phys Rev B 39:5005
21. Kwon Y, Ceperley DM, Martin RM (1993) Phys Rev B 48:12037
22. Ortiz G, Ballone P (1994) Phys Rev B 50:1391
23. Rapisarda F, Senatore G (1996) Aust J Phys 49:161
24. Kwon Y, Ceperley DM, Martin RM (1998) Phys Rev B 58:6800
25. Ortiz G, Harris M, Ballone P (1999) Phys Rev Lett 82:5317
26. Attaccalite C, Moroni S, Gori-Giorgi P, Bachelet GB (2002) Phys Rev Lett 88:256601
27. Zong FH, Lin C, Ceperley DM (2002) Phys Rev E 66:036703
28. Drummond ND, Needs RJ (2009) Phys Rev Lett 102:126402
29. Drummond ND, Needs RJ (2009) Phys Rev B 79:085414
30. Perdew JP, Wang Y (1986) Phys Rev B 33:8800
31. Gill PMW, Johnson BG, Pople JA, Frisch MJ (1992) Int J Quantum Chem Symp 26:319
32. Becke AD (1993) J Chem Phys 98:5648
33. Tau J, Perdew JP, Staroverov VN, Scuseria GE (2003) Phys Rev Lett 91:146401
34. Curtiss LA, Raghavachari K, Redfern PC, Pople JA (2000) J Chem Phys 112:7374
35. Woodcock HL III, Schaefer HF III, Schreiner PR (2002) J Phys Chem A 106:11923
36. Dreuw A, Head-Gordon M (2004) J Am Chem Soc 126:4007
37. Wodrich MD, Corminboeuf C, Schleyer P (2006) Org Lett 8:3631
38. Cohen AJ, Mori-Sanchez P, Yang W (2008) Science 321:792
39. Brittain DRB, Lin CY, Gilbert ATB, Izgorodina EI, Gill PMW, Coote ML (2009) Phys Chem Chem Phys 11:1138
40. Gill PMW, O'Neill DP, Besley NA (2003) Theor Chem Acc 109:241
41. Gill PMW, Crittenden DL, O'Neill DP, Besley NA (2006) Phys Chem Chem Phys 8:15
42. Gill PMW (2011) Annu Rep Prog Chem Sect C 107:229
43. Dumont EE, Crittenden DL, Gill PMW (2007) Phys Chem Chem Phys 9:5340
44. Crittenden DL, Dumont EE, Gill PMW (2007) J Chem Phys 127:141103

45. Crittenden DL, Gill PMW (2007) J Chem Phys 127:014101
46. Bernard YA, Crittenden DL, Gill PMW (2008) Phys Chem Chem Phys 10:3447
47. Pearson JK, Crittenden DL, Gill PMW (2009) J Chem Phys 130:164110
48. Hollett JW, Gill PMW (2011) Phys Chem Chem Phys 13:2972
49. Loos PF, Gill PMW (2009) Phys Rev A 79:062517
50. Loos PF, Gill PMW (2009) J Chem Phys 131:241101
51. Loos PF, Gill PMW (2009) Phys Rev Lett 103:123008
52. Loos PF, Gill PMW (2010) Phys Rev A 81:052510
53. Loos PF (2010) Phys Rev A 81:032510
54. Loos PF, Gill PMW (2010) Phys Rev Lett 105:113001
55. Loos PF, Gill PMW (2010) Mol Phys 108:2527
56. Loos PF, Gill PMW (2010) Chem Phys Lett 500:1
57. Loos PF, Gill PMW (2011) J Chem Phys 135:214111
58. Seidl M (2007) Phys Rev A 75:062506
59. http://mathworld.wolfram.com/Glome.html
60. Avery J (1989) Hyperspherical harmonics: applications in quantum theory. Kluwer Academic, Dordrecht
61. Kestner NR, Sinanoglu O (1962) Phys Rev 128:2687
62. Kais S, Herschbach DR, Levine RD (1989) J Chem Phys 91:7791
63. Taut M (1993) Phys Rev A 48:3561
64. Olver, FWJ, Lozier, DW, Boisvert, RF, Clark, CW (eds) (2010) NIST handbook of mathematical functions. Cambridge University Press, New York
65. Hartree DR (1928) Proc Cam Phil Soc 24:89
66. Fock V (1930) Z Phys 61:126
67. Kohn W, Sham LJ (1965) Phys Rev 140:A1133
68. Glasser L, Boersma J (1983) SIAM J Appl Math 43:535
69. Perdew JP, Wang Y (1992) Phys Rev B 45:13244
70. Coulson CA, Neilson AH (1961) Proc Phys Soc (London) 78:831
71. Gori-Giorgi P, Moroni S, Bachelet GB (2004) Phys Rev B 70:115102
72. Kato T (1957) Comm Pure Appl Math 10:151
73. Gill PMW (2001) Aust J Chem 54:661
74. Gill PMW (1994) Adv Quantum Chem 25:141
75. Varganov SA, Gilbert ATB, Deplazes E, Gill PMW (2008) J Chem Phys 128:201104
76. Gill PMW, Gilbert ATB (2009) Chem Phys 356:86
77. Limpanuparb T, Gill PMW (2009) Phys Chem Chem Phys 11:9176
78. Limpanuparb T, Gilbert ATB, Gill PMW (2011) J Chem Theory Comput 7:830
79. Limpanuparb T, Gill PMW (2011) J Chem Theory Comput 7:2353
80. Ghosh S, Gill PMW (2005) J Chem Phys 122:154108
81. Slater JC (1929) Phys Rev 34:1293
82. Møller C, Plesset MS (1934) Phys Rev 46:681
83. Purvis GD, Bartlett RJ (1982) J Chem Phys 76:1910
84. Hollett JW, Gill PMW (2011) J Chem Phys 134:114111
85. Hollett JW, McKemmish LK, Gill PMW (2011) J Chem Phys 134:224103
86. Kutzelnigg W, Klopper W (1991) J Chem Phys 94:1985
87. Nakatsuji H (2004) Phys Rev Lett 93:030403
88. Nakatsuji H, Nakashima H, Kurokawa Y, Ishikawa A (2007) Phys Rev Lett 99:240402
89. Lüchow A, Anderson JB (2000) Annu Rev Phys Chem 51:501
90. Foulkes WMC, Mitas L, Needs RJ, Rajagopal G (2001) Rev Mod Phys 73:33
91. Mitas L (2006) Phys Rev Lett 96:240402
92. Drummond ND, Towler MD, Needs RJ (2004) Phys Rev B 70:235119
93. Drummond ND, Needs RJ, Sorouri A, Foulkes WMC (2008) Phys Rev B 78:125106

Theor Chem Acc (2012) 131:1070
DOI 10.1007/s00214-011-1070-1

REGULAR ARTICLE

# Explicitly correlated wave functions: summary and perspective

**Seiichiro Ten-no**

Received: 14 June 2011 / Accepted: 8 August 2011 / Published online: 13 January 2012
© Springer-Verlag 2012

**Abstract** We summarize explicitly correlated electronic structure theory in perspective of future work in the field. Earlier stages of approaches with different *Ansätze* in physics and chemistry are described. We then discuss recent advances focusing on explicitly correlated wave functions using cusp conditions. Removal of Coulomb singularities in terms of the rational generator is brought out from the viewpoint of many-body perturbation theory. On the basis of decomposition schemes for many-electron integrals in R12 and F12 methods, we further discuss the possibility of increasing the accuracy of molecular numerical integration and massively parallel calculations of explicitly correlated methods.

**Keywords** Cusp conditions · R12 theory · F12 theory · SP *Ansatz* · Slater-type geminals · Many-electron integrals · Resolution of the identity · Auxiliary basis set · Numerical quadratures · Pseudospectral reweighting · Hybrid parallelization

## 1 Introduction

Explicitly correlated wave function treatment is the most strengthened strategy for accurate electronic structure calculations in modern quantum chemistry. The success of ab initio theory is certainly due to the popularity of

Published as part of the special collection of articles celebrating the 50th anniversary of Theoretical Chemistry Accounts/Theoretica Chimica Acta.

S. Ten-no (✉)
Graduate School of System Informatics,
Kobe University, Nada-ku, Kobe 657-8501, Japan
e-mail: tenno@cs.kobe-u.ac.jp

Gaussian-type basis functions, in which efficient algorithms for molecular integrals are readily available. Nevertheless, all standard post Hartree–Fock (HF) calculations in many-body perturbation theory (MBPT), configuration interaction (CI), and coupled–cluster (CC) theory suffer from a slow convergence of electron correlation even with one-electronic basis sets specifically designed for electron correlation [1–7] to minimize the basis set error with respect to the limiting energy for the same wave function model.

The use of extrapolation schemes has successfully filled up the deficiency of basis set convergence. The two-point extrapolation formula of Helgaker et al. [8, 9],

$$E_X = E_\infty + AX^{-3}, \tag{1}$$

for the correlation consistent polarized hierarchy of Dunning, cc-pV$X$Z with the cardinal number $X$, has been applied for accurate molecular equilibrium structures, atomization energies, and reaction enthalpies [10, 11], yet the information about wave function in near basis set limit is missing. Before long, a wide interest in the direct treatment of Coulomb holes arose in the community, and several research groups have been engaged in explicitly correlated electronic structure theory extensively for the last decade. The recent advances with novel ideas crucial for practical applications are largely indebted to the signpost paper of Kutzelnigg [12] published at the halfway point of the 50 year history of the present journal. Many review articles have appeared recently [13–18], and the readers should refer to them for theoretical details.

In the present perspective, we first outline earlier stages of explicitly correlated wave functions in the following section. We shall start from the Hylleraas expansion for the ground state He atom, the uses of possible wave functions forms for explicit electron correlation in transcorrelated,

Reprinted from the journal

⚡ Springer

quantum Monte Carlo, Gaussian-type geminal (GTG), and R12 methods are described. In Sect. 3, we outline explicitly correlated theory using cusp conditions. In terms of the rational generator with a short-range correlation factor as the Slater-type geminal function, we investigate the expansion of explicitly correlated wave functions perturbationally. In Sect. 4, summary and perspective on computational methods for many-electron integrals are presented.

## 2 Earlier stages of explicitly correlated methods

The Hylleraas expansion for the ground state of He [19],

$$\psi = e^{-\zeta s} \sum_{nlm} c_{nlm} s^n t^l u^m, \tag{2}$$

with the coordinates, $s = r_1 + r_2$, $t = -r_1 + r_2$, and $u = r_{12}$, has exhibited that the inclusion of the inter-electronic distance $u$ is significantly important. The use of only three terms 1, $u$, and $t^2$ in $s^n t^l u^m$, is accurate to 0.035 eV [20]. More general expansion of Kinoshita with negative powers of variables $s$ and $u$, which suffices for a formal solution of the wave function, has given extremely accurate energies [21]. Calculations with analogous expansions were performed for the $H_2$ molecule [22, 23].

The exact wave function must obey cusp conditions that prescribe the proper discontinuity behavior at the coalescence point. We shall focus on the solution of the Schrödinger equation of two-electrons letting $r$, $\theta$, and $\varphi$ spherical polar coordinates of the inter-electronic distance vector $\mathbf{r}_{12}$. The wave function around the coalescence point written as

$$\psi = \sum_{l=0}^{\infty} \sum_{m=-l}^{l} r^l f_{lm}(r) Y_l^m(\theta, \varphi), \tag{3}$$

satisfies the cusp conditions [24, 25],

$$f_{lm} = \left[1 + \frac{\lambda r}{2(l+1)} + O(r^2)\right] f_{lm}^{(0)}, \tag{4}$$

where $f_{lm}$ is a function of the center of mass coordinate, $\mathbf{R} = \frac{\mathbf{r}_1 + \mathbf{r}_2}{2}$, $\lambda$ is a strength parameter of the Coulomb interaction in the electron repulsion part, and $f_{lm}^{(0)}$ is the first term of the expansion, $f_{lm} = \sum_{k=0}^{\infty} r^k f_{lm}^{(k)}$. For solutions of a spin-free Hamiltonian, the spatial wave function is either symmetric (singlet) or antisymmetric (triplet) to interchange of the particles, and thus the expansion contains only terms either with even or odd $l$ as [25]

$$\psi^{(S)} = \hat{\psi}_0^{(S)} \left(1 + \frac{\lambda}{2} r\right) + O(r^2), \tag{5}$$

$$\psi^{(T)} = \mathbf{r}_{12} \cdot \mathbf{w}_{12} \left(1 + \frac{\lambda}{4} r\right) + O(r^3), \tag{6}$$

where $\hat{\psi}_0^{(S)} = f_{00}^{(0)} Y_0^0$ and $\mathbf{r}_{12} \cdot \mathbf{w}_{12} = \sum_{l=-1}^{1} f_{1m}^{(0)} Y_1^m r$. (5) is rewritten as

$$\lim_{r \to 0} \frac{\partial \psi^{(S)}}{\partial r} = \frac{\lambda}{2} \hat{\psi}_0^{(S)} (r = 0), \tag{7}$$

whose spherical average is equivalent to the cusp condition of Kato [24]. In either case, it is essential that the cusp behavior assures the regularity of the entire Schrödinger equation at the coalescence point with Coulomb singularity.

Kutzelnigg and Morgan suggested that a further expansion of $\mathbf{R}$ in spherical harmonics leads to unnatural parity states [26], and the alternation rules that account for most of violations of Hund's first rule are associated with the parity cases [27]. The leading term of $f_{00}^{(0)}$ is absent in unnatural parity singlet states, which follow a cusp condition different from (5) and (6). It is also noted that the exact solution at the triple coalescence point is more complicated with logarithmic terms as $\rho^2 ln \rho$ for $\rho = (r_1^2 + r_2^2)^{1/2}$ [28–30], and the importance of such a term for a improved convergence was shown numerically [31].

For correlated $N$-electronic wave functions, the most popular expansion is in the form,

$$\psi = C \sum_k \mathrm{d}_k D_k, \tag{8}$$

where $D_k$ are Slater determinants of spin-orbitals, and $C$ is a product of $n$-electron correlation factors $J_n$,

$$C = \prod_{n=2}^{N} J_n. \tag{9}$$

The simplest choice of this type is the Jastrow–Slater wave function consisting of two-body correlation factors [32] and a single Slater-determinant,

$$\psi = \prod_{i > j} J_2(\mathbf{r}_i, \mathbf{r}_j) D(\phi_1, \phi_2, \ldots, \phi_N). \tag{10}$$

Various function forms have been proposed for $J_n$ using power series of electron–nucleus and electron–electron distances and their scaled variables as $r_{ij}/(a + r_{ij})$. However, the difficulty of calculating the multidimensional integrals arising from $C\hat{H}C$ in a variational treatment has hindered the use of the wave function *Ansatz* in ordinary electronic structure calculations of molecules. The variational Monte Carlo (VMC) techniques permit the use of the correlated wave function (8) by minimizing the energy variance [33].

Boys and Handy [34] introduced a method to determine the variables $C$ and $D$ with the transcorrelated Hamiltonian,

$$\hat{H}_T = C^{-1} \hat{H} C. \tag{11}$$

As $C$ is commutable with any operators in $\hat{H}$ except for the kinetic energy one, only a few additional terms arise from the similarity transformation. For $C$ of only two-body correlation factor $J_2(\mathbf{r}_i, \mathbf{r}_j) = \exp(f_{ij})$, the transcorrelated Hamiltonian $C^{-1}\hat{H}C$ contains extra two-electron operators from the commutators, $[\nabla_i^2, f_{ij}]$ and $\left[[\nabla_i^2, f_{ij}], f_{ij}\right]$, and three-electron operators from $\left[[\nabla_i^2, f_{ij}], f_{ik}\right]$. Besides the purpose to determine $C$ and $D$ (which still requires the computation of awkward three-electron integrals), the use of a similarity transformation has drawn attention to eliminate the Coulomb singularities from the many-electron Hamiltonian to accelerate the convergence of a CI expansion. Hirschfelder [35] showed the explicit form of the similarity transformed Hamiltonian for the spherically symmetric correlation factor,

$$C = \prod_{i > j} \left[ 1 + \frac{1}{2} r_{ij} \exp(-\alpha r_{ij}) \right], \tag{12}$$

and the system of equation involving new operators as

$$-\frac{\mathbf{r}_{ij}}{r_{ij}} \cdot (\nabla_i - \nabla_j), \tag{13}$$

for cusp-less wave functions can be made regular. Similar discussions were made more recently [36–38], and second-order perturbation [39] and approximated coupled cluster methods [40, 41] were developed either with the usual HF reference or biorthogonal one-particle states for the similarity transformed Hamiltonian [39, 40]. It is also noted that the use of the transcorrelated Hamiltonian $\hat{H}_T$ in VMC (TC-VMC) is advantageous permitting the nodal optimization achieved by the self-consistent field calculation for $\hat{H}_T$ [42, 43]. (See also the recent paper of Luo et al. [44] for more applications of $\hat{H}_T$ in quantum Monte Carlo calculations.)

The use of explicitly correlated Gaussian (ECG) functions permits the computation of many-electron integrals inevitable in explicitly correlated calculations [45, 46]. Especially, on the basis of pair correlation theory of Sinanoğlu [47–49], various many-body perturbation theories have been developed using GTG basis [50–57] in the form,

$$\hat{Q}_{12} \exp\left(-\alpha r_{12}^2\right) \phi_i(1) \phi_j(2), \tag{14}$$

where $\hat{Q}_{12}$ is the strong orthogonality (SO) projection operator,

$$\hat{Q}_{12} = \left(1 - \hat{O}_1\right)\left(1 - \hat{O}_2\right), \tag{15}$$

with projectors onto the occupied orbitals,

$$\hat{O}_n = \sum_{i=1}^{\text{occ.}} |\phi_i(n)\rangle\langle\phi_i(n)|, \tag{16}$$

and the orbitals $\phi$ are also expanded into Gaussian-type functions. Several approaches with different functional forms instead of the original use of SO were proposed to eliminate the requirement of high-dimensional objects like four-electron integrals. (For details involving ECG, see the recent review article on GTG [17].) Apparently, it is necessary to take a linear combination of GTG functions to approximate the cusp behavior as the first derivative of GTG is zero at the coalescence.

In the seminal paper of Kutzelnigg [12], partial wave expansion analyses for the $S$-state of He-like atoms were performed using the simple *Ansatz*,

$$\psi = \left(1 + \frac{1}{2} r_{12}\right)\psi_0 + \chi, \tag{17}$$

with a bare-nucleus wave function $\psi_0$ and a component expanded in products of one-particle functions, $\chi$. It was demonstrated that the leading term of the partial wave increment of the correlation energy obtained with a linear combination of determinants alone [58, 59] going as $(l + \frac{1}{2})^{-4}$ is due to the basis set expansion of

$$\frac{1}{2} = \sum_{l=0}^{\infty} \langle \psi_0 | (r_{12}^{-1})_l \left(\frac{1}{2} r_{12}\right)_l | \psi_0 \rangle, \tag{18}$$

and a closed summation of the slowly converging contributions to the partial wave expansion can be accomplished by the introduction of the term $1 + \frac{1}{2} r_{12}$. This notion was realized for general molecules by the introduction of the resolution of the identity (RI) approximations for many-electron integrals [60]. (See the following section for RI approximations, and Sect. 3.3 of the review article [13] for elaborative discussions of RI.) Kutzelnigg and Klopper developed the so-called standard approximation (SA) for intermediates on the basis of the analysis of partial wave expansion in R12 theory [61]. It is noted that the linear $r_{12}$ behavior is not favorable for large distances, and Klopper introduced the unitary invariant *Ansatz* to mitigate this feature [62],

$$|w_{ij}\rangle = \frac{1}{2} \sum_{kl} c_{kl}^{ij} \hat{Q}_{12} r_{12} |\{kl\}\rangle, \tag{19}$$

where $w_{ij}$ is the explicitly correlated part of the pair function for the noninteracting $\{ij\}$. (See Table 1 for definitions of orbitals indices.) The GTG and original R12 methods retain only the diagonal amplitudes $c_{ij}^{ij}$ in (19) for spin-adapted pair functions, and the linear $r_{12}$ *Ansatz* with delocalized orbitals leads to $c_{ij}^{ij} = 0$ in the dissociation limit. This problem can be avoided by introducing the off-diagonal amplitudes $c_{kl}^{ij}$. Furthermore, Noga et al. [63–65] developed coupled–cluster methods with the R12 *Ansatz* (CC-R12), which provides highly accurate results [66] at the same scaling of the standard CC calculations.

**Table 1** Convention of orbital indices

| | |
|---|---|
| $p, q, r, ...$ | Orbitals in the given basis set (GBS) |
| $i, j, k, ...$ | Occupied orbitals |
| $a, b, c, ...$ | Virtual orbitals in GB |
| $\kappa, \lambda, \mu, ...$ | Orbitals in the complete basis set (CBS) |
| $\alpha, \beta, \gamma, ...$ | Virtual orbitals in CBS |
| $\alpha', \beta', \gamma', ...$ | Orthogonal complement of GBS |
| $p', q', r', ...$ | Auxiliary basis set (ABS) |

Corresponding capital indices are used for spatial orbitals

## 3 Explicitly correlated theory using cusp conditions

The linear $r_{12}$ correlation factor is valid in the vicinity of the coalescence point where the cusp conditions hold. However, one can easily confirm that the behavior cannot be applied for distant pairs. The second-order interaction energy of the He dimer with the HF reference in the dissociation limit is

$$U^{(2)} = -4 \sum_{\substack{m \neq 0 \\ n \neq 0}} \frac{|\langle n p_A m p_B | r_{12}^{-1} | 1s_A 1s_B \rangle|^2}{E_m^A + E_n^B - E_{1s}^A - E_{1s}^B}. \tag{20}$$

The standard binomial Taylor expansion of the Coulomb interaction $|\mathbf{R} + \mathbf{a} - \mathbf{b}|^{-1}$ gives the well-known limit of the first-order amplitude,

$$\psi_{n p_A m p_B}^{(1)} = -\frac{\langle n p_A m p_B | r_{12}^{-1} | 1s_A 1s_B \rangle}{E_m^A + E_n^B - E_{1s}^A - E_{1s}^B} \propto R^{-3}, \tag{21}$$

while the expansion of $|\mathbf{R} + \mathbf{a} - \mathbf{b}|$ leads to a different asymptotic behavior,

$$\langle n p_A m p_B | r_{12} | 1s_A 1s_B \rangle \propto R^{-1}. \tag{22}$$

Fortunately, the usual CI expansion is effective for dispersion coefficients because the virtual orbitals only with $\Delta l = \pm 1$ with respect to the occupied orbitals contribute to the $C_6$ coefficients.

There are two ways to eliminate the unphysical long-range behavior for distant pairs. One is to employ extra geminals that cancel the unfavorable contribution as in the unitary invariant *Ansatz* (19) (See also more general GG$n$ ($n = 0, 1, 2$) *Ansätze* [67] including virtual orbitals in pair functions), and the other is to introduce a short-range correlation factor $f_{12}$ in place of $r_{12}$. The introduction of the RI approximation and other decomposition methods for many-electron integrals allows us to employ more efficient correlation factors, as far as the two-electron integrals over $f_{12}, f_{12}/r_{12}, f_{12}^2, [\hat{T}_1, f_{12}]$, and $\frac{1}{2}[[\hat{T}_1, f_{12}], f_{12}]$ can be calculated. Short-range correlation factors were parameterized with a contracted GTG (CGTG),

$$f_{12}^{(CGTG)} = \sum_G^{N_G} c_G \exp(-\zeta_G r_{12}^2). \tag{23}$$

approximating the linear $r_{12}$ behavior at short distances with suitable weight functions [68–72]. It has been well known in the community that the necessary integrals can be calculated efficiently with $F_m(T)$-based algorithms such as McMurchie–Davidson and Obara–Saika ones [73–77] for a correlation factor represented by any linear combination of GTG functions (23). Samson et al. also examined the Gaussian-damped $r_{12}$ factor aiming at low-scaling R12 calculations [78, 79]. The Slater-type geminal (STG) function [80],

$$f_{12}^{(STG)} = -\frac{1}{\gamma} \exp(-\gamma r_{12}), \tag{24}$$

has been exclusively employed in F12 methods. (F12 refers to the choice of non-linear correlation factor instead of R12.) The necessary integrals can be calculated either by approximating $f_{12}^{(STG)}$ in terms of an $n$-component CGTG (STG-$n$G) or using analytic expressions involving the generalized Boys function [80, 81],

$$G_m(T, U) = \int_0^1 t^{2m} \exp[-Tt^2 + U(1 - t^{-2})]dt, \tag{25}$$

in place of $F_m(T)$ for the usual electron repulsion integrals. (Recently, the Dupuis–Rys–King algorithm [82] was extended to calculate integrals over $e^{-\gamma r_{12}}$ and $\frac{e^{-\gamma r_{12}}}{r_{12}}$ for F12 [83].) The efficiency of STG has been confirmed in MP2-F12 [84–87], and CC-F12 methods have been developed within SA [88], beyond SA with the linear approximation CCSD(F12) [89, 90], perturbative inclusion of explicitly correlated terms [91], and using automated code synthesis [92–95]. The short-range correlation factors are relevant to the explicitly correlated local methods [96–100], and F12 basis sets have been developed based on STG as well [101–103].

Once a correlation factor with correct asymptotical behaviors both in the short and long distance limits becomes available, a transparent *Ansatz* of wave function is conceivable. The cusp conditions (3) and (4) through $l = 1$ are expressed in terms of $f_{12} = r_{12} + O(r_{12}^2)$ as

$$\psi = (1 + \hat{\mathcal{R}}_{12})\psi_0, \tag{26}$$

where $\psi_0$ is the unperturbed wave function ($\lambda = 0$), and $\hat{\mathcal{R}}_{12}$ is the rational generator [71],

$$\hat{\mathcal{R}}_{12} = f_{12} \left( \frac{3}{8} + \frac{1}{8} \hat{p}_{12} \right). \tag{27}$$

with a permutation operator $\hat{p}_{12}$ of electronic positions, i.e. $\hat{p}_{12}\psi = \psi$ for singlet, and $= -\psi$ for triplet. It is noted that the usual spin-dependent Jastrow factor $J_n$ without $\hat{p}_{12}$ does not satisfy the $s$- and $p$-wave cusp conditions simultaneously to lead to a spin-contaminated solution. The most important indication of the rational generator in Eq. (26) is the cancelation of the Coulomb singularity in the Schrödinger equation as

$$\left[ -\frac{1}{2}\left(\nabla_1^2 + \nabla_2^2\right)\hat{\mathcal{R}}_{12} + r_{12}^{-1} \right]\psi_0 = O(r_{12}^0)\psi_0, \qquad (28)$$

for regular singlet and triplet pairs of any block of orbital products $\psi_0 = \{ij\}$, $\{i\alpha\}$, and $\{\alpha\beta\}$.

General applications of the rational generator in MBPT can be formulated with the partitioning,

$$\hat{H} = \hat{H}_0 + \hat{V}, \qquad (29)$$

where the zeroth-order Hamiltonian is

$$\hat{H}_0 = \sum_{i=1}^{N}\left[ -\frac{1}{2}\nabla_i^2 + v(\mathbf{r}_i) + \hat{U}_i \right], \qquad (30)$$

with the external field $v(\mathbf{r}_i)$, and the potential $\hat{U}_i$ can be either local or nonlocal as in the Møller–Plesset perturbation theory. Wave operators are obtained by means of the linked diagram theorem [104, 105],

$$\left[ \hat{H}_0, \hat{\Omega}^{(n)} \right] = -\hat{Q}\left\{ \hat{V}\hat{\Omega}^{(n-1)} \right\}_{\text{linked}}, \qquad (31)$$

i.e., the wave operator diagrams of the order $n$ are obtained by operating on the wave operator diagrams of $n-1$ with the perturbation $\hat{V}$ keeping only linked diagrams, where $\hat{Q}$ is the projector onto the space orthogonal to the unperturbed wave function $\hat{Q} = 1 - |\Psi_0\rangle\langle\Psi_0|$. We introduce the *Ansatz* for the wave operator

$$\hat{\Omega}^{(n)} = \hat{\omega}^{(n)} + \hat{\chi}^{(n)}, \qquad (32)$$

which is a sum of the explicitly correlated part in the form,

$$\hat{\omega}^{(n)} = \hat{Q}\left\{ \hat{\mathcal{R}}\hat{\Omega}^{(n-1)} \right\}_{\text{C}}, \qquad (33)$$

and its is residual $\hat{\chi}^{(n)}$, where C denotes connected diagrams. $\hat{\mathcal{R}}$ is the particle block of the rational generator,

$$\hat{\mathcal{R}} = \frac{1}{4}\sum_{\kappa\lambda}\sum_{\alpha\beta}\bar{\mathcal{R}}_{\alpha\beta}^{\kappa\lambda}\left\{ \hat{a}_\alpha^\dagger \hat{a}_\beta^\dagger \hat{a}_\lambda \hat{a}_\kappa \right\} \qquad (34)$$

with $\mathcal{R}_{\alpha\beta}^{\kappa\lambda} = \langle\alpha\beta|\hat{\mathcal{R}}_{12}|\kappa\lambda\rangle$, the overbar stands for anti-symmetrization $\bar{\mathcal{R}}_{\alpha\beta}^{\kappa\lambda} = \mathcal{R}_{\alpha\beta}^{\kappa\lambda} - \mathcal{R}_{\beta\alpha}^{\kappa\lambda}$, and the curly brackets {} denote normal ordering. Substituting $\hat{\Omega}^{(n)}$ of (33) into (31) and using the relation $[\hat{H}_0, \hat{Q}] = 0$, we establish the setoff between the Coulomb singularity in the connected terms of $\hat{Q}\{\hat{V}\hat{\Omega}^{(n-1)}\}_{\text{linked}}$ and $\hat{H}_0\{\hat{\mathcal{R}}\hat{\Omega}^{(n-1)}\}_{\text{C}}$ in each order in an analogous manner to the two-electronic case. (Thus, $\hat{\mathcal{R}}$ is first order in $\hat{V}$.) Starting from the initial condition,

$\hat{\omega}^{(0)} = 0$ and $\hat{\chi}^{(0)} = 1$, we can perform an order-by-order expansion of the wave operator.

The above formalism does not lead to a proper factorization of the wave operator like the absence of $\frac{1}{2}(\hat{\omega}^{(1)})^2$ in $\hat{\Omega}^{(2)}$ with a finite basis expansion of $\hat{\chi}^{(n)}$, albeit the correct energy scaling is guaranteed by the linked diagram expansion. This problem can be easily settled by exponentiating the *Ansatz*,

$$\hat{\Omega} = \exp(\hat{S}) = \exp(\hat{T}' + \hat{T}), \qquad (35)$$

with the explicitly correlated cluster operator as indicated by Köhn [115]

$$\hat{T}' = (\hat{\mathcal{R}}\hat{\Omega})_{\text{C}}, \qquad (36)$$

in which we retain only particle-hole excitation operators. In Fig. 1, we show representative diagrams of $\hat{T}'$. For the first-order wave operator $\hat{\Omega}^{(1)}$, d2:00 is the only term in $\hat{T}'$, and the $s$- and $p$-wave cusp conditions are automatically fulfilled by the rational generator for all pair functions (SP *Ansatz*) [71, 81]. The SP *Ansatz* is sometimes referred to as the diagonal orbital-invariant (DOI) or FIXED amplitude *Ansatz* and has led to efficient perturbation and CC-F12 methods [106–112]. For the ground state around the equilibrium geometry, the contribution of other diagrams associated with $\hat{S}$ would be small with a reasonable size of basis set. d2:20 and d3:20 appear in $\hat{\Omega}^{(2)}$ to contribute to the third- and fourth-order energies, respectively. The explicitly correlated connected triples with d3:20 have been investigated by Köhn for the first time under the name of the extended SP (XSP) *Ansatz* [113], and the geminal contributions are analyzed perturbationally [114]. The diagrams d2:10 and d2:11 will be significant when $\hat{T}_1$ is large as in a non-HF reference. For excitation energies in linear response theory, dominant occupations are induced in the virtual space, and the inclusion of d2:10 is crucial [115, 116]. Alternatively, the cusp behavior in the virtual-occupied block can be dealt with by a partial augmentation of the geminal basis within the unitary invariant *Ansatz* [117].

Another important consequence of the rational generator is on the treatment of open-shell molecules. The permutation operator in (27) causes a spin-flipped geminal basis (SFG),

$$f_{12}\hat{p}_{12}\{I_\uparrow J'_\downarrow\} = f_{12}\{J'_\uparrow I_\downarrow\}. \qquad (37)$$

SFG, which is absent in traditional geminal *Ansatz* as (19), should be included explicitly to satisfy the cusp conditions for different orbitals for different spins. The MP2-F12 method with SFG was implemented for the unrestricted HF (UHF) reference [118], and simplified CCSD(T)-F12x methods were extended to the case of restricted open-shell

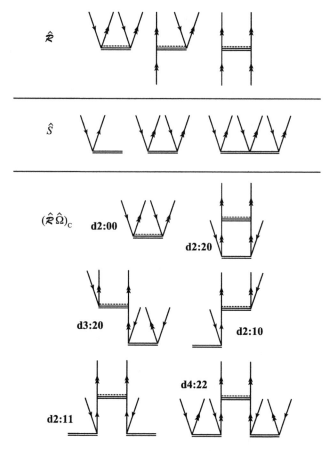

**Fig. 1** Diagrammatic representations of $\hat{\mathcal{R}}$, $\hat{S}$, and representative operators in $(\hat{\mathcal{R}}\Omega)_C$. The diagram for $\hat{T}'_\kappa$ of the contraction $(\hat{\mathcal{R}}\hat{S}_\mu\hat{S}_\nu)_C$ is labeled as d$\kappa$:$\mu\nu$. The *single* and *double up-going arrows* denote particle lines in GBS and CBS, respectively

HF (ROHF) reference functions similarly [119]. More recently, a Z-averaged perturbation theory with the rational generator was studied by Wilke et al. [120], and several *Ansätze* for open-shell molecules have been discussed in detail by Tew et al. [121]. For more general open-shell systems, multireference (MR) treatments are required, and the internally contracted geminal basis $\hat{\mathcal{R}}_{ext}|\text{CAS}\rangle$ was employed in explicitly correlated multireference perturbation theory [122], where $\hat{\mathcal{R}}_{ext}$ is the external excitation part of the rational generator on the complete active space self-consistent field (CASSCF) reference function [123] regarded as a generalized closed-shell vacuum. MR perturbation and MRCI methods with internally contracted basis have been developed [124–127]. The use of the internally contracted basis is more advantageous computationally than the earlier implementation of MRCI-R12 [128, 129], in which the unitary invariant *Ansatz* is applied for each determinantal state in the reference wave function. More recently, explicitly correlated Brillouin–Wigner (BW) MRCC methods [130, 131] with a modified Jeziorski–Monkhorst *Ansatz* [132] have been reported. Although a standard MRCC model has not been established in this field, the use of the similarity transformed Hamiltonian with the rational generator,

$$\hat{H}_R = \exp(-\hat{\mathcal{R}}_{ext})\hat{H}\exp(\hat{\mathcal{R}}_{ext}). \quad (38)$$

on top of MRCC would be promising for the inclusion of F12 effects. Although the problem of redundant variables is known in internally contracted MRCC [133], the use of the SP *Ansatz* bypasses the determination of F12 amplitudes. $\hat{H}_R$ with the rational generator is more advantageous than the transcorrelated Hamiltonian, $\hat{H}_{TC}$ in view of the cusp conditions. In this case, the correlation factor, $f_{12}$, in the rational generator is not necessarily a spherically symmetric function [71]. Even more drastic approximation is the replacement of $\hat{\mathcal{R}}_{ext}$ by

$$\hat{\mathcal{R}}'_{ext} = \frac{1}{4}\sum_{pq}\sum_{\alpha'\beta'}\bar{\mathcal{R}}^{pq}_{\alpha'\beta'}\left\{\hat{a}^\dagger_{\alpha'}\hat{a}^\dagger_{\beta'}\hat{a}_q\hat{a}_p\right\} \quad (39)$$

in $\hat{H}_R$, which gives rise to only the double contraction $\overset{\frown}{\hat{H}\hat{\mathcal{R}}'_{ext}}$ consuming the complementary orbitals $\alpha'$, $\beta'$. This similarity transformed Hamiltonian has been employed to accelerate the basis set convergence in a projector Monte Carlo (PMC) method based on Slater determinants [Ohtsuka (in preparation), 134]. Despite its simple implementation, $\hat{\mathcal{R}}'_{ext}$ resembles *Ansatz* 1 of standard MP2-R12 and CC-R12, which is not as accurate as *Ansatz* 2. This feature was also noted by Torheyden and Valeev in their work on MR R12 method [124].

In Table 2, we compare all-electron correlation energies of Ne from CCSD(F12), CCSD(T)(F12), VMC, and diffusion MC (DMC) calculations. VMC recovery of the correlation energy is 91% despite the inclusion of the four-body three-electron-nucleus (e3-n) Jastrow factor. This relatively poor performance is partially attributed to the deficient form of the Jastrow factor for cusp conditions. The error in DMC from the fixed-node approximation amounts to a few % of the correlation energy. One can obtain highly accurate correlation energies with CC-F12 methods. The CC-F12 energy with connected triples in aug-cc-pCVQZ basis set [5–7] is accurate to 1 $mE_h$, and a similar accuracy can be attained for general polyatomic molecules.

On closing this section, some thoughts to refine the use of cusp conditions in explicitly correlated wave functions would be mentioned. Firstly, an optimum way to determine the length-scale parameter $\gamma$ in STG has not been established, albeit it has been known that the result is less sensitive to $\gamma$ especially for valence electronic states. Within the SP *Ansatz*, $\gamma$ is connected to the second-order coefficients of cusp conditions, and Rassolov–Chipman

**Table 2** All-electron correlation energies of the Ne atom obtained in CC-F12 and quantum Monte Carlo methods

| Method | $E_c$ ($mE_h$) | (%) |
| --- | --- | --- |
| VMC[a] | 355.8 | 91.1 |
| DMC[a] | 377.2 | 96.6 |
| CCSD(F12)[b] | 382.9 | 98.1 |
| CCSD(T)(F12)[b] | 389.2 | 99.7 |
| CCSDT-R12[c] | 390.1 | 99.9 |
| Exact[d] | 390.5 | 100.0 |

[a] QMC results with Jastrow factor through 3e-n, Ref. [135]

[b] F12 calculation with aug-cc-pCVQZ basis.

[c] F12 calculation with aug-cc-pCV5Z basis, Ref. [95]

[d] Estimated exact correlation energy, Ref. [136]

[137] and Tew [138] investigated the behavior of the coalescence conditions. In either case, energy dependences enter the second-order coefficients. More recently, MP2-F12 and CCSD(T)-F12 methods with orbital-pair-specific Slater-type geminals have been proposed by Werner et al. [139]. Different geminal exponents are used for different correlations of core–core, core–valence and valence–valence orbital pairs for improved correlation energies and molecular properties for a number of diatomic molecules involving $d$-shell correlation. Secondly, the F12 methods have exclusively employed spherically symmetric correlation factors. Nevertheless, the logarithmic behavior at the triple-point coalescence point [28–30] indicates the possibility of a partial introduction of position dependence in $f_{12}$ for an improved accuracy. Investigation will be also made for unnatural parity singlet pairs [26, 27] whose rate of convergence is much faster than the natural pairs. Relativistic effects are of importance as well since the partial wave increment of the leading term of the relativistic correction goes as $(l + \frac{1}{2})^{-2}$; the slow rate can be also removed by the F12 *Ansatz* [140].

## 4 Computational aspects for many-electron integrals

The penalty in explicitly correlated theory is the introduction of many-electron integrals. Efficient treatment of the many-electron integrals is the key for wide applicability of theory. The geminal-based methods require three-electron integrals over $f_{12} r_{13}^{-1}$, $f_{12} r_{23}^{-1} f_{13}$, and $[\nabla_1^2 + \nabla_2^2, f_{12}] f_{13}$, and four-electron integrals over the operators as $r_{12}^{-1} f_{13} f_{14}$, and $r_{12}^{-1} f_{23} f_{34}$. All many-electron integrals can be expressed in closed form algebraic expressions using the Gaussian correlation factor [50, 51]. The three- and four-electron integrals are so numerous that the key to any practical method lies in the efficient handling of the integrals. The weak-orthogonality

projector was invented to avoid four-electron integrals in the GTG theory. The use of the transcorrelated Hamiltonian also requires only three-electron integrals over $(\nabla_1 f_{12}) \cdot (\nabla_1 f_{13})$. Recurrence relations have been developed for efficient evaluation of three-electron integrals [38].

RI introduced in the pioneering work of Kutzelnigg and Klopper [60, 61] approximates the primary three-electron integrals as

$$\langle ijk|f_{12}r_{13}^{-1}|lmn\rangle \simeq \sum_{p'} F_{ij}^{p'm} G_{p'k}^{ln}, \tag{40}$$

$$F_{ij}^{p'm} = \langle ij|f_{12}|p'm\rangle, \tag{41}$$

$$G_{p'k}^{ln} = \langle p'k|r_{12}^{-1}|ln\rangle. \tag{42}$$

In the atomic case, the expansion is saturated at the RI basis $p'$ up to the angular momentum quantum number $l = 3l_{\text{occ.}}$, because each of the two-electron integrals $F_{ij}^{p'm}$ and $G_{p'k}^{ln}$ vanishes beyond the range for the spherically symmetric operators $f_{12}$ and $r_{12}^{-1}$. Although the expansion of the cyclic integrals,

$$\langle ijk|f_{12}r_{23}^{-1}f_{13}|lmn\rangle \simeq \sum_{p'q'r'} F_{ij}^{p'q'} G_{q'k}^{mr'} F_{p'r'}^{ln}, \tag{43}$$

does not terminate due to the two RI indices in each 4-index object, the partial wave increment of the integrals goes as $(l + \frac{1}{2})^{-6}$ which is regarded as "an acceptable rate of convergence" [12]. As a result, slowly convergent two-electron objects as $\langle ij|f_{12}r_{12}^{-1}|kl\rangle$ are calculated analytically, and all many-electron integrals are approximated by RI in R12/F12 theories. The RI approximation is also employed for the three-electron integrals in the transcorrelated method [37], and it was shown that the commutator form is more accurate than the simple application of RI to the vector product $(\nabla_1 f_{12}) \cdot (\nabla_1 f_{13})$ [41]. The standard approximation (SA) for various correlated methods in earlier R12 theory employs the same basis set for orbital expansion and RI. More flexible expansions using an auxiliary basis set (ABS) were introduced by Klopper and Samson in 2002 [141], and Valeev proposed the use of a union of the given basis for orbitals and its orthogonal component, complementary ABS (CABS) for RI to improve the accuracy of ABS [142]. The ABS and CABS approaches enabled us to use a compact basis set in R12/F12 methods, and various methods beyond SA have been developed for the last decade.

The computational cost for the requirement of a large basis set in ABS and CABS approaches has been reduced by the use of density fitting (DF). DF expands an orbital product into a fitting basis,

$$\rho_{pq}(\mathbf{r}) = \phi_p(\mathbf{r})\phi_q(\mathbf{r})$$
$$\simeq \tilde{\rho}_{pq}(\mathbf{r}) = \sum_A c_{A,pq} \Xi_A(\mathbf{r}), \tag{44}$$

and has provided fast self-consistent field (SCF) and correlated methods [143–149]. (DF is often referred to as RI, but we use the terminology of DF to avoid the confusion with the RI in R12/F12 methods.) Manby introduced DF in explicitly correlated theory [150, 72]. All two-electron integrals are replaced by those of the robust formula, the error in which is quadratic in the errors in the fitted densities [151], and efficient methods have been developed in conjunction with local approximations [96–100]. DF was also employed to increase the accuracy of the RI approximation [152]. "A momentum transfer" by grouping the orbitals at the coordinate joining operators in three-electron integrals before the application of RI,

$$\langle ijk|f_{12}r_{13}^{-1}|lmn\rangle \simeq \langle \tilde{\rho}_{il}jk|f_{12}r_{13}^{-1}|1mn\rangle$$
$$\simeq \sum_{p'A} c_{A,il} F_{Aj}^{p'm} G_{p'k}^{1n}, \qquad (45)$$

reduces the requirement of the RI basis from $3l_{\mathrm{occ.}}$ to $2l_{\mathrm{occ.}}$, where 1 denotes unity instead of an orbital. This reduction will be significant for molecules with heavy elements especially of occupied $f$-shells. Further development would be possible to eliminate the remaining four-index integrals as $F_{Aj}^{p'm}$ using DF with the robust formula.

The central idea of breaking up many-electron integrals in the contemporary explicitly correlated electronic structure methods was actually suggested about 20 years earlier than the emergence of RI. Boys and Handy pointed out that a couple of 3-D integrals in the special kind of 9-D three-electron integrals of the transcorrelated Hamiltonian,

$$\int F(1,2)G(1,3)\mathrm{d}\mathbf{r}_1\mathrm{d}\mathbf{r}_2\mathrm{d}\mathbf{r}_3 \qquad (46)$$

can be evaluated without interference, and the final 3-D integration can be performed on the results of them [34]. Moreover, efficient 3-D numerical quadratures (QD) have been developed for exchange-correlation contribution in density functional theory [153–157].

QD introduced in F12 theory [71, 81] decomposes the three-electron integrals

$$\langle ijk|f_{12}r_{13}^{-1}|lmn\rangle \simeq \sum_{g} \bar{\phi}_i(\mathbf{r}_g)\phi_l(\mathbf{r}_g)F_{jm}^{g}G_{kn}^{g}, \qquad (47)$$

with only three-index objects,

$$F_{jm}^{g} = \int \mathrm{d}\mathbf{r}\phi_j(\mathbf{r})f_{12}\phi_m(\mathbf{r}), \qquad (48)$$

$$G_{kn}^{g} = \int \mathrm{d}\mathbf{r}\phi_k(\mathbf{r})r_{12}^{-1}\phi_n(\mathbf{r}), \qquad (49)$$

and orbital amplitudes, $\bar{\phi}_i(\mathbf{r}_g)$ and $\bar{\phi}_i(\mathbf{r}_g) = \phi_i(\mathbf{r}_g)w_g$ with the quadrature weights $w_g$. "The momentum transfer" as in DF holds in (47) from the beginning since the orbital

amplitudes are decoupled at a grid point. Thus, the numerical integration is saturated at $2l_{\mathrm{occ.}}$ of the angular grid in the atomic case. The four-electron and cyclic three-electron integrals can be treated by the hybrid QD/RI approximation [81] analogous to the ABS/RI approach of Klopper [158]. The Coulomb interaction in those many-electron integrals is arising from the exchange operator in the Fockian as,

$$\langle ijk|\hat{K}_1 f_{12}f_{13}|lmn\rangle \simeq \sum_{p'g} \langle i|\hat{K}_1|p'\rangle \bar{\phi}_{p'}(\mathbf{r}_g)\phi_l(\mathbf{r}_g)F_{jm}^{g}F_{kn}^{g}. \qquad (50)$$

The summation over $p'$ can be replaced by the given basis set (GBS) since the RI expansion should converge much faster than $l = 3l_{\mathrm{occ.}}$ as indicated from the rare gas case. In this case, the four-electron integrals in the special form can be calculated explicitly without the hybrid approximation,

$$\langle ijk|\hat{K}_1 f_{12}f_{13}|lmn\rangle \simeq \sum_{g} K_i(\mathbf{r}_g)\bar{\phi}_l(\mathbf{r}_g)F_{jm}^{g}F_{kn}^{g}, \qquad (51)$$

in terms of the exchange operator in the physical space,

$$K_i(\mathbf{r}_g) = \sum_{j} \bar{\phi}_j(\mathbf{r}_g)G_{ij}^{g}. \qquad (52)$$

The cyclic integrals are canceled out by other terms within the hybrid QD/RI approximation, otherwise the RI basis can be replaced by ABS. One can calculate even two-electron integrals using QD optionally, and the reduction of the scaling of intermediates from quartic to cubic is effective for large-scale correlated calculations. The best MP2 energy of Benzene in the near basis set limit, $-1.0575(5)$ $E_h$, was obtained by performing MP2-F12 calculations with this method through aug-cc-pV6Z basis set in a fractional computational time of the SCF calculation [159].

We explore possibilities of further improved methods for many-electron integrals. Is it possible to mitigate the numerical integration using ABS? We introduce the overlap metric over ABS,

$$s_{p'q'} = \sum_{g} \phi_{p'}(\mathbf{r}_g)w_g\phi_{q'}(\mathbf{r}_g). \qquad (53)$$

The matrix $\mathbf{s}$ is positively definite for the usual numerical quadratures and is an identity matrix in the limit of the exact numerical integration. Using the symmetric orthogonalization matrix for $\mathbf{s}$,

$$\mathbf{x} = \mathbf{s}^{-\frac{1}{2}} = \mathbf{U}\mathbf{d}^{-\frac{1}{2}}\mathbf{U}^{+}, \qquad (54)$$

we define the renormalized orbital amplitudes

$$R_{gp} = \sum_{p'q'} \phi_{p'}(\mathbf{r}_g)x_{p'q'}\langle q'|p\rangle, \qquad (55)$$

where **d** is the diagonal matrix with the eigen values of the metric **s**, and $\langle q' | p \rangle = \delta_{q'p}$ for ABS consisting of GB and its orthogonal components (CABS). The numerical integration with the renormalized quadrature (rQD) is given by

$$\langle p|\hat{A}|q\rangle \simeq \sum_g \bar{R}_{gp}\hat{A}(\mathbf{r}_g)R_{gq}, \qquad (56)$$

with $\bar{R}_{gp} = R_{gp}w_g$. One can easily conform that rQD integrates ABS exactly for $\hat{A}(\mathbf{r}_g) = 1$.

As a preliminary application, we show errors in numerical MP2 energies of water approximating electron repulsion integrals $(ai|bj)$ using QD and rQD in Fig. 2. The cc-pVXZ and aug-cc-pVXZ basis sets [5–7] labeled by the cardinal numbers as X and aX for GB are used, and ABS are formed from the GB and the uncontracted cc-pVQZ basis set as CABS. The numbers of quadrature points are 1,440 (72 angular and 20 radial) and 9,216 (268 angular and 32 radial) per atom for the coarse and medium grids, respectively. In either case, the rQD errors are one order of magnitude smaller than the corresponding ones of QD. rQD with the medium grid is accurate to 1 $\mu E_h$. The use of rQD in F12 methods is straightforward. The calculation of derivatives of ABS amplitudes required for the B-intermediate can be bypassed by the use of the approximation C [160]. The additional computational costs for the application of rQD to F12 methods are negligibly small since the most time-consuming steps involving three-index objects coincide with the original QD method. The detail of the use of rQD in F12 theory will be discussed in more detail elsewhere. It is noted that rQD is closely related with the pseudospectral (PS) methods [161–165], which employ more parametrized grids and dealiasing functions. PS typically uses only ca. 100 grid points with a fair accuracy. This fact indicates that there is a possibility of further improvement of the performance of rQD using integral corrections and length scales algorithms employed in PS [163].

More recently, a message passing interface (MPI) and open multi-processing (OpenMP) hybrid parallel algorithm for MP2 calculations using QD have been developed [166]. High parallel efficiency is attained by distributing QD points and virtual molecular orbital pairs. The communication of the three-index objects and the assembly of electron repulsion integrals of them scale as $OVG$ and $O^2V^2G$, for the numbers of occupied MOs, virtual MOs, and quadrature points, $O$, $V$, and $G$, respectively, and the parallel efficiency is increased for large molecules. The CPU timing of all-electron correlated calculations for coronene with the aug-cc-pCVTZ basis set is shown in Fig. 3. Calculations were performed on SGI Altix ICE 8400EX (CPU: Intel Xeon X5570 2.93GH, 8 CPU cores/node, Memory: 24 GB/node, Network: Enhanced Hypercube with 4 × QDR InfiniBand) at the Supercomputer Center, Institute for Solid State Physics, University of Tokyo with eight threads per process generated in hybrid parallel calculations. Another calculation of fullerene ($C_{60}$) with the same basis set (3,540 basis functions) was also completed in ca. 4.8 h on 8,192 cores without invoking molecular symmetry. The hybrid parallel algorithm can be applied to QD-based F12 methods, and massively parallel algorithms involving higher-order correlation will be presented in the near future.

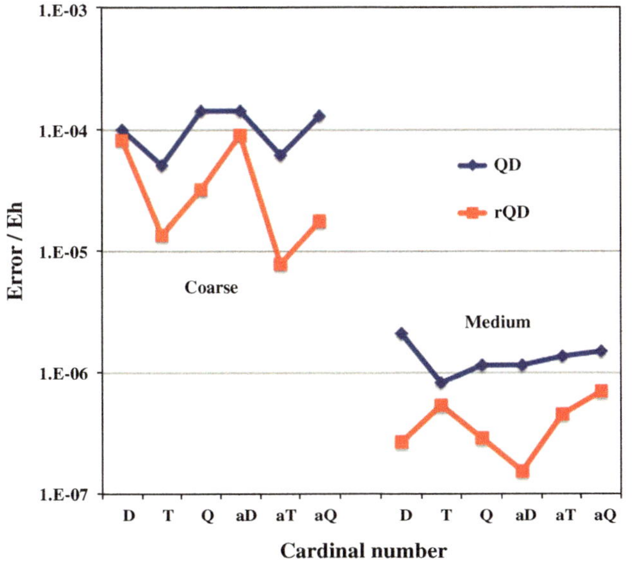

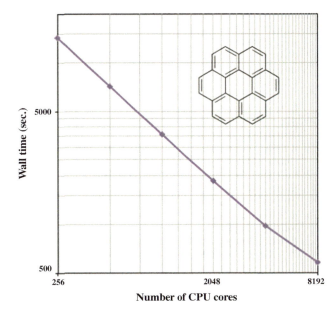

**Fig. 2** Errors in numerical MP2 energoes of QD and rQD methods for the water molecule ($E_h$)

**Fig. 3** Parallel efficiency of numerical MP2 calculation of coronene. All electrons are correlated, and the number of grid points per atom is 9,216. Point group symmetries are not employed

## 5 Summary

In this overview, we have summarized milestones in the history of explicitly correlated wave functions since the Hylleraas expansion for the He ground state in 1929. The latest methods utilize short-range correlation factor, mostly the Slater-type geminal function, which enables us to use the rational generator for cusp conditions to eliminate the electron–electron Coulomb singularity from the entire Schrödinger equation to grant improved basis set convergence. There remain issues to hone the present strategy on the inclusions of higher-order cusp conditions, triple-point coalescence conditions, and relativistic effects, which will be investigated extensively before long. At the same time, efficient computational schemes to realize accurate calculations of explicitly correlated methods rest with the treatment of many-electron integrals. This aspect has been also discussed with a perspective.

**Acknowledgments** This work is partly supported by the Grant-in-Aids for Scientific Research (B) (No. 00270471) from the Japan Society for the Promotion of Science (JSPS), and the 25th Grant-in-Aid of Tokyo Ohka Foundation the promotion of Science and Technology.

## References

1. Almlöf J, Taylor PR (1987) J Chem Phys 86:4070–4077
2. Widmark P-O, Malmqvist P-Å, Roos BO (1990) Theor Chim Acta 77:291–306
3. Widmark P-O, Persson BJ, Roos BO (1991) Theor Chim Acta 79:419–432
4. Pou-Amerigo R, Merchan M, Nebot-Gil I, Widmark P-O, Roos BO (1995) Theor Chim Acta 92:149–181
5. Dunning TH Jr. (1989) J Chem Phys 90:1007–1023
6. Kendall RA, Dunning TH Jr., Harrison RJ (1992) J Chem Phys 96:6796–6806
7. Woon DE, Dunning TH Jr. (1995) J Chem Phys 103:4572–4585
8. Helgaker T, Klopper W, Koch H, Noga J (1997) J Chem Phys 106:9639
9. Halkier A, Helgaker T, Jørgensen P, Klopper W, Koch H, Olsen J, Wilson AK (1998) Chem Phys Lett 286:243–252
10. Bak KL, Jørgensen P, Olsen J, Helgaker T, Klopper W (2000) J Chem Phys 112:9229–9242
11. Bak KL, Gauss J, Jørgensen P, Olsen J, Helgaker T, Stanton JF (2001) J Chem Phys 114:6548–6556
12. Kutzelnigg W (1985) Theor Chim Acta 68:445–469
13. Klopper W, Manby FR, Ten-no S, Valeev EF (2006) Int Rev Phys Chem 25:427–468
14. Helgaker T, Klopper W, Tew DP (2008) Mol Phys 106:2107–2143
15. Tew DP, Hättig C, Bachorz RA, Klopper W, (2010) In: Čársky P, Paldus J, Pittner J (eds), Recent progress in coupled cluster methods. Springer, Berlin pp 535–572
16. Werner H-J, Adler TB, Knizia G, Manby FR (2010) In: Čársky P, Paldus J, Pittner J (eds), Recent progress in coupled cluster methods. Springer, Berlin pp 573–620
17. Szalewicz K (2010) Mol Phys 108:3091–3103
18. Ten-no S, Noga J. Wiley Interdisciplinary Reviews: Computational Molecular Science, in press
19. Hylleraas EA (1929) Z Phys. 54:347–366
20. Hylleraas EA (1964) Adv Quantum Chem 1:1–33
21. Kinoshita T (1957) Phys Rev 105:1490–1502
22. James HM, Coolidge AS (1933) J Chem Phys 1:825
23. Kołos W, Wolniewicz L (1968) J Chem Phys 49:404
24. Kato T (1957) Commun Pure Appl Math 10:151–177
25. Pack RT, Brown WB (1966) J Chem Phys 45:556–559
26. Kutzelnigg W, Morgan W (1992) J Chem Phys 96:4484–4508
27. Moragan W, Kutzelnigg W (1992) J Phys Chem 97:2425–2434
28. Bartlett JH (1937) Phys Rev 51:661–669
29. Fock VA (1954) Izv Akad Nauk SSSR, Ser. Fiz 18:161
30. Fock VA (1958) D Kngl Norske Videnskab Selsk Forh 31:138
31. Frankowski K, Pekeris CL (1966) Phys Rev 146:46
32. Jastrow R (1955) Phys Rev 98:1479–1484
33. Umrigar CJ, Wilson KG, Wilkins JW (1988) Phys Rev Lett 60:1719–1722
34. Boys SF, Handy NC (1969) Proc R Soc A 310:43–61
35. Hirschfelder JO (1963) J Chem Phys 39:3145–3146
36. Nooijen M, Bartlett RJ (1998) J Chem Phys 109:8232–8240
37. Ten-no S (2000) Chem Phys Lett 330:169–174
38. Ten-no S (2000) Chem Phys Lett 330:175–179
39. Hino O, Tanimura Y, Ten-no S (2001) J Chem Phys 115:7865–7871
40. Hino O, Tanimura Y, Ten-no S (2002) Chem Phys Lett 353:317–323
41. Ten-no S, Hino O (2002) Int J Mol Sci 3:459–474
42. Umezawa N, Tsuneyuki S (2003) J Chem Phys 119:10015
43. Umezawa N, Tsuneyuki S (2004) J Chem Phys 121:7070
44. Luo HJ, Hackbusch W, Frad HJ (2010) Mol Phys 108:425
45. Boys SF (1960) Proc Roy Soc London Ser A 258:402
46. Singer K (1960) Proc Roy Soc London Ser A 258:412
47. Sinanoğlu O (1961) Phys Rev 122:493
47. Sinanoğlu O (1962) J Chem Phys 36:3198
49. Sinanoğlu O (1964) Adv Chem Phys 6:315
50. Lester WA, Krauss M (1965) J Chem Phys 41:1407
51. Lester WA, Krauss M (1965) J Chem Phys 42:2990
52. Pan K-C, King HF (1970) J Chem Phys 53:4397–4399
53. Szalewicz K, Jeziorski B, Monkhorst HJ, Zabolitsky JG (1983) J Chem Phys 78:1420
54. Szalewicz K, Jeziorski B, Monkhorst HJ, Zabolitsky JG (1983) J Chem Phys 79:5543
55. Jeziorski B, Monkhorst HJ, Szalewicz K, Zabolitzky JG (1984) J Chem Phys 81:368
56. Szalewicz K, Jeziorski B, Monkhorst HJ, Zabolitsky JG (1984) J Chem Phys 81:2723
57. Wenzel KB, Zabolitzky JG, Szalewicz K, Jeziorski B, Monkhorst HJ (1986) J Chem Phys 85:3964
58. Schwartz C (1962) Phys Rev 126:1015
59. Hill RN (1985) J Chem Phys 83:1173
60. Klopper W, Kutzelnigg W (1987) Chem Phys Lett 134:17–22
61. Kutzelnigg W, Klopper W (1991) J Chem Phys 94:1985–2001
62. Klopper W (1991) Chem Phys Lett 186:583–585
63. Noga J, Kutzelnigg W, Klopper W (1992) Chem Phys Lett 199:497–504
64. Noga J, Kutzelnigg W (1994) J Chem Phys 101:7738–7762
65. Noga J, Kutzelnigg W, Klopper W (1997) In: R. J. Bartlett (ed), Recent advances in computational chemistry, vol 3. World Scientific, Singapore, pp 1–48
66. Noga J, Valiron P, Klopper W (2001) J Chem Phys 115:2022
67. Dahle P, Helgaker T, Jonsson D, Taylor PR (2007) Phys Chem Chem Phys 9:3112–3126
68. Persson BJ, Taylor PR (1996) J Chem Phys 105:5915–5926
69. Persson BJ, Taylor PR (1997) Theor Chem Acc 97:240–250
70. Ten-no S (2003) Lect Notes Comput Sci 2660:152–158
71. Ten-no S (2004) J Chem Phys 121:117–129
72. May AJ, Manby FR (2004) J Chem Phys 121:4479–4485

73. McMurchie LE, Davidson ER (1978) J Comput Phys 26:218–231
74. Obara S, Saika A (1986) J Chem Phys 84:3963–3974
75. Obara S, Saika A (1988) J Chem Phys 89:1540–1559
76. Head-Gordon M, Pople JA (1988) J Chem Phys 89:5777
77. Ahlrichs R (2006) Phys Chem Chem Phys 8:3072–3077
78. Samson CCM, Klopper W, Helgaker T (2002) Comput Phys Commun 149:1–10
79. Samson CCM (2004) Highly accurate treatment of dynamical electron correlation through R12 methods and extrapolation techniques (PhD thesis, University of Utrecht)
80. Ten-no S (2004) Chem Phys Lett 398:56–61
81. Ten-no S (2007) J Chem Phys 126:014108
82. Dupuis M, Rys J, King HF (1976) J Chem Phys 65:111–116
83. Shiozaki T (2009) Chem Phys Lett 479:160–164
84. May AJ, Valeev E, Polly R, Manby FR (2005) Phys Chem Chem Phys 7:2710–2713
85. Tew DP, Klopper W (2005) J Chem Phys 123:074101
86. Valeev EF (2006) J Chem Phys 125:244106
87. Werner H-J, Adler TB, Manby FR (2007) J Chem Phys 126:164102
88. Noga J, Kedžuch S, Šimunek J, Ten-no S (2008) J Chem Phys 128:174103; Erratum: (2009) J Chem Phys 130:029901
89. Fliegl H, Hättig C, Klopper W (2006) J Chem Phys 124:044112
90. Tew DP, Klopper W, Neiss C, Hättig C (2007) Phys Chem Chem Phys 9:1921–1930
91. Valeev EF (2008) Phys Chem Chem Phys 10:106
92. Shiozaki T, Kamiya M, Hirata S, Valeev EF (2008) Phys Chem Chem Phys 10:3358
93. Shiozaki T, Kamiya M, Hirata S, Valeev EF (2008) J Chem Phys 129:071101
94. Köhn A, Richings GW, Tew DP (2008) J Chem Phys 129:201103
95. Shiozaki T, Kamiya M, Hirata S, Valeev EF (2009) J Chem Phys 130:054101
96. Werner H-J, Manby FR (2006) J Chem Phys 124:054114
97. Manby FR, Werner H-J, Adler TB, May AJ (2006) J Chem Phys 124:094103
98. Werner HJ (2008) J Chem Phys 129:101103
99. Adler TB, Werner H-J, Manby FR (2009) J Chem Phys 130:054106
100. Adler TB, Werner H-J (2009) J Chem Phys 130:241101
101. Peterson KA, Adler TB, Werner H-J (2008) J Chem Phys 128:084102
102. Yousaf KE, Peterson KA (2008) J Chem Phys 129:184108
103. Yousaf KE, Peterson KA (2009) Chem Phys Lett 476:303
104. Brandow BH (1967) Rev Mod Phys 39:771
105. Lindgren I, Morrison M (1986) Atomic many-body theory, 2nd edn. Springer, Berlin
106. Tew DP, Klopper W, Hättig C (2008) Chem Phys Lett 452:326
107. Bokhan D, Ten-no S, Noga J (2008) Phys Chem Chem Phys 10:3320–3326
108. Torheyden M, Valeev EF (2008) Phys Chem Chem Phys 10:3410
109. Alder TB, Knizia G, Werner H-J (2007) J Chem Phys 127:221106
110. Bokhan D, Bernadotte S, Ten-no S (2009) Chem Phys Lett 469:214
111. Hättig C, Tew DP, Köhn A (2010) J Chem Phys 132:231102
112. Köhn A, Tew DP (2010) J Chem Phys 133:174117
113. Köhn A (2009) J Chem Phys 130:131101
114. Köhn A (2010) J Chem Phys 133:174118
115. Köhn A (2009) J Chem Phys 130:104104
116. Bokhan D, Ten-no S (2010) unpublished
117. Neiss C, Hättig C, Klopper W (2006) J Chem Phys 125:064111
118. Bokhan D, Bernadotte S, Ten-no S (2009) J. Chem Phys 131:084105
119. Knizia G, Adler TB, Werner H-J (2009) J Chem Phys 130:054104
120. Wilke JJ, Schaefer HF (2009) J Chem Phys 131:244116
121. Tew DP, Klopper W (2010) Mol Phys 108:315–325
122. Ten-no S (2007) Chem Phys Lett 447:175–179
123. Roos BO, Linse P, Siegbahn PEM, Blomberg MRA (1981) Chem Phys 66:197
124. Torheyden M, Valeev EF (2009) J Chem Phys131:171103
125. Shiozaki T, Werner H-J (2010) J Chem Phys 133:141103
126. Shiozaki T, Knizia G, Werner H-J (2011) J Chem Phys 134:034113
127. Shiozaki T, Werner H-J (2011) J Chem Phys 134:184104
128. Gdanitz RJ (1993) Chem Phys Lett 210:253
129. Gdanitz RJ (1998) Chem Phys Lett 283:253
130. Kedžuch S, Demel O, Pittner J, Noga J (2010) In: Čársky P, Paldus J, Pittner J (eds), Recent progress in coupled cluster methods. Springer, Berlin, pp 251–266
131. Kedžuch S, Demel O, Pittner J, Ten-no S, Noga J (2011) Chem Phys Lett 511:418–423
132. Jeziorski B, Monkhorst HJ (1981) Phys Rev A 24:1668–1681
133. Banerjee A, Simons J (2981) Int J Quantum Chem 19:207–216
134. Ohtsuka Y, Nagase S (2010) Chem Phys Lett 485:367–370
135. Huang C-J, Umrigar CJ, Nightingale MP (1997) J Chem Phys 107:3007–3013
136. Chakravorty SJ, Gwaltney SR, Davidson ER (1993) Phys Rev A 47:3649–3670
137. Rassolov VA, Chipman DM (1996) J Chem Phys 104:9908–9912
138. Tew DP (2008) J Chem Phys 129:014104
139. Werner H-J, Knizia G, Manby FR (2011) Mol Phys 109:407–417
140. Kutzelnigg W (2008) Intern J Quantum Chem 108:2280–2290
141. Klopper W, Samson CCM (2002) J Chem Phys 116:6397–6410
142. Valeev EF (2004) Chem Phys Lett 395:190–195
143. Alsenoy CV (1988) J Comput Chem 9:620
144. Vahtras O, Almlof J, Feyreisen MW (1993) Chem Phys Lett 213:514
145. Feyreisen M, Fitzgerald G, Komornicki A (1993) Chem Phys Lett 208:359
146. Bernholdt DE, Harrison RJ (1996) Chem Phys Lett 250:477
147. Ten-no S, Iwata S (1995) Chem Phys Lett 240:578
148. Ten-no S, Iwata S (1996) J Chem Phys 105:3604
149. Rendell AP, Lee TJ (1994) J Chem Phys 101:400
150. Manby FR (2003) J Chem Phys 119:4607–4613
151. Dunlap BI (2000) Phys Chem Chem Phys 2:2113
152. Ten-no S, Manby FR (2003) J Chem Phys 119:5358–5363
153. Becke AD (1987) J Chem Phys 88:2547
154. Murray CW, Handy NC, Laming GJ (1993) Mol Phys 78:997
155. Gill PMW, Johnson BG, Pople JA (1993) Chem Phys Lett 209:506
156. Treutler O, Ahlrichs R (1995) J Chem Phys 102:346
157. Krack M, Köster AM (1988) J Chem Phys 108:3226
158. Klopper W (2004) J Chem Phys 120:10890
159. Yamaki D, Koch H, Ten-no S (2007) J Chem Phys 127:144104
160. Kedžuch S, Milko M, Noga J (2005) Int J Quantum Chem 105:929–436
161. Friesner RA (1985) Chem Phys Lett 116:39
162. Friesner RA (1988) J Chem Phys 92:3091
163. Murphy RB, Friesner RA, Ringnalda MN, Goddard WA III (1994) J Chem Phys 101:2986
164. Martinez TJ, Mehta A, Carter EA (1992) J Chem Phys 97:1876
165. Martinez TJ, Carter EA (1994) J Chem Phys 100:3631
166. Ishimura K, Ten-no S (2011) Theor Chem Acc 130:317–321

Theor Chem Acc (2012) 131:1086
DOI 10.1007/s00214-011-1086-6

REGULAR ARTICLE

# Theoretical chemistry: current applications to photochemistry and thermochemistry

**Fernando R. Ornellas**

Received: 7 August 2011 / Accepted: 11 August 2011 / Published online: 11 January 2012
© Springer-Verlag 2012

**Abstract** A historical perspective is given contrasting challenges and advances in theoretical chemistry at the time the first issue of Theoretical Chemistry Accounts appeared in 1962 and the progress achieved since then as expressed in current state-of-the-art applications in photochemistry and thermochemistry.

**Keywords** Quantum chemistry and photochemistry · Quantum chemistry and thermochemistry · Applied quantum chemistry: future perspectives

The fertile change in paradigm brought forth by the quantum mechanical study of atoms and molecules provided many of the new tools and concepts, and it also presented a continued challenge to scientists to express the chemical concepts in the more mathematical language being developed. The early application of quantum mechanics to the ground state of He by Hylleraas [1] and to the ground state of $H_2$ by James and Coolidge, [2] although bringing some excitement about the accuracy one could obtain for these systems, has led to some frustration, since it was also soon realized that the explicit incorporation of the interelectronic distance in calculations of larger systems was not readily applicable.

Published as part of the special collection of articles celebrating the 50th anniversary of Theoretical Chemistry Accounts/Theoretica Chimica Acta.

F. R. Ornellas (✉)
Departamento de Química Fundamental, Instituto de Química,
Universidade de São Paulo, Av. Lineu Prestes, 748, São Paulo,
São Paulo 05508-000, Brazil
e-mail: frornell@usp.br

In the meantime, since these earlier calculations appeared until 1962, when the first issue of Theoretica Chimica Acta was published, methodological advances like the Hartree–Fock and configuration interaction (CI) methods had already been made. Perturbational approaches like the one developed by Møller and Plesset, [3] although published as early as 1934, was largely unexploited until the work of Pople, Binkley and Seeger [4] later in 1976, and a general and straightforward application of multiconfigurational self-consistent field wavefunctions would only appear in the literature a decade later in the work of Das and Wahl [5].

Ab initio calculations were mostly restricted to atoms and diatomic molecules, since the evaluation of multicenter integrals over Slater functions was a major problem at the time, until the Gaussian-type functions introduced by Boys in 1950 replaced the Slater-type functions in calculations [6]. The contribution of Roothaan [7] where the molecular orbitals of the Hartree–Fock formalism were expressed in terms of a finite set of analytic functions leading to a rigorous matrix representation of the Hartree–Fock equation was almost simultaneous with the appearance of the first automatic computing machines in the academic environment. This combination of more easily implemented theoretical models and automatic computation seemed to indicate a brighter future for quantum chemistry. The attainment of the Hartree–Fock limit, thus, became a goal for some schools.

By 1960, a feeling of some of the hot issues in question can be appreciated in the collected papers from the Conference on Molecular Quantum Mechanics held at the University of Colorado the previous year [8]. It is interesting to recall that in his banquet speech summarizing the then present state of molecular structure calculations presented in this conference, Coulson [9], besides expressing

Reprinted from the journal

Ⓓ Springer

his opinion "that the whole group of theoretical chemists is on the point of splitting into two parts … almost alien to each other", now known as the ab initio and semi-empirical theoretical chemists, also prophesized that "… the speeding up of calculations, and the design of even faster machines should enable us to extend the range of effectively exact solutions. I am inclined to think that the range of 6–20 electrons belongs to this picture." Even considering the potential role that the large-scale use of electronic computers would have in the future, he again reemphasized that "It looks as if somewhere around 20 electrons there is an upper limit to the size of a molecule for which accurate calculations are ever likely to become practicable." A few years later, the investigation of Pitzer and Lipscomb [10] on the barrier to internal rotational in ethane, performed at the self-consistent Hartree–Fock level of theory with a minimal basis represented a major step in obtaining reasonable accuracy and physical insights on the origin of the barrier. As part of Pitzer's Ph.D. thesis, this project involved the development of an SCF program from scratch especially designed for this molecule [11]. Also, an early example of the predictive role that quantum chemical calculations would play in spectroscopy in the years to come was related to the structures and energy splitting of the singlet and triplet states in $CH_2$ [12]. For a critical review of the apparent conflict between theory and experiment, resolved in favor of quantum chemistry, the reader is referred to the survey of Shavitt [13]; a more recent account of studies on this system is given by Neugebauer and Häfelinger [14].

In this short and necessarily incomplete overview of some of the progress theoretical/computational chemistry had attained by the time the first issue of TCA came to be published, I am trying to convey to the younger generation of practitioners in this area that for a full appreciation of the progress we have achieved so far in the application of the methods of electronic structure, as presented in the articles that follow, they would benefit very much by getting acquainted with some of the landmark papers in the area. Fortunately for them, in the short monograph by Schaefer, a period of about 50 years since the very beginning of quantum chemistry has been covered in which his personal selection of 149 papers guides the reader to what he expresses as part of a necessarily "vigorous program of theoretical education" [15]. In this context, the young practitioner might be surprised to learn that in the 1970–1980s special installations were required to house huge mainframes with a computing power and disk storage much smaller than his or her desktop computer. This limitation of hardware required ingenuity to implement the various theoretical methods of electronic structure calculations, of which the determination of the eigenvalues of very large Hamiltonian matrices, the direct approach to the

SCF-HF and CI methods, the transformation of atomic to molecular integrals, and the calculation of gradients of the energy with respect to the nuclear coordinates are a few examples. The new generation of vectorial computers appearing in the early 1980s brought new excitement to the quantum chemistry community, and since then the steep increase in computing power opened new frontiers to explore. Concomitantly, the progress made in the development of methods to solve the electronic Schrödinger equation and the availability of easy-to-use software packages made computational quantum chemistry a necessary field of study in every chemistry department.

In this section of the 50th anniversary issue of *TCA*, two contributions showing state-of-the-art applications of quantum chemistry to present day research make evident the enormous progress we have achieved in the last 50 years. The first paper by Lischka and collaborators [16] brings us to a complex, beautiful, and exciting universe never dreamed by Coulson in his 1959 speech. How to describe the significant physico-chemical changes that follow photoexcitation of molecular systems, what is the basic physics underlying these phenomena, and what are the quantum chemistry tools that can help us to reliably describe the excited state phenomenology are topics that only very recently could be successfully approached and that are treated in a state-of-the-art way by the authors. A survey of current possibilities and future challenges include topics related to nucleobases and nucleic acid fragments, model systems of retinal (the chromophore of rhodopsins), the fields of photovoltaics, photodevices, phototriggers, and molecular devices. For a realization of the immense advances made so far, one should contrast these subjects with the early problems of the internal rotation in $C_2H_6$, and the structure and energy splitting in $CH_2$, for example. Directions of as yet to be explored new territories are also presented as a challenge for future generations.

How much have we progressed in attaining increasingly accurate results since the first issue of Theoretica Chimica Acta came into press? The focus of the contribution of Peterson, Feller, and Dixon [17] on "chemical accuracy" leads us to an analysis and discussion of the intrinsic accuracy of highly correlated wavefunction-based electronic structure methods as applied in ab initio thermochemistry and spectroscopy. What to expect of increasing levels correlation treatment coupled with extensions of the 1-particle basis sets to the complete basis set limit in the characterization of molecular systems? How apparently small effects like scalar relativistic, spin–orbit (oute)core correlation effects, and zero point energies affect thermochemical properties like atomization energies and vibrational frequencies? What is an acceptable transition boundary between tractable and untractable systems for pushing the attainment of chemical and spectroscopic accuracy to their limits? How can we investigate these latter systems? And how "composite approaches" come to our rescue? By answering these questions, the authors conduct the readers through the state-of-the-art in this field in an authoritative way. Interesting to see in this arena, as in the old days, is the resurgence of the interelectronic coordinate ($r_{12}$) in explicitly correlated methods as an alternative approach to improve convergence, thus reducing the requirements of very large basis sets. Challenges to be overcome in the near future are finally presented for the accuracy seekers.

As a final remark in this introductory section, one should note that implicit in the calculations surveyed in these two contributions is the motto "Getting the Right Answer for the Right Reason" that should serve as a guide to all practitioners in the area.

## References

1. Hylleraas EA (1928) Z Physik 48:469
2. James HM, Coolidge AS (1933) J Chem Phys 1:825
3. Møller C, Plesset MS (1934) Phys Rev 46:618
4. Pople JA, Binkley JS, Seeger R (1976) Int J Quantum Chem Symp 10:1
5. Das G, Wahl AC (1976) J Chem Phys 56:1769
6. Boys SF (1950) Proc Roy Soc Lond A200:542
7. Roothaan CCJ (1951) Rev Mod Phys 23:69
8. Parr RG (1960) Rev Mod Phys 32:169
9. Coulson CA (1960) Rev Mod Phys 32:170
10. Pitzer RM, Lipscomb WN (1963) J Chem Phys 39:1995
11. Pitzer RM (2011) J Chem Theory Comput 7:2346
12. Jordan PCH, Longuet-Higgins HC (1962) Mol Phys 5:121
13. Shavitt I (1985) Tetrahedron 41:1531
14. Neugebauer A, Häfelinger G (2005) Int J Mol Sci 6:157
15. Schaefer HF (1984) Quantum chemistry—the development of ab initio methods in molecular electronic structure theory. Oxford, New York
16. Plasser F, Barbatti, M, Aquino AJA, Lischka H (2012) Theor Chem Acc 131:1073
17. Peterson KA, Feller D, Dixon DA (2012) Theor Chem Acc 131:1079

REGULAR ARTICLE

# Electronically excited states and photodynamics: a continuing challenge

Felix Plasser · Mario Barbatti · Adélia J. A. Aquino · Hans Lischka

Received: 12 July 2011 / Accepted: 1 August 2011 / Published online: 11 January 2012
© Springer-Verlag 2012

**Abstract** The purpose of this contribution is the description of the progress in theoretical investigations on electronically excited states in connection with photodynamical simulations made within the last years and to provide an outlook on the scope of future applications and challenges. An overview over excited-state phenomenology is given and the applicability of different computational methods is discussed. Both electronic structure- and dynamics methods are considered. The examples presented comprise the explanation of the photostability of individual DNA nucleobases, the photodynamics of DNA including excitonic and charge-transfer processes, the primary processes of vision and the broad field of photovoltaics, photodevices, and molecular machines.

**Keywords** Electronic structure · Excited states · Photodynamics · Nonadiabatic phenomena

---

Published as part of the special collection of articles celebrating the 50th anniversary of Theoretical Chemistry Accounts/Theoretica Chimica Acta.

F. Plasser · A. J. A. Aquino · H. Lischka (✉)
Institute for Theoretical Chemistry, University of Vienna,
Waehringerstrasse 17, 1090 Vienna, Austria
e-mail: hans.lischka@univie.ac.at

M. Barbatti (✉)
Max-Planck-Institut für Kohlenforschung,
Kaiser-Wilhelm-Platz 1, 45470 Mülheim an der Ruhr, Germany
e-mail: barbatti@kofo.mpg.de

A. J. A. Aquino · H. Lischka
Department of Chemistry and Biochemistry,
Texas Tech University, Lubbock, TX 79409-1061, USA

A. J. A. Aquino
Institute of Soil Research, University of Natural Resources and Applied Life Sciences Vienna, Peter-Jordan-Straße 82, 1190 Vienna, Austria

## 1 Introduction

Photoinduced phenomena in molecules play an important role in many scientific and technological fields. In biological sciences, they are related to photoaging and photodamage [1–3], to vision and light detection [4, 5], to photosynthesis and light harvesting [6–8]. In technology, they are central for photocatalysis [9, 10], photovoltaics [11, 12], imaging [13, 14], photodevices, [15, 16], conventional [17, 18] and time-resolved spectroscopy [19, 20]. It would, certainly, be unrealistic to attempt examining such huge variety of fields in one review. Instead, we intend to provide an account of recent theoretical investigations, which, although restricted to a more modest variety of topics, will still illustrate a broad range of recent achievements, as well as current limitations of quantum chemical investigations of molecular excited states.

Special focus will be laid on the interactions between chromophores in π-systems, which have recently attracted substantial interest for computational studies because of many challenging questions still to be answered. Note that as a consequence of progress in computer hardware and quantum chemical algorithms, high-level quantum chemical treatment of many of the examples discussed below has become accessible only in the last years. A large number of systems of biological and technological interest exist, where interactions between chromophores play an important role. In particular, the photophysics and charge migration dynamics of DNA fragments have attracted widespread interest and many open questions remain [2, 21]. Exciton dynamics in photosynthetic complexes and other aggregates are studied to understand how the precise arrangement of the chromophores affects light harvesting efficiency and what is the role of quantum coherence [7, 22, 23]. In particular, the technique of two-dimensional optical spectroscopy provides fascinating signatures of the quantum nature of these processes but often poses questions that cannot be answered without modeling [7, 23]. Excitonic and charge dynamics is fundamental in organic electronics. Not only migration properties of excitons and charges are of highest interest, but especially the charge separation and charge recombination steps have a crucial influence on the efficiency of photovoltaic and electroluminescent devices [24]. Another widely used application that is based on the specific properties of excitation energy transfer (EET) is the analytical technique of fluorescence resonance energy transfer, which is used to obtain time-dependent structural information on macromolecules [25].

Several physical phenomena play a role in photoexcited molecules and molecular aggregates. Electronic excitations correspond to significant changes in the electronic structure, which may in turn lead to ultrafast non-equilibrium phenomena. For a successful modeling of excited states, it is important to understand the basic physics of different classes of excited states, as well as the available computational methods and their advantages and problems with respect to the different questions to be asked. We will start in Sect. 2 by outlining excited-state phenomenology, where aside from excited states of single molecules special focus will be laid on changes that occur due to interactions between chromophores. In Sect. 3, a number of quantum chemical methods that can be used for excited states and their advantages and limitations will be discussed. Possibilities for considering dynamical phenomena and couplings between electronic and nuclear degrees of freedom will be outlined in Sect. 4. An overview of possibilities for computing excited states in dimers and aggregates will be presented in Sect. 5. Finally, Sect. 6 will feature a number of examples selected from biology and technology and special attention will be given to the applicability of different methods for these classes of systems.

## 2 Excited-state phenomenology

### 2.1 Photoinduced phenomena in molecules

Molecules can be electronically excited by irradiation at UV or visible wavelengths. The electronic excitation promotes the molecule to a non-equilibrium state, which triggers a sequence of relaxation processes. As illustrated in Fig. 1, these processes may correspond to simple vibrational relaxation on a single potential energy surface, or may involve radiationless crossings to other adiabatic states with the same spin multiplicity (internal conversion) or different spin multiplicities (intersystem crossing); alternatively, they may involve radiative transitions to other states by fluorescent or phosphorescent processes.

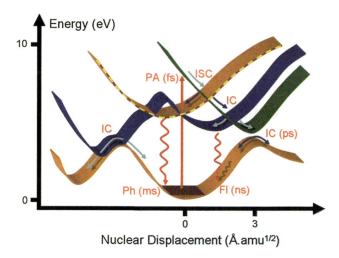

**Fig. 1** Illustration of photodynamical processes occurring on ground and excited potential energy *curves* of a molecule. *PA* photoabsorption, *ISC* Intersystem crossing, *IC* internal conversion, *Fl* Fluorescence, *Ph* Phosphorescence

In any case, between the fastest processes—vibrational relaxation taking place within few tens of femtoseconds (fs), and slowest processes—phosphorescence occurring within milliseconds (ms), a span in time of more than ten orders of magnitude is found. This large variability of processes and time scales imposes, naturally, a great challenge to theoreticians, who should be prepared to employ many different approaches and methods, tailored for each special case.

Theoretical investigations of molecular excited states usually start with the determination of the vertical excitation spectrum for the ground-state minimum geometry and continue with the determination of geometries and energies of excited-state stationary structures, of coupling terms for state crossings, and of reaction pathways connecting all those structures. In comparison with ground-state research, excited-state investigations impose a new level of challenges due to the high density of closely lying states possessing different delocalization character (localized valence states, delocalized excitons, separated charge-transfer states, diffuse Rydberg states). Moreover, rather than having nearly harmonic wells separated by high energy barriers—as usually found in the ground state—excited-state surfaces are often anharmonic, with multiple wells separated by low energy barriers, allowing the molecule to reach geometrical conformations far from chemical intuition.

The development of experimental femtosecond spectroscopic techniques [26, 27] has been a driving force pushing theoretical research in excited states beyond simple assignment of vertically excited spectra. The need for theoretical models helping to deconvolute time-dependent spectra has continually motivated the research of excited-state reaction pathways and excited-state dynamics simulations. In this context, internal conversion processes have been a main field of theoretical research in excited states. After photoexcitation, the molecule can reach regions of the configuration space where the Born–Oppenheimer approximation breaks down (see Fig. 2). In particular, degeneracy between states of the same spin multiplicity, also known as conical intersections [28, 29], creates an efficient funnel for radiationless transfer between adiabatic states. Seams of conical intersections are ubiquitous [30]; the main question to be answered is whether a specific molecule excited at a determined wavelength can reach the intersection seam or not.

Algorithms for localization of conical intersection have been developed [31–33], making their search comparable to the conventional search for stationary points. They are reviewed and benchmarked in Ref. [34]. The research of internal conversion pathways performed for a large variety of molecules has shown that the structures at conical intersections often keep close structural resemblance even between very different molecules. For instance, two examples of $S_1/S_0$ conical intersections are shown for two distinct molecules in Fig. 3. The substructure that causes the degeneracy between the two states is a twist-pyramidalized configuration (indicated by lines in bold) in both cases. The twist has the effect of stabilizing the $\pi\pi^*$ state and destabilizing the closed shell (cs) state, bringing them to an avoided crossing. Pyramidalization ($sp^2$ to $sp^3$ hybridization change) tunes the degeneracy [35]. Five of the most common motifs giving rise to $S_1/S_0$ conical

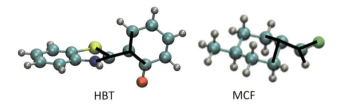

**Fig. 3** Examples of twisted pyramidalized (or ethylenic) $S_1/S_0$ conical intersections in 2-(2′-hydroxyphenyl)-benzothiazole (HBT) [36] and 4-methylcyclohexylidene-fluoromethane (MCF) [37]. The *lines* in **bold** indicate the origin of the "primitive conical intersection"

intersections in organic molecules are collected and illustrated in Table 1.

In spite of the large variety of types of processes observed in molecular photodynamics, these motifs for conical intersections indicate that there are common patterns followed by molecules. For instance, Fig. 4 shows the state occupation during dynamics of pyrrole [38], adenine [39], and pyridone [40] according to surface hopping simulations. These state occupations belong to three distinct classes of excited-state pathways commonly observed in dynamics of organic molecules. For pyrrole, a very fast (sub-100 fs) dynamics occurs following the NH dissociation along the $\pi\sigma^*$ state. For adenine, ring puckering toward an ethylenic conical intersection along the $\pi\pi^*$ state controls the dynamics. For pyridone, the conical intersection cannot be reached efficiently, turning this molecule into a fluorescent species.

### 2.2 Charge transfer and excitonic interactions

If two or more chromophores interact with each other, new intriguing phenomena come into play. Strong interactions may significantly alter the character of the excited states by delocalizing them among several fragments. Additionally, defect migration plays a role even at much weaker coupling strengths and energy transfer occurs up to a spatial separation of several nanometers between chromophores [25]. Electronic defects in aggregated or bulk systems are commonly described in terms of excess electrons and holes relative to the neutral ground state. If the electron and the hole are in close contact, possibly delocalized over a few fragments, the term Frenkel exciton is commonly used. Typically, only Frenkel excitons are active in absorption and emission processes. Later, they may split into charge-transfer states, i.e., an electron and a hole on separate fragments bound by a strong mutual Coulomb interaction. Finally, complete charge separation can occur where the electron and hole migrate independently. The complete description of these phenomena is quite challenging as it is not only necessary to describe the electronic structure at

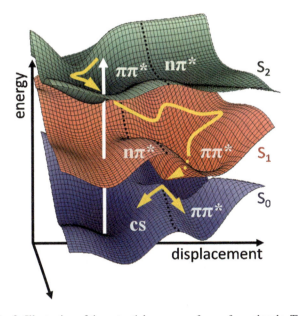

**Fig. 2** Illustration of the potential energy surfaces of a molecule. The *arrows* indicate the photoabsorption (*vertical*) and the relaxation processes. Each adiabatic surface ($S_0$, $S_1$, $S_2$) can have different diabatic characters (cs (*closed shell*), $n\pi^*$, $\pi\pi^*$) depending on the nuclear geometry

**Table 1** Common motifs or "primitive conical intersections" giving rise to conical intersections between the first excited and the ground singlet states of organic molecules

| Conical intersection | Primitive structure | Examples |
| --- | --- | --- |
| Twisted; Twisted-pyramidalized ($\pi\pi^*$/cs) | $R_1R_2X-CR_3R_4$ | Conjugated chains (ethylene [189], polar substituted ethylenes [190], protonated Schiff bases [191], stilbene [192], azobenzene [193]) Aromatic rings (benzene [194], purines [195], pyrimidines [196]) |
| H-migration/carbene ($\pi\pi^*$/cs) | $C:$ with $H$, $R_1$, $R_2$, $R_3$ | Ethylidene [197] Cyclohexene [198] |
| Out-of-plane O ($\pi_O\pi^*$/cs) | $R_1R_2C-O$ | Formamide [199] Rings with carbonyl groups (pyridone [40], pyrimidines [200], guanine [201]) |
| Bond breaking; ring opening ($\pi\sigma_{XY}^*$/cs) | $R_1R_2X\text{-----}Y$ | Heteroaromatic rings (pyrrole [202, 203], purines [204], thiophene [205], furan [206], imidazole [207]) |
| Proton transfer ($\pi\pi^*$/cs) | $X\text{-----}\rightarrow H$ with $R_1$, $R_2$ | Watson–Crick base pairs [118] |

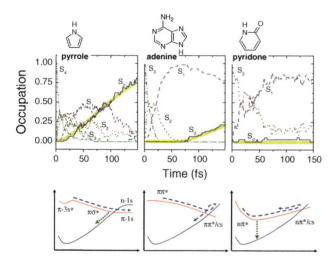

**Fig. 4** State occupation during dynamics of pyrrole [38], adenine [39], and pyridone [40] and corresponding main reaction pathways. The highlighted *curve* indicates the evolution of the ground-state population

high level, but it is also essential to include the coupling between electronic and nuclear degrees of freedom. The computational description can either occur by finding the parameters for global theories or by a direct dynamical description.

In Fig. 5 the model shape of a potential energy surface (PES) of a defect located on two interacting fragments is shown, cf. Ref. [41]. The diabatic states $H_{ii}^{\text{diab}}$ and $H_{ff}^{\text{diab}}$ where the defect is localized on the *initial* or *final* fragments are represented by displaced parabolas. These are characterized by the reorganization energy $\lambda$, the energy required to move from one minimum geometry to the other while staying on the same diabatic surface. These diabatic curves are modulated in this example by a constant coupling $H_{if}^{\text{diab}}$ to yield the adiabatic energy curves $E_1^{\text{ad}}$ and $E_2^{\text{ad}}$. If the two fragments are different, then a non-vanishing reaction free energy $\Delta_r G$ for the defect transfer reaction will be observed. If the coupling is on the order of the reorganization energy, then the resulting adiabatic states are strongly mixed and a unique delocalized minimum is formed, Fig. 5a. If the coupling is significantly smaller, then the adiabatic states remain very similar in character to that of the localized diabatic states and only at the transition region an interaction between the states occurs, Fig. 5b. The essence of Marcus theory is to combine $H_{if}^{\text{diab}}$, $\lambda$, and $\Delta_r G$ to form a rate equation including non-adiabatic effects for weak coupling cases [42]. Förster theory allows connecting spectroscopically available quantities to form a rate equation of excitation energy transfer [43].

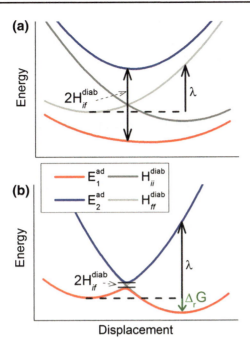

**Fig. 5** Model potential energy curves for a defect delocalized over two fragments in the (**a**) strong coupling and (**b**) weak coupling cases

## 3 Quantum chemical method for excited states

### 3.1 Ab initio methods

Many ab initio methods are currently available for exited-state calculations. Some single-reference methods, such as the second-order approximate coupled-cluster (CC2) [44], algebraic diagrammatic construction (ADC) [45], and the equation-of-motion coupled-cluster (EOM-CC) [46] methods, have proven to be reliable and affordable sources for vertical excitation energy calculations, with data quality considerably superior to former common approximations such as the configuration interaction with single excitations (CIS) [47]. It should be noted that combination of CC2 and ADC(2) with the resolution-of-the-identity (RI) approximation [48, 49] has significantly increased the applicability of these methods (see below).

At strongly distorted molecular geometries or conical intersections with the ground state, the Hartree–Fock method does not provide a satisfactory reference wave function. In these cases, the usage of multiconfigurational and multireference methods is necessary, and the multi-configurational self-consistent field (MCSCF) [50, 51], the multireference configuration interaction (MRCI) [52–54], or the complete active space perturbation theory to second-order (CASPT2) [55] can be applied. Analytical gradients are available for all these methods [56–58]. Analytical nonadiabatic couplings terms are available for MCSCF and MRCI methods [59].

**Table 2** Recent reviews and benchmark investigations on excited-state calculations

| Description | Refs. |
| --- | --- |
| Benchmark of excited states: ab initio methods | [208–211] |
| Benchmark of excited states: DFT-based methods | [82, 212–217] |
| Benchmark of excited states: semiempirical methods | [88] |
| Benchmark of conical intersections | [34] |
| Review on quantum chemical methods | [83] |
| Review on the QM/MM approach | [218, 219] |
| Review on surface hopping | [100, 220] |
| Review on excited-state phenomena | [157, 221–223] |

Besides the high computational costs common to all these methods, another handicap affecting especially the MCSCF and the MRCI methods is their dependence on the choice of active and reference spaces [60]. Of particular concern is that larger active spaces, which are often used when a large number of orbitals is to be considered, start to include dynamic electron correlation effects for some of the electrons, which causes an imbalance by leaving the other electrons uncorrelated [61]. Additionally, convergence problems may increase computation times, and certain orbital rotations may appear for specific geometries changing the character of the MCSCF wavefunction that can lead to discontinuities in potential energy surfaces. MCSCF and MRCI often overestimate the energy of ionic states in the Franck–Condon region [62, 63]. CASPT2, on its turn, tends to underestimate the vertical spectrum [64, 65]. A collection of references to investigations benchmarking results for excited-state calculations at ab initio levels are given in Table 2.

### 3.2 DFT-based methods

Currently, excited-state properties can be obtained with several density functional (DFT) methods. Besides the popular time-dependent (TD) DFT [66, 67], usually computed within the linear-response approximation, other DFT-based methods, such as the restricted open-shell Kohn–Sham (ROKS) [68], restricted ensemble-reference Kohn–Sham (REKS) [69], time-dependent density functional tight-binding (TD-DFTB) [70], and DFT/MRCI [71, 72], can provide excitation energies and other properties. Among these methods, analytic energy gradients and analytic and numeric nonadiabatic couplings terms are available for TDDFT [73–78].

The result of a TDDFT calculation critically depends on the functional used. Because of its superior quality in ground-state calculations, the hybrid B3-LYP [79] functional is often used as a starting point. In a similar sense, the generalized gradient approximation functional PBE [80] is often applied. PBE0 [81] is a parameter free hybrid extension of PBE, which was formulated with a special focus on molecular properties and excited states. Highly parameterized functionals like Truhlar's M06 series [82] give remarkable results in many cases, but have the disadvantage that the physical understanding of the functional form becomes more difficult. In practical applications, usually several functionals would be considered and computed results compared to reference values (obtained, e.g., from experiment or a high-level ab initio method) thus allowing to find an appropriate functional. A difficulty for applying TDDFT is the large number of available functionals, from which it is not trivial to find the best one for each specific case. Additionally, TDDFT lacks the possibility of systematically improving results, which is one of the major characteristics of ab initio methods.

One of the main problems with DFT-based methods has been the description of charge-transfer states [83], which can be strongly overstabilized by conventional functionals. This problem, however, has been largely attenuated by the development of asymptotically corrected functionals [82, 84, 85], but more experience may still be needed to properly evaluate these approaches. In the case of TDDFT, the single-reference character and the linear-response approximation also turn the computation of potential energy surfaces near conical intersections between the ground and the excited states problematic, if not impossible. These intersections can be described by ROKS (within a simple two-state approximation), REKS, and DFT/MRCI methods.

A collection of references to works benchmarking results for excited-state calculations at DFT-based levels are given in Table 2.

## 3.3 Semiempirical-based methods

As for ground-state problems, one promising option to overcome the limitations imposed by computational costs is the use of semiempirical methods. Semiempirical-based methods have been also developed for excited-state calculations. In particular, multireference problems have received special attention with the development of semiempirical configuration interaction methods employing floating-occupation of molecular orbitals (FOMO/CI) [86] and MRCI based on the graphic unitary group approach (GUGA) [87].

Results of mixed quality have been reported from the application of these methods. Good results, for instance, can be obtained for vertical excitations with the OM2/MRCI method [88]. The excited-state relaxation dynamics, however, has been often in contradiction with results obtained at ab initio levels (see, for instance, the discussion about cytosine in Sect. 6.1). Transferable, large-scale

parametrization over the whole periodic table and including excited-state stationary structures and intersections is still necessary to make these methods fully reliable for excited-state calculations.

## 4 Methods for photodynamical simulations

### 4.1 Wavepacket dynamics

The full time-dependent Schrödinger equation for a molecule can be partitioned between self-consistent sets of equations for the electronic and nuclear systems [28, 89]. Then, the nuclear wavepacket of an electronically excited molecule can be propagated by grid-projection techniques [90, 91]. The problem with such methods is that only a limited small number of nuclear coordinates can be included in the calculations. Even though the situation is improved by time-dependent wavefunction expansions, like in the multiconfigurational time-dependent Hartree (MCTDH) method [89], computational costs and difficulties involved in building multidimensional potential energy and coupling surfaces still limit wavepacket propagation to small subsets of nuclear coordinates.

In spite of such limitations, excited-state wavepacket dynamics is an important tool for obtaining highly accurate spectroscopic results [89, 92] or verifying the quality of predictions obtained with trajectory-based methods [93, 94].

### 4.2 Trajectory-based approaches

Trajectory-based approaches for nonadiabatic dynamics simulations have been popularized in the last decade. They overcome the main problem of nuclear wavepacket propagation—the limited dimensionality due to computational costs and to difficulties of building multidimensional potential energy surfaces—by adopting a local approximation. Within the local approximation, energies, energy gradients, and coupling terms need to be computed only along a classical trajectory, rather than over the full space as the time-dependent Schrödinger equation requests. Delocalization of the nuclear wavepacket is partially recovered by the propagation of multiple independent trajectories. Nonadiabatic behavior is recovered by different approaches. Two of the most well-tested trajectory-based methods are multiple spawning [13, 95] and surface hopping [96].

While multiple spawning propagates Gaussian wavepackets centered at classical trajectories and can spawn new trajectories at nonadiabatic crossing regions, surface hopping propagates classical trajectories on the energy surface of a single state, allowing each independent trajectory to switch between states during the propagation according to a stochastic algorithm. The multiple spawning technique has the advantage of including quantum effects for the nuclear motion more rigorously. Surface hopping is widely used because of its methodological simplicity and ease of interpretation. In many cases, multiple spawning and surface hopping should lead to similar results at similar computational costs if a proper integration of the time-dependent Schrödinger equation and some additional consistency corrections [97] are performed in the surface hopping dynamics. A recent review about the multiple spawning method can be found in Ref. [98]. Surface hopping dynamics has been reviewed in Refs. [99, 100] and a general program implementation is described in Ref. [101].

## 5 Computational considerations for dimers and aggregates

A significant amount of attention has been devoted to computing the electronic coupling between chromophores as the central descriptive quantity for the defect transfer process, cf. Fig. 5. There are two general approaches for computing these couplings: direct supermolecular computations and computations of interaction matrix elements between molecules considered independently. In the supermolecular approach, the coupling is obtained as half the energy splitting at resonance conditions [102]. Resonance can either be enforced by symmetry considerations, cf. Refs. [103, 104], by scanning of geometries, or by applying an external electric field [105]. It has been pointed out that simply calculating the gap without checking for resonance may significantly overestimate the coupling [103]. Alternatively, localized states in supermolecular structures may be found by property-based diabatization schemes [102, 105]. In a third approach, the parameters are obtained by considering repeat units [106]. If the two chromophores are considered separately, the task is to compute the interaction matrix element between the two excited states. If there is no overlap between donor and acceptor wave functions, exchange can be neglected and the coupling is given as the Coulomb interaction between the transition densities [107]. This electrostatic interaction can be computed according to the transition density cube method [108] or using analytical Coulomb integrals [107]. For larger separations between the chromophores it is often enough to consider the interaction of transition dipoles, which is the basis for Förster theory. If charge migration rather than energy migration properties are to be computed in a fragment-based approach, the coupling can be estimated as the matrix element between the highest occupied molecular orbitals of the fragments [103, 109]. If the couplings and other descriptive quantities are available, they can provide the background for global theories or

multi-scale techniques to describe larger systems [22, 41, 110, 111].

In many cases, it is advantageous to go beyond the simple rate equations provided by Marcus and Förster theory. This was done by adapting the existing theories through inclusion of vibronic terms [41] or by explicitly considering molecular aggregates [111]. But atomistic ab initio non-adiabatic dynamics simulations, as described above, should give a direct unbiased view of the processes occurring and can naturally be applied to intermediate coupling situations where Marcus and Förster theory cannot be applied. An examination of the applicability of surface hopping dynamics for defect transport and its connection to Marcus theory is given in Ref. [104]; the underlying physics are outlined in Sect. 2 of Ref. [112]. The atomistic picture allows for an inclusion of environmental effects through QM/MM coupling schemes. In particular, electrostatic embedding (see Ref. [113] and further references therein) allows coupling the electronic polarization of the core system directly to the orientational polarization of the solvent. The strong influence of such an environmental polarization on the charge-transfer properties of DNA have been discussed in Ref. [109].

## 6 Applications showing current possibilities and future aspects

### 6.1 Nucleobases

A very active field of theoretical research in excited states has been the investigation of the behavior of nucleobases [1, 114, 115] and other nucleic-acids fragments [116–118] after UV irradiation. The main motivation has been to understand how this radiation can damage the genetic code and what intrinsic protection mechanisms DNA has developed against it. Experimental research has revealed that all nucleobases can efficiently get rid of the photoenergy at the picosecond time scale [119, 120]. Theoretical research of reaction pathways and conical intersections of nucleobases has identified the main internal conversion channels available for each nucleobase [1, 121–124]. Dynamics simulations have provided information about the efficiency of each of those reaction pathways [20, 114, 115, 125, 126].

Although the reaction pathways described by most of theoretical methods are qualitatively equivalent, results of the dynamics simulations can be quite dependent on the specific properties of the respective potential energy surfaces. Divergent results about the importance of each pathway have been found for all nucleobases. This situation is illustrated for cytosine in Table 3, which surveys the results of simulations from Refs. [114, 115, 117, 126–129].

Table 3 Gas phase dynamics results for UV-excited cytosine obtained with different methods

| Dynamics | Electronic structure | Main pathway | $\tau$ (ps) | Refs. |
|---|---|---|---|---|
| MS | CAS (2, 2) | oop-NH$_2$ (65%) | ~0.8 | [127] |
| SH | OM2/MRCI | C6-puck (100%) | 0.37 | [115] |
| SH | CAS (10, 8) | Quasi-planar (64%) | 0.69 | [114, 128] |
| SH | CAS (12, 9) | Quasi-planar | ~0.5[a] | [129] |
| SH | FOMO/AM1 (PM3) | C6-puck (77%)[b] | 0.09 (0.17) | [117] |
|  |  | oop-NH$_2$ (81%)[c] | – |  |
| SH | ROKS | C6-puck | 0.7 | [126] |

*MS* multiple spawning, *SH* surface hopping

[a] Extrapolation of the reported data with a single exponential fitting function

[b] Cytosine + sugar

[c] Isolated cytosine

It is well established that cytosine has three main reaction pathways for internal conversion after excitation into the first bright $\pi\pi^*$ singlet state [1, 124, 127]. One pathway is characterized by puckering at the C6 atom, another one is characterized by out-of-plane displacement of the amino group, and the third one is characterized by a quasi-planar distortion bringing cytosine to near a three-state conical intersection. As shown in Table 3, although all methods predict ultrafast deactivation, there is still no consensus about which pathway is the most important one.

Semi-classical simulations of the UV-photo absorption spectrum at the RI-CC2 level were carried out for each nucleobase, adenine, guanine, cytosine, thymine, and uracil in gas phase [18]. The simulation of the absorption cross section was approached by constructing a nuclear phase space distribution in the electronic ground state and then projecting it onto the electronic excited states. The ground-state distribution was prepared by means of a probabilistic sampling of the Wigner distribution for the ground-state quantum harmonic oscillator [32, 130, 131]. The spectra of adenine, guanine, thymine, and uracil (Fig. 6) showed a common characteristic by the presence of a two-band structure separated by a low intensity region. On the other hand, the cytosine spectrum is formed by a succession of three bands of increasing intensity (Fig. 6). For all five nucleobases, the bands are formed mostly by absorption into $^1\pi\pi^*$ states.

### 6.2 DNA fragments

Whereas most of the photophysics of isolated nucleobases is now quite well understood, the photodynamics of several interacting nucleobases still poses many challenges. Research on the excitonic interactions between nucleobases is driven by the observation that the photophysics of DNA is significantly different from that of isolated nucleobases. Whereas all the isolated bases decay on a

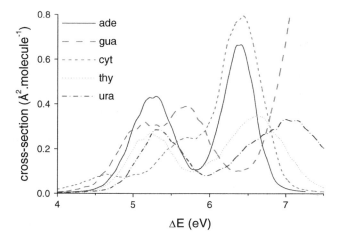

Fig. 6 Simulated absorption spectra of nucleobases using the RI-CC2 method

picosecond time scale [119, 120], long-lived transients up to 10–100 ps [132–134] are observed in DNA, which are strongly dependent on sequence and structure [135]. In principle, both inter- and intrastrand interactions could play a significant role. It is now believed that intrastrand stacking interactions are the major factor for the increase in life time [2]. It has been suggested that UV absorption occurs into Frenkel excitonic states with a delocalization length of 3–4 bases in a homo-adenine strand [134] or probably less in more heterogenous sequences [2].

Computational studies on DNA base stacks are aimed at elucidating the processes described above. Computational challenges arise from the large system sizes to be treated, the importance of the environment, and the need of considering dynamical effects. In principle, TDDFT would be favorable for describing systems of such a size and base tetramers have been successfully treated [136], but the accuracy of TDDFT, at least in the case of standard functionals without long-range corrections, has been questioned [137]. Ab initio methods have been used for slightly smaller systems. RI-CC2 has been used for system sizes up to base trimers [106], and the CASPT2 method could be applied to dimers [138]. It is a particular challenge to assess the involvement of charge-transfer states. TDDFT studies using a polarized continuum model emphasized the importance of charge-transfer states even when long-range corrected functionals were used [139, 140]. By contrast, a CASPT2 study concluded that Frenkel excitonic states stabilized by resonant interactions should be the most stable trapping sites [138]. Clarifications are still needed to understand these essential photophysical processes. More information may be provided by ab initio methods combined with QM/MM electrostatic embedding treatment of the surrounding DNA to treat both the electronic structure and the polarization of the environment at a high level.

Extended benchmark calculations for excitation energies, oscillator strengths, and characters of the low-lying singlet excited states of the stacked nucleobase pairs adenine–thymine (AT) and guanine-cytosine (GC) were performed by means of a selected set of density functional, B3-LYP and PBE0, and the recently developed M06-2X and M06-HF within the time-dependent density functional theory (TD-DFT) [141]. The resolution-of-the-identity second-order algebraic diagrammatic construction (RI-ADC(2)) method served as reference approach for comparison. For the AT dimer at its ground-state optimized geometry, the charge-transfer state at RI-ADC(2)/TZVP level corresponds to the excited state $S_6$, which is an excitation from a $\pi$ orbital in adenine to a $\pi^*$ in thymine. For the GC dimer, the RI-ADC(2)/TZVP the charge-transfer state corresponds to $S_4$, which is a $\pi-\pi^*$ excitation from guanine to cytosine. Comparison of DFT and ADC(2) results shows that the M06-2X version provides a relatively good reproduction of the reference results. It avoids the serious overstabilization and overcrowding of the spectrum with charge-transfer states found with the B3LYP functional. Solvent effect was also investigated for the AT and GC complexes. The main observation is that the amount of charge-transfer character increases with the polarity of the solvent.

Non-adiabatic dynamics simulations have been performed for one quantum mechanically treated base in an MM base stack [142] and a short double helix [143]. The study illustrated that purely mechanical restrictions do not play a large role and that the monomer likely decay pathways are not hindered by the stacking interaction.

Interactions between Watson–Crick base pairs in DNA could lead to electron or proton transfer processes between the strands as well. The possibility of an electron-driven proton transfer in a guanine-cytosine base pair has been suggested from a static computational study [118] and further examined by dynamics simulations [116]. The effect has been found experimentally in isolated base pairs [144]. But it is now believed that interstrand interactions do not play a significant role in the photophysics of DNA strands [2, 136].

A significant amount of attention has been devoted to charge migration dynamics of DNA [21, 109, 145]. Also in this case, the question of delocalization and of transport properties arises. Aside from biological interests, charge transfer in DNA has been studied in the context of nanotechnology as a model self-assembling system with molecular recognition properties. The task in this context is to modify the properties in order to get sufficient electric conductivity [21]. It has been theoretically suggested [146] and experimentally verified [145] that charge transfer may proceed through two distinct mechanisms: direct superexchange up to a separation of about three bases and multistep hopping for larger distances. Several strategies exist for getting a detailed picture of different aspects of these processes. Models from solid state physics are able to capture the periodicity of the system but have difficulties in properly considering structural fluctuations and molecular motions [21, 147]. High-level ab initio supermolecular calculations (CASSCF, CASPT2, and MRCI) give detailed descriptions of the interactions between stacked bases [148, 149] and even non-adiabatic effects in connection with dynamics simulations, cf. Ref. [104]. However, these methods can only be applied selectively because of high computational requirements. In contrast, semiempirical DFTB calculations in connection with QM/MM inclusion of the environment allow for an efficient sampling of the conformational space and for estimating the importance of solvent degrees of freedom [109].

## 6.3 Retinal and model systems

Retinal, the chromophore of rhodopsins, has attracted attention from theoreticians since the 1970s [150]. It plays a main function in the transduction of a light signal into a chemical impulse via photoisomerization inside the protein cage. Theoretical research of retinal has been focused on two main aspects, first, solvatochromic effects, which allow its absorption to be tuned to different wavelengths depending on the protein-specific environment [151–154]; second, determination of isomerization mechanism occurring after photoexcitation [4, 5, 155]. Along the last decade, most of the investigations in this field dealt with retinal models [156–161], usually protonated Schiff bases neglecting the cyclohexene ring. Recently, research considering the full length retinal at quantum mechanical level has been reported [4, 5, 155].

The photoisomerization mechanism is still an issue under debate. While ab initio (CASSCF, CASPT2, MRCI) [155–157] and semiempirical (OM2/MRCI) [159] methods predict that torsion around formal double bonds is the main mechanism bringing the molecule to the conical intersection, TDDFT results predict a different scenario with a major role played by torsion around formal single bonds [162]. Although the agreement about the torsional mode coming from most of methods seems to indicate a failure of TDDFT to deal with this molecule, recent CASPT2 and quantum Monte Carlo evaluations of excited-state potential energy surfaces of a retinal model indicated that the changes in bond length alternation in the excited state should be much smaller than that predicted by CASSCF and even MRCI, which would favor to some extent the TDDFT predictions [163].

### 6.4 Photovoltaics

Electronic devices based on organic chemistry have many potential advantageous properties compared to silicon-based devices. These include lower production costs, ease of processing, and mechanical flexibility [24]. In particular, the field of photovoltaics has been intensively studied with the aim of producing photovoltaic energy at a comparable or lower cost as traditional non-renewable energy sources. Computational studies are aimed at elucidating the complex nature of the excited states involved and to examine the vibronic coupling that leads to interconversion between them, partially on an ultrafast time scale. A variety of interesting classes of solar cells have been suggested. In dye-sensitized solar cells, an organic dye is used as a chromophore in direct contact with an inorganic framework and charge separation occurs at the interface between the dye molecules and inorganic nanoparticles [164, 165]. If an organic bulk phase is to be used, then the exciton migration properties play a decisive role. The poly (p-phenylene vinylene) (PPV) system and its methoxy derivative have been extensively studied in this context. A special focus was laid on coherences in energy migration [166], which were explained in terms of excitons surfing on the polymer backbone [167]. A DFT study illustrated the strong electron–phonon coupling along the bond length alternation coordinates [168], which could serve as a basis for the surfing process. Whereas bond length alternation is expected to affect on-site excitation energies, torsions may modulate the electronic couplings between the different monomer units. Using full quantum dynamics, these nuclear coordinates can be coupled with electronic degrees of freedom and valuable information about electron–phonon coupling and possible quantum coherences could be obtained, cf. the preliminary exploration of Ref. [110].

After exciton migration, charge separation has to occur at the heterojunction. This process is not well understood yet [12], but quantum dynamical studies may give insight into the electron–phonon coupling processes involved. After charge separation, the charge migration properties become the decisive feature. Computational studies on charge transport in organic systems have been concerned

with the estimation of transfer parameters [103] as well as with non-adiabatic dynamics simulations [169]. A number of interesting extensions exists for making organic solar cells more efficient. Self-assembled nanoscale heterojunctions may be used for decreasing exciton migration lengths [170]. Alternatively, supramolecular assemblies with an electron donating group, a chromophore, and an electron accepting group covalently connected can eliminate the need of diffuse electron migration altogether. The properties of such a triad have been studied by DFT [171] and TDDFT [172]. Moreover, the process of multiple exciton generation from one photon can be used to increase the fundamental limits of conversion efficiency [173]. In organic molecules, this can be realized through singlet fission, i.e., one singlet exciton is split into two triplet excitons [173]. The process has been examined computationally for a pentacene dimer [174]. It is suggested that the photoexcitation occurs into a bright single excitation $S_2$ state, which can cross with a dark doubly excited lower lying state. After this state is reached the two molecules repel each other, and as the system dissociates the doubly excited state obtains the nature of two singlet coupled triplets [174, 175]. To describe these complex excited states, multi-reference perturbation theory was applied with a triple zeta basis set [174]. Biomimetic solar cells using porphyrin aggregates are another promising strategy [176]. In this case, a connection between large-scale theories [22] and smaller-scale parameter calculations may be suitable to provide a solid theoretical background. In general, it can be said that computational studies provide a foundation for understanding the signatures of complex excited-state processes, and may therefore aid in the design of new efficient solar cells in the different categories described.

### 6.5 Photodevices, phototriggers, and molecular machines

In analogy to the processes occurring in nature for retinal (see Sect. 6.3), molecular photodevices can be used for a clean and ultrafast way to control reactions and trigger processes [177]. Theoretical research in this field has been focusing on laser control of reactions, such as to induce photoisomerization [15, 178, 179], photodissociation [9, 180], and proton transfer processes [181–184]. Light-driven motors have been considered as well [185].

One of the main examples in this field is the photoisomerization of azo-molecules, especially of azobenzene [15, 178, 186, 187], whose molecular length is strongly affected by photoexcitation. In fact, theoretical investigations have played a major role in this field by means of dynamics simulations, which have revealed the intrinsically multidimensional character of the photoisomerization process, rather than a simple unidimensional one, as commonly thought.

Photoexcitation may induce electron or proton transfer within the molecule or between molecules. This process leads to a strong red shift of the emission wavelength in comparison with the absorption wavelength. Theoretical studies of excited-state intramolecular proton transfer (ESIPT) have been useful to support experimental ultrafast spectroscopic investigations. These studies have revealed, for instance, details on the transfer mechanism [181–183] and how it may trigger internal conversion processes [36, 188]. Another interesting theoretical suggestion was a reversible molecular switch based on ESIPT [184].

# 7 Conclusion and outlook

After photoexcitation, molecules may undergo a variety of different processes. The excited states are characterized by a significant shift in electron density, which may, for example, lead to ionic or diffuse character. In addition, excited states can interact, resulting in irregular shapes of potential energy surfaces. Ultrafast non-equilibrium processes further complicate the situation. For these reasons, excited-state calculations are still quite challenging in spite of the large effort spent in developing new computational methods. Whereas ground-state thermochemistry has evolved into a field where many questions can be routinely addressed by non-specialists, excited-state calculations usually still require a careful assessment of available methods. Not only considerations of computational efficiency play a role, but methods have to be chosen that are appropriate to describe the different kinds of excited states and processes of interest. The availability of analytic energy gradients and nonadiabatic coupling elements is also an important selection criterion.

A major goal in the development of new excited-state methods will be the increase of applicability and stability of the different approaches. Many coupled-cluster-based single-reference methods exist, which offer high-quality results at geometries close to the ground-state minimum. However, the description of conical intersections and strongly distorted geometries is still highly challenging. It would be highly desirable to have a unified approach for these multi-reference cases, rather than having to carefully choose the methodological details as is currently the case for many such problems. TDDFT serves as an attractive method because of its computational efficiency, but it suffers from different systematic errors, most prominently the overstabilization of charge-transfer states. Long-range corrected functionals offer a promising solution. Semiempirical methods offer an even computationally faster approach and appropriate multireference methods are available, but careful parameterization is usually necessary and questions about transferability arise. When choosing a dynamics method, it is necessary to compromise between computational efficiency and the level of approximation, i.e., whether nonadiabatic effects and all internal degrees of freedom have to be considered or not. Dimers and aggregates offer new challenges that may have to be addressed with efficient multi-scale methodology. While there are many successful applications of present methods, there are still many outstanding problems remaining and the development of new approaches to electronic structure calculations will play an important role.

**Acknowledgments** This work has been supported by the Austrian Science Fund within the framework of the Special Research Program and F41 Vienna Computational Materials Laboratory (ViCoM). FP is a recipient of a DOC fellowship of the Austrian Academy of Sciences. The calculations were performed in part at the Vienna Scientific Cluster (project nos. 70019 and 70151) of the Vienna University Computer Centre. This work was also performed as part of research supported by the National Science Foundation Partnership in International Research and Education (PIRE) Grant No. OISE-0730114; support was also provided by the Robert A. Welch Foundation under Grant No. D-0005.

# References

1. Serrano-Andres L, Merchan M (2009) J Photochem Photobiol C-Photochem Rev 10:21
2. Middleton C, de La Harpe K, Su C, Law Y, Crespo-Hernandez C, Kohler B (2009) Annu Rev Phys Chem 60:217
3. Barbatti M (2011) Phys Chem Chem Phys 13:4686
4. Polli D, Altoe P, Weingart O, Spillane KM, Manzoni C, Brida D, Tomasello G, Orlandi G, Kukura P, Mathies RA, Garavelli M, Cerullo G (2010) Nature 467:440
5. Hayashi S, Taikhorshid E, Schulten K (2009) Biophys J 96:403
6. Olaso-Gonzalez G, Merchán M, Serrano-Andrés L (2006) J Phys Chem B 110:24734
7. Cheng Y, Fleming G (2009) Annu Rev Phys Chem 60:241
8. Dreuw A, Harbach P, Mewes J, Wormit M (2010) Theor Chem Acc 125:419
9. Daniel C, Full J, González L, Lupulescu C, Manz J, Merli A, Vajda Š, Wöste L (2003) Science 299:536
10. Duncan WR, Stier WM, Prezhdo OV (2005) J Am Chem Soc 127:7941
11. Shang Y, Li Q, Meng L, Wang D, Shuai Z (2011) Theor Chem Acc 129:291
12. Brédas J-L, Norton JE, Cornil J, Coropceanu V (2009) Acc Chem Res 42:1691
13. Virshup AM, Punwong C, Pogorelov TV, Lindquist BA, Ko C, Martinez TJ (2009) J Phys Chem B 113:3280
14. Groenhof G, Bouxin-Cademartory M, Hess B, deVisser SP, Berendsen HJC, Olivucci M, Robb MA (2004) J Am Chem Soc 126:4228
15. Böckmann M, Doltsinis NL, Marx D (2010) Angew Chem 122:3454
16. Habenicht BF, Prezhdo OV (2009) J Phys Chem C 113:14067
17. Petrenko T, Neese F (2007) J Chem Phys 127:164319
18. Barbatti M, Aquino AJA, Lischka H (2010) Phys Chem Chem Phys 12:4959
19. Stenrup M, Larson A (2011) Chem Phys 379:6
20. Hudock HR, Levine BG, Thompson AL, Satzger H, Townsend D, Gador N, Ullrich S, Stolow A, Martinez TJ (2007) J Phys Chem A 111:8500

21. Endres R, Cox D, Singh R (2004) Rev Mod Phys 76:195
22. Prokhorenko V, Steensgaard D, Holzwarth A (2003) Biophys J 85:3173
23. Nemeth A, Milota F, Sperling J, Abramavicius D, Mukamel S, Kauffmann H (2009) Chem Phys Lett 469:130
24. Bredas J, Beljonne D, Coropceanu V, Cornil J (2004) Chem Rev 104:4971
25. Selvin P (2000) Nat Struct Biol 7:730
26. Hertel IV, Radloff W (2006) Rep Prog Phys 69:1897
27. Zewail AH (2000) Angewandte Chemie (International Edition) 39:2586
28. Koppel H, Domcke W, Cederbaum LS (1984) Adv Chem Phys 57:59
29. Atchity GJ, Xantheas SS, Ruedenberg K (1991) J Chem Phys 95:1862
30. Truhlar DG, Mead CA (2003) Phys Rev A 68:032501
31. Bearpark MJ, Robb MA, Schlegel HB (1994) Chem Phys Lett 223:269
32. Ciminelli C, Granucci G, Persico M (2004) Chem Eur J 10:2327
33. Dallos M, Lischka H, Shepard R, Yarkony DR, Szalay PG (2004) J Chem Phys 120:7330
34. Keal TW, Koslowski A, Thiel W (2007) Theor Chem Acc 118:837
35. Michl J, Bonačić-Koutecký V (1990) Electronic aspects of organic photochemistry. Wiley-Interscience, New York
36. Barbatti M, Aquino AJA, Lischka H, Schriever C, Lochbrunner S, Riedle E (2009) Phys Chem Chem Phys 11:1406
37. Schreiber M, Barbatti M, Zilberg S, Lischka H, Gonzalez L (2007) J Phys Chem A 111:238
38. Vazdar M, Eckert-Maksic M, Barbatti M, Lischka H (2009) Mol Phys 107:845
39. Barbatti M, Lischka H (2008) J Am Chem Soc 130:6831
40. Barbatti M, Aquino AJA, Lischka H (2008) Chem Phys 349:278
41. Barbara P, Meyer T, Ratner M (1996) J Phys Chem 100:13148
42. Marcus RA (1993) Angewandte Chemie (International Edition in English) 32:1111
43. Forster T (1948) Annalen Der Physik 2:55
44. Christiansen O, Koch H, Jorgensen P (1995) Chem Phys Lett 243:409
45. Trofimov AB, Schirmer J (1995) J Phys B At Mol Opt Phys 28:2299
46. Stanton JF, Bartlett RJ (1993) J Chem Phys 98:7029
47. Foresman JB, Head-Gordon M, Pople JA, Frisch MJ (1992) J Phys Chem 96:135
48. Hättig C, Weigend F (2000) J Chem Phys 113:5154
49. Hättig C, Köhn A (2002) J Chem Phys 117:6939
50. Schmidt MW, Gordon MS (1998) Annu Rev Phys Chem 49:233
51. Shepard R (1987) The multiconfiguration self-consistent field method. In: Lawley KP (ed) Ab Initio methods in quantum chemistry II. Advances in chemical physics. Wiley, New York, pp 63
52. Buenker RJ, Peyerimhoff SD, Butscher W (1978) Mol Phys 35:771
53. Karwowski JA, Shavitt I (2003) Configuration interaction. In: Wilson S (ed) Handbook of molecular physics and quantum chemistry. Wiley, Chichester, p 227
54. Werner H-J, Knowles PJ (1988) J Chem Phys 89:5803
55. Finley J, Malmqvist PA, Roos BO, Serrano-Andrés L (1998) Chem Phys Lett 288:299
56. Lischka H, Dallos M, Shepard R (2002) Mol Phys 100:1647
57. Lengsfield BH, Saxe P, Yarkony DR (1984) J Chem Phys 81:4549
58. Celani P, Werner HJ (2003) J Chem Phys 119:5044
59. Lischka H, Dallos M, Szalay PG, Yarkony DR, Shepard R (2004) J Chem Phys 120:7322
60. Szymczak JJ, Barbatti M, Lischka H (2011) Int J Quantum Chem 111:3307

61. Slavicek P, Martinez T (2010) J Chem Phys 132:234102
62. Müller T, Dallos M, Lischka H (1999) J Chem Phys 110:7176
63. Angeli C (2009) J Comput Chem 30:1319
64. Ghigo G, Roos BO, Malmqvist P-A (2004) Chem Phys Lett 396:142
65. Bomble YJ, Sattelmeyer KW, Stanton JF, Gauss J (2004) J Chem Phys 121:5236
66. Bauernschmitt R, Ahlrichs R (1996) Chem Phys Lett 256:454
67. Casida M (1995) Time-dependent density functional response theory for molecules. In: Chong D (ed) Recent advances in density functional methods. Part I. World Scientific, Singapore, p 155
68. Frank I, Hutter J, Marx D, Parrinello M (1998) J Chem Phys 108:4060
69. Filatov M, Shaik S (1999) Chem Phys Lett 304:429
70. Elstner M (2006) Theor Chem Acc 116:316
71. Grimme S, Waletzke M (1999) J Chem Phys 111:5645
72. Beck EV, Stahlberg EA, Burggraf LW, Blaudeau J-P (2008) Chem Phys 349:158
73. Furche F, Ahlrichs R (2002) J Chem Phys 117:7433
74. Send R, Furche F (2010) J Chem Phys 132:044107
75. Tavernelli I, Curchod BFE, Rothlisberger U (2009) J Chem Phys 131:196101
76. Werner U, Mitric R, Suzuki T, Bonacic-Koutecký V (2008) Chem Phys 349:319
77. Tapavicza E, Tavernelli I, Rothlisberger U (2007) Phys Rev Lett 98:023001
78. Barbatti M, Pittner J, Pederzoli M, Werner U, Mitric R, Bonacic-Koutecky V, Lischka H (2010) Chem Phys 375:26
79. Becke A (1993) J Chem Phys 98:5648
80. Perdew J, Burke K, Ernzerhof M (1996) Phys Rev Lett 77:3865
81. Adamo C, Barone V (1999) J Chem Phys 110:6158
82. Zhao Y, Truhlar D (2008) Theor Chem Acc 119:525
83. Dreuw A, Head-Gordon M (2005) Chem Rev 105:4009
84. Peach MJG, Helgaker T, Salek P, Keal TW, Lutnaes OB, Tozer DJ, Handy NC (2005) Phys Chem Chem Phys 8:558
85. Yanai T, Tew DP, Handy NC (2004) Chem Phys Lett 393:51
86. Granucci G, Persico M, Toniolo A (2001) J Chem Phys 114:10608
87. Koslowski A, Beck ME, Thiel W (2003) J Comput Chem 24:714
88. Silva-Junior MR, Thiel W (2010) JCTC 6:1546
89. Worth GA, Cederbaum LS (2004) Annu Rev Phys Chem 55:127
90. Kosloff R (1994) Annu Rev Phys Chem 45:145
91. Schneider R, Domcke W, Koppel H (1990) J Chem Phys 92:1045
92. Worth GA, Meyer HD, Cederbaum LS (2004) Conical intersections: electronic structure, dynamics and spectroscopy. Advanced series in physical chemistry. World Scientific Publishing Company, Singapore
93. Worth GA, Hunt P, Robb MA (2003) J Phys Chem A 107:621
94. Santoro F, Lami A, Olivucci M (2007) Theor Chem Acc 117:1061
95. Ben-Nun M, Quenneville J, Martínez TJ (2000) J Phys Chem A 104:5161
96. Tully JC (1990) J Chem Phys 93:1061
97. Granucci G, Persico M (2007) J Chem Phys 126:134114
98. Martínez TJ (2006) Acc Chem Res 39:119
99. Barbatti M, Shepard R, Lischka H (2011) Computational and methodological elements for nonadiabatic trajectory dynamics simulations of molecules. In: Domcke W, Yarkony DR, Köppel H (eds) Conical Intersections: theory, computation and experiment. advanced series in physical chemistry, vol 17. World Scientific, Singapore
100. Barbatti M (2011) WIREs Comp Mol Sci 1:620
101. Barbatti M, Granucci G, Persico M, Ruckenbauer M, Vazdar M, Eckert-Maksic M, Lischka H (2007) J Photochem Photobiol. A 190:228

102. Hsu C (2009) Acc Chem Res 42:509
103. Valeev E, Coropceanu V, da Silva D, Salman S, Bredas J (2006) J Am Chem Soc 128:9882
104. Plasser F, Lischka H (2011) J Chem Phys 134:034309
105. Voityuk A, Rosch N (2002) J Chem Phys 117:5607
106. Nachtigallova D, Hobza P, Ritze H (2008) Phys Chem Chem Phys 10:5689
107. Tamura H, Mallet J, Oheim M, Burghardt I (2009) J Phys Chem C 113:7548
108. Krueger B, Scholes G, Fleming G (1998) J Phys Chem B 102:5378
109. Kubar T, Elstner M (2008) J Phys Chem B 112:8788
110. Sterpone F, Martinazzo R, Panda A, Burghardt I (2011) Zeitschrift fuer Physikalische Chemie 225:541
111. Scholes G, Jordanides X, Fleming G (2001) J Phys Chem B 105:1640
112. Brunschwig BS, Logan J, Newton MD, Sutin N (1980) J Am Chem Soc 102:5798
113. Ruckenbauer M, Barbatti M, Muller T, Lischka H (2010) J Phys Chem A 114:6757
114. Barbatti M, Aquino AJA, Szymczak JJ, Nachtigallová D, Hobza P, Lischka H (2010) Proc Natl Acad Sci USA 107:21453
115. Lan Z, Fabiano E, Thiel W (2009) J Phys Chem B 113:3548
116. Groenhof G, Schafer LV, Boggio-Pasqua M, Goette M, Grubmuller H, Robb MA (2007) J Am Chem Soc 129:6812
117. Alexandrova AN, Tully JC, Granucci G (2010) J Phys Chem B 114:12116
118. Sobolewski AL, Domcke W, Hättig C (2005) Proc Natl Acad Sci USA 102:17903
119. Crespo-Hernández CE, Cohen B, Hare PM, Kohler B (2004) Chem Rev 104:1977
120. Canuel C, Mons M, Piuzzi F, Tardivel B, Dimicoli I, Elhanine M (2005) J Chem Phys 122:074316
121. Marian CM (2007) J Phys Chem A 111:1545
122. Kistler KA, Matsika S (2007) J Phys Chem A 111:2650
123. Yamazaki S, Domcke W, Sobolewski AL (2008) J Phys Chem A 112:11965
124. Blancafort L, Bearpark MJ, Robb MA (2008) Computational modeling of cytosine photophysics and photochemistry: from the gas phase to DNA. In: Shukla MK, Leszczynski J (eds) Radiation induced molecular phenomena in nucleic acid. Challenges and advances in computational chemistry and physics. Springer, The Netherlands
125. Picconi D, Barone V, Lami A, Santoro F, Improta R (2011) Chem Phys Chem 12:1957
126. Doltsinis NL, Markwick PRL, Nieber H, Langer H (2008) Ultrafast radiationless decay in nucleic acids: insights from nonadiabatic ab initio molecular dynamics. In: Shukla MK, Leszczynski J (eds) Radiation induced molecular phenomena in nucleic acids, vol 5. Challenges and advances in computational chemistry and physics. Springer, Netherlands, p 265
127. Hudock HR, Martinez TJ (2008) Chemphyschem 9:2486
128. Barbatti M, Aquino AJA, Szymczak JJ, Nachtigallova D, Lischka H (2011) Phys Chem Chem Phys 13:6145
129. González-Vázquez J, González L (2010) Chemphyschem 11:3617
130. Bergsma JP, Berens PH, Wilson KR, Fredkin DR, Heller EJ (1984) J Phys Chem 88:612
131. Lukes V, Solc R, Milota F, Sperling J, Kauffmann HF (2008) Chem Phys 349:226
132. Crespo-Hernandez C, Cohen B, Kohler B (2005) Nature 436:1141
133. Vaya I, Miannay F, Gustavsson T, Markovitsi D (2010) Chemphyschem 11:987
134. Buchvarov I, Wang Q, Raytchev M, Trifonov A, Fiebig T (2007) Proc Natl Acad Sci USA 104:4794

135. Schwalb N, Temps F (2008) Science 322:243
136. Santoro F, Barone V, Improta R (2009) J Am Chem Soc 131:15232
137. Kozak C, Kistler K, Lu Z, Matsika S (2010) J Phys Chem B 114:1674
138. Olaso-Gonzalez G, Merchan M, Serrano-Andres L (2009) J Am Chem Soc 131:4368
139. Lange A, Herbert J (2009) J Am Chem Soc 131:3913
140. Santoro F, Barone V, Lami A, Improta R (2010) Phys Chem Chem Phys 12:4934
141. Aquino AJA, Nachtigallova D, Hobza P, Truhlar DG, Hattig C, Lischka H (2011) J Comput Chem 32:1217
142. Nachtigallova D, Zeleny T, Ruckenbauer M, Muller T, Barbatti M, Hobza P, Lischka H (2010) J Am Chem Soc 132:8261
143. Zeleny T, Hobza P, Nachtigallova D, Ruckenbauer M, Lischka H (2011) Collect Czech Chem Commun 76:631
144. Abo-Riziq A, Grace L, Nir E, Kabelac M, Hobza P, de Vries M (2005) Proc Natl Acad Sci USA 102:20
145. Giese B, Amaudrut J, Kohler A, Spormann M, Wessely S (2001) Nature 412:318
146. Jortner J, Bixon M, Langenbacher T, Michel-Beyerle M (1998) Proc Natl Acad Sci USA 95:12759
147. Ye Y, Chen R, Martinez A, Otto P, Ladik J (1999) Solid State Commun 112:139
148. Blancafort L, Voityuk A (2006) J Phys Chem A 110:6426
149. Roca-Sanjuan D, Olaso-Gonzalez G, Coto P, Merchan M, Serrano-Andres L (2010) Theor Chem Acc 126:177
150. Warshel A (1976) Nature 260:679
151. Hoffmann M, Wanko M, Strodel P, Konig PH, Frauenheim T, Schulten K, Thiel W, Tajkhorshid E, Elstner M (2006) J Am Chem Soc 128:10808
152. Wanko M, Hoffmann M, Strodet P, Koslowski A, Thiel W, Neese F, Frauenheim T, Elstner M (2005) J Phys Chem B 109:3606
153. Rohrig UF, Guidoni L, Rothlisberger U (2005) Chemphyschem 6:1836
154. Altun A, Morokuma K, Yokoyama S (2011) ACS Chem Biol 6:775
155. Altoè P, Cembran A, Olivucci M, Garavelli M (2010) Proc Natl Acad Sci 107:20172
156. Szymczak JJ, Barbatti M, Lischka H (2009) J Phys Chem A 113:11907
157. Garavelli M (2006) Theor Chem Acc 116:87
158. Sumita M, Ryazantsev MN, Saito K (2009) Phys Chem Chem Phys 11:6406
159. Keal T, Wanko M, Thiel W (2009) Theor Chem Acc 123:145
160. Weingart O, Schapiro I, Buss V (2006) J Mol Model 12:713
161. Burghardt I, Cederbaum LS, Hynes JT (2004) Faraday Discuss 127:395
162. Send R, Sundholm D (2007) J Phys Chem A 111:8766
163. Valsson O, Filippi C (2010) JCTC 6:1275
164. Oregan B, Gratzel M (1991) Nature 353:737
165. Odobel F, Le Pleux L, Pellegrin Y, Blart E (2010) Acc Chem Res 43:1063
166. Collini E, Scholes G (2009) Science 323:369
167. Bredas J, Silbey R (2009) Science 323:348
168. Nayyar I, Batista E, Tretiak S, Saxena A, Smith D, Martin R (2011) J Phys Chem Lett 2:566
169. Jones G, Carpenter B, Paddon-Row M (1998) J Am Chem Soc 120:5499
170. Park S, Roy A, Beaupre S, Cho S, Coates N, Moon J, Moses D, Leclerc M, Lee K, Heeger A (2009) Nat Photonics 3:297
171. Baruah T, Pederson M (2009) J Chem Theory Comput 5:834
172. Spallanzani N, Rozzi C, Varsano D, Baruah T, Pederson M, Manghi F, Rubio A (2009) J Phys Chem B 113:5345
173. Hanna M, Nozik A (2006) J Appl Phys 100:074510

174. Zimmerman P, Zhang Z, Musgrave C (2010) Nat Chem 2:648
175. Siebbeles L (2010) Nat Chem 2:608
176. Balaban T (2005) Acc Chem Res 38:612
177. Zhao Y, Ikeda T (2009) Smart light-responsive materials: azobenzene-containing polymers and liquid crystals. Wiley-Interscience, Hoboken
178. Carstensen O, Sielk J, Schonborn JB, Granucci G, Hartke B (2010) J Chem Phys 133:124305
179. Manz J, Sundermann K, de Vivie-Riedle R (1998) Chem Phys Lett 290:415
180. Crespo-Otero R, Barbatti M (2011) J Chem Phys 134:164305
181. Schriever C, Barbatti M, Stock K, Aquino AJA, Tunega D, Lochbrunner S, Riedle E, de Vivie-Riedle R, Lischka H (2008) Chem Phys 347:446
182. Plasser F, Barbatti M, Aquino AJA, Lischka H (2009) J Phys Chem A 113:8490
183. Aquino AJA, Lischka H, Hattig C (2005) J Phys Chem A 109:3201
184. Sobolewski A (2008) Phys Chem Chem Phys 10:1243
185. Koumura N, Geertsema E, van Gelder M, Meetsma A, Feringa B (2002) J Am Chem Soc 124:5037
186. Weingart O, Lan Z, Koslowski A, Thiel W (2011) J Phys Chem Lett 2:1506
187. Pederzoli M, Pittner J, Barbatti M, Lischka H (2011) J Phys Chem A 115:11136
188. Sobolewski AL, Domcke W, Hattig C (2006) J Phys Chem A 110:6301
189. Ben-Nun M, Martínez TJ (2000) Chem Phys 259:237
190. Barbatti M, Aquino AJA, Lischka H (2006) Mol Phys 104:1053
191. Ruiz DS, Cembran A, Garavelli M, Olivucci M, Fuss W (2002) Photochem Photobiol 76:622
192. Bearpark MJ, Bernardi F, Clifford S, Olivucci M, Robb MA, Vreven T (1997) J Phys Chem A 101:3841
193. Cattaneo P, Persico M (1999) Phys Chem Chem Phys 1:4739
194. Li Q, Mendive-Tapia D, Paterson MJ, Migani A, Bearpark MJ, Robb MA, Blancafort L (2010) Chem Phys 377:60
195. Marian CM (2005) J Chem Phys 122:104314
196. Merchan M, Gonzalez-Luque R, Climent T, Serrano-Andres L, Rodriuguez E, Reguero M, Pelaez D (2006) J Phys Chem B 110:26471
197. Barbatti M, Paier J, Lischka H (2004) J Chem Phys 121:11614
198. Wilsey S, Houk KN (2002) J Am Chem Soc 124:11182
199. Antol I, Eckert-Maksic M, Barbatti M, Lischka H (2007) J Chem Phys 127:234303

200. Perun S, Sobolewski AL, Domcke W (2006) J Phys Chem A 110:13238
201. Lan ZG, Fabiano E, Thiel W (2009) Chemphyschem 10:1225
202. Barbatti M, Vazdar M, Aquino AJA, Eckert-Maksic M, Lischka H (2006) J Chem Phys 125:164323
203. Sobolewski AL, Domcke W (2000) Chem Phys 259:181
204. Perun S, Sobolewski AL, Domcke W (2005) Chem Phys 313:107
205. Salzmann S, Kleinschmidt M, Tatchen J, Weinkauf R, Marian CM (2008) Phys Chem Chem Phys 10:380
206. Fuji T, Suzuki Y-I, Horio T, Suzuki T, Mitric R, Werner U, Bonacic-Koutecky V (2010) J Chem Phys 133:234303
207. Barbatti M, Lischka H, Salzmann S, Marian CM (2009) J Chem Phys 130:034305
208. Schreiber M, Silva MR, Sauer SPA, Thiel W (2008) J Chem Phys 128:134110
209. Silva-Junior MR, Schreiber M, Sauer SPA, Thiel W (2010) J Chem Phys 133:174318
210. Fleig T, Knecht S, Hättig C (2007) J Phys Chem A 111:5482
211. Parac M, Grimme S (2002) J Phys Chem A 106:6844
212. Silva-Junior MR, Schreiber M, Sauer SPA, Thiel W (2008) J Chem Phys 129:104103
213. Peach MJG, Benfield P, Helgaker T, Tozer DJ (2008) J Chem Phys 128:044118
214. Zhao Y, Truhlar DG (2006) J Phys Chem A 110:13126
215. Perdew JP, Ruzsinszky A, Constantin LA, Sun JW, Csonka GI (2009) JCTC 5:902
216. Jacquemin D, Wathelet Vr, Perpète EA, Adamo C (2009) JCTC 5:2420
217. Parac M, Grimme S (2003) Chem Phys 292:11
218. Lin H, Truhlar DG (2007) Theor Chem Acc 117:185
219. Senn HM, Thiel W (2009) Angew Chem Int Ed Engl 48:1198
220. Doltsinis NL (2002) Nonadiabatic dynamics: mean-field and surface hopping. In: Grotendorst J, Marx D, Muramatsu A (eds) Quantum simulations of complex many-body systems: from theory to algorithms, vol 10. NIC Series. John von Neumann Institute for Computing, Julich
221. Levine BG, Martínez TJ (2007) Annu Rev Phys Chem 58:613
222. Barbatti M, Ruckenbauer M, Szymczak JJ, Aquino AJA, Lischka H (2008) Phys Chem Chem Phys 10:482
223. Migani A, Olivucci M (2004) Conical intersections and organic reaction mechanisms. In: Domcke W, Yarkony DR, Köppel H (eds) conical intersections: electronic structure, dynamics and spectroscopy. World Scientific Publishing Company, Singapore

Theor Chem Acc (2012) 131:1079
DOI 10.1007/s00214-011-1079-5

REGULAR ARTICLE

# Chemical accuracy in ab initio thermochemistry and spectroscopy: current strategies and future challenges

**Kirk A. Peterson · David Feller · David A. Dixon**

Received: 4 July 2011 / Accepted: 25 August 2011 / Published online: 11 January 2012
© Springer-Verlag 2012

**Abstract** The current state of the art in wavefunction-based electronic structure methods is illustrated via discussions of the most important effects incorporated into a selection of high-accuracy methods chosen from the chemical literature. If one starts with a high-quality correlation treatment, such as provided by the CCSD(T) coupled cluster method, the leading effects include convergence of the results with respect to the 1-particle basis set, (outer) core/valence correlation, scalar relativistic effects and a number of smaller effects. For thermochemical properties such as the heat of formation, the zero-point vibrational energy also becomes important, introducing its own set of difficulties to the computational approach. Changes in the various components as the chemical systems incorporate heavier elements and as the size of the systems grows are also considered. Finally, challenges arising from the desire to extend existing methods to transition metal and heavier elements are considered.

**Keywords** Ab initio thermochemistry ·
Ab initio spectroscopy · Chemical accuracy ·
Composite methods · Coupled cluster

Published as part of the special collection of articles celebrating the 50th anniversary of Theoretical Chemistry Accounts/Theoretica Chimica Acta.

K. A. Peterson (✉) · D. Feller
Department of Chemistry, Washington State University,
Pullman, WA 99164-4630, USA
e-mail: kipeters@wsu.edu

D. A. Dixon
Department of Chemistry, The University of Alabama,
Shelby Hall, Box 870336, Tuscaloosa, AL 35487-0336, USA

## 1 Introduction

The last 20 years or so have been marked by phenomenal progress in the area of quantitatively accurate quantum chemistry. Arguably, the greatest impact due to these advances has been in the area of ab initio thermochemistry and to a lesser extent ab initio spectroscopy. The steep increase in computing power over these years has certainly been a major driver for these advances, but these gains have also been matched by significant methodological developments in solving the electronic Schrödinger equation. Simultaneously with both advances, researchers have benefited from the implementation of these methods in easy-to-use software packages that perform well on high-performance computers. In the current paper, the emphasis is not on density functional theory but necessarily on wavefunction-based methods, e.g., the hierarchy of coupled cluster techniques, since they represent a scheme that can systematically approach the exact solution of the Schrödinger equation in a given basis set. The other main approximation that strongly influences the accuracy of a quantum chemistry calculation is the choice of 1-particle basis set used to describe the underlying molecular orbitals. The introduction of the systematically convergent (to the complete basis set (CBS) limit) correlation consistent basis sets by Dunning [1] in 1989, cc-pV$n$Z where $n$ = D, T, Q, 5, etc., has led to a paradigm shift in strategies for accurate solutions of the electronic Schrödinger equation. Using sequences of these basis sets with a chosen approximation method, e.g., coupled cluster singles and doubles with perturbative triples [2], CCSD(T), the CBS limit can be accurately estimated. This capability effectively eliminates the basis set incompleteness error from the calculation. In addition, it can minimize issues with basis set superposition error (BSSE). This strategy is now a common thread in all

Reprinted from the journal

161

🖄 Springer

high-accuracy ab initio thermochemistry and spectroscopy methodologies whether coupled cluster or multireference configuration interaction (MRCI) methods are used to solve the electronic Schrödinger equation.

Our goal is to design methods that provide at least chemical accuracy in terms of thermochemistry (and spectroscopy). So what does one mean by "chemical accuracy"? In the thermochemistry literature, this is almost universally interpreted as 1 kcal/mol or about 4 kJ/mol (1 kcal/mol = 4.184 kJ/mol). In the computational chemistry literature, a mean absolute deviation (MAD), or analogously a mean unsigned deviation (MUD), of 1 kcal/mol has been used to establish the criterion of chemical accuracy for a given thermochemistry methodology. As has been previously noted in the literature, however, an MUD of 1 kcal/mol does not actually represent chemical accuracy analogous to experiment since for experimentalists, the latter is generally based on a 95% confidence level or about 2 standard deviations. Hence, a more stringent MUD of about 0.5 kcal/mol is actually required to achieve an experimentalist's notion of chemical accuracy in a quantum chemistry methodology. The situation is a bit more ambiguous in terms of spectroscopic properties, such as equilibrium structures and vibrational frequencies. The term "spectroscopic accuracy" in the quantum chemistry literature generally refers to vibrational frequencies accurate to within $\pm 1$ cm$^{-1}$. While not typically defined for equilibrium bond lengths (which in any event are not directly observable quantities), this high level of accuracy for vibrational frequencies can be loosely equated to bond lengths accurate to better than $\pm 0.0005$ Å. This represents a serious challenge for computational quantum chemistry, and from a thermochemistry perspective, such error bars are more in line with kJ/mol or sub-kJ/mol accuracy. For the purposes of the present contribution, which is focused on *chemical* accuracy, the rather arbitrary definition of $\pm 0.005$ Å and $\pm 15$ cm$^{-1}$ for structures and frequencies, respectively, will be used.

Attempting to achieve reliable chemical accuracy for a general polyatomic molecule from a single quantum chemistry calculation is a daunting if not impossible task. As will be demonstrated in detail below, results of this quality require the use of methods capable of recovering significant amounts of electron correlation together with 1-particle basis sets near the CBS limit. For example, the CCSD(T) method, which often has an inherent accuracy (defined as the limiting accuracy at the CBS limit) equivalent to chemical accuracy, at least for thermochemical properties, has a computational expense that scales roughly as $n^3 N^4$, where $n$ is the number of occupied orbitals and $N$ the number of virtual orbitals (this is loosely given as $M^7$ where $M$ is the number of basis functions, although for a fixed molecule, the cost actually scales as just $N^4$).

To directly achieve chemical accuracy with CCSD(T) without extrapolation would require correlation of both the valence and outer-core electrons with a basis set of at least aug-cc-pCV6Z or aug-cc-pCV7Z quality [3]. For very small molecules, this is not especially difficult in terms of the computational cost, but for molecules of any size, this strategy becomes impractical, especially for the determination of the equilibrium geometry and the vibrational frequencies. Furthermore, if the goal is to push beyond chemical accuracy toward spectroscopic accuracy, this approach becomes completely intractable.

The solution to this problem is the use of so-called composite approaches, which are based on additivity arguments. The dominant sources of error in the calculation are treated as accurately as possible, and the remaining energy contributions are tackled with lower levels of approximation. For molecular systems with wavefunctions dominated by a single Slater determinant, one of the best foundations for such a composite scheme is based on the frozen-core CCSD(T) method. In order to mitigate the steep basis set requirements of these calculations, the systematic convergence characteristics of the correlation consistent family of basis sets of Dunning and co-workers can be very effectively exploited via simple extrapolation to the CBS limit. As detailed below, extrapolation of the results of two or more basis sets of increasing size can in many cases reduce the computational burden from requiring a very large sextuple-$\zeta$ or septuple-$\zeta$ calculation (often including augmenting diffuse functions) to one of only quadruple-$\zeta$. As also discussed below, this can be alternatively treated by using explicitly correlated methods whereby functions with explicit $r_{12}$ dependence are included in the basis set.

In the present work, Sect. 2 briefly reviews some of the more common composite thermochemistry methodologies in use to date in the quantum chemistry literature, while Sect. 3 follows with a detailed discussion of the strategies involved in effectively calculating the major components of these schemes. In Sect. 4, any modifications of these thermochemical strategies in order to calculate "chemically accurate" spectroscopic properties will be addressed with future challenges and a short summary following thereafter in Sects. 5 and 6, respectively.

# 2 Theoretical thermochemical procedures from the literature

Electronic structure methods designed to achieve "high accuracy" in predictions of thermochemical properties have proliferated over the past several decades. In the following section, we briefly discuss a select subset of methods taken from the chemical literature. This list is representative of commonly used methods and not meant to be exhaustive.

## 2.1 Gaussian-n

In 1989, Pople and co-workers introduced a composite procedure, a so-called model chemistry, named "Gaussian-1 theory" (G1) that was intended to predict properties that depended upon energy differences, such as total atomization energies, ionization potentials, as well as proton and electron affinities [4]. G1 attempted to approximate a time-consuming, high-level calculation [QCISD(T)/6-311+G(2df,p)] at a greatly reduced computational cost. It consisted of a fixed collection of well-defined steps that were easily implemented in the popular Gaussian program (it was readily invoked in the Gaussian program with the command line option G1) [5]. G1 blended high-level methods, as judged by the standards of the time, with a correction (based on the $H_2$ molecule and H atom) to achieve a target accuracy of $\pm 2$ kcal/mol in atomization energies for most of a 31-molecule test set containing elements taken from the first row of the Periodic Table. Errors for seven molecules exceeded 2 kcal/mol, with a maximum error of 4.0 kcal/mol. Subsequent extension to second-row elements yielded an accuracy of approximately $\pm 3$ kcal/mol [6]. It is important to note that the need to approach the complete basis set limit to deal with issues in the 1-particle space was not well understood when G1 was designed since the correlation consistent basis sets had just been developed [1].

Due to the "black box" nature of G1, it could be readily adopted by researchers who preferred to focus on chemical questions using an off-the-shelf, calibrated procedure. The more time-consuming alternative, which was common practice at the time, was to begin most research projects by investing resources in deciding which theoretical method (or combination of methods) was best suited for that project. In this regard, G1 represented a departure from the norm for high-accuracy methods in the 1980s but was in line with Pople's concepts about model chemistries that he had been developing for more than 30 years [7]. Because techniques were seldom applied to large numbers of atoms or molecules, drawing conclusions about the robustness and generality of the method were difficult to reach. The term "G1" served as a convenient abbreviation for a complex, 7-step procedure. However, what began as a widely used communication shorthand was heavily eroded by the explosion of model chemistries in the following decades. It became nearly impossible for the casual user to assimilate the myriad acronyms and oftentimes complex sequence of steps they represented.

Due to G1's focus on energy differences, the developers felt justified in performing most of the individual computational steps at reference geometries obtained from a much lower level of theory (MP2(full)/6-31G*, where "full" means all electrons are correlated). Although the 6-31G* basis set is inherently not capable of describing core/valence correlation effects, the choice of MP2(full) was predicated on the availability of analytic first derivatives for MP2 in the version of Gaussian available at that time. The harmonic frequencies that are needed for computing zero-point vibrational energies were acquired from scaled HF/6-31G* values. The practice of using reference geometries obtained at a much lower level of theory than the ultimate desired level, as well as frequencies (often scaled) from a lower-level calculation, is a general characteristic shared by nearly all model chemistries that were to follow.

The desire for improved accuracy led to the introduction of Gaussian-2 theory (G2) by Pople and co-workers in 1991 [8]. It started with the G1 energy and added a two-part MP2 correction designed to account for the error arising from an additivity assumption in G1 and a basis set correction involving a third $d$ function on non-hydrogen atoms and a second set of $p$ functions on hydrogen. Finally, an empirical, "higher-level correction" (HLC) based on the number of electron pairs was added. This term was determined from a best fit to the experimental atomization energies of 55 molecules for which the experimental values were well established. An important concept that arose at the time of the G2 method was the need to develop an experimental data set of energetic properties that are known to $\pm 1$ kcal/mol for testing their approach, as well as for the necessary parameterization. For a set of 39 small first- and second-row compounds, the MUD was reduced from 1.4 kcal/mol (G1) to 0.9 kcal/mol (G2). Differences exceeding 5 kcal/mol were still observed in some atomization energies. Modifications to one or more of the steps in G2 intended to reduce the computing requirements led to a number of related methods, including: G2(MP2) and G2(MP3) [9], G2(B3LYP/MP2/CC) [10], G2(MP2,SVP) [11] and G2(MP2,SV) [12].

The next major modification, Gaussian-3 theory (G3), appeared in 1998 [13]. It replaced several steps in the G2 procedure and incorporated a new higher-level correction (still empirical) as well as added corrections for spin–orbit effects in atoms and a core/valence correlation correction. For a test set of 148 experimental enthalpies of formation at 298 K, G3 reduced the MUD from 1.56 (G2) to 0.94 kcal/mol. Nonetheless, numerous cases were found where the unsigned deviations were much larger, e.g., $C_2F_4 = 4.9$, $PF_3 = 4.8$, $Na_2 = 4.0$ and $SO_2 = 3.8$ kcal/mol. As with G2, there were a number of derivative model chemistries based on a range of smaller modifications to the procedure, such as G3(MP2) [14], G3//B3LYP, G3(MP2)//B3LYP

[15], G3(CCSD), G3(MP2,CCSD) [16], G3X, G3X(MP2), G3X(MP3), G3SX [17], G3-RAD and G3X-RAD [18].

In 2007, the development of the Gaussian-4 (G4) procedure was reported by Curtiss et al. [19]. It incorporated five modifications to G3: (1) the Hartree–Fock (HF) component was extrapolated to the basis set limit, (2) an additional set of $d$ functions was added to the basis set, (3) CCSD(T) replaced QCISD(T), (4) B3LYP was used for determining the geometries and zero-point vibrational energies and (5) two new higher-level corrections replaced the former ones. However, the correction for scalar relativistic effects was removed and is now included in the HLCs. On the G3/05 test set of 454 experimental energy differences, G4 yielded a MUD of 0.83 kcal/mol versus 1.13 kcal/mol for G3. It should be noted that the experimental energy test set included many ionization potentials, electron affinities and proton affinities, which are in principle easier to calculate than heats of formation from total atomization energies. The root mean square (RMS) deviations were 1.19 kcal/mol (G4) versus 1.67 kcal/mol (G3). The number of molecules where the error exceeded 2 kcal/mol was 35, and the maximum error was 8.9 kcal/mol (ionization potential of $B_2F_4$). A follow-up paper that same year reported variants G4(MP2) and G4(MP3) whose performance on the G3/05 test set was only slightly worse than G4, e.g., $\varepsilon_{MAD}[G4(MP2)] = 1.03$ kcal/mol [20]. Thus, we see that the original G1 procedure eventually spawned a family of 22 different model chemistries based on four primary forms, each with their own strengths and weaknesses. This pattern of proliferating models will be seen to repeat itself with the other approaches discussed below.

## 2.2 Petersson-style CBS models

Petersson and co-workers have produced a large body of work focused on procedures for estimating the correlation energy at the complete basis set limit. Their publications cover more than three decades [21]. One of the earliest model chemistries to incorporate the core ideas was the $CBS^{(\infty,3)}$(Full)/DZ+P atomic pair natural orbital model, which was limited to closed-shell systems [22]. An RMS accuracy of $\pm 0.0014$ $E_h$ (0.9 kcal/mol) in total energies was claimed with a very modest double-zeta plus polarization quality basis set. A subsequent unrestricted Hartree–Fock-quadratic-CI-based method, CBS-CI(full)/(14s9p4d 2f,6s3pld)/[6s6p3d2f,4s2p1d] was reported to yield total energies accurate to $\pm 0.0012$ $E_h$ (0.8 kcal/mol) [23].

Further development work led to several new procedures that were applied to energy differences. The CBS2(FC)/6-311+G**-QCI/6-311+G†† model was found to produce dissociation energies, electron affinities and ionization energies that were only slightly worse than G1 values while being 10 times faster [24]. These conclusions were based on a small set of first-row atoms and diatomic molecules. A further factor of ten reduction in computer time was found with the CBS2(FC)/6-311+G†† model, but the RMS error in dissociation energies more than doubled (1.91 → 4.23 kcal/mol), and the error in electron affinities increased by a factor of 9 (0.18 → 1.58 eV).

Following the release of the previously discussed Gaussian-2 method, further refinements in the approach of Petersson and co-workers led to the CBS-QCI/APNO model [25]. It included a higher-order correction based on QCISD(T)/6-311++G(2df,p) calculations and a size-consistent empirical correction. Tests on a collection of 64 first-row compounds (the method was only defined for first-row elements) showed a MUD of 0.53 kcal/mol for various energy differences (dissociation energies, ionization potentials and electron affinities). Although maximum errors were not explicitly given, several errors exceeding 1 kcal/mol were evident in the included Tables. Additional tests on 20 hydrocarbon bond energies showed an RMS error of 0.9 kcal/mol versus 2.2 kcal/mol for G2 and a maximum error of 2.6 kcal/mol.

A year later, three new methods (CBS-4, CBS-q and CBS-Q) were introduced in an attempt to increase the size of the systems that could be handled to six non-hydrogen atoms by significantly lowering the computational cost [26]. The mean unsigned deviations measured with respect to the 125-member G2 test set were 2.0, 1.7 and 1.0 kcal/mol, respectively, for the three methods. Maximum errors of 5.4, 4.6 and 2.3 kcal/mol were obtained for atomization energies. These three methods and the CBS-QCI/APNO method all exploit the asymptotic convergence of the second-order Møller-Plesset (MP2) pair energies. CBS-Q was later modified to use B3LYP geometries and frequencies, leading to a method labeled CBS-QB3 [27]. That modification produced a reduction in the maximum error for the entire G2 test set from 3.9 to 2.8 kcal/mol. More recently, a modification was made to the CBS-QB3 method to allow the spin correction to be removed. In terms of heats of formation, the resulting ROCBS-QB3 method did not perform quite as well as the method it was meant to replace, $\varepsilon_{MUD} = 0.91$ kcal/mol versus 0.69 kcal/mol for CBS-QB3, based on the G2/97 test set [28].

Radom and co-workers proposed a modification to the CBS-Q method in which the geometries and zero-point vibrational energies were obtained from QCISD/6-31G(d) calculations and the highest level of theory involved CCSD(T) rather than QCI [29]. This method, which was intended for free radical thermochemistry, was called CBS-RAD. It included an empirical correction of 0.30 kcal/mol per triple or double bond.

Like the Gaussian-n methods, the collection of Petersson-style CBS model chemistries has undergone a continual evolution that has resulted in many different implementations. The CBS chemistries are also readily invoked by keywords in various versions of the Gaussian program system. We are unaware of any direct comparisons between the most recent Gaussian-n and Petersson-style methods across a large body of experimental data.

## 2.3 Weizmann-n

A decade following the appearance of the Gaussian-1 method, Martin and co-workers introduced the Weizmann-1 (W1) and Weizmann-2 (W2) "black box" model chemistries, adopting a naming convention analogous to the Gn sequence [30]. It was intended that the Wn family of models include all terms that contribute $\sim 0.24$ kcal/mol (1 kJ/mol) to atomization energies and that all of the components individually be demonstrated to have converged with respect to the 1-particle basis set to approximately this same level of accuracy. Experimentally derived empirical parameters were not to be used. Some of the earliest Wn implementations did not conform to these admittedly high standards due to the computational cost of certain terms.

Although atomic electron affinities were used to help guide the selection of the best CBS extrapolation formula, the Wn models were primarily intended for the prediction of heats of formation. W1 used B3LYP/cc-pVTZ+1 geometries and zero-point energies, with the latter being scaled by 0.985. The "+1" notation signified the addition of a tight d function for second-row elements. Separate CBS extrapolations were performed on the SCF, CCSD and (T) energy components. The most expensive single-point CCSD(T) calculation involved an aug-cc-pVTZ+2d1f basis set, while the lower-level CCSD calculation used a larger aug-cc-pVQZ+2d1f basis set.

The more expensive W2 included up through h-type basis functions. W1 included a molecule-independent, empirical parameter not present in W2. The former was reported to yield a mean unsigned deviation of 0.30 kcal/mol for atomization energies, while the latter method reduced that to 0.23 kcal/mol (or 0.18 kcal/mol for strongly single configuration dominant species). The maximum errors were 1.01 (W1) and 0.64 kcal/mol (W2). These statistical metrics were based on a collection of 28 small, well-behaved, first- and second-row molecules. For comparison purposes, the corresponding CBS-QB3 and G3 mean unsigned deviations were 0.91 and 0.86 kcal/mol. Using a different test set of 30 molecules, including the more difficult-to-describe $O_3$, W2 had a MUD of 0.40 kcal/mol and a maximum error of 3.0 kcal/mol [31]. It becomes readily apparent that the error metrics can be quite different from one small collection of molecules to the next. A number of W1 and W2 variants were subsequently introduced, including W1U, W1USc, W1BD and W1RO [32] and W1w, W2w and W2.2 [33]. Mintz et al. discussed

additional modifications labeled W2C, W2C-CAS-ACPF and W2C-CAS-AQCC methods [34].

W3, the next major Weizmann-n revision, appeared 5 years later [31]. For W3, the contribution of triple excitations was determined from CCSDT calculations with the cc-pVDZ and cc-pVTZ basis sets. The effects of quadruple excitations were based on CCSDTQ/cc-pVDZ calculations scaled by 1.2532, a value derived from a small training set. Results included atomization energies, ionization potentials and electron affinities. For the same set of 30 small molecules used with the W2 method, W3 provided a smaller mean unsigned deviation (0.22 vs. 0.40 kcal/mol) and reduced the maximum error from 3.00 to $-0.78$ kcal/mol. A number of variations have also been discussed, including W3a, W4a and W4b.

Karton et al. introduced the Weizmann-4 (W4) model for first- and second-row compounds 2 years later [33]. In this implementation, reference geometries were based on CCSD(T)(FC)/cc-pVQZ calculations. The vibrational zero-point energies were taken from the best available sources, which often translated to a combination of experimental fundamentals and theoretical anharmonic corrections. For a test set of 26 small molecules for which active thermochemical table (ATcT) [35] atomization energies (which are in principle highly accurate) were available, W4 yielded a MUD of 0.15 kcal/mol with an RMS of 0.09 kcal/mol. When the test set was expanded to 30 molecules, both error metrics became 0.15 kcal/mol. Individual cases that exceeded 0.24 kcal/mol (1 kJ/mol) included ClCN ($-0.63$), $PH_3$ ($+0.31$) and $O_3$ ($-0.24$). Table I in the Karton et al.'s [33]. paper included a total of 36 molecules, but some of these were excluded from the final statistical analysis. Among these was $C_2$ ($^1\Sigma_g^+$) which has an even larger higher-order correction than $O_3$ and for which an ATcT value is available [35]. As with the earlier Wn models, a number of related models were subsequently proposed, including W4lite, W4.2, W4.3 [36] W3.2 [33] and W4.4 [36]. Thus, we see that while the Weizmann-n model chemistries are much younger than either the Gaussian-n or the Petersson-style CBS methods, they nonetheless have spawned many variations as well.

Karton and Martin have also discussed using the W4 electronic energies evaluated at fixed geometries, i.e., not the reference CCSD(T)(FC)/cc-pVQZ geometries used in the standard application of W4, to map out portions of the potential energy surface in order to predict high-accuracy spectroscopic properties such as bond lengths, harmonic frequencies and anharmonicities [37].

## 2.4 High-accuracy extrapolated ab initio thermochemistry

In 2004, Tajti et al. introduced the high-accuracy extrapolated ab initio thermochemistry (HEAT) model which was intended to "achieve high accuracy for enthalpies of formation of atoms and small molecules" without resorting to empirical scale factors [38]. Reference geometries were determined at the CCSD(T)(CV)/cc-pVQZ level of theory, and open-shell systems were described with unrestricted Hartree–Fock (UHF) wavefunctions. The CCSD(T)(CV) basis set limit was estimated by a $1/l_{max}^3$ extrapolation [39] of aug-cc-pCVQZ and aug-cc-pCV5Z energies. A frozen-core higher-order correction was determined by combining a $1/l_{max}^3$ extrapolation of CCSDT/cc-pVTZ and CCSDT/cc-pVQZ energies with a CCSDTQ/cc-pVDZ energy. Anharmonic zero-point vibrational energies were evaluated at the CCSD(T)(CV)/cc-pVQZ level of theory. For a test set of 26 small molecules involving 5 atoms (H, C, N, O and F), HEAT produced a mean unsigned deviation of 0.09 kcal/mol relative to well-established experimental $\Delta H_f(0\ K)$ values. The maximum observed error was 0.33 kcal/mol ($C_2H_2$).

Two years later, Bomble et al. discussed a number of improvements that lead to the HEAT345-Q, HEAT345-(Q), HEAT345-Q(P) and HEAT345-Q(P) methods [40]. When applied to a set of 18 small molecules, these variants yielded mean unsigned deviations that were marginally better than the original HEAT method. The most accurate was HEAT345-QP which had a MUD that was 0.01 kcal/mol smaller than the original HEAT. The maximum error declined from 0.17 kcal/mol (HEAT) to 0.12 kcal/mol (HEAT345-QP).

In 2008, Harding et al. explored the impact of extending the basis sets used in the HEAT protocol up through aug-cc-pCV6Z and including a diagonal Born–Oppenheimer correction. This led to the HEAT456-Q and HEAT-456QP methods [41]. A statistical comparison based on a collection of 18 small molecules for which ATcT values were available showed small *increases* in the RMS errors for the larger basis set variants, e.g., 0.09 (HEAT345-Q) versus 0.14 (HEAT456-Q) kcal/mol. The maximum errors were also found to increase slightly, e.g., 0.17 (HEAT345-Q) versus 0.29 (HEAT456-Q) kcal/mol. This suggests that the earlier implementations may have benefited from a fortuitous cancellation of errors, as have many of the other methods described in the current paper. The authors commented that HEAT was not intended strictly for the purpose of computing atomization energies, but "rather for estimating the total energies of molecules".

### 2.5 Correlation consistent composite approach

The MP2-based correlation consistent composite approach (ccCA) of Cundari, Wilson and co-workers was proposed considerably later than the first four families of model chemistries as an alternative to the Gaussian-n methods [42]. The design goal was to achieve a mean unsigned deviation of $\pm 1$ kcal/mol for the same energetic quantities targeted by Gn (enthalpies of formation, electron affinities, ionization potentials and proton affinities) but without reliance on empirical corrections. Compared to the Wn models, ccCA is computationally faster and therefore can be applied to much larger systems. There have already been many variations of this basic approach. The first paper in this series discussed five different implementations (ccCA-DZ, ccCA-TZ, ccCA-aTZ, ccCA CBS-1 and ccCA CBS-2). All of them used reference geometries and scaled harmonic frequencies obtained at the B3LYP/6-31(d) level of theory. The first three of these begin with an MP4(FC)/cc-pVnZ ($n$ = D,T) or MP4(FC)/aug-cc-pVTZ calculation. The final two variations of ccCA involve an extrapolation to the MP2(FC) complete basis set limit using aug-cc-pVnZ ($n$ = D,T,Q) basis sets with either an exponential [43] or mixed Gaussian/exponential formula [44]. Finally, a QCISD(T)/cc-pVTZ correction was added to better account for correlation effects. On a collection of 28 molecules taken from the G2-2 and G3/99 test sets, the ccCA-CBS2 method yielded a mean unsigned deviation of 1.16 kcal/mol after incorporating additional corrections for atomic spin–orbit effects and scalar relativistic effects based on first-order stationary direct perturbation theory values reported by Kedziora et al. [45]. In several cases, the error in $\Delta H_f$ exceeded 3 kcal/mol, e.g., $AlCl_3$ = $-3.8$, $SF_6$ = $-3.8$ and $C_2F_4$ = $-3.5$ kcal/mol.

Subsequent modifications of ccCA (ccCA-F, ccCA-P, ccCA-S4, ccCA-S3 and ccCA-WD) explored the use of different MP2/CBS extrapolation formulas [46]. Other changes included the following: (1) replacing QCISD(T) with CCSD(T), (2) using MP2 Douglas–Kroll calculations for computing scalar relativistic corrections and (3) using cc-pV($n$ + $d$)Z basis sets for second-row elements. The MUD for the complete G3/99 test set (223 enthalpies of formation, 88 ionization potentials, 58 electron affinities and 9 proton affinities) was 0.96 and 0.97 kcal/mol with the mixed Gaussian/exponential and $1/(l_{max} + 1/2)^4$ formulas, respectively. This compares to 0.95 kcal/mol for G3X. In cases where the ccCA errors for $\Delta H_f$ were unusually large, e.g., $B_2F_4^+$ = 7.7, $C_{10}H_8$ (azulene) = $-5.3$ kcal/mol, the authors recommended a re-examination of the experimental data.

A year after the ccCA method was announced, further changes were introduced in order to handle transition metal compounds, leading to the ccCA-TM variant [47, 48]. A much looser target of "transition metal chemical accuracy", corresponding to $\pm 3$ kcal/mol, was adopted. Further modifications intended to allow the method to describe potential energy surfaces were introduced in the multireference ccCA (MR-ccCA) method [49]. By replacing the standard self-consistent field and MP2 steps with the corresponding resolution of the identity (RI) counterparts,

Prascher et al. reported improved computational efficiency with the development of the RI-ccCA and RI-ccCA+L methods [50]. The +L notation indicates that the density fitting, local CCSD(T) method of Schütz and Werner was used [51, 52]. The average savings in computer time for the RI-ccCA method was 72%. With the +L alternative, the computational savings increased slightly to 76%. However, as might be expected, both approximations introduced additional error relative to ccCA. For RI-ccCA, the average unsigned error relative to ccCA is 0.27 kcal/mol, compared to 2.63 kcal/mol for RI-ccCA+L on a test set of 120 molecules. The maximum positive and negative errors are as follows: RI-ccCA = 3.81 and −3.56 kcal/mol, RI-ccCA+L = 9.20 and −1.77 kcal/mol.

Five more modifications to ccCA were proposed in 2009, including changes to the basis set used in the B3LYP calculations, separate extrapolations of the SCF and MP2 energies and the use of new vibrational scale factors [53]. When applied to the G3/05 test set (454 comparisons), the latest version (ccCA-PS3) produced a mean unsigned deviation of 1.01 kcal/mol. The corresponding G4(MP2) value was 1.04 kcal/mol. However, maximum errors were not reported.

## 2.6 Focal-point analysis

All of the approaches considered to this point fall under the category of model chemistries. Like the original G1 method, they consist of a multi-step, fixed recipe implicitly defined for a subset of the periodic table. In many cases, this meant hydrogen and the main group elements from the first two rows. The focal-point approach (FPA) of Allen, Császár and co-workers differs in that it is more of a flexible strategy that can be tailored for particular research interests [54, 55]. In the initial 1993 report, the method was applied to the determination of the heats of formation of NCO ($^2\Pi$) (cyanato radical) and HCNO (isocyanic acid) using a mixture of experimental and theoretical reference geometries. For NCO, a CISD/DZ(d,p) optimized structure was adopted, although the CO bond length differed substantially from the experimental $r_0$ value. Five independent reactions, involving a mixture of experimental and theoretical data, were used to determine the NCO $\Delta H_f(0\ K)$. A variety of basis sets, the largest of which was an [13s,8p,6d,4f/8s,6p,4d] contraction, were combined with various levels of perturbation theory and coupled cluster theory, all of which utilized the frozen-core approximation. Essentially, CCSD(T)/(2d,1f/2p,1d) reaction energies were adjusted by MP2 energies evaluated with the largest basis set, allowing the very expensive CCSD(T) calculations in the large basis set to be avoided. The analysis was carried out using two sequences of methods: MP2 → MP3 → MP4(SDTQ) → MP5(SDTQ) and MP3 → CCSD → CCSD(T).

Calibration calculations on some prototypical bonds, e.g., $CO_2 \rightarrow O + CO$ and $N_2 \rightarrow N + N$, showed that the FPA level of theory underestimated bond strengths by as much as 6.4 kcal/mol ($N_2$). Consequently, bond additivity corrections were adopted for the four (out of five) reactions involving homolytic bond cleavage.

A 1998 focal-point study of conformational energies (e.g., the torsional barrier in ethane) replaced the basis sets used in earlier studies with members of the correlation consistent basis set family up through cc-pV6Z, aug-cc-pV5Z and cc-pCVQZ [56]. In some cases, it was necessary to use a hybrid cc-pVTZ(C,N,O)/cc-pVDZ(H) basis set. Correlation recovery was accomplished through fifth-order perturbation theory or through coupled cluster calculations at the CCSD, CCSD(T) and CCSDT levels. Geometries were optimized with CCSD(T)(CV)/cc-pVTZ or CCSD(T)(CV)/aug-cc-pVTZ calculations. The largest basis set used in a CCSD(T) calculation varied from system to system. In the case of ethane, the largest basis set was aug-cc-pVTZ. Relativistic corrections were determined with first-order perturbation theory using the mass velocity and one-electron Darwin terms [57]. The diagonal Born–Oppenheimer correction was computed at the Hartree–Fock level of theory [58]. Once again, there were two sequences of correlation recovery procedures, one based on perturbation theory through fifth order and another based on coupled cluster theory through CCSDT. A Padé approximant was used to extrapolate the perturbation theory sequence to infinite order. The authors described this as a two-dimensional extrapolation grid in the 1-particle and n-particle expansions.

There have been a substantial number of studies involving the focal-point method, but a comprehensive survey is beyond the scope of the present work. As an illustration of a more up-to-date application of the method, we consider a 2004 study which re-examined the heats of formation of NCO and HNCO by Allen and co-workers [59]. An early focal-point study had reported $\Delta H_f(0\ K)$ values of 35.3 kcal/mol (NCO) and −26.1 kcal/mol (HNCO) [60]. Another study that same year by the same authors offered revised values of $31.4 \pm 0.5$ kcal/mol (NCO) and $-27.5 \pm 0.5$ kcal/mol (HNCO) [55]. In the 2004 investigation, the heat of formation of NCO was established via two formation reactions. The corresponding heat of formation for HNCO involved seven additional reactions. After accurately determining the reaction energies, reference heats of formation for component species were taken from experiment or in a few cases from theory. This approach seeks to avoid computing total atomization energies in favor of the use of isogyric reactions. The MPn ($n = 2$–5) scheme used in previous FPA studies was abandoned in favor of the HF → MP2 → CCSD → CCSD(T) → CCSDT sequence of methods. The largest

CCSD(T) calculation employed the cc-pV5Z basis set. Separate extrapolations were performed on the Hartree–Fock and correlation energy components. Core/valence correlation effects were treated with up to CCSD(T)(CV)/cc-pCVQZ and MP2(CV)/cc-pCV5Z calculations, followed by extrapolation to the CBS limit.

For closed-shell systems, the MP2/CBS limit was estimated with explicitly correlated MP2-R12/A calculations performed with an uncontracted cc-pV5Z basis set augmented with additional tight and diffuse functions. Open-shell systems were described with conventional UHF-MP2(FC)/cc-pVnZ ($n$ = D–5) followed by extrapolation to the CBS limit. In their previous studies, the highest level of correlation recovery was obtained from CCSDT computations. For the newer work, the impact of quadruple excitations was explored via a series of homolytic bond breaking reactions, three of which involved a single bond and four of which involved a multiple bond. For this purpose, the authors chose the Brueckner orbital coupled cluster theory, BD(TQ) and CCSD(2) methods. Although the impact of quadruple excitations was sometimes found to be "not negligible", the two approaches to including quadruples gave "disconcerting inconsistencies" and an "anomalously large correction" in certain cases. Therefore, they recommended not attempting to correct for quadruple excitations until full-blown CCSDTQ calculations were affordable for all of the species in their formation reactions. The final recommended 0 K heats of formation were $30.5 \pm 0.2$ kcal/mol (NCO) and $-27.6 \pm 0.2$ kcal/mol (HNCO). The latter is in essentially exact agreement with the 1993 value of $-27.5 \pm 0.5$ kcal/mol, and the former falls slightly outside the error bars of the earlier value.

To the best of our knowledge, there have been no surveys comparing the performance of the focal-point approach to other approaches for high-accuracy thermochemistry across a broad range of molecules.

## 2.7 Feller-Peterson-Dixon procedure

The general strategy of this approach developed by the current authors for the studies of thermochemical or spectroscopic properties such as molecular structures and vibrational frequencies is similar in spirit to the focal-point analysis technique of Allen, Császár and co-workers in that it is a flexible, multi-step approach intended to address all of the major sources of error, as opposed to a fixed recipe model chemistry. For the sake of this discussion, the approach will be referred to as the Feller–Peterson–Dixon (FPD) procedure that has been under development for almost 15 years. The overall goal is to include all physically significant effects that contribute to the property of interest at a level that will guarantee results within the target accuracy. For example, if the goal is computing the dissociation energy of a heavy element diatomic like HI ($^1\Sigma^+$) to an accuracy of $\leq 0.5$ kcal/mol, a second-order molecular spin–orbit correction would be included even though for lighter elements the contribution is negligible [61, 62]. Consequently, the basic approach can be easily modified to include effects that contribute significantly in some cases and not in others. This characteristic enables it to address many properties across a large portion of the periodic table. An important feature of the approach is that, whenever possible, an attempt is made to gauge the uncertainty in each of the (up to) eight components required to achieve a final value. Even a rough knowledge of the uncertainty in the individual components improves our ability to avoid combining pieces with markedly different accuracies, so that computational resources can be directed to the most problematic areas where they can be used most effectively. While this approach differs in detail from FPA and the other high-level approaches already discussed, it also shares many common features.

For thermochemical studies, the initial step in the procedure involves a series of frozen-core CCSD(T) geometry optimizations using the diffuse function–augmented correlation consistent sets, denoted aug-cc-pVnZ and aug-cc-pV($n$ + d)Z, $n$ = D, T, Q, 5,…,10. The wide variation in available basis set sizes allows the treatment of fairly large systems, such as $C_8H_{18}$, [63] with up through quadruple-$\zeta$ quality sets or the ability to employ sets of 10-$\zeta$ quality for atoms [64, 65]. The largest basis set possible for the system of interest is used in order to reduce the reliance on basis set extrapolation procedures, e.g., use of aug-cc-pV6Z through aug-cc-V8Z whenever they are computationally tractable [3, 62, 66–68]. In the FPD approach, a combination of up to five CBS extrapolation formulas is utilized, taking the average as the best estimate and half the spread in the values as a crude measure of the associated uncertainty. The application of these formulas have been discussed in detail elsewhere [62]. A very recent study of the effectiveness of these formulas across a collection of 141 small-to-medium size chemical systems for which reliable estimates of the CBS limit were available supports the accuracy of this approach (see below) [3]. Open-shell calculations are based on the R/UCCSD(T) method, which is based on restricted open-shell Hartree–Fock (ROHF) orbitals but allows a small amount of spin contamination in the solution of the CCSD equations for open-shell systems. Calculations on isolated atoms imposed full atomic symmetry on the orbitals. As in the HEAT or Wn methods, higher-order (HO) correlation effects are handled with the CCSDT and CCSDTQ or CCSDT(Q) sequence of methods. Combinations such as CCSDT/cc-pVQZ+CCSDTQ/cc-pVTZ or CCSDT/cc-pVTZ+CCSDTQ/cc-pVDZ are typically used to provide a balanced treatment. When both combinations are affordable, the difference between them provides a measure of the uncertainty in the HO correction. In a limited number of cases, CCSDTQ5 or even FCI calculations have been performed. In most other instances, the FCI result has been estimated by using a continued fraction (cf) approximant that involves the CCSD, CCSDT and CCSDTQ sequence of energies [62, 69].

Corrections for outer core/valence (CV) correlation effects are treated with the weighted core/valence basis sets [70], cc-pwCV$n$Z, $n$ = D, T, Q and 5 with extrapolation to the CBS limit using the same procedure as used for the frozen-core energies. It is also possible to compute the CV correction using higher-order correlation methods, but the calculations are extremely time-consuming. For molecules with 4–6 "heavy" atoms, it seems that generally the HO CV correction typically falls into the 0.1–0.2 kcal/mol range. Thus, they are only required in studies aiming for the highest possible accuracy. Scalar relativistic (SR) effects are described with second-order Douglas-Kroll-Hess (DKH) CCSD(T)(FC) calculations [71, 72] using the cc-pV$n$Z-DK basis sets [73]. Spin–orbit coupling effects in the atoms are taken from the experimental zero-field splittings, while, when applicable, molecular SO corrections are taken either from experiment or calculated with the state-interacting approach using CI wavefunctions.

Anharmonic zero-point vibrational energies (ZPVEs) for diatomic molecules are obtained from sixth-degree Dunham fits of the potential energy curves [74]. For triatomic systems fits to potential energy surfaces are often used with ZPVEs calculated using second-order vibrational perturbation theory. Finally, anharmonic ZPEs for general polyatomic species are determined by combining CCSD(T) harmonic frequencies with anharmonic corrections obtained from second-order Møller-Plesset perturbation theory (MP2). For example, CCSD(T)(FC)/aug-cc-pVQZ harmonic frequencies might be combined with MP2(FC)/aug-cc-pVTZ anharmonic corrections. Whenever possible, the degree of convergence in the ZPE is tracked as the underlying basis sets are improved.

Another factor considered in studies aiming for very high accuracy, especially when the molecule contains hydrogen atoms, is the effect of the diagonal Born–Oppenheimer correction (see, e.g., Ref. [75]). Typically, the FPD approach uses the aug-cc-pVTZ basis set at either the CISD or CCSD levels of theory to evaluate this component.

As an indication of the performance of the FPD procedure for atomization energies, a mean signed deviation of $-0.04$ kcal/mol, RMS = 0.28 kcal/mol and MAD = 0.17 kcal/mol is found when comparing against 121 molecules whose experimental uncertainties are $\pm 1$ kcal/mol or less [62]. If the comparison is restricted to molecules with an experimental uncertainty of $\pm 0.3$ kcal/mol or less, the MAD falls to 0.10 kcal/mol (61 comparisons).

Molecular structures are also accurately predicted with this approach. Mean unsigned deviations for CCSD(T)(FC) bond lengths between non-hydrogen atoms drop from 0.029 Å (281 comparisons) with the aug-cc-pVDZ basis set to 0.012 Å with aug-cc-pVTZ and 0.007 Å with aug-cc-pVQZ [62]. At the estimated basis set limit and including all of the minor corrections discussed above except the DBOC, a MAD value of 0.0006 Å is found (103 comparisons). Specific comparisons with accurate semi-experimental bond lengths for polyatomic molecules have been discussed elsewhere [76].

## 3 Components of an accurate composite methodology

Among the composite methods reviewed above, many include a common collection of similar contributions. In this section, the most important are discussed with an emphasis on the calculation of molecular atomization energies (needed for heats of formation) since these are among the most demanding properties to predict. In general, the conclusions based on this particular property also carry over to equilibrium geometries and vibrational frequencies.

### 3.1 Estimating the frozen-core, CCSD(T) complete basis set limit

Notwithstanding molecules containing alkali and alkaline earth metals, the basis set incompleteness error in the correlation of the *valence* electrons tends to be one of the largest sources of error in the calculation of an atomization energy, particularly when the accurate CCSD(T) method is used. A detailed discussion of the error introduced separating the frozen-core and core correlation contributions is given in the next section. Figure 1 shows the basis set convergence of the total energies and dissociation energies toward the CBS limit for the diatomic series $N_2$ through $Bi_2$ using a sequence of correlation consistent basis sets. For these results the aug-cc-pV$n$Z sets [1, 77] were used for $N_2$, aug-cc-pV($n$ + $d$)Z for $P_2$, [78] and aug-cc-pV$n$Z-PP was used [79] for $As_2$ through $Bi_2$. The -PP (pseudopotential) basis sets are accompanied by a Stuttgart-Köln, energy consistent, small-core effective core potential that replaces 10 electrons in As, 28 electrons in Sb and 60 electrons in Bi. The CBS limits were obtained by extrapolation of the total energies ($n$ = 4 and 5 for QZ and 5Z, respectively) by the formula (see below)

$$E(n) = E(CBS) + A\left(n + \frac{1}{2}\right)^{-4}. \tag{1}$$

Whereas there is considerable spread in the basis set error at the double-zeta level, smooth regular convergence is

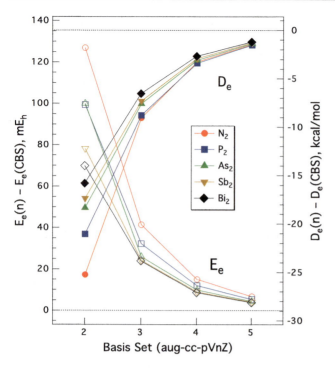

**Fig. 1** Convergence of the frozen-core CCSD(T) total and dissociation energies with respect to the CBS limit of $N_2$ through $Bi_2$ as a function of the correlation consistent basis set size (aug-cc-pV$n$Z, aug-cc-pV$(n+d)$Z, or aug-cc-pV$n$Z-PP)

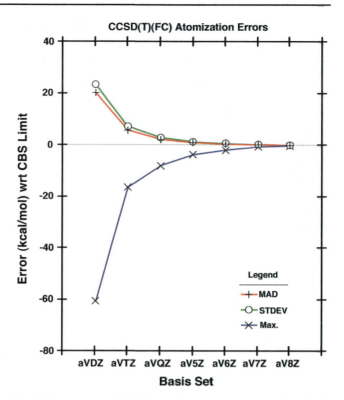

**Fig. 2** Frozen-core CCSD(T) atomization energy errors (kcal/mol) with respect to the estimated complete basis set limit as a function of the correlation consistent basis set size (aV$n$Z = aug-cc-pV$n$Z and/or aug-cc-pV$(n+d)$Z) for 141 molecules

observed. Curiously, at least for these simple non-polar molecules, the convergence rate is the slowest for $N_2$ and fastest for $Bi_2$. Presumably this is related to the decrease in the correlation energy between these species. To the extent one can generalize from the data in this figure, it seems likely that to obtain an accuracy of 1 kcal/mol in the atomization energy from a calculation with a single finite basis set, one would need a set of at least aug-cc-pV6Z quality. This conjecture is strengthened by Fig. 2, which demonstrates the slow convergence of atomization energies for 141 molecules of various sizes, mostly consisting of elements of the first and second rows. Beginning with the aug-cc-pVTZ basis set, each increment in basis set index cuts the error metrics by approximately a factor of two. Depending on which error metric one chooses to employ, achieving an accuracy of ±0.5 kcal/mol or less in this component without the use of extrapolation techniques to the CBS limit would require the use of basis sets on the order of aug-cc-pV7Z. Such calculations are currently only possible for very small molecules. This analysis was based on data contained in the Computational Results Database (CRDB) [80]. The basis set errors can also be strongly influenced by the molecular size. Figure 3 shows the basis set errors in the CCSD(T) atomization energies for a series of simple alkanes from $C_2H_6$ to $C_6H_{14}$ [68]. While the results for the smallest member of this series mimics the behavior seen in Figs. 1 and 2, there is a strong size dependence capped by the nearly 10 kcal/mol error for $C_6H_{14}$ at the aug-cc-pVQZ level. Whereas it is clear that for a size-extensive method like CCSD(T), the errors of the individual energies should scale with system size, it is perhaps a bit disconcerting that the relative energy should also display such a strong size dependence. This would appear to indicate that as the molecules increase in size not only is the correlation treatment more computationally intensive due to the growth in the number of electrons, but a larger basis set must also be used to maintain the same level of accuracy in the atomization energy. As discussed in detail below, this can be mitigated by effective basis set extrapolation strategies.

In lieu of basis set extrapolation, there is a more fundamental technique to improve the basis set convergence rate at the correlated level, namely the use of explicitly correlated methods. The slow convergence of the correlation energy with basis set size is due to the poor description of the electron coalescence cusp by products of one-electron functions when the interelectronic distance $r_{12}$ becomes small. Recently developed F12 methods [81, 82] utilize a Slater-type correlation factor, $F_{12} = e^{-\gamma r_{12}}$, and exhibit much faster convergence toward the CBS limit. Figure 4 compares the convergence of conventional CCSD(T) and CCSD(T)-F12b atomization energies toward

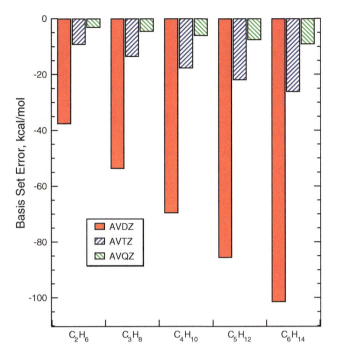

**Fig. 3** Convergence toward the CBS limit of the frozen-core CCSD(T) atomization energies for a series of simple alkanes (AV$n$Z = aug-cc-pV$n$Z, $n$ = D, T, Q)

the CBS limit for a set of 51 small molecules [3]. The F12b calculations [83] have used the cc-pV$n$Z-F12 sequence of basis sets [84], which are similar in size to the standard aug-cc-pV$n$Z [aug-cc-pV($n$ + $d$)Z for the second row] sets. From Fig. 4, it is clear that the use of F12 methods improves the basis set convergence rate by about two basis set index ($n$) levels, e.g., the cc-pVDZ-F12 set has errors similar to conventional calculations utilizing aug-cc-pVQZ basis sets and the cc-pVQZ-F12 results are very similar to those obtained from computationally expensive aug-cc-pV6Z calculations. Some of the benefits of F12 calculations exhibited in Fig. 4 are accomplished by a reduction in the HF basis set error by the effective use of the auxiliary basis set required in the explicitly correlated treatment [85]. The evaluation of additional multi electron integrals does result in an increased computational cost for F12 calculations compared to the analogous conventional cases, but, with the CCSD(T)-F12b method, the conventional (T) correction begins to again dominate the calculation already at the cc-pVTZ-F12 level. From the results shown in Fig. 4, chemical accuracy in the atomization energies of these small molecules is obtained on average with just the cc-pVTZ-F12 basis set and in the worst case with cc-pVQZ-F12. As shown in Fig. 5 for the same saturated hydrocarbon series as in Fig. 3, while a similar increase in basis set error with system size is evident (although perhaps now approaching a maximum at $C_5H_{12}$), even the $C_6H_{14}$ molecule is nearly converged to within $\sim$1 kcal/mol at the CCSD(T)-F12b/cc-pVQZ-F12 level of theory.

Given the regular convergence patterns clearly observed in Figs. 1, 2, 3, 4, and 5, it is not surprising that there have been a tremendous number of publications proposing and testing various basis set extrapolation procedures, nearly exclusively using correlation consistent basis sets. It is outside the scope of the present contribution to review this literature, but the reader is referred to the recent comprehensive study of Ref. [3]. A total of 5 commonly used CBS extrapolation formulas were studied, as well as an average of the results obtained when these formulas are each used on the total energies. The latter approach has generally

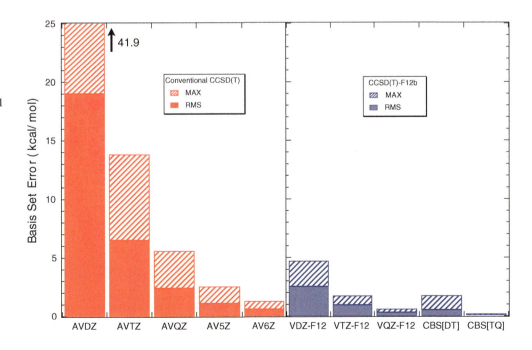

**Fig. 4** Comparison of the convergence toward the CBS limit of frozen-core conventional CCSD(T) and explicitly correlated CCSD(T)-F12b calculations of the atomization energies of 51 small molecules. Also shown are the F12b CBS extrapolated values, CBS[DT] and CBS[TQ]. See the text. (AV$n$Z = aug-cc-pV$n$Z or aug-cc-pV($n$ + $d$)Z and V$n$Z-F12 = cc-pV$n$Z-F12, $n$ = D, T, Q)

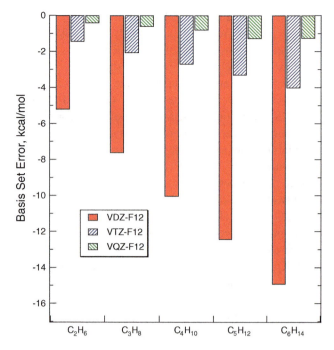

**Fig. 5** Convergence toward the CBS limit of the frozen-core CCSD(T)-F12b atomization energies for a series of simple alkanes

HF energy using the Karton and Martin relation [86] followed by an extrapolation of the correlation energy by $E(n) = E(CBS) + An^{-3}$. Whereas the RMS error is very similar to that of Eq. 1, the maximum error is noticeably larger and was generally positive in sign. Compared to the basis set errors of Fig. 2, all of the formulas shown in Fig. 6 are very successful at reducing the basis set incompleteness error; both the RMS and mean unsigned errors decrease by factors of 3–4. Given the smooth convergence with basis set shown in Fig. 4 for the CCSD(T)-F12b method, it is not surprising that CBS extrapolations can be beneficial in these calculations as well. Also shown in Fig. 4 are the results of extrapolating both DZ/TZ and TZ/QZ basis set pairs using the Schwenke-style approach [87] of Hill et al. [88]. For the small DZ/TZ pair, the RMS error is decreased by a factor of two compared to the raw cc-pVTZ-F12 result, but no improvement is observed for the maximum error. On the other hand, extrapolation of the TZ/QZ pair yields results on par with the estimated accuracy of the reference data, i.e., about $\sim 0.1$ kcal/mol. Both conventional and F12 extrapolations were used previously on the series of alkanes in Figs. 3 and 5. In these cases, the DZ/TZ extrapolations from the F12 calculations were in excellent agreement with the more computationally intensive F12 TZ/QZ results, as well as the conventional TZ/QZ extrapolations, although the latter had relatively large estimated uncertainties.

In summary, in order to reliably converge the frozen-core CCSD(T) atomization energies to better than 1 kcal/mol one has at least three choices:

1. perform a conventional CCSD(T) calculation with basis sets of aug-cc-pV6Z to aug-cc-pV7Z quality,

been used in the FPD composite method outlined above. With a reference set of 141 relatively small molecules, Eq. 1 applied to the total energies and the 5-formula average exhibited the best statistical results on the whole. Both the MUD and RMS across the entire 141 molecule dataset are shown in Fig. 6 for four of the approaches described in Ref. [3]. In Fig. 6, the results for the popular $\ell_{max}^{-3}$ formula [39] involved a separate extrapolation of the

**Fig. 6** A comparison of the effectiveness of various extrapolation formulas. Only the $\ell_{max}^{-3}$ results employed separate extrapolations of the HF and correlation energy components. Except for the mixed exponential/gaussian formula, only the largest two basis sets in each group were used in the extrapolations. See the text

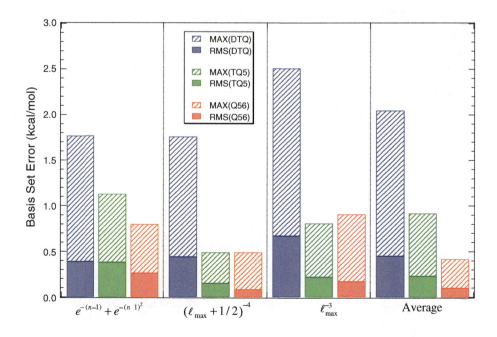

2. use conventional basis sets of at least aug-cc-pVQZ and aug-cc-pV5Z quality to be used with a CBS extrapolation with a formula such as Eq. 1, or
3. perform a CCSD(T)-F12b/cc-pVQZ-F12 calculation.

In the above summary, it is assumed that the outcome of any given calculation might result in the maximum possible errors that are shown in Figs. 2, 4 and 6. To obtain accuracies in the 1–2 kJ/mol range, it seems clear that either a basis set extrapolation using conventional CCSD(T) with sets of at least aug-cc-pV6Z quality must be used or a TZ/QZ extrapolation at the CCSD(T)-F12b level of theory using cc-pV$n$Z-F12 basis sets.

### 3.2 Contributions due to correlation of outer-core electrons

Since the estimation of the frozen-core CBS limit is presumed to be the dominant source of error in this CCSD(T)-based procedure, it is important that this and all subsequent contributions be converged with respect to basis set and level of theory to much better than 1 kcal/mol so that the final composite result will achieve chemical accuracy. For the recovery of core correlation effects, only the outer-core shell of electrons is usually treated, i.e., the principal shell of electrons lying just below the valence electrons ($1s$ for 1st row, $2s2p$ for 2nd row, etc.). Basis sets specially designed for correlating these types of electrons, e.g., cc-pwCV$n$Z or cc-pCV$n$Z [70, 89], must be used in order to minimize spurious basis set superposition errors. In principle, the combination of core and valence correlation could be combined into a single contribution by using a sequence of core/valence correlation consistent basis sets. Based on the discussion in the previous section, large sets of at least aug-cc-pwCV5Z sizes would be required for chemical accuracy (with CBS extrapolations), while very large cc-pwCV6Z or larger basis sets would be needed for more accurate work. As mentioned above, this is the procedure utilized in the HEAT method. All other composite schemes choose to separate the two contributions, and the rationale relies on the fact that the effect of core correlation on properties such as atomization energies generally converges much more rapidly with basis set than the frozen-core contribution.

As a simple example, consider again the diatomic series N$_2$ through Bi$_2$. In these cases, the definition of outer-core correlation ranges from including just the (4) $1s$ electrons in N$_2$ to the (20) $5d$ electrons in Bi$_2$ (it should be noted that generally the $(n-1)sp$ electrons are included in the outer-core definition for the post-$d$ elements but these relatively small contributions have been neglected in this discussion). Figure 7 shows the convergence with basis set of the CCSD(T) core correlation effects on the dissociation energies, i.e., $\Delta D_e = D_e(\text{valence} + \text{core}) - D_e(\text{valence})$ using aug-cc-pwCV$n$Z basis sets [70] (aug-cc-pwCV$n$Z-PP for As, Sb, and Bi) [90] on *both* the frozen-core (valence) and core correlated (val + core) calculations. The CBS limits were obtained via extrapolation with Eq. 1. The use of core/valence basis sets for both calculations (FC and core correlated) is essential since the tight functions that are included in the core/valence basis sets do make small contributions to the frozen-core correlation energy. Using a basis set such as cc-pwCV$n$Z for valence + core correlation but using only cc-pV$n$Z in the frozen-core calculation will result in an overestimation of the core correlation contribution. Even if the results calculated in this way are extrapolated to the CBS limit, the problem can persist since the accuracy of the frozen-core extrapolation will depend on whether valence or core/valence basis sets are used. Returning to Fig. 7 and comparing these results to those of Fig. 1 clearly show that the core correlation contributions to $D_e$ are much more rapidly convergent with basis set than the frozen-core dissociation energies. In the former case, convergence to within 1 kcal/mol is achieved with core/valence basis sets two levels smaller in the basis set index than what was required for the FC dissociation energy. It should be stressed, however, that the cc-pwCV$n$Z series of core/valence basis sets converge this contribution

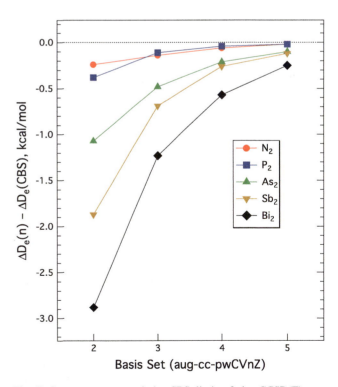

**Fig. 7** Convergence toward the CBS limit of the CCSD(T) core correlation contributions to the dissociation energies of N$_2$ through Bi$_2$ as a function of basis set (aug-cc-pwCV$n$Z or aug-cc-pwCV$n$Z-PP)

considerably faster than the cc-pCVnZ basis sets, so, if the latter are used, the comparison may not be nearly as dramatic [41]. Except for the Bi$_2$ case, even basis sets of aug-cc-pwCVTZ(-PP) level are sufficient for chemical accuracy. It is obvious, however, that accurate recovery of core correlation for molecules containing post-$d$ elements is much more challenging than in the case of molecules containing just first- and second-row elements. The core correlation effects are much larger as well; at the CBS limit, the core correlation contributions to $D_e$ for these 5 diatomics are 0.80 (N$_2$), 0.77 (P$_2$), 2.39 (As$_2$), 3.16 (Sb$_2$) and 4.34 (Bi$_2$) kcal/mol. It should be noted that for these cases, the "core d" orbitals are not active in the valence space and that compounds such as AsF$_3$, SbF$_3$ and BiF$_3$ may require these d orbitals to be treated in the valence space since the F 2s orbitals are of comparable energy to the "core d" orbitals. Of course the core correlation contribution also grows with the molecular size, e.g., the CCSD(T)/CBS limit values for the core correlation contributions to the atomization energies of C$_2$H$_6$, C$_3$H$_6$ and C$_4$H$_6$ have been reported to be (in kcal/mol) 2.42, 3.72, and 4.78, respectively [62]. For a system as large as C$_8$H$_{18}$ (n-octane), it is has been calculated to be at least 8.60 kcal/mol [63].

Given the smooth convergence with basis set for the core/valence effects shown in Fig. 7, it is not surprising that extrapolation is also effective in this case as well. Using Eq. 1 with $n = 3$ and 4 with the aug-cc-pwCVnZ series of basis sets (TZ and QZ, respectively) yields results nearly identical to the CBS limits obtained with $n = 4$ and 5, except for Bi$_2$, where the [TQ] extrapolation underestimated the more accurate [Q5] CBS limit by 0.18 kcal/mol. It should be noted that a CCSD(T) [TQ] extrapolation of aug-cc-pwCVnZ basis sets is the same level of theory used in the W4 method [33] to accurately account for core correlation effects. For the series of molecules of Fig. 7, even a [DT] extrapolation is reasonably effective, yielding a factor of 2 or better improvement compared to the raw aug-cc-pwCVTZ results.

Given the dramatic improvement in the basis set convergence rate when explicitly correlated F12 methods are used, it might be expected that similar behavior will occur for core correlation effects. This is indeed the case and CCSD(T)-F12b results analogous to those shown for conventional CCSD(T) in Fig. 7 for N$_2$ through Bi$_2$ are shown in Fig. 8. These results used the cc-pCVnZ-F12 basis sets of Hill et al. [91]. for N$_2$ and P$_2$ and the newly developed cc-pVnZ-F12-PP sets of Peterson et al. [92] for the heavier elements. In the F12 cases, convergence to within about 0.2–0.5 kcal/mol is observed already at the DZ level, while at least 0.2 kcal/mol accuracy is obtained in all cases at just the TZ level. Some small non-monotonic behavior for Bi$_2$ is observed, but it is only on the order of 0.2 kcal/mol. The

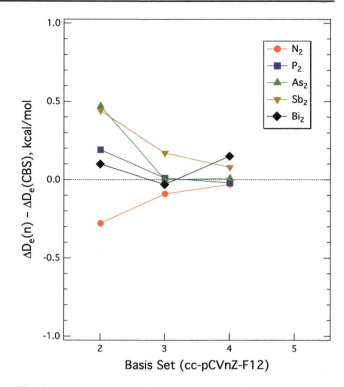

Fig. 8 Convergence toward the CBS limit of the core correlation contributions to the dissociation energies of N$_2$ through Bi$_2$ calculated with CCSD(T)-F12b as a function of basis set (cc-pCVnZ-F12 or cc-pVnZ-F12-PP)

latter can probably be attributed to the cc-pVnZ-F12-PP basis sets, which were only minimally augmented for outer-core correlation. In any event, it is obvious that F12 methods have the potential of providing a very efficient means to accurately account for core correlation effects on atomization energies.

3.3 Accounting for relativistic effects

*3.3.1 Scalar relativity*

For molecules containing elements from H through Ar, the effects of scalar relativity on atomization energies is generally small, on the order of a few to just several tenths of a kcal/mol for small molecules, and this will increase with molecular size. There are exceptions, of course, such as the sulfur oxides SO$_2$ and SO$_3$, where the scalar relativistic contributions have been calculated to be −0.78 and −1.76 kcal/mol, respectively [62], as well as a contribution of −2.3 kcal/mol for C$_8$H$_{18}$ [63]. The first-row transition metals already exhibit strong effects due to scalar relativity, for example the contribution to the dissociation energy of CuH amounts to 2.5 kcal/mol [93]. For molecules containing post-3$d$ elements, the effect can also be very large. For example, the scalar relativistic effect on the atomization energy of GaCl$_3$ has been calculated to be 6 kcal/mol

[94]. Most composite methodologies have focused on three techniques for recovery of scalar relativistic effects: the 1-electron mass velocity and Darwin operators from the Breit-Pauli Hamiltonian in first-order perturbation theory (MVD) [57, 95], the 2nd-order (or sometimes $3^{rd}$ order) Douglas-Kroll-Hess (DKH) Hamiltonian [71, 96], and in a few cases stationary direct perturbation theory (DPT) [97, 98]. The MVD approach is fairly accurate for light elements but can lead to small errors compared to more accurate treatments even for molecules containing second-row elements like $SO_2$ [99]. In addition, the MVD approach is much more sensitive to the underlying basis set than other methods, and generally requires uncontracting at least the $s$ functions of the basis set for the most accurate results [100, 101]. The workhouse for ab initio thermochemistry and spectroscopy, however, is the 1-electron DKH method, typically truncated at 2nd-order (DKH2). After recontracting the underlying basis set for DKH, e.g., cc-pVnZ-DK [73], the scalar relativistic correction is simply obtained as the difference in two energy evaluations, one with and one without the DKH hamiltonian, e.g., DK-CCSD(T)/cc-pVTZ-DK–CCSD(T)/cc-pVTZ. For relative energies, DKH3 does not seem to be required until one encounters elements as heavy as $Z = 54$ (Xe) [102]. The basis set dependence of the scalar relativistic contributions evaluated with DKH2 is not overly strong. Convergence to within $\sim 0.1$ kcal/mol is often reached at the TZ level and is converged to within a few hundredths of a kcal/mol with aug-cc-pVQZ-DK [99].

For molecules containing heavy elements, scalar relativistic effects should be included from the beginning (see Sec. III A and B above), either in all-electron calculations with the DKH Hamiltonian or using accurate relativistic effective core potentials (ECPs) or pseudopotentials (PPs). The family of correlation consistent basis sets has been extended to much of the periodic table in conjunction with newly optimized Stuttgart-style PPs, e.g., cc-pVnZ-PP [79, 103–106]. When the latter approach is taken, scalar relativistic effects are accurately taken into account for the heavy elements, but scalar relativity on any remaining light elements is not accounted for in addition to the small errors due to using the pseudopotential approximation. To this end, a DKH correction has been used, generally using TZ-level correlation consistent basis sets, i.e., $\Delta DK = DK$-CCSD(T)/cc-pVTZ-DK–CCSD(T)/cc-pVTZ-PP, where in the 2nd term standard cc-pVTZ basis sets are used on any light elements that do not employ PPs. In a series of diatomic molecules containing $5d$ metal atoms, this "PP correction" ranged from 4.9 kcal/mol for HfO to just $-0.07$ kcal/mol for IrN [107]. In the case of the halogen oxides BrO and IO, this contribution amounted to at most 0.17 kcal/mol [108]. It should be noted at this point that the PPs used in the cc-pVnZ-PP basis sets have been adjusted

to data that contain contributions not included at the DKH level of theory, namely the Breit interaction, which can become important for heavy elements, e.g., $5d$ and post-$5d$ elements. The above DKH correction effectively removes these contributions.

### 3.3.2 Spin–orbit effects

For molecules involving only light elements, spin–orbit (SO) effects on atomization energies are generally limited to the spin–orbit fine structure of the atoms, which can be easily extracted from experiment [109] for lighter atoms by simply $J$-averaging over the levels of the ground-state term:

$$\Delta E_{SO} = -\frac{\sum_J (2J+1)E_J}{\sum_J (2J+1)} \qquad (2)$$

For heavier atoms, it may not be possible to use this simple expression due to mixing of different spin/spatial states with the ground state. For diatomics or linear molecules with non-zero spin and orbital angular momentum, e.g., the $X^2\Pi$ state of NO, the spin–orbit contribution is given as one-half the spin–orbit constant obtained from experiment. In either case, these quantities can also be accurately calculated at the configuration interaction (CI) level using 1st-order perturbation theory and the Breit-Pauli Hamiltonian [110]. For heavier elements, even starting with the post-$3d$ elements, 2nd-order spin–orbit molecular effects begin to have non-negligible contributions. For example, molecular SO effects on the closed-shell BrF and $Br_2$ molecules contribute 0.3 and 0.4 kcal/mol, respectively, to their dissociation energies [61]. In the case of $I_2$, this effect increases to 1.6–2.0 kcal/mol [61, 102]. These contributions can be recovered using a state-interacting approach using either the all-electron Breit-Pauli Hamiltonian or spin–orbit potentials from relativistic effective core potentials (ECPs), but the large number of electronic states that need to be included to accurately recover these 2nd-order effects can be cumbersome [111]. In the case of small molecules, this can be accurately overcome by carrying out a double-group relativistic CI calculation with ECPs [112]. Perhaps the most promising avenue is the use of 2-component Kramers-restricted CCSD using SO ECPs. This approach has seen only very limited use, however, in composite schemes due in part to its lack of availability in commonly used software [113].

### 3.4 Electron correlation beyond CCSD(T)

For molecules whose wavefunctions are dominated by the HF determinant, the contributions discussed to this point are usually sufficient to achieve chemical accuracy for an atomization energy at the equilibrium geometry. For cases with multireference character or when more accurate results (<1 kcal/mol) are required, extension of the correlation treatment beyond CCSD(T) is required. Estimating the remaining contributions between full CI (FCI) and CCSD(T) is generally carried out by performing calculations with higher-order coupled cluster methods (CCSDT, CCSDT(Q), CCSDTQ, etc.). In the majority of cases, these higher-order correlation corrections are carried out at the frozen-core level of theory with relatively small basis sets since the computational expense scales very strongly with both the number of correlated electrons and basis functions. Studies have also shown that smaller basis sets are justified with higher-order correlated methods because the corrections due to each step up in the excitation level converges much more rapidly in terms of the basis set size than the previous excitation level [64]. It also has been generally observed in the literature that CCSD(T) atomization energies (and often bond lengths) are fortuitously closer to the FCI limit than the CCSDT values. Hence, any coupled-cluster-based correction for higher-order correlation should also include some estimate of connected quadruple excitations. To make matters somewhat more complicated, in the majority of cases the CCSDT correction, defined as CCSDT–CCSD(T), and the quadruples correction, e.g., CCSDTQ–CCSDT, enter with opposite signs and often with comparable magnitudes. So extra care must be taken to balance the convergence with respect to basis set of the two contributions, e.g., a CBS extrapolation of the CCSDT contribution combined with a CCSDTQ or CCSDT(Q) correction using just a cc-pVDZ basis set is not generally recommended [62]. A balanced approach might involve CBS extrapolation of both components, e.g., using TZ/QZ for CCSDT and DZ/TZ for CCSDTQ or CCSDT(Q), or by simply utilizing cc-pV$(n + 1)$Z for the CCSDT correction and cc-pV$n$Z for the quadruples since CCSDT has a stronger observed basis set dependence than CCSDTQ [33, 36, 114].

Going beyond corrections for connected quadruple excitations is often necessary for high accuracy. However, extending the procedure to pentuples, i.e., CCSDTQ5–CCSDTQ, is applicable to only the smallest of molecules. Another avenue that has proven fruitful is to carry out a continued fraction (cf) extrapolation using the sequence CCSD, CCSDT and CCSDTQ [62]. Although the results from this procedure can be a bit erratic, we reported that in 42 out of 49 comparisons with FCI, it improved atomization energies based on CCSDTQ [62]. In cases where large cf corrections are predicted, the results should be viewed with caution.

## 3.5 Miscellaneous contributions

For molecules composed of light elements, especially hydrogen, the commonly employed Born–Oppenheimer approximation, which separates electronic and nuclear motions, introduces a potentially significant error into the calculation of molecular properties when judged by the goals of high-accuracy studies. The diagonal correction to the BO approximation (DBOC) has been shown to contribute up to a few tenths of a kcal/mol to the atomization energies of hydrogen-containing molecules [115]. Calculating this contribution at the HF level of theory can lead to overestimations compared to more accurate CCSD calculations with cc-pCVTZ basis sets with outer-core correlation [116]. Even the use of cc-pCVDZ basis sets in the latter case yields nearly converged results. Including DBOC corrections for molecules that do not include hydrogen does not seem to be warranted at this time for the accuracies discussed above.

Related to the scalar relativistic correction discussed above is the leading quantum electrodynamic (QED) effect, the Lamb shift. While generally neglected in most composite methodologies, it has been shown previously by Dyall et al. [117]. that the contribution of the Lamb shift to the atomization energy can approach 3–5% of the scalar relativistic contribution. In cases where the scalar relativistic correction is large, for example when the valence $s$ orbital occupation strongly changes, the Lamb shift can contribute $\sim 0.1$ to 0.2 kcal/mol to the atomization energy for even relatively light systems like $AlF_3$ and $GaF_3$. This latter work calculated the contributions using perturbation theory. In the work of Shepler et al. [102]. involving small molecules containing iodine and mercury, the Lamb shift was calculated using the model potential approach of Pyykkö and Zhao [118] in DKH calculations. For the various mercury-containing species reported in Ref. [102], the Lamb shift contributed 0.3–0.7 kcal/mol to the atomization energies. Molecules such as $I_2$ or IBr exhibited contributions of only about 0.1 kcal/mol.

## 3.6 Zero-point vibrational corrections

The situation with zero-point vibrational energy (ZPVE) contributions in many ways parallels the situation with the atomization energies themselves. The largest contribution comes from estimating the harmonic value at something approaching the CCSD(T)(FC)/CBS limit, followed by a number of corrections of differing signs that may largely cancel but nevertheless need to be included for very high-accuracy studies. Anharmonic contributions to the ZPVE seem to rarely exceed 0.5 kcal/mol for small molecules [62, 119], but can increase strongly with the size of the molecule and number of bonds to hydrogen. Several of the model chemistries outlined in Sect. 2 utilize scaled harmonic frequencies with scale factors empirically chosen for a given method and basis set combination. Others use quartic force fields together with 2nd-order vibrational perturbation theory. Whereas both the HEAT and Wn approaches calculate these at the CCSD(T) level of theory, the FPD approach currently exploits the fact that the anharmonic contribution can be accurately determined by lower levels of theory, so accurate CCSD(T) harmonic frequencies (at least with cc-pVQZ) are combined with anharmonic effects calculated at the MP2/aug-cc-pVDZ level of theory. It is recommended that core correlation contributions be neglected in the calculation of the CCSD(T) harmonic frequencies for ZPVEs if higher-order correlation effects are neglected (see below). Otherwise, the resulting ZPVEs tend to be too large. Certainly the error associated with the computed ZPVE is often overlooked in many composite thermochemistry methodologies, and its error can easily grow to the 1 kcal/mol range for small molecules and can be much larger for large molecules, especially if they contain many hydrogen atoms [62, 120]. For example, the ZVPE is estimated to be 151.8 kcal/mol for $C_8H_{18}$ of which the anharmonic correction could be on the order of at least 10 kcal/mol [63].

## 4 Additional considerations for ab initio spectroscopy

In general, the application of an accurate composite methodology originally designed for thermochemistry applications needs little modification for use in the accurate prediction of spectroscopic properties such as rotational constants and vibrational band origins. Compared to calculating an atomization energy, the biggest challenge is simply one of dimensionality—all of the contributions from Sects. 3.1–3.6 must now be applied (in principle) on many near-equilibrium geometries of the 3N-6 dimensional potential energy surface (PES) instead of just at the minimum. For diatomic and even triatomic molecules it is relatively straightforward to explicitly sample the near-equilibrium part of the PES, to obtain an analytical representation and to evaluate spectroscopic constants from the derivatives of the PES using perturbation theory or variational methods. Each point sampled on the PES is constructed from the several contributions of the composite approach just as in thermochemical applications. From such small molecule benchmarks, it is clear that the dominant contribution to the overall accuracy is still the estimation of the frozen-core CCSD(T)/CBS values. For molecules containing only 1st- and 2nd-row elements, the resulting FC-CCSD(T)/CBS equilibrium bond lengths exhibit typical average errors with respect to experiment of about +0.003 Å (too long) and harmonic frequencies too large by an average of about 6 cm$^{-1}$ [37, 98, 121]. The addition of outer-core correlation reduces the error in the equilibrium structures by about a factor of two but leaves the distances uniformly too short on average. The errors in

the harmonic frequencies exhibit a further increase, tending to be even further away from experiment. Both of these deficiencies are unfortunately not ameliorated until the contributions of connected quadruples are included into the methodology (the inclusion of full iterative triples also tends to worsen the agreement with experiment compared to CCSD(T) alone). So, it is clear that the success of the FC-CCSD(T)/CBS level of theory is to some extent due to a fortuitous cancellation of errors. For equilibrium structures, it is clear that correlation of the outer-core electrons is desirable, particularly as shown recently for post-$d$ elements where $d$-electron correlation can lead to substantial bond shortening [90]. On the other hand for harmonic frequencies, if core correlation contributions are included, then some treatment of relativistic and higher-order correlation effects must also be addressed.

For polyatomic molecules, a common strategy is to assume that the various contributions of the composite methodology are additive for the structural parameters and harmonic frequencies [62]. The CCSD(T)(FC)/CBS limits can be estimated by either large basis set conventional calculations, explicitly correlated CCSD(T), or, in the case of bond lengths, by direct CBS extrapolation [62, 122]. Of course it is also possible to incorporate the CBS extrapolation scheme into a numerical or analytical derivative calculation [121, 123]. Overall, the conclusions reached by the small molecule studies have been shown to be applicable to larger polyatomics as well [62]. In particular for harmonic frequencies, however, the importance of quadruple excitations makes the accurate calculation (better than 10–15 cm$^{-1}$) of harmonic frequencies (and subsequent prediction of anharmonic band origins) very computationally challenging for polyatomic molecules [123].

Of course neither equilibrium geometries nor harmonic frequencies are directly comparable to experiment; hence, some calculation of the contributions of vibrational anharmonicity must be considered in order to be truly predictive. Fortunately, the anharmonic part of the PES is not nearly as sensitive to the electron correlation method or basis set as compared to the harmonic portion. For polyatomic molecules, this can be exploited by a multilayer approach to the calculation of vibrational frequencies, whereby the harmonic frequencies are calculated as accurately as possible, perhaps also with the diagonal anharmonicities, and then the remaining anharmonic contributions are reliably calculated at some lower level of theory, e.g., MP2 or DFT [62, 124, 125]. Generally, the anharmonic contributions are calculated from quartic force fields and 2nd-order vibrational perturbation theory [126, 127] or by explicit sampling of the PES with multimode normal coordinate expansions and subsequent vibrational SCF or vibrational CI calculations (see, e.g., Ref [128]). For the prediction of ground-state rotational constants, vibrationally averaged structures can be obtained from these same calculations from the cubic force field, which defines the vibration–rotation interaction constants [129]. These latter ab initio values can also be combined with experimental ground-state rotational constants to produce accurate semi-experimental equilibrium geometries [76, 130–132], which are more directly comparable to bottom of the well values obtained from theory [133].

## 5 Future challenges

For molecules consisting of relatively light elements from the first few rows of the periodic table, sophisticated composite ab initio methodologies have demonstrated a potential for very high accuracy in both thermochemistry and spectroscopy. Given the important role of the FC-CCSD(T)/CBS limit in these schemes, further development and implementation of explicitly correlated methods, e.g., CCSD(T)-F12x, will bring immediate dividends to these methods, especially in terms of the size of molecules that then can be accurately treated. Particularly for the accurate calculation of vibrational frequencies beyond the FC-CCSD(T) level of theory, more efficient methods for recovering higher-order correlation corrections are essential in order to carry out more reliable calculations on polyatomic molecules. Finally, the biggest challenge is the extension of these ideas to molecules containing transition metals and other heavy elements. For molecules that are dominated by the HF determinant, the contributions outlined above in Sec. IIIA-E are perhaps sufficient [47, 134–137], but, in particular for the transition metals, the existence of multireference character is common, which puts the current composite schemes that are based on the single-reference CCSD(T) method in jeopardy. Recently attempts have been made to include multireference components, for example the MR-ccCA variant [138], which includes CASPT2 and MRCI contributions. It remains to be seen, however, whether this approach will provide sufficient accuracy or can be made sufficiently general for larger systems. Also as accurate composite calculations are pushed to heavier systems, relativistic effects begin to be extremely important. For the most part, scalar relativistic effects appear to be straightforward to include, but accurate recovery of spin–orbit effects remains a challenge. This, in part, is due to difficulties in calculating the correct energies even for the atoms, which can have very complicated spin–orbit fine structure [109].

## 6 Summary

Current strategies for the calculation of molecular thermochemical and spectroscopic properties have been reviewed with particular attention to composite schemes designed to obtain chemical accuracy and beyond. The most important consideration in these methodologies is the accurate estimation of the frozen-core complete basis set limit, either at the equilibrium geometry in the case of thermochemistry applications or as a function of the 3N-6 internal coordinates for the calculation of spectroscopic properties. Practical solutions to this problem include effective basis set extrapolation strategies using sequences of correlation consistent basis sets or newly developed F12 explicitly correlated methods. The effects of correlating the outer-core electrons are also discussed. These latter correlation corrections are essential for accurate thermochemistry, e.g., atomization energies or heats of formation, as well as decreasing the errors in calculated equilibrium bond lengths. In the case of vibrational frequencies determined at the CCSD(T) level, correlating the outer-core electrons tends to increase the errors with respect to experiment and should generally be accompanied by higher-order electron correlation contributions in order to improve upon the accuracy already attained with just the valence electrons correlated. Relativistic effects can also make important contributions to both thermochemical and spectroscopic properties, even in molecules containing only elements of the first few rows of the periodic table, and scalar relativistic effects need to be included from the outset in heavier systems. The treatment of chemical systems with significant multireference character, as well as the efficient and accurate recovery of spin–orbit coupling effects for molecules containing heavy elements, remains challenge for the area of chemically accurate thermochemistry.

**Acknowledgments** K. A. P. would like to acknowledge the funding of the National Science Foundation (CHE-0723997). D. A. D. was supported by the Chemical Sciences, Geosciences and Biosciences Division, Office of Basic Energy Sciences, U.S. Department of Energy (DOE) under the catalysis center program. D. A. D. thanks the Robert Ramsay Fund of the University of Alabama and Argonne National Laboratory for partial support.

## References

1. Dunning TH Jr (1989) J Chem Phys 90:1007
2. Raghavachari K, Trucks GW, Pople JA, Head-Gordon M (1989) Chem Phys Lett 157:479
3. Feller D, Peterson KA, Hill JG (2011) J Chem Phys 135:044102
4. Pople JA, Head-Gordon M, Fox DJ, Raghavachari K, Curtiss LA (1989) J Chem Phys 90:5622
5. Frisch MJ, Trucks GW, Head-Gordon M, Gill PMW, Wong MW, Foresman JB et al (1992) Gaussian 92. Gaussian, Inc., Pittsburg
6. Curtiss LA, Jones C, Trucks GW, Raghavachari K, Pople JA (1990) J Chem Phys 93:2537
7. Hehre WJ, Radom L, Schleyer PvR, Pople JA (1986) Ab initio molecular orbital theory. Wiley-Interscience, New York
8. Curtiss LA, Raghavachari K, Trucks GW, Pople JA (1991) J Chem Phys 94:7221

9. Curtiss LA, Raghavachari K, Pople JA (1993) J Chem Phys 98:1293
10. Bauschlicher CW Jr, Partridge H (1995) J Chem Phys 103:1788
11. Smith BJ, Radon L (1995) J Phys Chem 99:6468
12. Curtiss LA, Redfern PC, Smith BJ, Radom L (1996) J Chem Phys 104:5148
13. Curtiss LA, Raghavachari K, Redfern PC, Pople JA (1998) J Chem Phys 109:7764
14. Curtiss LA, Redfern PC, Raghavachari K, Rassolov V, Pople JA (1999) J Chem Phys 110:4703
15. Baboul AG, Curtiss LA, Redfern PC, Raghavachari K (1999) J Chem Phys 110:7650
16. Curtiss LA, Raghavachari K, Redfern PC, Baboul AG, Pople JA (1999) Chem Phys Lett 314:101
17. Curtiss LA, Redfern PC, Raghavachari K, Pople JA (2001) J Chem Phys 114:108
18. Henry DJ, Sullivan MB, Radom L (2003) J Chem Phys 118:4849
19. Curtiss LA, Redfern PC, Raghavachari K (2007) J Chem Phys 126:084108
20. Curtiss LA, Redfern PC, Raghavachari K (2007) J Chem Phys 127:124105
21. Nyden MR, Petersson GA (1981) J Chem Phys 75:1843
22. Petersson GA, Bennett A, Tensfeldt TG, Al-Laham MA, Shirley WA, Mantzaris J (1988) J Chem Phys 89:2193
23. Petersson GA, Al-Laham MA (1991) J Chem Phys 94:6081
24. Petersson GA, Tensfeldt TG, Montgomery JA Jr (1991) J Chem Phys 94:6091
25. Montgomery JA Jr, Ochterski JW, Petersson GA (1994) J Chem Phys 101:5900
26. Ochterski JW, Petersson GA, Montgomery JA Jr (1996) J Chem Phys 104:2598
27. Montgomery JA Jr, Frisch MJ, Ochterski JW, Petersson GA (1999) J Chem Phys 110:2822
28. Wood GPF, Radom L, Petersson GA, Barnes EC, Frisch MJ, Montgomery JA Jr (2006) J Chem Phys 125:094106
29. Mayer PM, Parkinson CJ, Smith DM, Radom L (1998) J Chem Phys 108:604
30. Martin JML, de Oliveira G (1999) J Chem Phys 111:1843
31. Boese AD, Oren M, Atasoylu O, Martin JML, Kallay M, Gauss J (2004) J Chem Phys 120:4129
32. Barnes EC, Petersson GA, Montgomery JJA, Frisch MJ, Martin JML (2009) J Chem Theory Comput 5:2687
33. Karton A, Rabinovich E, Martin JML, Ruscic B (2006) J Chem Phys 125:144108
34. Mintz B, Chan B, Sullivan MB, Buesgen T, Scott AP, Kass SR, Radom L, Wilson AK (2009) J Phys Chem A 113:9501
35. Ruscic B (unpublished) Results obtained from the active thermochemical tables, ATcT ver. 1.25 and the Core (Argonne) thermochemical network ver. 1.056 2006
36. Karton A, Taylor PR, Martin JML (2007) J Chem Phys 127:064104
37. Karton A, Martin JML (2010) J Chem Phys 133:144102
38. Tajti A, Szalay PG, Császár AG, Kállay M, Gauss J, Valeev EF, Flowers BA, Vázquez J, Stanton JF (2004) J Chem Phys 121:11599
39. Helgaker T, Klopper W, Koch H, Noga J (1997) J Chem Phys 106:9639
40. Bomble YJ, Vázquez J, Kállay M, Michauk C, Szalay PG, Császár AG, Gauss J, Stanton JF (2006) J Chem Phys 125:064108
41. Harding ME, Vazquez J, Ruscic B, Wilson AK, Gauss J, Stanton JF (2008) J Chem Phys 128:114111
42. DeYonker NJ, Cundari TR, Wilson AK (2006) J Chem Phys 124:114104
43. Feller D (1992) J Chem Phys 96:6104

44. Peterson KA, Woon DE, Dunning TH Jr (1994) J Chem Phys 100:7410
45. Kedziora GS, Pople JA, Ratner MA, Redfern PC, Curtiss LA (2001) J Chem Phys 115:718
46. DeYonker NJ, Grimes T, Yokel S, Dinescu A, Mintz B, Cundari TR, Wilson AK (2006) J Chem Phys 125:104111
47. DeYonker NJ, Peterson KA, Steyl G, Wilson AK, Cundari TR (2007) J Phys Chem A 111:11269
48. DeYonker NJ, Williams TG, Imel AE, Cundari TR, Wilson AK (2009) J Chem Phys 131:024106
49. Mintz B, Williams TG, Howard L, Wilson AK (2009) J Chem Phys 130:234104
50. Prascher BP, Lai JD, Wilson AK (2009) J Chem Phys 131:044130
51. Schütz M, Chem J (2000) Phys 113:9986
52. Schütz M, Werner H-J (2000) Chem Phys Lett 318:370
53. DeYonker NJ, Wilson BR, Pierpont AW, Cundari TR, Wilson AK (2009) Mol Phys 107:1107
54. East ALL, Allen WD, Császár AG (1993) In: Laane J, van der Veken B, Overhammer H (eds) Structure and conformations of non-rigid molecules. Kluwer, Dordrecht, p 343
55. East ALL, Allen WD (1993) J Chem Phys 99:4638
56. Császár AG, Allen WD, Schaefer HF, II I (1998) J Chem Phys 108:9751
57. Cowan RD, Griffin M (1976) J Opt Soc Am 66:1010
58. Handy NC, Yamaguchi Y, Schaefer HF (1986) J Chem Phys 84:4481
59. Schuurman MS, Muir SR, Allen WD, Schaefer HF, II I (2004) J Chem Phys 120:11586
60. East ALL, Johnson CS, Allen WD (1993) J Chem Phys 98:1299
61. Feller D, Peterson KA, de Jong WA, Dixon DA (2003) J Chem Phys 118:3510
62. Feller D, Peterson KA, Dixon DA (2008) J Chem Phys 129:204105
63. Pollack L, Windus TL, de Jong WA, Dixon DA (2005) J Phys Chem A 109:6934
64. Feller D, Peterson KA, Crawford TD (2006) J Chem Phys 124:054107
65. Barnes EC, Petersson GA, Feller D, Peterson KA (2008) J Chem Phys 129:194115
66. Feller D, Peterson KA, Dixon DA (2010) J Phys Chem A 114:61
67. Feller D, Peterson KA, Dixon DA (2011) J Phys Chem A 115:1440
68. Feller D, Peterson KA, Hill JG (2010) J Chem Phys 133:184102
69. Feller D, Peterson KA (2007) J Chem Phys 126:114105
70. Peterson KA, Dunning TH Jr (2002) J Chem Phys 117:10548
71. Douglas M, Kroll NM (1974) Ann Phys NY 82:89
72. Jansen G, Hess BA (1989) Phys Rev A 39:6016
73. de Jong WA, Harrison RJ, Dixon DA (2001) J Chem Phys 114:48
74. Dunham JL (1932) Phys Rev 41:713
75. Kutzelnigg W (1997) Mol Phys 90:909
76. Feller D, Craig NC, Matlin AR (2008) J Phys Chem A 112:2131
77. Kendall RA, Dunning Jr TH, Harrison RJ (1992) J Chem Phys 96:6796
78. Woon DE, Dunning TH Jr (1993) J Chem Phys 98:1358
79. Peterson KA (2003) J Chem Phys 119:11099
80. Feller D (1996) J Comp Chem 17:1571
81. Werner H-J, Knizia G, Adler TB, Marchetti O (2010) Z Phys Chem 224:493
82. Klopper W, Manby FR, Ten-no S, Valeev EF (2006) Int Rev Phys Chem 25:427
83. Knizia G, Adler TB, Werner H-J (2009) J Chem Phys 130:054104
84. Peterson KA, Adler TB, Werner H-J (2008) J Chem Phys 128:084102

85. Knizia G, Werner H-J (2008) J Chem Phys 128:154103
86. Karton A, Martin JML (2006) Theor Chem Acc 115:330
87. Schwenke DW (2005) J Chem Phys 122:014107
88. Hill JG, Peterson KA, Knizia G, Werner H-J (2009) J Chem Phys 131:194105
89. Woon DE, Dunning TH Jr (1995) J Chem Phys 103:4572
90. Peterson KA, Yousaf KE (2010) J Chem Phys 133:174116
91. Hill JG, Mazumder S, Peterson KA (2010) J Chem Phys 132:054108
92. Peterson KA, Hill JG (2011) (in preparation)
93. Balabanov NB, Peterson KA (2005) J Chem Phys 123:064107
94. Bauschlicher Jr CW (1998) J Phys Chem A 102:10424
95. Martin RL (1983) J Phys Chem 87:750
96. Hess BA (1986) Phys Rev A 33:3742
97. Kutzelnigg W, Ottschofski E, Franke R (1995) J Chem Phys 102:1740
98. Tew DP, Klopper W, Heckert M, Gauss J (2007) J Phys Chem A 111:11242
99. Boese AD, Oren M, Atasoylu O, Martin JML, Kallay M, Gauss J (2004) J Chem Phys 120:4129
100. Bauschlicher CW Jr, Martin JML, Taylor PR (1999) J Phys Chem A 103:7715
101. Michauk C, Gauss J (2007) J Chem Phys 127:044106
102. Shepler BC, Balabanov NB, Peterson KA (2005) J Phys Chem A 109:10363
103. Figgen D, Peterson KA, Dolg M, Stoll H (2009) J Chem Phys 130:164108
104. Peterson KA, Figgen D, Dolg M, Stoll H (2007) J Chem Phys 126:124101
105. Peterson KA, Puzzarini C (2005) Theor Chem Acc 114:283
106. Peterson KA, Figgen D, Goll E, Stoll H, Dolg M (2003) J Chem Phys 119:11113
107. Spohn B, Goll E, Stoll H, Figgen D, Peterson KA (2009) J Phys Chem A 113:12478
108. Peterson KA, Shepler BC, Figgen D, Stoll H (2006) J Phys Chem A 110:13877
109. Moore CE, (1971) Vol Natl Stand Ref Data Ser, National Bureau of Standards (US), Washington 35
110. Hess BA, Marian CM, Peyerimhoff SD (1995) In: Yarkony DR (ed) Modern electronic structure theory, part I, vol 2. World Scientific, Singapore, p 152
111. Shepler BC, Peterson KA (2003) J Phys Chem A 107:1783
112. Yabushita S, Zhang Z, Pitzer RM (1999) J Phys Chem A 103:5791
113. Cho WK, Choi YJ, Lee YS (2005) Mol Phys 103:925
114. Ruden TA, Helgaker T, Jorgensen P, Olsen J (2003) Chem Phys Lett 371:62
115. Ruscic B, Wagner AF, Harding LB, Asher RL, Feller D, Dixon DA, Peterson KA, Song Y, Qian XM, Ng CY, Liu JB, Chen WW (2002) J Phys Chem A 106:2727
116. Gauss J, Tajti A, Kallay M, Stanton JF, Szalay PG (2006) J Chem Phys 125:4111
117. Dyall KG, Bauschlicher Jr CW, Schwenke DW, Pyykkö P (2001) Chem Phys Lett 348:497
118. Pyykkö P, Zhao U (2003) J Phys B 36:1469
119. Barone V (2004) J Chem Phys 120:3059
120. Karton A, Ruscic B, Martin JML (2007) J Mol Struct 811:345
121. Heckert M, Kallay M, Tew DP, Klopper W, Gauss J (2006) J Chem Phys 125:4108
122. Puzzarini C (2009) J Phys Chem A 113:14530
123. Peterson KA, Francisco JS (2011) J Chem Phys 134:084308
124. Rauhut G, Knizia G, Werner H-J (2009) J Chem Phys 130:054105
125. Puzzarini C, Barone V (2008) Phys Chem Chem Phys 2008:6991
126. Barone V (2005) J Chem Phys 122:014108
127. Clabo DA Jr, Schaefer HF III (1987) Int J Quantum Chem 31:429
128. Rauhut G (2004) J Chem Phys 121:9313
129. Puzzarini C, Heckert M, Gauss J (2008) J Chem Phys 128:194108
130. Craig NC, Feller D, Groner P, Hsin HY, McKean DC, Nemchick DJ (2007) J Phys Chem A 111:2498
131. Feller D, Craig NC (2009) J Phys Chem A 113:1601
132. Feller D, Craig NC, Groner P, McKean DC (2011) J Phys Chem A 115:94
133. Puzzarini C, Cazzoli G, Gambi A, Gauss J (2006) J Chem Phys 125:054307
134. Craciun R, Picone D, Long RT, Li S, Dixon DA, Peterson KA, Christe KO (2010) Inorg Chem 49:1056
135. Li S, Hennigan JM, Dixon DA, Peterson KA (2009) J Phys Chem A 113:7861
136. Puzzarini C, Peterson KA (2005) Chem Phys 311:177
137. Pahl E, Figgen D, Borschevsky A, Peterson KA, Schwerdtfeger P (2011) Theor Chem Acc 129:651
138. Jiang W, Wilson AK (2011) J Chem Phys 134:034101

REGULAR ARTICLE

# Negative energy states in relativistic quantum chemistry

Christoph van Wüllen

Received: 1 August 2011 / Accepted: 31 August 2011 / Published online: 11 January 2012
© Springer-Verlag 2012

**Abstract** The role of the negative energy states in relativistic quantum chemistry is shortly discussed. They must be included in a sum over states formula that arises for second-order properties. Relativistic calculations for the electric dipole polarizability and the diamagnetic susceptibility (with on-center and displaced gauge origins) for hydrogen-like ions are presented to illustrate the problem. Relativistic electron correlation calculations mostly use a configuration space Dirac-Coulomb operator together with the no-pair approximation, which excludes the negative energy states from the correlation treatment. Despite all efforts, no consistent theoretical description between this and a full QED treatment seems to exist.

**Keywords** Dirac operator · Electron correlation · Magnetic shielding · Negative energy states · Polarizability · Relativistic

## 1 Introduction

It is clear that the Schrödinger equation cannot be "true" because its classical limit is Newtonian mechanics, and since Einstein's seminal work on special relativity, we know that this can only be approximately true at velocities small compared with the speed of light. For a long time,

Published as part of the special collection of articles celebrating the 50th anniversary of Theoretical Chemistry Accounts/Theoretica Chimica Acta.

C. van Wüllen (✉)
Fachbereich Chemie and Forschungszentrum OPTIMAS,
Technische Universität Kaiserslautern,
Erwin-Schrödinger-Straße, 67663 Kaiserslautern, Germany
e-mail: vanwullen@chemie.uni-kl.de

this has not been regarded as a problem for quantum chemistry, basically because the typical energy scale of chemical processes (few electron volts) is very small compared with the rest energy of an electron (half a mega electron volt). Around 1970, this viewpoint changed and increasing evidence was found [1–5] that also valence electrons in atomic or molecular systems (especially those containing heavy elements) are significantly affected by the so-called *relativistic effects*. According to Pyykkö [5], a relativistic effect is *anything arising from the finite speed of light*, so it is not something we can measure, but an error we make if we stick to nonrelativistic quantum mechanics.

Dirac has derived a relativistic wave equation for a single electron aready in 1928, and exact solutions are known for a free electron as well as for an electron moving in the electric field of an infinitely heavy point-like nucleus (we only note in passing that taking into account the finite nuclear mass is more complicated than in the nonrelativistic case [6]). The Dirac operator is a configuration space Hamiltonian, its spectrum for a free particle contains two branches, one continuum for positive energies $E > mc^2$ (with $m$ the rest mass of the electron and $c$ the speed of light), and one for negative energies $E < mc^2$. For an electron bound by the electric field of a nucleus, there are additionally discrete positive energy levels with $E < mc^2$. For very heavy nuclei, there may be discrete levels with negative energy, but these are bound states that can easily be distinguished from the negative-energy continuum containing only scattering states. If one had to summarize all the fundamental (as opposed to technical) problems of relativistic quantum chemistry—both from a historical perspective and looking at the current status of the field!— into a single question, then it would most likely be "*what should we do with the negative energy eigenstates of the Dirac operator?*". On one hand, these states do not represent physical solutions for the electron: their existence is associated with charge conjugation and they acquire a physical meaning in a field theory. On the other hand, we cannot just forget about them because they do form (in a mathematical sense) a part of the spectrum of the Dirac operator. Here, we just name three areas of relativistic quantum chemistry where this question arises, namely the electron correlation problem, the relativistic calculation of magnetic properties, and the quest for accurate *quasirelativistic* Hamiltonians which is an attempt to get rid of the negative energy states at an early stage.

## 2 Negative energy states and second-order properties

If one describes a one-electron system with the configuration space Dirac operator, the calculation of a second-order property leads to a sum-over-states (SOS) formula that encompasses the whole spectrum of the unperturbed Dirac operator, including the negative energy states. For the electric dipole polarizability, for example, one gets

$$\alpha = 2 \sum_{p \neq 0} \frac{\langle \psi_0 | \hat{h}' | \psi_p \rangle \langle \psi_p | \hat{h}' | \psi_0 \rangle}{\epsilon_p - \epsilon_0} \tag{1}$$

where $\psi_0$ is the ground state and the summation goes over states $\psi_p$ which are either positive-energy excited states or negative energy states. It should be noted that if we use the exact states, the SOS expression is meant to include the continuum. In a finite basis set calculation this is not an issue, since everything is discrete while the continuum is still well represented for large basis sets. For an electric field in the $z$ direction, the operator describing the perturbation is (for the standard representation for the Dirac operator, and with $I_2$ the $2 \times 2$ unit matrix)

$$\hat{h}' = \begin{pmatrix} zI_2 & 0 \\ 0 & zI_2 \end{pmatrix} \tag{2}$$

There is no question that one must include the negative energy states in the SOS expression (Eq. 1) to get a (mathematically) correct result. Electric perturbations change the diagonal blocks of the Dirac operator in the standard representation, and therefore, the contribution of the negative energy states is usually small and vanishes in the nonrelativistic limit $c \to \infty$: their energy denominators in the SOS expression are large in absolute value (of the order $2\,mc^2$ and larger) while the numerators are of order unity. This is demonstrated by numerical calculations on the hydrogen atom, presented in Table 1. Here, a very large and dense even-tempered Gaussian basis set has been used (100 exponents, smallest exponent 0.00001, largest exponent $2.71...*10^{12}$, ratio of "adjacent" exponents is $\frac{3}{2}$). These exponents have been used to generate upper

**Table 1** Polarizability ($\alpha$) of a relativistic hydrogen atom, calculated for different values $c$ of the speed of light

| $c$ | $\alpha_+$ | $\alpha_-$ | $\alpha$ |
|---|---|---|---|
| 137,036 | 4.5000 | −2.599 [−21] | 4.5000 |
| 137.036 | 4.4996 | −2.599 [−9] | 4.4996 |
| $\frac{137.036}{10}$ | 4.4628 | −2.574 [−5] | 4.4628 |
| $\frac{137.036}{50}$ | 3.6243 | −0.0132 | 3.6112 |
| $\frac{137.036}{100}$ | 1.5849 | −0.1224 | 1.4625 |

All entries in Hartree atomic units such that $c^{-1}$ is the fine structure constant. Numbers in brackets indicate the power of 10 by which the number has to be multiplied. $\alpha_+$ is the contributions from the positive energy states in the sum over states expression, while $\alpha_-$ come from the negative energy states. The exact nonrelativistic value is $\alpha = \frac{9}{2}$

component basis functions of $s_{\frac{1}{2}}, p_{\frac{1}{2}}, p_{\frac{3}{2}}$ and $d_{\frac{3}{2}}$ symmetry. Basis functions for the lower components have been generated by applying $\boldsymbol{\sigma}\mathbf{p}$ to each of the upper component basis functions and then normalizing the resulting functions for better numerical stability. Here, $\boldsymbol{\sigma}$ is the vector of the Pauli spin matrices and $\mathbf{p}$ is the momentum operator. This procedure is known as *restricted kinetic balance*, the number of upper and lower component basis functions is the same. The speed of light, $c$, has been varied between 137036 and 1.37036 atomic units, the physical value being close to 137.036. This numerical experiment is not as unphysical as it may seem: the first line is the weakly relativistic limit, the second line is the relativistic result for the hydrogen atom, and the results of the next three lines can also be obtained using the physical value of $c$ for hydrogen-like ions with $Z = 10, 50, 100$, and scaling the calculated polarizability $\alpha$ with $Z^4$. All calculations have been performed with the MATHEMATICA program (a product of Wolfram Research, Inc., Version 7.0) using a numerical precision of 30 digits.

One clearly sees that the contribution of the negative energy states to the polarizability vanishes in the nonrelativistic limit with $\mathcal{O}(c^{-4})$. The polarizability decreases for a more and more relativistic hydrogen atom, most likely because the $1s$ ground state is lowered in energy and becomes more compact. Note that this decrease is stronger than calculated by Kaneko [7] some time ago. But even for the last line (corresponding to $Z = 100$), the contribution of the negative energy states is less than 10% of the total value. The situation is very different for magnetic perturbations. Let us for example calculate the diamagnetic susceptibility. Here we have

$$\hat{h}' = \begin{pmatrix} 0 & c\boldsymbol{\sigma}\mathbf{A} \\ c\boldsymbol{\sigma}\mathbf{A} & 0 \end{pmatrix} \tag{3}$$

(and an additional minus sign in Eq. 1 because of the sign convention for the susceptibility, and $\mathbf{A}$ is the vector potential describing the homogeneous magnetic field). We choose $\mathbf{A} = \frac{1}{2}\mathbf{B} \times (\mathbf{r} - \mathbf{R})$ with the gauge origin $\mathbf{R}$ either at the position of the nucleus or displaced by 1 bohr. It should be noted that formally there is no contribution which looks like the diamagnetic contribution in nonrelativistic theory [8]. But because $\hat{h}'$ couples upper and lower components of the atomic Dirac spinors, the numerator in the negative energy contributions of the SOS expression can be very large, of the order $c^2$. The contribution of the negative energy states represents diamagnetism in a relativistic theory [8] and survives in the nonrelativistic limit. This analysis is made more transparent by our hydrogenic calculations which are presented in Table 2. If the gauge origin is at the position of the nucleus, there is no paramagnetic contribution to the susceptibility in nonrelativistic theory. Accordingly, the contribution of the positive energy states $\chi_+$ vanishes in the nonrelativistic limit, it clearly goes as $\mathcal{O}(c^{-4})$. The contribution of the negative energy states just recovers the nonrelativistic diamagnetic contribution in the weakly relativistic limit. For the strongly relativistic hydrogen atoms, the susceptibility decreases (most likely again, because the $1s$ orbital becomes more compact), and this is mostly due to $\chi_-$. This analysis is corroborated by the calculations for an off-center gauge origin. While the total magnetic susceptibility is not affected by the change of the gauge origin, the contributions from positive and negative energy eigenstates change, and they do so in the same manner as the nonrelativistic paramagnetic and diamagnetic contributions do. It should be noted that the results of the last three lines in Table 2 correspond to calculations on hydrogen-like ions with the physical value of $c = 137.036$ and $Z = 10, 50, 100$, provided that the gauge origin displacement is scaled with $Z^{-1}$ and the calculated magnetizabilities are scaled with $Z^2$. Our results for the total magnetizabilities agree with the semi-analytical values given by Poszwa and Rutkowski (Ref. [9], Table III).

It is interesting to note the symmetry of the eigenstates $\psi_p$ in the SOS formula Eq. 1 which actually give nonzero contributions, since this dictates the basis set requirements of the calculation. To calculate the polarizability, only $p_{\frac{1}{2}}$ and $p_{\frac{3}{2}}$ spinors contribute, and this is as in the nonrelativistic case: to calculate the polarizability of the hydrogen atom, one needs a sufficiently flexible $p$ basis set. For a nonrelativistic calculation of the magnetizability, one only needs $s$ functions if the gauge origin is at the nucleus (there is only a diamagnetic term) and a flexible $p$ basis set for a displaced gauge origin. In the relativistic case, this is different: If the gauge origin is at the nucleus, $s_{\frac{1}{2}}$ as well as $d_{\frac{3}{2}}$ spinors contribute to the diamagnetic susceptibility. This is so because the linear response of the $1s$ orbital to the magnetic field is dominated by a $p_{\frac{1}{2}}$ and $p_{\frac{3}{2}}$ symmetry contribution to the lower component, and these are generated from $s_{\frac{1}{2}}$ and $d_{\frac{3}{2}}$ large component basis functions in the restricted kinetic balance scheme. The $d$ functions contribute one-third of the total value of $\chi$. These $s$ and $d$ contributions do not change their numerical value if the gauge origin is displaced, in this case additional nonzero terms from the $p_{\frac{1}{2}}$ and $p_{\frac{3}{2}}$ spinors arise which sum up to zero, their contribution to $\chi_-$ and $\chi_+$ has the same absolute value with opposite sign. This means that a relativistic calculation of the magnetic susceptibility for a hydrogen atom and possibly an off-center gauge origin requires a flexible $p$ and $d$ basis set, and this translates to very severe basis set requirements for the calculation of molecular magnetic properties if one takes the SOS formula as it stands. Therefore, a theory specifically designed to the calculation of second-order magnetic properties has to be designed, and the article by Xiao, Sun and Liu in this volume [10] will recapitulate the status of the field as well as current developments.

**Table 2** Diamagnetic susceptibility ($\chi$) of a relativistic hydrogen atom, calculated for different values $c$ of the speed of light, and the sign convention for $\chi$ is such that negative values indicate diamagnetism

| $c$ | Gauge origin at (0,0,0) | | | Gauge origin at (1,0,0) | | |
|---|---|---|---|---|---|---|
| | $\chi_+$ | $\chi_-$ | $\chi$ | $\chi_+$ | $\chi_-$ | $\chi$ |
| 137,036 | 8.231 [−22] | −0.5000 | −0.5000 | 0.2500 | −0.7500 | −0.5000 |
| 137.036 | 8.229 [−10] | −0.5000 | −0.5000 | 0.2500 | −0.7500 | −0.5000 |
| $\frac{137.036}{10}$ | 8.065 [−6] | −0.4965 | −0.4965 | 0.2489 | −0.7454 | −0.4965 |
| $\frac{137.036}{50}$ | 0.0039 | −0.4169 | −0.4130 | 0.2305 | −0.6435 | −0.4130 |
| $\frac{137.036}{100}$ | 0.0405 | −0.2155 | −0.1750 | 0.2123 | −0.3873 | −0.1750 |

All entries in Hartree atomic units such that $c^{-1}$ is the fine structure constant. Numbers in brackets indicate the power of 10 by which the number has to be multiplied. $\chi_+$ is the contributions from the positive energy states in the sum over states expression, while $\chi_-$ come from the negative energy states. The magnetic field is in $z$ direction, and the gauge origin is either at the position of the nucleus, or displaced by 1 $a_0$ along the $x$ axis. The exact nonrelativistic value is $\chi = -\frac{1}{2}$, with a diamagnetic contribution of $-\frac{3}{4}$ and a paramagnetic contribution of $+\frac{1}{4}$ in the case of the displaced gauge origin

## 3 Relativistic quantum chemistry without negative energy states: quasirelativistic approaches

The problems mentioned so far have nothing to do with the *physical* (as opposed to mathematical) relevance of the negative-energy states of the Dirac operator, and in many cases, it is the consequence of their pure existence that complicates the calculation. For example, they prevent using the variational principle in a naive way, because this would require a spectrum bounded from below. Because of such problems, the development of quasirelativistic Hamiltonians has a long and rich history. The label "quasirelativistic" is used with different meanings in the literature but here it indicates that one disregards the charge conjugation degrees of freedom, loosely speaking, one gets rid of the negative energy eigenstates. Apart from that, everything including the spin-orbit interaction is still included. Such a Hamiltonian "for electrons only" [11] can be obtained by applying a unitary transformation to the Dirac operator such that it becomes block diagonal, and each block as a two-component operator contains one branch (positive or negative) branch of the spectrum. The block representing the positive branch or the spectrum is then the two-component quasirelativistic "electronic" operator $\hat{h}_+$. Such a decoupling of the Dirac equation can easily be written down for a free electron. First attempts to do so for a bound electron resulted in an iterative procedure which yields an expansion of $\hat{h}_+$ in powers of $c^{-2}$, named Foldy-Wouthuysen expansion after the authors [12]. Any truncation of this expansion (say, after the $c^{-2}$ term) yields highly singular operators, so an alternative route, the Douglas-Kroll-Hess transformation (for recent reviews, see [13, 14]) has become very popular. Here, $\hat{h}_+$ is expanded in the strength of the external potential, and low-order truncations of this series already yield very useful operators which cover a large parts of the relativistic effects. Other low-order approximation such as the ZORA Hamiltonian [15, 16] have also found much use, but the Douglas-Kroll series has been systematically tweaked to very high accuracy for one-electron systems. It should be noted that including the electron interaction requires approximations in order not to loose the computational efficiency [17–19]. More recently, manipulating the matrix representation of the Dirac operator rather than performing the decoupling at operator level has been strongly advocated [20–25] because at matrix level, achieving exact decoupling is rather straightforward. The article by Peng and Reiher in this volume [26] on the exact decoupling of the relativistic Fock operator reviews this field and reports its latest developments.

## 4 The electron correlation problem

A complete relativistic treatment of interacting electrons is established by a field theory that encompasses the photon (radiation) field and the electron-positron field; this theory is known as quantum electrodynamics (QED) or rather bound-state QED if the electrons move in the electric field of the nuclear framework. Electrons and positrons are described simultaneously because the particle number is not a conserved quantity: if the radiation field can deliver energies $>2mc^2$, (real) electron-positron pairs can be created. Some QED effects such as vacuum polarization can be understood through the creation of short-lived or *virtual* electron-positron pairs. In such a field theory, the negative-energy states of the Dirac operator do not only have a physical meaning but are *necessary* to describe both electrons and positrons on the same footing. For atomic and molecular systems with more than just a few electrons, a complete bound-state QED treatment seems impossible at the moment, so relativistic quantum chemistry is essentially based on the Dirac-Coulomb *configuration space* Hamiltonian

$$\hat{H}_{DC} = \sum_i \hat{h}_D(i) + \sum_{i<j} \frac{1}{r_{ij}} \tag{4}$$

where $\hat{h}_D$ is the one-electron Dirac operator in the electric field of the nuclear framework, and the last term is the nonrelativistic Coulomb interaction between the electrons. This operator may be augmented by the two-particle Breit term which describes the magnetic interaction between the electrons as well as retardation. For most valence properties of heavy-element compounds, the Breit interaction gives rise to minor corrections (about one percent of the total relativistic effect), but it contributes strongly to relativistic corrections for light elements, and also has a significant contribution to spin-orbit splittings as it is the origin of the spin-other-orbit interaction. For most of the following discussion, it is of little importance whether one uses the Dirac-Coulomb operator as it stands here or whether one adds the Breit interaction. It is a long-standing debate whether one can actually *derive* the Dirac-Coulomb operator from QED in some way, or whether one must regard it as an ad hoc semi-classical relativistic many-electron operator. With the second viewpoint, we at least do not leave safe ground and can pragmatically justify the Dirac-Coulomb operator with the success story of relativistic quantum chemistry. Everything beyond is then summarized as "QED effects", in the same way as we do this for everything that is beyond the Dirac description of hydrogenic ions (the Lamb shift being the first and most prominent example).

If we proceed as in nonrelativistic theory, we construct Slater determinants from the eigenspinors of $\hat{h}_D$ or some mean-field variant of it like the four-component (Dirac) extension of Hartree-Fock. The reference function is constructed using the aufbau principle, applied to the positive-energy (bound) eigenstates, leaving the negative energy states unoccupied ("empty Dirac sea"). Optimizing the orbitals for such a reference is usually not a problem, this is done in the Dirac-Hartree-Fock (DHF) step. If one now starts a configuration interaction type calculation, one finds a large number of configurations quasi degenerate with the reference configuration, for example with double replacements where one electron is excited high into the positive continuum, while the other electron is "excited" down into the negative energy continuum. Using a complete basis set, we would even find infinitely many such configurations exactly degenerate with the reference configuration. Because of the electron interaction, the matrix element with the reference function does not vanish and one should therefore expect that such "doubly excited" configurations will heavily mix with the reference configuration, until the weight of the latter is essentially gone. This (in)famous problem has been termed "Brown-Ravenhall disease" (after the authors [27]) or "continuum dissolution", and in particular Sucher has warned the community not to use the Dirac-Coulomb operator in such a way [28–30]. As has been pointed out by Kutzelnigg [31] recently, the fact that the bound states are immersed in a continuum of non-physical scattering states already arises in the absence of the electron-electron interaction (without electron interaction there is still no coupling between them). It was therefore generally accepted that one had to use projection operators onto a set of positive energy one-particle states to be put around the Hamiltonian. This procedure is called the *no pair* or *no virtual pair* approximation. One usually uses the DHF orbitals to define the projection (this was first advocated by Mittleman [32]), and then the no pair approximation is 'implemented' by just dropping the negative-energy orbitals from the calculation after the DHF step. As has been pointed out by Sapirstein [33], even a complete configuration interaction ("*full CI*") calculation is only complete within the positive-energy orbital space, and therefore, the energy depends on the choice of the orbitals, in contrast to the nonrelativistic case. Therefore it has been tried to include the negative-energy orbitals in the correlation treatment. For a complete basis set, we know that continuum dissolution *does* happen [30], but in a finite basis this is not necessarily the case. Including the negative-energy states in the correlation treatment of helium-like ions *increases* the total energy [33, 34]. In nonrelativistic theory, the total energy decreases when the CI space is enlarged, but here it was explained that the CI energy raises because the state of interest must remain orthogonal to the non-physical scattering states around it that arise when including the negative energy states in the

correlation treatment [34]. The effect of including the negative energy states has also been investigated theoretically by Brown [35] and here the energy raising was related to the picture that if two paired electrons in the $1s$ orbital come very close, the exclusion principle prevents production of a virtual electron-positron pair ("*Pauli blocking*").

It is not clear what one actually calculates if one includes the negative-energy states in the correlation treatment of the configuration space Dirac-Coulomb operator. This also requires a certain amount of trust in having good luck because one relies on the absence of degeneracies which are accidental in a finite basis set but will arise if the basis set becomes complete. Most likely, this procedure does not include QED effects on the correlation energy in a systematic way. Kutzelnigg [31] condemns it as "*illegitimate*" and only accepts the Dirac-Coulomb Hamiltonian together with the no-pair approximation, because this can be viewed as a simplification of the Fock space formulation he pushes forward. In this formulation, the "empty Dirac sea" of the Dirac-Coulomb Hamiltonian is filled by introducing a physical vacuum (together with normal ordering), so continuum dissolution cannot happen. But here as well the question what one really gets when including the negative energy states in a Fock space formulation is not answered satisfactorily. Since the radiation field is not quantized in the Fock space approach, important low-order QED radiative contributions such as the self-energy are missing from the outset. Liu has analyzed the Fock space approach recently [36] and found here as well, that the negative energy states are *anti-correlating* (i.e., energy raising when included in the correlation treatment) and that this should not be the case. The question whether there is a consistent theoretical level between no-pair Dirac-Coulomb and full QED is therefore still open.

Although some fundamental problems of the relativistic electron correlation problem are not yet solved, we expect that the Dirac-Coulomb operator, together with the no-pair approximation, will remain the cornerstone of wave function–based relativistic quantum chemistry in the foreseeable future. Then, the correlation treatment becomes formally equivalent to what we do in a nonrelativistic framework. Recent reviews on the four-component relativistic correlation methodology are available [37, 38]. An important difference to the nonrelativistic case is that the symmetry (spatial and spin) is reduced in the relativistic case, which increases the numerical effort by about one order or magnitude. This applies to the computational steps that follow the integral transformation, but the latter does not dominate the overall effort for very accurate calculations. Beyond the largely technical problem of making four-component correlation calculations fast and efficient, there are two remaining issues left. First, the target molecules for relativistic quantum chemical calculations frequently are heavy-element compounds, and here, one usually does not get very far with single-reference methods. In many applications, one must calculate the dynamical correlation energy based on multiconfiguration reference functions, and this is a hard and not yet satisfactorily solved problem even in nonrelativistic theory. Second, the convergence of the correlation energy with increasing basis set size is *much* slower than in the nonrelativistic case [39–41]. The truncation error is proportional to $(L + \frac{1}{2})^{-1}$ for a basis set saturated up to an angular momentum $L$, this means that there is practically no hope obtaining converged results from conventional methods that expand the many-electron wave function in Slater determinants composed of one-particle spinors. Recently, there has been a first general attempt to overcome this problem using wave functions that explicitly depend on the interelectronic distances [42]. This approach is based on the nonrelativistic cusp condition, which also governs the behavior of the relativistic wave function for small (but not too small) interelectronic distances. It should be noted that for the Dirac-Coulomb operator, the wave function has a (weak) singularity at the coalescence of two electrons, but it remains finite if the Gaunt or Breit interaction is included [43]. A rigorous treatment of the relativistic correlation cusp (beyond imposing the nonrelativistic cusp condition) seems impossible since we simply do not know what a configuration space Hamiltonian for the electron interaction at very short distances looks like. These two remarks show that even if we forget about all the fundamental problems and take the no-pair Dirac-Coulomb Hamiltonian for granted, the relativistic correlation problem is much more challenging than in the nonrelativistic case, and more than can be anticipated from the symmetry reduction alone.

## References

1. Desclaux JP (1973) At Data Nucl Data Tables 12:311
2. Pitzer KS (1979) Acc Chem Res 12:271
3. Pyykko P, Desclaux JP (1979) Acc Chem Res 12:276
4. Christiansen PA, Ermler WC, Pitzer KS (1985) Annu Rev Phys Chem 36:407
5. Pyykko P (1988) Chem Rev 88:563
6. Breit G, Brown GE (1948) Phys Rev 74:1278
7. Kaneko S (1977) J Phys B At Mol Opt Phys 10:3347
8. Kutzelnigg W (2003) Phys Rev A 67:032109
9. Poszwa A, Rutkowski A (2007) Phys Rev A 75:033402
10. Xiao Y, Sun Q, Liu W (2012) Fully relativistic theories and methods for NMR parameters, article in this volume
11. Kutzelnigg W (1997) Chem Phys 225:203
12. Foldy LL, Wouthuysen SA (1950) Phys Rev 78:29
13. Reiher M (2006) Theor Chem Acc 116:241
14. Nakajima T, Hirao K (2011) Chem Rev. doi:10.1021/cr200040s
15. Vanlenthe E, Baerends EJ, Snijders JG (1993) J Chem Phys 99:4597
16. Sadlej AJ, Snijders JG, van Lenthe E, Baerends EJ (1995) J Chem Phys 102:1758
17. van Wüllen C, Michauk C (2005) J Chem Phys 123:204113
18. Peng DL, Liu WJ, Xiao YL, Cheng L (2007) J Chem Phys 127:104106
19. Liu WJ (2010) Mol Phys 108:1679
20. Barysz M, Sadlej AJ, Snijders JG (1997) Int J Quantum Chem 65:225
21. Barysz M, Sadlej AJ (2002) J Chem Phys 116:2696
22. Kutzelnigg W, Liu WJ (2005) J Chem Phys 123:241102
23. Kutzelnigg W, Liu WJ (2006) Mol Phys 104:2225
24. Liu WJ, Peng DL (2006) J Chem Phys 125:044102
25. Liu WJ, Kutzelnigg W (2007) J Chem Phys 126:114107
26. Peng D, Reiher M (2012) Exact decoupling of the relativistic fock operator, article in this volume
27. Brown GE, Ravenhall DG (1951) Proc Roy Soc A 208:552
28. Sucher J (1980) Phys Rev A 22:348
29. Sucher J (1984) Int J Quantum Chem 25:3
30. Sucher J (1985) Phys Rev Lett 55:1033
31. Kutzelnigg W (2011) Chem Phys. doi:10.1016/j.chemphys.2011.06.001
32. Mittleman MH (1981) Phys Rev A 24:1167
33. Sapirstein J, Cheng KT, Chen MH (1999) Phys Rev A 59:259
34. Watanabe Y, Nakano H, Tatewaki H (2007) J Chem Phys 126:174105
35. Brown GE (1987) Phys Scr 36:71
36. Liu W (2012) Phys Chem Chem Phys 14:35
37. Visscher L (2002) In: Schwerdtfeger P (ed) Relativistic electronic structure theory: Part 1 fundamentals. Elsevier, Amsterdam, pp 291
38. Fleig T (2011) Chem Phys. doi:10.1016/j.chemphys.2011.06.032
39. Ottschofski E, Kutzelnigg W (1997) J Chem Phys 106:6634
40. Halkier A, Helgaker T, Klopper W, Olsen J (2000) Chem Phys Lett 319:287
41. Kutzelnigg W (2008) Int J Quantum Chem 108:2280
42. Bischoff FA, Valeev EF, Klopper W, Janssen CL (2010) J Chem Phys 132:214104
43. Kutzelnigg W (1989) In: Mukherjee D (ed) Aspects of many body effects in molecules and extended systems, (Lecture Notes in Chemistry, vol 50). Springer, Berlin, p 353

Theor Chem Acc (2012) 131:1080
DOI 10.1007/s00214-011-1080-z

REGULAR ARTICLE

# Fully relativistic theories and methods for NMR parameters

**Yunlong Xiao · Qiming Sun · Wenjian Liu**

Received: 13 August 2011 / Accepted: 30 August 2011 / Published online: 11 January 2012
© Springer-Verlag 2012

**Abstract** Nuclear magnetic shielding and spin–spin coupling constants are intrinsically all-electron relativistic properties and demand in principle fully relativistic treatments. Here, the magnetic balance (MB) condition plays an essential role, both conceptually and computationally. The various formulations can be unified in terms of the idea of "orbital decomposition." Further combined with the ansatz of "gauge-including atomic orbitals" (GIAO) for distributed gauge origins leads to very efficient four-component relativistic methods at both the mean-field and correlated levels. To illustrate the latter, the no-pair MB-GIAO-MP2 expressions for nuclear shieldings are derived in two different ways, one with the derivative technique and the other through the induced current. Due to the non-variational nature of MP2, the two expressions are not identical. The current-dependent expression is much simpler and appears more natural in view of the experimental measurement.

**Keywords** NMR parameters · Four-component relativistic theories · Orbital decomposition · Magnetic balance · Gauge-including atomic orbitals

Published as part of the special collection of articles celebrating the 50th anniversary of Theoretical Chemistry Accounts/Theoretica Chimica Acta.

Y. Xiao · Q. Sun · W. Liu (✉)
Beijing National Laboratory for Molecular Sciences, Institute of Theoretical and Computational Chemistry, State Key Laboratory of Rare Earth Materials Chemistry and Applications, College of Chemistry and Molecular Engineering, and Center for Computational Science and Engineering, Peking University, Beijing 100871, People's Republic of China
e-mail: liuwjbdf@gmail.com

## 1 Introduction

Nuclear magnetic resonance (NMR) spectroscopy is the most powerful experimental tool for determining the electronic and molecular structures of various chemical systems. What is really measured in an NMR experiment is the transition frequencies $v_L$ between the quantized spin states $\Phi_i$ of an active nucleus $K$, viz.,

$$E_i = \langle \Phi_i | -\vec{M}_K(\vec{R}) \cdot \vec{B}_{eff}(\vec{R})\delta(\vec{R} - \vec{R}_K) | \Phi_i \rangle$$
$$= -g_K \mu_N M_i |\vec{B}_{eff}(\vec{R}_K)|, \tag{1}$$

$$\vec{M}_K = g_K \mu_N \vec{I}_B, \quad \vec{I}_B |\Phi_i\rangle = M_i |\Phi_i\rangle, \quad |\Phi_i(\vec{R}_K)|^2 = 1, \tag{2}$$

where $\vec{M}_K$ is the point-like nuclear magnetic momentum operator with the (dimensionless) spin angular momentum $\vec{I}_B$ quantized along the direction of the effective magnetic field $\vec{B}_{eff}$ at the position $\vec{R}_K$ of the nucleus. It is seen that the energy levels $E_i$ are proportional to the strength of $\vec{B}_{eff}(\vec{R}_K)$. They are equally separated for the absolute difference between the spin projections $M_{i+1}$ and $M_i$ is always one. Therefore, the transition frequency $v_L$ between any two adjacent states reads

$$\Delta E = g_K \mu_N |\vec{B}_{eff}(\vec{R}_K)| = hv_L. \tag{3}$$

As the nuclear magneton $\mu_N = \frac{e\hbar}{2m_p}$ is 1836 times smaller than the Bohr magneton $\mu_B = \frac{e\hbar}{2m_e}$ and the magnetic field $\vec{B}_{eff}(\vec{R}_K)$ is very weak even in modern experimental setup, the energy separation $\Delta E$ is in the radiofrequency domain, much smaller than the electronic excitation energies of a closed-shell molecule. That is, what is described here is a nucleus–electron composite system where the (quantized) nuclei are in the hyperfine excited states while the electrons in the ground state. Note in passing that, in this "experimental picture," the nuclei are considered as "internal particles," whereas the electrons as "external particles." This is different from the "theoretical picture" where the role of these two types of particles is reversed. To see the equivalence of the two pictures, we realize that the effective field $\vec{B}_{eff}(\vec{R}_K)$ is composed of the applied (homogeneous) external magnetic field $\vec{B}^{10}$ and the magnetic field $\vec{B}_{ind}(\vec{R}_K)$ produced by the induced electronic current $-\vec{j}_{ind}(\vec{r})$ in the molecule:

$$\vec{B}_{eff}(\vec{R}_K) = \vec{B}^{10} + \vec{B}_{ind}(\vec{R}_K), \tag{4}$$

$$\vec{B}_{ind}(\vec{R}_K) = \frac{1}{c^2} \int \frac{\vec{j}_{ind}(\vec{r}) \times (\vec{r} - \vec{R}_K)}{|\vec{r} - \vec{R}_K|^3} d\vec{r}, \tag{5}$$

$$\vec{j}_{ind}(\vec{r}) = \Psi_0^\dagger(\vec{B}^{10}) c\vec{\alpha}\Psi_0(\vec{B}^{10}), \tag{6}$$

where the SI-based atomic units have been used and $c$ is the speed of light. Moreover, $\Psi_0(\vec{B}^{10})$ is the relativistic electronic ground state in the presence of the external magnetic field $\vec{B}^{10}$ and $\vec{\alpha}$ is the $4 \times 4$ Dirac matrix. Substituting $\vec{j}_{ind}(\vec{r})$ (6) into Eq. 5 leads to

$$\vec{B}_{ind}(\vec{R}_K) = \langle\Psi_0(\vec{B}^{10})| - \frac{\partial D^{01}}{\partial \vec{\mu}_K}|\Psi_0(\vec{B}^{10})\rangle, \tag{7}$$

$$D^{01} = c\vec{\alpha} \cdot \vec{A}_K^{01}, \tag{8}$$

$$\vec{A}_K^{01}(\vec{r}) = \frac{1}{c^2} \frac{\vec{\mu}_K(\vec{R}_K) \times (\vec{r} - \vec{R}_K)}{|\vec{r} - \vec{R}_K|^3}, \tag{9}$$

where $\vec{A}_K^{01}(\vec{r})$ is recognized as the vector potential produced by the *classical* nuclear magnetic moment $\vec{\mu}_K$ located at $\vec{R}_K$ and $D^{01}$ represents hence the hyperfine magnetic interaction. In this way, the standard "theoretical picture" has been recovered, where the quantized electrons move in the classical fields of the clamped nuclei. As a good approximation, the electronic wave function $\Psi_0(\vec{B}^{10})$ can be truncated up to first order in $\vec{B}^{10}$, i.e.,

$$\Psi_0(\vec{B}^{10}) \approx \Psi_0^{00} + \Psi_0^{10}, \quad \Psi_0^{10} = \frac{d\Psi_0}{d\vec{B}^{10}}|_{\vec{B}^{10}=0} \cdot \vec{B}^{10}. \tag{10}$$

The induced magnetic field $\vec{B}_{ind}(\vec{R}_K)$ then reads

$$B_{ind,i}(\vec{R}_K) = \langle\Psi_0^{10}| - \frac{\partial D^{01}}{\partial \mu_{K,i}}|\Psi_0^{00}\rangle + \langle\Psi_0^{00}| - \frac{\partial D^{01}}{\partial \mu_{K,i}}|\Psi_0^{10}\rangle,$$
$$i = x, y, z. \tag{11}$$

This arises from the fact that the zeroth-order current $\vec{j}^{00}(\vec{r})$ vanishes pointwise for non-degenerate closed-shell molecules under consideration, such that the term

$$\langle\Psi_0^{00}| - \frac{\partial D^{01}}{\partial \vec{\mu}_K}|\Psi_0^{00}\rangle = -\frac{\partial}{\partial \vec{\mu}_K}\int \vec{j}^{00}(\vec{r}) \cdot \vec{A}_K^{01} d\vec{r} \tag{12}$$

in the expansion of $\vec{B}_{ind}(\vec{R}_K)$ also vanishes. One can further define a mixed second-order energy

$$E^{11}(\vec{\mu}_K, \vec{B}^{10}) = \langle\Psi_0^{10}|D^{01}|\Psi_0^{00}\rangle + \langle\Psi_0^{00}|D^{01}|\Psi_0^{10}\rangle \tag{13}$$

such that

$$B_{ind,i}(\vec{R}_K) = -\sum_j \sigma_{K,ij}B_j^{10}, \quad i,j = x,y,z, \tag{14}$$

$$\sigma_{K,ij} = \frac{\partial^2 E^{11}}{\partial B_j^{10}\partial \mu_{K,i}}, \tag{15}$$

where $\sigma_K$ is just the so-called nuclear shielding tensor. Note that Eq. 13 can also be written as

$$E^{11}(\vec{\mu}_K, \vec{B}^{10}) = \int \vec{j}^{10}(\vec{r}) \cdot \vec{A}_K^{01}(\vec{r}) d\vec{r} \tag{16}$$

in terms of the first-order current

$$\vec{j}^{10}(\vec{r}) = \Psi_0^{00} c\vec{\alpha}\Psi_0^{10} + c.c. \tag{17}$$

In the above discussions, the induced current density $\vec{j}_{ind}(\vec{r})$ (6) has been assumed to arise only from the applied homogeneous magnetic field $\vec{B}^{10}$. In principle, it should also encompass contributions from the magnetic moments of all the nuclei of nonzero spin. However, the latter lead to different resonant transitions and can hence be treated separately. Factually, they will give rise just to the so-called (indirect) nuclear spin–spin coupling tensor $J_{ij}$, the derivations of which are precisely the same as the above. It is just that the $\vec{B}^{10}$ is to be replaced with the magnetic momentum $\vec{\mu}_L$ of a different nucleus $L$. Likewise, the magnetizability tensor $\chi_{ij}$ can be obtained by replacing $\vec{\mu}_K$ with $\vec{B}^{10}$. Because of such similarities, we will in the following focus only on the nuclear shieldings in terms of the scalar quantity $E^{11}$ rather than of the tensor $\sigma_K$. For a single-determinantal electronic wave function, $E^{11}$ (13) becomes

$$E^{11}(\vec{\mu}_K, \vec{B}^{10}) = \sum_i^{N_{occ}}[\langle\psi_i^{10}|D^{01}|\psi_i^{00}\rangle + \langle\psi_i^{00}|D^{01}|\psi_i^{10}\rangle]$$
$$= \int \vec{j}^{10}(\vec{r}) \cdot \vec{A}_K^{01}(\vec{r}) d\vec{r}, \tag{18}$$

$$\vec{j}^{10}(\vec{r}) = \psi_i^{00} c\vec{\alpha}\psi_i^{10} + c.c. \tag{19}$$

where $\psi_i^{00}$ and $\psi_i^{10}$ are the zeroth- and first-order occupied Dirac bispinors.

A number of points can be made here. First, the expression $E^{11}$ (13) or (18) for the mixed second-order energy is *asymmetric* with respect to the perturbations $\vec{\mu}_K$ and $\vec{B}^{10}$. The latter is considered here as the primary perturbation, a natural option in view of the "experimental picture". A symmetric formulation is still possible, where the two types of perturbations are treated on an equal footing. However, for the nuclear shieldings, the asymmetric formulation is much preferred: Once the first-order orbitals $\psi_i^{10}$ due to the three components of $\vec{B}^{10}$ have been obtained, the shieldings of all the active nuclei can immediately be calculated. In contrast, in the symmetric formulation, the first-order orbitals $\psi_i^{01}$ due to $3N$ components of $\vec{\mu}_K (K = 1, \ldots N)$ have to be constructed as well, which is obviously more expensive. In addition, special care has to be taken of the singularities [1, 2] arising from the nuclear vector potential $\vec{A}_K^{01}$ (9). Of course, if one wants to calculate the nuclear shieldings and spin–spin couplings in the same time, it is the $\vec{\mu}_K$ that should be taken as the primary perturbation. Second, the nuclear vector potential $\vec{A}_K^{01} \sim |\vec{r} - \vec{R}_K|^{-2}$ is very local and samples the electronic wave function nearby the vicinity of the nucleus. As such, the nuclear shieldings and spin–spin couplings are intrinsically *all-electron relativistic* properties. Any approximate treatment for relativity would fail to reproduce accurately the absolute shieldings and spin–spin couplings of heavy atoms such as Cl in HCl [3], as all the approximations inherent therein stem solely from the atomic cores, precisely the region sampled by the parameters. Third, as it stands, the expression $E^{11}$ (13) or (18) is composed only of a single term, the so-called paramagnetism due to the polarization of the electron cloud under the influence of the external magnetic field. The familiar diamagnetism [4] representing the rigid motion of the electron cloud is hence "missing" in such a standard four-component relativistic linear response theory (LRT) [5–10]. This is not only a conceptual problem but also a computational problem: The computation is extremely demanding [11, 12], particularly when the restricted kinetic balance (RKB) [13] prescription is used for constructing the basis sets. Even worse, the nonrelativistic limit (nrl) cannot be obtained correctly with a finite RKB basis as the contributions of negative energy states (NES) are of order $c^0$ and hence survive to the nrl [14]. The computation is partly expedited by using the so-called unrestricted kinetic balance (UKB) scheme [15]. However, this is true only for a Gaussian-type basis but not for, e.g., a Slater-type basis. Even with a Gaussian basis, the particular advantage of UKB holds only for atomic or nearly atomic systems [16], yet with the price of enhancing the risk of linear dependency in the basis set. Also, the UKB does not seem to converge faster to the basis set limit than the RKB [17]. The way out of these conceptual and computational difficulties is to explicitly build in the magnetic balance (MB) [15, 18, 19] between the small and large components of the Dirac bispinors (see Sect. 2)

Fourth, the first-order orbitals $\psi_i^{10}$ are determined by the Zeeman magnetic interaction $D^{10}$

$$D^{10} = c\vec{\alpha} \cdot \vec{A}_g^{10}, \tag{20}$$

$$\vec{A}_g^{10}(\vec{r}) = \frac{1}{2}\vec{B}^{10} \times (\vec{r} - \vec{R}_g), \tag{21}$$

$$\nabla \cdot \vec{A}_g^{10} = 0, \quad \vec{B}^{10} = \vec{\nabla} \times \vec{A}_g^{10}. \tag{22}$$

The problem here is that the gauge origin $\vec{R}_g$ cannot uniquely be determined by the magnetic field $\vec{B}^{10}$. This is easily seen as follows:

$$\vec{A}_\mu^{10}(\vec{r}) = \vec{A}_g^{10}(\vec{r}) - \vec{\nabla}\Lambda_\mu^g, \quad \Lambda_\mu^g = \frac{1}{2}(\vec{B}^{10} \times \vec{R}_\mu^g) \cdot \vec{r},$$
$$\vec{R}_\mu^g = \vec{R}_\mu - \vec{R}_g, \quad \vec{B}^{10} = \vec{\nabla} \times \vec{A}_\mu^{10}. \tag{23}$$

Both $\vec{A}_g^{10}(\vec{r})$ and $\vec{A}_\mu^{10}(\vec{r})$ correspond to the same homogeneous magnetic field $\vec{B}^{10}$. A change of the gauge origin is effectively a unitary gauge transformation of the Hamiltonian and the wave function under which all observables should be invariant. However, this is not the case for calculations with finite bases: the calculated nuclear shieldings may depend strongly on the choice of gauge origin for a regular basis set is not flexible enough to describe the corresponding changes in the wave function. An efficient way of eliminating the problem is to use gauge-including atomic orbitals (GIAO) [20–23]

$$\omega_\mu(\vec{R}_g) = e^{-i\Lambda_\mu^g}\chi_\mu(\vec{R}_\mu). \tag{24}$$

Effectively, the gauge origin $\vec{R}_g$ is transferred to the optimum location, the center $\vec{R}_\mu$ of the basis function $\chi_\mu$. To first order, this reads

$$\omega_\mu(\vec{R}_g) = Z_g^{10}\chi_\mu(\vec{R}_\mu) + \chi_\mu(\vec{R}_\mu), \tag{25}$$

$$Z_g^{10} = -i\Lambda_\mu^g. \tag{26}$$

If one stops here and takes $\chi_\mu$ as restricted or unrestricted kinetically balanced, one would obtain the standard GIAO-based LRT, which is still very demanding on the basis sets due to the lack of an explicit diamagnetic term. This can be resolved by further using the MB condition $Z_m^{10}$ (vide post), leading to

$$\omega_\mu(\vec{R}_g) = Z^{10}\chi_\mu(\vec{R}_\mu) + \chi_\mu(\vec{R}_\mu), \tag{27}$$

$$Z^{10} = Z_g^{10} + Z_m^{10}. \tag{28}$$

The $\chi_\mu$ in Eq. 27 is then simply restricted kinetically balanced. This ansatz has been dubbed as the "magnetically balanced gauge-including atomic orbitals" (MB-GIAO) [24] to emphasize that the gauge origins of the external vector potential $\vec{A}^{10}$ in both the Hamiltonian and the MB condition are distributed among the individual atomic orbitals.

Now, it is clear that the four-component relativistic methods for nuclear shieldings of molecular systems can be classified into four categories: (1) with neither the MB nor the distributed gauge origins [9], i.e., $Z^{10}$ is set to zero in Eq. 27. As pointed out by Mannien and Vaara [25], except for very few solid calculations [26–29], most of the so-obtained results suffer either from severe basis set incompleteness errors [30–33] or crude approximations for the contributions of NES [34]; (2) with the distributed gauge origins but without the MB [35–39], see Eq. 25; (3) with the MB but without the distributed gauge origins [2, 3, 16, 40, 41], i.e., only the second term of Eq. 28 is used in Eq. 27; (4) with both the MB and the distributed gauge origins [24, 42], see Eq. 27. Compared with schemes (2) and (3), only marginal complexities have been introduced in scheme (4) but the requirements on the basis sets are greatly reduced. For a recent review on this topic, yet from a different perspective, see Ref. [43].

The remaining account is planned as follows. The concept of orbital decomposition [44, 45] for incorporating the MB is discussed in Sect. 2. The modified Dirac equation is then introduced in Sect. 3 for alternative notations. Section 4 is devoted to the coupled-perturbed equation for determining the first-order orbitals. To demonstrate that the MB-GIAO ansatz can be combined with wave function-based correlation methods, the second-order Møller-Plesset perturbation theory (MP2) is taken as an example in Sect. 5, where the contributions of NES to correlation are also taken into account (i.e., without the no-pair approximation). The account ends with concluding remarks in Sect. 6.

## 2 The concept of orbital decomposition for magnetic balance

To expedite subsequent derivations, we first introduce the following conventions. All equations will be written in the SI-based atomic units in which $\hbar = e = m = 4\pi\epsilon_0 = 1$ and $c \approx 137$. The indices $\{i, j, k, l, \ldots\}$, $\{a, b, c, d, \ldots\}$ and $\{p, q, r, s, \ldots\}$ are reserved for the occupied, virtual, and unspecified spinors, respectively. The barred indices indicate the corresponding time-reversed Kramers partners. When necessary, the NES will be denoted explicitly as $\{\tilde{a}, \tilde{b}, \tilde{c}, \tilde{d}, \cdots\}$. The Einstein summation convention over repeated indices is always employed.

As the Dirac operator (see Eq. 30) is linear with respect to the vector potential, the mixed second-order energy $E^{11}$ (18) is purely paramagnetic, when the first-order orbitals are expanded in the full basis of unperturbed states. There

have been four different schemes to recapture the missing diamagnetism, viz. (a) by a separation of the positive and negative energy states [15, 46], (b) by the Gordon decomposition of the Dirac current density [47–50], (c) by a unitary transformation of the Dirac operator [2, 14], and (d) by the decomposition of the orbitals [2, 3, 44, 45]. In scheme (a), the contributions of NES are attributed to the diamagnetism. This amounts to a block-diagonal approximation to the orbital Hessian. Apart from the approximate nature, the interpretation of the diamagnetism in terms of NES is conceptually unacceptable as the nonrelativistic theory [4] never refers to NES. In addition, the computation is still very demanding on the basis sets. In contrast, the other three schemes recover the diamagnetism in a natural manner without recourse to NES or any approximation. The contributions of NES are reduced to order $c^{-2}$ or higher so as to guarantee the correct nrl even with a finite RKB basis. The basis set requirements are hence greatly alleviated [3]. Critical comparisons of the last three schemes have been made before [24] and will not be repeated here. Yet, some remarks are still necessary to clarify the misunderstandings still prevailing in the literature. The Kutzelnigg unitary transformation (KUT) [14] was formulated originally for the magnetizability of one-electron system subject to a homogeneous magnetic field. Given its elegance, considerable confusions have emerged as soon as it is extended to nuclear shieldings of many-electron systems due to two magnetic fields. The very first implementation [16] of the idea led to the point that the KUT is numerically very different from the LRT without the MB. Such numerical evidence was confirmed by a subsequent formal analysis [51] showing that the KUT and LRT agree only up to first order. Possible reasonings were further speculated [52] to "explain" the different performances of the KUT and LRT. Factually, all these were artificial, since the particular implementation [16] did not consider the full KUT but instead neglected the contributions of NES. It was shown later on that this implementation corresponds to a second-order coupling approximation (SOCA) [2], which turns out to be very crude. Moreover, it was also argued [39, 53] that the KUT has a serious drawback in that, it does not commute with the Gaunt two-electron interaction such that, unlike the Gaunt-free case, the desired decoupling between positive and negative energy states is no longer possible. This is also a misunderstanding, again stemming from the "curse of decoupling." The decoupling is not what one wants, for the contributions of NES are to be accounted for anyway. In other words, the decoupling is only needed if one wants to neglect the NES, which however may not be justified at all as shown by the SOCA. Even in the presence of the Gaunt interaction, the KUT need only be applied to the one-electron Dirac operator and truncated to a desired order (e.g., first and mixed second orders). The remaining residuals of the perturbed orbitals can then be expanded effectively in a standard basis set. In this way, all the above problems do not appear. That is, what is really important [2, 3, 44, 45] is not the decoupling *per se* but the MB. It is then understood that, among the last three schemes, scheme (d) is most generic, in terms of which the various strictly equivalent methods, including the orbital decomposition approach (ODA) [2], the external field-dependent unitary transformation at operator (EFUT) or matrix (EFUTm) level [2], the full field-dependent unitary transformation at matrix level (FFUTm) [2], and the restricted magnetic balance (RMB) [3, 40, 41, 54] can be obtained in the same manner [3]. It can also be employed to formulate the exact two-component (X2C) counterparts [55]. The last point to be made here is that FFUT, equivalent to the final result of Pyper [48] obtained by using two times Gordon decompositions of the current, suffers from severe numerical instabilities when used in a basis set expansion in conjunction with a point charge-point dipole description of the nucleus and therefore cannot be recommended. Yet, its matrix counterpart, FFUTm, is immune to such instabilities.

To illustrate the concept of orbital decomposition [44, 45], suffice it to consider the one-electron Dirac equation

$$D\psi_p = \epsilon_p \psi_p, \tag{29}$$

$$D = D^{00} + D^{01} + D^{10}, \tag{30}$$

$$D^{00} = c\vec{\alpha} \cdot \vec{p} + (\beta - 1)c^2 + V_N, \tag{31}$$

$$D^{01} = c\vec{\alpha} \cdot \vec{A}^{01}, \tag{32}$$

$$D^{10} = c\vec{\alpha} \cdot \vec{A}^{10}. \tag{33}$$

Equation 29 can be rewritten in block form

$$\begin{pmatrix} V & c\vec{\sigma} \cdot \vec{\pi} \\ c\vec{\sigma} \cdot \vec{\pi} & V - 2c^2 \end{pmatrix} \begin{pmatrix} \psi_p^L \\ \psi_p^S \end{pmatrix} = \epsilon_p \begin{pmatrix} \psi_p^L \\ \psi_p^S \end{pmatrix},$$
$$\vec{\pi} = \vec{p} + \vec{A}^{01} + \vec{A}^{10}, \tag{34}$$

from which one obtains the following exact relationship between the small and large components of $\psi_p$

$$\psi_p^S = X_p \psi_p^L, \tag{35}$$

$$X_p = \frac{1}{2c} R_p(\vec{r}) \vec{\sigma} \cdot \vec{\pi}, \tag{36}$$

$$R_p(\vec{r}) = \left[ 1 + \frac{\epsilon_p - V(\vec{r})}{2c^2} \right]^{-1} = \frac{2c^2}{2c^2 - V} \left[ 1 + \frac{\epsilon_p}{2c^2 - V} \right]^{-1}. \tag{37}$$

Regarding $\vec{A}^{10}$ as the primary perturbation, the orbital $\psi_p$ and the operator $X_p$ can be expanded up to first order

$$\psi_p^{10} = \begin{pmatrix} \psi_p^{L,10} \\ \psi_p^{S,10} \end{pmatrix}, \tag{38}$$

$$\psi_p^{S,10} = X_p^{10} \psi_p^{L,00} + X_p^{00} \psi_p^{L,10}, \tag{39}$$

$$X_p^{00} = \frac{1}{2c} R_p^{00}(\vec{r}) \vec{\sigma} \cdot \vec{p}, \tag{40}$$

$$R_p^{00}(\vec{r}) = \left[ 1 + \frac{\epsilon_p^{00} - V^{00}(\vec{r})}{2c^2} \right]^{-1}$$
$$= \frac{2c^2}{2c^2 - V^{00}} \left[ 1 + \frac{\epsilon_p^{00}}{2c^2 - V^{00}} \right]^{-1}, \tag{41}$$

$$X_p^{10} = \frac{1}{2c} R_p^{00}(\vec{r}) \vec{\sigma} \cdot \vec{A}^{10} - \frac{\epsilon_p^{10}}{4c^3} [R_p^{00}(\vec{r})]^2 \vec{\sigma} \cdot \vec{p}. \tag{42}$$

Note that, although $\vec{\sigma} \cdot \vec{p}$ and $\vec{\sigma} \cdot \vec{A}^{10}$ in $X_p^{10}$ are both of odd parity, their actions on atomic spinors are very different (for details see Appendix A in Ref. [3]. As a tensor operator of rank 0, $\vec{\sigma} \cdot \vec{p}$ does not change the $j$ and $m_j$ values of an atomic spinor $|ljm_j\rangle$. In contrast, as a tensor operator of rank 1, the action of $\vec{\sigma} \cdot \vec{A}^{10}$ on $|ljm_j\rangle$ leads to a linear combination of spinors $|l'j'm_j'\rangle$ with $l' = l \pm 1, j' = j - 1, j, j + 1$ and $m_j' = m_j - 1, m_j, m_j + 1$. This further implies that up to $l_{occ}^{max} + 2$-type RKB basis functions are required to get converged shieldings if $l_{occ}^{max}$ is the highest angular momentum among the occupied orbitals. Such an analysis shows that the two terms of $X_p^{10}$ (42) must be treated differently. As $R_p^{00}$ is totally symmetric, the second term of $X_p^{10}$ has the same symmetry as $X_p^{00}$, implying that its effect may be shifted to the second term of Eq. 39, if $\Psi_p^{L,00}$ and $\Psi_p^{L,10}$ are to be expanded in the same basis. In addition, it is of order $c^{-2}$ relative to the first term of Eq. 42. Therefore, $X_p^{10}$ can be approximated as

$$X_p^{10} \approx \frac{1}{2c} R_p^{00} \vec{\sigma} \cdot \vec{A}^{10} = \frac{1}{2c} \vec{\sigma} \cdot \vec{A}^{10} R_p^{00}. \tag{43}$$

The $X_p^{10}$ operator is then fixed by $\vec{\sigma} \cdot \vec{A}^{10}$ and the zeroth-order quantities in view of Eq. 41 for $R_p^{00}$. The deduction is that the first-order orbital $\Psi_p^{10}$ can formally be decomposed into a known magnetic field-dependent term $\Psi_{p,m}^{10}$ and a residual $\Psi_{p,r}^{10}$

$$\psi_p^{10} = \Psi_{p,m}^{10} + \psi_{p,r}^{10}, \tag{44}$$

$$\psi_{p,m}^{10} = Z_m^{10} \psi_p^{00}. \tag{45}$$

This is the origin of the name "orbital decomposition approach" (ODA) [44]. $Z_m^{10}$ will mimic closely the action of $X_p^{10}$ such that the residual $\Psi_{p,r}^{10}$ can effectively be expanded in the space of zeroth-order orbitals that are further expanded in a RKB basis, viz.,

$$\psi_{p,r}^{10} = \psi_q^{00} C_{qp}^{10}, \tag{46}$$

$$\psi_q^{00} = \chi_\mu C_{\mu p}^{00} = \begin{pmatrix} \psi_q^{L,00} \\ \psi_q^{S,00} \end{pmatrix} = \begin{pmatrix} g_\mu A_{\mu q}^{00} \\ \frac{\vec{\sigma}\cdot\vec{p}}{2c} g_\mu B_{\mu q}^{00} \end{pmatrix}. \tag{47}$$

A number of choices for the $Z_m^{10}$ operator are possible by simplifying the $R_p^{00}$ in Eq. 43. The simplest case is $R_p^{00} = 0$, leading to

$$Z_{m,LRT}^{10} = 0. \tag{48}$$

This is just the standard LRT, which does not have an explicit diamagnetic term and is hence computationally very demanding on the basis sets of high angular momenta. Another simple case is to choose $R_p^{00} = 1$, the nrl of the first equality of Eq. 41, which leads to

$$Z_{m,ODA}^{10} = \begin{pmatrix} 0 & 0 \\ \frac{\vec{\sigma}\cdot\vec{A}_\mu^{10}}{2c} & 0 \end{pmatrix}. \tag{49}$$

This is just the original ODA, first derived as a first-order (i.e., $O(c^{-2})$) no-pair approximation by neglecting the contributions of NES to the paramagnetic term [44] and then extended to the exact ODA [2]. The anti-Hermitian part of $Z_{m,ODA}^{10}$ leads to EFUT [2], which is obtained originally in a different way and completely equivalent to the Kutzelnigg unitary transformation for magnetizibility [14],

$$Z_{m,EFUT}^{10} = Z_{m,ODA}^{10} - Z_{m,ODA}^{10\dagger} = \begin{pmatrix} 0 & -\frac{\vec{\sigma}\cdot\vec{A}_\mu^{10}}{2c} \\ \frac{\vec{\sigma}\cdot\vec{A}_\mu^{10}}{2c} & 0 \end{pmatrix}. \tag{50}$$

The Hermitian part of $Z_{m,ODA}^{10}$ results in the same diamagnetic term as that proposed by Sternheim [46],

$$Z_{m,SD}^{10} = Z_{m,ODA}^{10} + Z_{m,ODA}^{10\dagger} = \begin{pmatrix} 0 & \frac{\vec{\sigma}\cdot\vec{A}_\mu^{10}}{2c} \\ \frac{\vec{\sigma}\cdot\vec{A}_\mu^{10}}{2c} & 0 \end{pmatrix}. \tag{51}$$

Furthermore, in the spirit of zeroth-order regular approximation (ZORA) for $R_p^{00}$ (see the second equality of Eq. 41), we have

$$Z_{m,ZORA}^{10} = \begin{pmatrix} 0 & 0 \\ \frac{c\vec{\sigma}\cdot\vec{A}_\mu^{10}}{2c^2 - V^{00}} & 0 \end{pmatrix}, \tag{52}$$

which however involves complicated integrals and additionally suffers from gauge dependence due to the appearance of the scalar potential in the denominator. This option is documented here only for completeness. Finally, we have the RMB ansatz [2, 40, 41, 54]

$$Z_{m,RMB}^{10} = \begin{pmatrix} 0 & 0 \\ 0 & \frac{\vec{\sigma}\cdot\vec{A}_\mu^{10}}{2c}\left(\frac{\vec{\sigma}\cdot\vec{p}}{2c}\right)^{-1} \end{pmatrix}. \tag{53}$$

Each of the above $Z_m^{10}$ can be combined with the diagonal GIAO factor defined in Eq. 26, leading to the MB-GIAO expansion of the first-order orbitals [24],

$$\psi_p^{10} = \psi_{p,b}^{10} + \psi_{p,r}^{10}, \tag{54}$$

$$\psi_{p,b}^{10} = Z^{10}\psi_p^{00} = Z_\mu^{10}\chi_\mu C_{\mu p}^{00}, \quad Z_\mu^{10} = -i\Lambda_\mu^g + Z_m^{10}. \tag{55}$$

Note that the action of $Z^{10}$ is on the RKB basis function $\chi_\mu$, whose center $\vec{R}_\mu$ is chosen to be the gauge origin of $\vec{A}_\mu^{10}$ in $Z_m^{10}$. The residual $\Psi_{p,r}^{10}$ has been defined in Eq. 46, while the known field-dependent part $\Psi_{p,b}^{10}$ can be written out explicitly in terms of the large component basis $\{g_\mu\}$ alone

$$\psi_{p,b,LRT}^{10} = \begin{pmatrix} -i\Lambda_\mu^g g_\mu \\ 0 \end{pmatrix} A_{\mu p}^{00} + \begin{pmatrix} 0 \\ -i\Lambda_\mu^g \frac{\vec{\sigma}\cdot\vec{p}}{2c} g_\mu \end{pmatrix} B_{\mu p}^{00}, \tag{56}$$

$$\psi_{p,b,ODA}^{10} = \begin{pmatrix} -i\Lambda_\mu^g g_\mu \\ \frac{\vec{\sigma}\cdot\vec{A}_\mu^{10}}{2c} g_\mu \end{pmatrix} A_{\mu p}^{00} + \begin{pmatrix} 0 \\ -i\Lambda_\mu^g \frac{\vec{\sigma}\cdot\vec{p}}{2c} g_\mu \end{pmatrix} B_{\mu p}^{00}, \tag{57}$$

$$\psi_{p,b,EFUT}^{10} = \begin{pmatrix} -i\Lambda_\mu^g g_\mu \\ \frac{\vec{\sigma}\cdot\vec{A}_\mu^{10}}{2c} g_\mu \end{pmatrix} A_{\mu p}^{00} + \begin{pmatrix} -\frac{\vec{\sigma}\cdot\vec{A}_\mu^{10}}{2c}\frac{\vec{\sigma}\cdot\vec{p}}{2c} g_\mu \\ -i\Lambda_\mu^g \frac{\vec{\sigma}\cdot\vec{p}}{2c} g_\mu \end{pmatrix} B_{\mu p}^{00}. \tag{58}$$

$$\psi_{p,b,SD}^{10} = \begin{pmatrix} -i\Lambda_\mu^g g_\mu \\ \frac{\vec{\sigma}\cdot\vec{A}_\mu^{10}}{2c} g_\mu \end{pmatrix} A_{\mu p}^{00} + \begin{pmatrix} \frac{\vec{\sigma}\cdot\vec{A}_\mu^{10}}{2c}\frac{\vec{\sigma}\cdot\vec{p}}{2c} g_\mu \\ -i\Lambda_\mu^g \frac{\vec{\sigma}\cdot\vec{p}}{2c} g_\mu \end{pmatrix} B_{\mu p}^{00}. \tag{59}$$

$$\psi_{p,b,RMB}^{10} = \begin{pmatrix} -i\Lambda_\mu^g g_\mu \\ 0 \end{pmatrix} A_{\mu p}^{00} + \begin{pmatrix} 0 \\ \left(\frac{\vec{\sigma}\cdot\vec{A}_\mu^{10}}{2c} - i\Lambda_\mu^g \frac{\vec{\sigma}\cdot\vec{p}}{2c}\right) g_\mu \end{pmatrix} B_{\mu p}^{00}. \tag{60}$$

Up to now, the RKB prescription [13] has been assumed to construct the basis set $\{\chi_\mu\}$, see Eq. 61. As well known, it is somewhat biased toward the positive energy states (PES) and does not guarantee the correct nrl of the NES [17]. In contrast, the DKB prescription [56] provides more balanced descriptions of both the PES and NES and even full variational safety [17]. In this prescription, the zeroth-order molecular orbitals are expanded as [57][1]

$$\begin{pmatrix} \psi_p^{L,00} \\ \psi_p^{S,00} \end{pmatrix} = \begin{pmatrix} g_\mu \\ \frac{1}{2c}\vec{\sigma}\cdot\vec{p} g_\mu \end{pmatrix} \mathbf{A}_{\mu p}^{00} + \begin{pmatrix} -\frac{1}{2c}\vec{\sigma}\cdot\vec{p} P g_\mu \\ P g_\mu \end{pmatrix} \mathbf{B}_{\mu p}^{00}, \tag{61}$$

where the operator $P$ changes the parity of a given atomic spinor $g_\mu = |ljm_j\rangle$, viz.,

$$P|ljm_j\rangle = \begin{cases} |(l+1)jm_j\rangle & \text{if } l<j \\ |(l-1)jm_j\rangle & \text{if } l>j \end{cases} \tag{62}$$

The corresponding first-order orbitals may be expanded as

$$\psi_{p,b,DMB}^{10} = (Z_g^{10} + Z_{m,DMB}^{10})\psi_p^{00} \tag{63}$$

---

[1] Equation 34 in Ref. [17] turns out to be inappropriate and should be replaced with the present Eq. 61.

$$
= \begin{pmatrix} -i\Lambda^g_\mu g_\mu \\ \left(\frac{\vec{\sigma}\cdot\vec{A}^{10}_\mu}{2c} - i\Lambda^g_\mu \frac{\vec{\sigma}\cdot\vec{p}}{2c}\right)g_\mu \end{pmatrix} A^{00}_{\mu p} + \begin{pmatrix} -\left(\frac{\vec{\sigma}\cdot\vec{A}^{10}_\mu}{2c} - i\Lambda^g_\mu \frac{\vec{\sigma}\cdot\vec{p}}{2c}\right)Pg_\mu \\ -i\Lambda^g_\mu Pg_\mu \end{pmatrix} B^{00}_{\mu p}.
$$

$$\tag{64}$$

This ansatz may be termed as "dual magnetic balance" (DMB) that has not been considered before. Note that, at variance with the other $Z^{10}_m$ operators, the $Z^{10}_{m,DMB}$ operator is only implicity but its action is well defined. While the DKB is considerably more involved than the RKB in field-free calculations [17], the DMB does not increase too much complexity compared with the other schemes for NMR parameters.

Substituting $\psi^{10}_i$ (54) into Eq. 18 leads to the mixed second-order energy

$$E^{11} = E^{11}_d + E^{11}_p, \tag{65}$$

$$E^{11}_d = \langle \psi^{00}_i | D^{01} | \psi^{10}_{i,b} \rangle + c.c. \tag{66}$$

$$= C^{00*}_{\mu i} \langle g_\mu | Z^{10\dagger}_\mu D^{01} + D^{01} Z^{10}_\nu | g_\nu \rangle C^{00}_{\nu i}, \tag{67}$$

$$E^{11}_p = \langle \psi^{00}_i | D^{01} | \psi^{00}_p \rangle C^{10}_{pi} + c.c. \tag{68}$$

$$= [\langle \psi^{00}_i | D^{01} | \psi^{00}_a \rangle C^{10}_{ai} + c.c.] - \langle \psi^{00}_i | D^{01} | \psi^{00}_j \rangle S^{10}_{ji}, \tag{69}$$

where $E^{11}_d$ and $E^{11}_p$ are the diamagnetic and paramagnetic terms, respectively. Equation 69 results from the unitary normalization condition

$$C^{10}_{pq} + C^{10*}_{qp} + S^{10}_{pq} = 0, \tag{70}$$

$$S^{10}_{pq} = \langle \psi^{00}_p | Z^{10\dagger} + Z^{10} | \psi^{00}_q \rangle = C^{00*}_{\mu p} \langle \chi_\mu | Z^{10\dagger}_\mu + Z^{10}_\nu | \chi_\nu \rangle C^{00}_{\nu q}. \tag{71}$$

It should be emphasized that all the above schemes are strictly equivalent to each other as they are just different decompositions of the same observable (total shielding) into the sum of two non-observables (paramagnetism and diamagnetism). As the schemes share the same gauge factor, their mutual relations can readily be understood by inspecting the $Z^{10}_m$ operators. More specifically, the diamagnetic term of ODA is just the average of those of EFUT and SD, as clearly seen from Eqs. 49, 50, and 51. The diamagnetic term of RMB is somewhat different, but this is only minor. As for the paramagnetic terms, the contributions of the first-order small components $\psi^{S,10}_p$ are the same for the SD, ODA, and EFUT approaches, whereas those of $\psi^{L,10}_p$ are the same for the LRT, ODA, and RMB ones. The latter also equal to the average of those by SD and EFUT. In particular, ODA and RMB differ only in the treatment for the first-order small components $\psi^{S,10}_p$ in view of Eqs. 49 and 53: $Z^{10}_{m,ODA}$ acts on directly the zeroth-order small component $\psi^{S,00}_p$ while $Z^{10}_{m,RMB}$ acts on the zeroth-order pseudo-large component $\widetilde{\psi}^{L,00}_p$ (see Sect. 3) It has been confirmed [2, 24] that the

ODA, EFUT, SD, and RMB schemes perform rather similarly for nuclear shieldings, with EFUT and RMB only marginally better, as far as the convergence to the basis set limit is concerned. Because of the inherent difference between DKB and RKB (see Eqs. 61 and 47), both $\psi^{L,10}_p$ and $\psi^{S,10}_p$ are treated differently in DMB. Its performance yet remains to be checked.

## 3 The modified Dirac equation

In the previous notation, the $Z^{10}_{m,RMB}$ operator (53) looks very peculiar and the $Z^{10}_{m,DMB}$ operator (63) even does not have an explicit form. To avoid this, we now introduce the modified Dirac equation (MDE) with a non-unit metric:

$$\widetilde{F}\widetilde{\psi}_p = \widetilde{S}\widetilde{\psi}_p, \tag{72}$$

$$\widetilde{F} = Y^\dagger DY, \quad Y = \begin{pmatrix} Y_{11} & Y_{12} \\ Y_{21} & Y_{22} \end{pmatrix}, \tag{73}$$

$$\widetilde{S} = Y^\dagger Y, \tag{74}$$

where the modified spinors $\widetilde{\psi}_p$ are to be expanded as

$$\widetilde{\psi}_p = g_\mu \widetilde{C}_{\mu p} = \widetilde{\psi}^{00}_q C_{qp}, \tag{75}$$

$$\widetilde{\psi}^{00}_p = \begin{pmatrix} \psi^{L,00}_p \\ \widetilde{\psi}^{L,00}_p \end{pmatrix} = \begin{pmatrix} g_\mu A^{00}_{\mu p} \\ g_\mu B^{00}_{\mu p} \end{pmatrix}. \tag{76}$$

The lower component $\widetilde{\psi}^L_p$ of $\widetilde{\psi}_p$ may be called "pseudo-large component" as it is of the same symmetry and nrl as the large component $\psi^L_p$ [2]. In the orthonormal basis of $\{\widetilde{\psi}^{00}_p\}$, the matrix representation of Eq. 72 reads

$$\widetilde{\mathbf{F}}\mathbf{C} = \widetilde{\mathbf{S}}\mathbf{C}\epsilon, \tag{77}$$

which can be expanded up to first order

$$\widetilde{\mathbf{F}}^{00} = \epsilon^{00}, \quad \widetilde{F}^{00} = Y^{00\dagger}D^{00}Y^{00}, \quad \widetilde{S}^{00} = Y^{00\dagger}Y^{00},$$
$$\mathbf{C}^{00} = \mathbf{I}, \tag{78}$$

$$\widetilde{\mathbf{F}}^{10} + \epsilon^{00}\mathbf{C}^{10} = \widetilde{\mathbf{S}}^{10}\epsilon^{00} + \widetilde{\mathbf{S}}^{00}\mathbf{C}^{10}\epsilon^{00} + \widetilde{\mathbf{S}}^{00}\epsilon^{10}, \tag{79}$$

$$\widetilde{F}^{10} = Y^{10\dagger}D^{00}Y^{00} + Y^{00\dagger}D^{10}Y^{00} + Y^{00\dagger}D^{00}Y^{10}, \tag{80}$$

$$\widetilde{S}^{10} = Y^{10\dagger}Y^{00} + Y^{00\dagger}Y^{10}, \tag{81}$$

where the matrix elements of the $Y^{00}$ and $Y^{10}$ operators are documented in Table 1. The $Y^{10}$ and the previous $Z^{10}$ operators are related by

$$Y^{10} = Z^{10}Y^{00}. \tag{82}$$

That is, the RKB/DKB condition $Y^{00}$, associated previously with the basis set $\{\chi_\mu\}$, is now folded into the Hamiltonian.

**Table 1** The $Y^{00}$ and $Y^{10}$ operators in Eqs. 78 and 80

| Method | $Y_{11}^{00}$ | $Y_{12}^{00}$ | $Y_{21}^{00}$ | $Y_{22}^{00}$ |
|---|---|---|---|---|
| RKB | 1 | 0 | 0 | $\frac{\vec{\sigma}\cdot\vec{p}}{2c}$ |
| DKB | 1 | $-\frac{\vec{\sigma}\cdot\vec{p}}{2c}P$ | $\frac{\vec{\sigma}\cdot\vec{p}}{2c}$ | $P$ |
| | $Y_{11}^{10}$ | $Y_{12}^{10}$ | $Y_{21}^{10}$ | $Y_{22}^{10}$ |
| LRT-GIAO[a] | $-i\Lambda_\mu^g$ | 0 | 0 | $-i\Lambda_\mu^g\frac{\vec{\sigma}\cdot\vec{p}}{2c}$ |
| ODA-GIAO[a] | $-i\Lambda_\mu^g$ | 0 | $\frac{\vec{\sigma}\cdot\vec{A}_\mu^{10}}{2c}$ | $-i\Lambda_\mu^g\frac{\vec{\sigma}\cdot\vec{p}}{2c}$ |
| EFUT-GIAO[a] | $-i\Lambda_\mu^g$ | $-\frac{\vec{\sigma}\cdot\vec{A}_\mu^{10}}{2c}\frac{\vec{\sigma}\cdot\vec{p}}{2c}$ | $\frac{\vec{\sigma}\cdot\vec{A}_\mu^{10}}{2c}$ | $-i\Lambda_\mu^g\frac{\vec{\sigma}\cdot\vec{p}}{2c}$ |
| SD-GIAO[a] | $-i\Lambda_\mu^g$ | $\frac{\vec{\sigma}\cdot\vec{A}_\mu^{10}}{2c}\frac{\vec{\sigma}\cdot\vec{p}}{2c}$ | $\frac{\vec{\sigma}\cdot\vec{A}_\mu^{10}}{2c}$ | $-i\Lambda_\mu^g\frac{\vec{\sigma}\cdot\vec{p}}{2c}$ |
| RMB-GIAO[a] | $-i\Lambda_\mu^g$ | 0 | 0 | $-i\Lambda_\mu^g\frac{\vec{\sigma}\cdot\vec{p}}{2c}+\frac{\vec{\sigma}\cdot\vec{A}_\mu^{10}}{2c}$ |
| DMB-GIAO[b] | $-i\Lambda_\mu^g$ | $\left(i\Lambda_\mu^g\frac{\vec{\sigma}\cdot\vec{p}}{2c}-\frac{\vec{\sigma}\cdot\vec{A}_\mu^{10}}{2c}\right)P$ | $-i\Lambda_\mu^g\frac{\vec{\sigma}\cdot\vec{p}}{2c}+\frac{\vec{\sigma}\cdot\vec{A}_\mu^{10}}{2c}$ | $-i\Lambda_\mu^g P$ |

The operator $P$ is defined in Eq. 62

[a] Associated with RKB

[b] Associated with DKB

The present MDE (72) amounts to a generalization of the original magnetic field-independent MDE in terms of the RKB only. It was first proposed by Kutzelnigg [57] but explored more thoroughly by Dyall [58]. Apart from the easy introduction of the RMB and DMB ansätze, one major advantage of the so-introduced MDE is that it allows the development of the X2C counterparts for NMR parameters [55] precisely in the same way as that of the X2C Hamiltonians for electronic structure calculations [59] (for a recent review see Ref. [60]). Nonetheless, in the subsequent sections, we still stick to the $Z^{10}$ operators to follow closely the previous works.

## 4 The coupled-perturbed DHF/DKS equation

It is seen from Eq. 69 that only the virtual-occupied coefficients $C_{bj}^{10}$ remain to be determined by the coupled-perturbed Dirac-Hartree-Fock (DHF)/Dirac-Kohn-Sham (DKS) equation

$$(\epsilon_a^{00} - \epsilon_i^{00})C_{ai}^{10} + K_{ai,bj}C_{bj}^{10} + K_{ai,jb}C_{bj}^{10*} = -G_{ai}^{10}, \tag{83}$$

whose dimension is $2N_v \times 2N_o \times 2$, with $N_v$ and $N_o$ being the respective virtual and occupied Kramers pairs. The additional factor of 2 arises from the fact that both $C_{bj}^{10}$ and $C_{bj}^{10*}$ are non-redundant, equivalent to the real and imaginary parts of $C_{bj}^{10}$ alone. In view of the following relation specific for closed-shell systems [24],

$$C_{\bar{a}\bar{i}}^{10} = -C_{ai}^{10*}, \tag{84}$$

Equation 83 can be rewritten as

$$\begin{aligned}(\epsilon_a^{00} - \epsilon_i^{00})C_{ai}^{10} + (V_{res}^{10})_{ai} &= -G_{ai}^{10}, \\ (V_{res}^{10})_{ai} &= (K_{ai,bj} - K_{ai,\bar{j}\bar{b}})C_{bj}^{10},\end{aligned} \tag{85}$$

whose dimension becomes $2N_v \times 2N_o$, half of the original one. The coupling matrix $K$ takes the following form

$$\begin{aligned}K_{pq,rs} = &-c_1(pr|sq) - c_2(p\vec{\alpha}q|s\vec{\alpha}r) + c_3(p\vec{\alpha}r|s\vec{\alpha}q) \\ &+ c_4(f_{xc})_{pq,rs},\end{aligned} \tag{86}$$

where the Mulliken notation has been adopted for the integrals. The first term arises from the exchange contributions of the Coulomb interaction, while the second and third terms from the respective direct and exchange contributions of the Gaunt interaction. Note that there are no direct Coulomb contributions here, as the electron density has no linear response to the magnetic fields [24]. The linear coefficients $c_i$ may be specified according to some hybrid density functionals. Under the non-collinear local density approximation for the density functional [61, 62]

$$\begin{aligned}E_{xc}[\rho, s] &= \int \epsilon(\rho, s)\,\mathrm{d}\vec{r}, \quad \rho = \psi_i^\dagger\psi_i, \quad s = |\vec{M}|, \\ \vec{M} &= \psi_i^\dagger\beta\vec{\Sigma}\psi_i, \quad \vec{\Sigma} = \begin{pmatrix} \vec{\sigma} & 0 \\ 0 & \vec{\sigma} \end{pmatrix},\end{aligned} \tag{87}$$

the exchange-correlation kernel reads

$$(f_{xc})_{pq,rs} = \left(\psi_p^{00}\beta\vec{\Sigma}\psi_q^{00}\Big|\frac{\partial^2\epsilon}{\partial s^2}[\rho^{00}, s^{00}]\delta(\vec{r} - \vec{r}')|\psi_s^{00}\beta\vec{\Sigma}\psi_r^{00}\right). \tag{88}$$

The forces $G_{ai}^{10}$ on the right-hand side of Eq. 85 take the following form

$$\begin{aligned}G_{ai}^{10} = &\tilde{D}_{ai}^{10} - S_{ai}^{10}\epsilon_i^{00} + c_1(V_{xx}^{10})_{ai} + c_2(V_{gc}^{10})_{ai} + c_3(V_{gx}^{10})_{ai} \\ &+ c_4(V_{xc}^{10})_{ai},\end{aligned} \tag{89}$$

$$\tilde{D}_{ai}^{10} = \langle a|D^{10} + Z^{10\dagger}F^{00} + F^{00}Z^{10}|i\rangle, \quad F^{00} = D^{00} + V_{scf}^{00}, \tag{90}$$

$$(V_{xx}^{10})_{ai} = -(aZ^{10}j|ji) - (aj|Z^{10}ji) + (aj|ki)S_{jk}^{10}, \tag{91}$$

$$V_{gc}^{10})_{ai} = -(a\vec{\alpha}i|Z^{10}j\vec{\alpha}j) - (a\vec{\alpha}i|j\vec{\alpha}Z^{10}j) + (a\vec{\alpha}i|k\vec{\alpha}j)S_{jk}^{10}, \tag{92}$$

$$(V_{gx}^{10})_{ai} = (a\vec{\alpha}j|Z^{10}j\vec{\alpha}i) + (a\vec{\alpha}Z^{10}j|j\vec{\alpha}i) - (a\vec{\alpha}j|k\vec{\alpha}i)S_{jk}^{10},$$
$$(93)$$

$$(V_{xc}^{10})_{ai} = (a\beta\vec{\Sigma}i|\frac{\partial^2\epsilon}{\partial s^2}[\rho^{00}, s^{00}]\delta(\vec{r} - \vec{r'})|\vec{M}_b^{10}),$$
$$(94)$$

$$\vec{M}_b^{10} = (\psi_i^{00\dagger}\beta\vec{\Sigma}Z^{10}\psi^{00} + c.c.) - \psi_i^{00\dagger}\beta\vec{\Sigma}\psi_j^{00}S_{ji}^{10}.$$
$$(95)$$

At this stage, it is of great interest to examine the contributions of NES (denoted as $\tilde{a}$ and $\tilde{b}$). It is easy to see that, in the uncoupled treatment (i.e., $K = 0$), the $C_{\tilde{a}j}^{10}$ are of order $c^{-1}$ in the LRT without the MB and $c^{-3}$ in the other schemes with the MB (see $C_{ai}^{10(0)}$ in Table 2). Therefore, the contributions of NES to the residual current density $\vec{j}_r^{10}$

$$\vec{j}_r^{10} = \psi_i^{00}c\vec{\alpha}\psi_a^{00}C_{ai}^{10} + c.c.$$
$$(96)$$

are of $c^0$ in the LRT and $c^{-2}$ in the other methods. The former implies that the contributions of NES to nuclear shieldings (18) survive even to the nrl in the LRT, precisely the reason why the LRT is so demanding on the quality of the basis sets. Likewise, the contributions of NES to the residual magnetization vector $\vec{M}_r^{10}$

$$\vec{M}_r^{10} = \psi_i^{00}\beta\vec{\Sigma}\psi_a^{00}C_{ai}^{10} + c.c.$$
$$(97)$$

are ,respectively, of $c^{-2}$ and $c^{-4}$ in the LRT and all the other methods. To further see the effects of NES to the couplings, it is first noticed from Table 3 that the Coulomb exchange as well as the Gaunt direct and exchange integrals are of order $c^0$ if both $a$ and $b$ are NES, whereas all the integrals in $K$ are of order $c^{-1}$ when either $a$ or $b$ is a NES. It is then seen from the first iteration $C_{ai}^{10(1)}$ of Eq. 85

$$C_{ai}^{10(1)} = C_{ai}^{10(0)} + \sum_b X_{ai,b}^{10(1)},$$
$$(98)$$

$$C_{ai}^{10(0)} = \frac{\tilde{D}_{ai}^{10} - S_{ai}^{10}\epsilon_i^{00}}{\epsilon_i^{00} - \epsilon_a^{00}},$$
$$(99)$$

**Table 2** Ordering of the first-order coefficients

| $Z_m^{10} = 0$ | | | | | $Z_m^{10} \neq 0$ | | | | |
|---|---|---|---|---|---|---|---|---|---|
| $a$ | $D_{ai}^{10}$ | $C_{ai}^{10(0)}$ | $ab$ | $X_{ai,b}^{10(1)}$ | $a$ | $\tilde{D}_{ai}^{10}$ | $C_{ai}^{10(0)}$ | $ab$ | $X_{ai,b}^{10(1)}$ |
| $+$ | $c^0$ | $c^0$ | $++$ | $c^0$ | $+$ | $c^0$ | $c^0$ | $++$ | $c^0$ |
| | | | $+-$ | $c^{-2}$ | | | | $+-$ | $c^{-4}$ |
| $-$ | $c^1$ | $c^{-1}$ | $-+$ | $c^{-3}$ | $-$ | $c^{-1}$ | $c^{-3}$ | $-+$ | $c^{-3}$ |
| | | | $--$ | $c^{-3}$ | | | | $--$ | $c^{-5}$ |

The signs $+$ and $-$ indicate positive and negative energy orbitals, respectively. $\tilde{D}^{10} = D^{10} + Z_m^{10\dagger}D^{00} + D^{00}Z_m^{10}$ is even and of $c^0$; $D^{10}$ is odd and of $c^1$; $C_{ai}^{10(0)}$ (99): Uncoupled; $X_{ai}^{10(1)}$ (100): The first iteration of Eq. 85

**Table 3** Ordering of the integrals needed in Eq. 85

| Integrals | Expression | $ab$ | | | |
|---|---|---|---|---|---|
| | | $++$ | $+-$ | $--$ | $-+$ |
| Coulomb direct[a]/$f_{xc}$ | $[(ai|jb) - (ai|\bar{b}j)]$ | $c^0$ | $c^{-1}$ | $c^{-1}$ | $c^{-2}$ |
| Coulomb exchange | $[(ab|ji) - (a\bar{j}|\bar{b}i)]$ | $c^0$ | $c^{-1}$ | $c^{-1}$ | $c^0$ |
| Gaunt direct | $[(a\vec{\alpha}i|j\vec{\alpha}b) - (a\vec{\alpha}i|\bar{b}\vec{\alpha}\bar{j})]$ | $c^{-2}$ | $c^{-1}$ | $c^{-1}$ | $c^0$ |
| Gaunt exchange | $[(a\vec{\alpha}b|j\vec{\alpha}i) - (a\vec{\alpha}\bar{j}|\bar{b}\vec{\alpha}i)]$ | $c^{-2}$ | $c^{-1}$ | $c^{-1}$ | $c^0$ |
| coupling $K$ | $K_{ai,bj} - K_{ai,\bar{j}b}$ | $c^0$ | $c^{-1}$ | $c^{-1}$ | $c^0$ |

The signs $+$ and $-$ indicate positive and negative energy orbitals, respectively

[a] Not really needed

$$X_{ai,b}^{10(1)} = \sum_j \frac{(K_{ai,bj} - K_{ai,\bar{j}b})C_{bj}^{10(0)}}{\epsilon_i^{00} - \epsilon_a^{00}}$$
$$(100)$$

that the corrections $X_{ai,b}^{10(1)}$ are in the LRT of order $c^{-2}$ relative to $C_{ai}^{10(0)}$ when at least one virtual orbital is of negative energy (cf. Table 2). The situation is even better for the other schemes: The $C_{\tilde{a}i}^{10(0)}$ themselves are already of order $c^{-3}$ and the corrections $X_{ai,\tilde{b}}^{10(1)}$ to $C_{ai}^{10(0)}$ ($\epsilon_a > 0$) are only of order $c^{-4}$. These findings suggest immediately the following algorithms:

1. The $C_{\tilde{a}i}^{10}$ for the NES are calculated simply according to Eq. 99. The indices $a$ and $b$ in Eq. 85 are then restricted only to the PES, which is computationally very much the same as the two-component counterpart.

2. Alternatively, the $C_{\tilde{a}i}^{10}$ (99) can be included when solving Eq. 85 for the PES. After convergence, the $C_{\tilde{a}i}^{10}$ for the NES can be recalculated as

$$C_{\tilde{a}i}^{10} = \frac{\tilde{G}_{ai}^{10} + (V_{res}^{10})_{ai}}{\epsilon_i^{00} - \epsilon_a^{00}}$$
$$(101)$$

with the induced potential $V_{res}^{10}$ from the PES.

3. Equation 85 can be solved exactly by further iterating the procedure in scheme (2).

It turns out that scheme (1) is already sufficiently accurate [24]. Note in passing that only $C_{ai}^{10}$ and $C_{\tilde{a}i}^{10}$ ($=C_{ai}^{10*}$) are independent coefficients in view of Eq. 84. The Gaunt term does not contribute to $V_{scf}^{00}$ and contributes to nuclear shieldings only indirectly through the coupled-perturbed equation. As only the Gaunt integrals involving the NES are significant (see Table 3) and only radially compact, atomic-like NES are relevant for total shieldings [2], only one-centered Gaunt integrals need to be retained. Therefore, the inclusion of the Gaunt interaction does not cost much computational overhead.

# 5 The MP2 method for nuclear shielding

Apart from relativistic effects, correlation effects also play an important role in accurate calculations of NMR parameters. To demonstrate that the MB-GIAO ansatz can also be combined with wave function-based correlation methods, we consider here the second-order Møller-Plesset perturbation theory (MP2) and derive the expressions for nuclear shielding in two different ways, one with the derivative technique and the other through the induced current. It turns out that the latter is much simpler.

## 5.1 $\mu$-MP2: nuclear shielding as second-order derivative of MP2 energy

The point of departure is the following partitioning of the configuration space Dirac-Coulomb (DC) Hamiltonian

$$H = H_0 + V, \tag{102}$$

$$H_0 = \sum_i^N h(i), \quad h = D^{00} + D^{01} + D^{10} + V_{HF}[\vec{\mu}, \vec{B}^{10}], \tag{103}$$

$$V = H - H_0 = \frac{1}{2} \sum_{ij}^N \frac{1}{r_{ij}} - \sum_i^N V_{HF}[\vec{\mu}, \vec{B}^{10}](i). \tag{104}$$

That is, the two external perturbations $D^{01}$ and $D^{10}$ are to be included in the variational DHF calculation. Such a partitioning facilitates the subsequent derivation of the MP2 energy $E_2$ according to the bound-state QED prescription (W. Liu, 2011, unpublished)

$$E_2 = E_{2+} + E_{2-}, \tag{105}$$

$$E_{2+} = \frac{1}{4} \bar{g}_{ij}^{ab} \bar{t}_{ab}^{ij}, \tag{106}$$

$$E_{2-} = -\frac{1}{4} \bar{g}_{ij}^{\bar{a}\bar{b}} \bar{t}_{\bar{a}\bar{b}}^{ij} - \bar{g}_{ij}^{\bar{a}j} \bar{t}_{\bar{a}k}^{ik} + g_{aj}^{i\bar{b}} t_{i\bar{b}}^{aj} + g_{ib}^{ai} t_{aj}^{j\bar{b}}, \tag{107}$$

$$t_{pq}^{rs} = g_{pq}^{rs} / \Delta_{pq}^{rs} = -(t_{rs}^{pq})^*, \tag{108}$$

$$\Delta_{xyz\cdots}^{pqrs\cdots} = \epsilon_{up} - \epsilon_{down} = (\epsilon_p + \epsilon_q + \cdots) - (\epsilon_x + \epsilon_y + \cdots), \tag{109}$$

$$\bar{t}_{pq}^{rs} = t_{pq}^{rs} - t_{pq}^{sr} = -(\bar{t}_{rs}^{pq})^*, \tag{110}$$

$$g_{pq}^{rs} = (pr|qs) = (g_{rs}^{pq})^*, \tag{111}$$

$$\bar{g}_{pq}^{rs} = g_{pq}^{rs} - g_{pq}^{sr} = (\bar{g}_{rs}^{pq})^*, \tag{112}$$

where $E_{2+}$ (106) represents the second-order correlation energy within the PES manifold, whereas $E_{2-}$ (107) arises from the contributions of the NES. It should be emphasized that the latter *cannot* be obtained correctly by the configuration space (associated with the empty Dirac picture) or the Fock space [63] (associated with the filled Dirac picture) approaches, where, e.g., the first term of $E_{2-}$ (107) appears as $\frac{1}{4} \bar{g}_{ij}^{\bar{a}\bar{b}} \bar{t}_{\bar{a}\bar{b}}^{ij}$, which is positively valued and hence "anti-correlating," resulting in a reduction in the correlation energy although the correlation space is enlarged. The contributions of $E_{2-}$ (107) to NMR parameters can readily be taken into account. However, the $E_{2-}$ term is of the same order of $(Z/c)^3$ as the leading radiative QED correction (Lamb shift). They should hence be treated on an equal footing, viz., either both included or both neglected. As the QED correction is not under concern here, the $E_{2-}$ term will not further be considered as well. We are then left with the $E_{2+}$ term only, the no-pair approximation.

So far the DHF orbitals involved here and hence the energy (105) are correct to infinite order in external perturbations. Yet, it is the mixed second-order energy $E_{2+}^{11}$ with respect to two external perturbations that is of interest here. The first-order derivative of $E_{2+}$ with respect to the first perturbation (e.g., the nuclear magnetic dipole $\vec{\mu}$) reads [64]

$$E_{2+}^1 = E_{2+}^{[1]} + \frac{\partial E_{2+}}{\partial \mathbf{C}} \mathbf{C}^1 + \frac{\partial E_{2+}}{\partial \epsilon} \epsilon^1, \tag{113}$$

$$E_{2+}^{[1]} = \frac{1}{4} \bar{g}_{ij}^{ab[1]} \bar{t}_{ab}^{ij} + c.c., \tag{114}$$

$$g_{pq}^{rs[1]} = (Z^1 pr||qs) + (pZ^1 r||qs) + (pr||Z^1 qs) + (pr||qZ^1 s), \tag{115}$$

$$\frac{\partial E_{2+}}{\partial \mathbf{C}} \mathbf{C}^1 = \frac{1}{2} (C_{pi}^{1*} \bar{g}_{pj}^{ab} \bar{t}_{ab}^{ij} + C_{pa}^1 \bar{g}_{ij}^{pb} \bar{t}_{ab}^{ij}) + c.c., \tag{116}$$

$$\frac{\partial E_{2+}}{\partial \epsilon} \epsilon^1 = \frac{1}{2} \bar{t}_{ij}^{ab} \bar{t}_{ab}^{ij} (\epsilon_i^1 - \epsilon_a^1). \tag{117}$$

Under the unitary normalization condition

$$C_{qp}^{1*} + S_{pq}^1 + C_{pq}^1 = 0, \tag{118}$$

$$S_{pq}^1 = \langle p|(Z^{1\dagger} + Z^1)|q \rangle, \tag{119}$$

Equation 116 may be rewritten as

$$\frac{\partial E_{MP2}}{\partial \mathbf{C}} \mathbf{C}^1 = \frac{1}{2} \left[ (C_{ci}^{1*} \bar{g}_{cj}^{ab} \bar{t}_{ab}^{ij} - C_{ak}^{1*} \bar{g}_{ij}^{kb} \bar{t}_{ab}^{ij} - S_{ka}^1 \bar{g}_{ij}^{kb} \bar{t}_{ab}^{ij}) + c.c. \right]$$
$$+ \frac{1}{2} (S_{ac}^1 \bar{g}_{cb}^{ij} \bar{t}_{ij}^{ab} - S_{ik}^1 \bar{g}_{kj}^{ab} \bar{t}_{ab}^{ij})$$
$$+ \frac{1}{2} (C_{ca}^1 (\epsilon_c^0 - \epsilon_a^0) \bar{t}_{ij}^{cb} \bar{t}_{ab}^{ij} - C_{ik}^1 (\epsilon_i^0 - \epsilon_k^0) \bar{t}_{kj}^{ab} \bar{t}_{ab}^{ij}). \tag{120}$$

Noticeably, Eq. 117 involves the first order orbital energies $\epsilon_p^1$ which requires a perturbation-adapted basis for each degenerate manifold. This can be avoided by realizing that MP2 is invariant under unitary transformations within the occupied and virtual subspaces. That is, Eq. 117 can be rewritten as [65]

$$\frac{\partial E_{2+}}{\partial \epsilon}\epsilon^1 = \gamma_{ji}\epsilon_{ij}^1 + \gamma_{ba}\epsilon_{ab}^1, \tag{121}$$

where

$$\gamma_{ji} = \frac{1}{2}\bar{t}_{jk}^{ab}\bar{t}_{ab}^{ik} = \gamma_{ij}^*, \tag{122}$$

$$\gamma_{ba} = -\frac{1}{2}\bar{t}_{bc}^{ij}\bar{t}_{ij}^{ac} = \gamma_{ab}^*, \tag{123}$$

$$\epsilon_{pq}^1 = G_{pq}^1 + C_{pq}^1(\epsilon_p^0 - \epsilon_q^0) + (C_{ak}^1\bar{g}_{pk}^{qa} + C_{ak}^{1*}\bar{g}_{pa}^{qk}), \tag{124}$$

$$G_{pq}^1 = \tilde{D}_{pq}^1 - S_{pq}^1\epsilon_q^0 - \bar{g}_{pl}^{qk}S_{kl}^1 + (pq||Z^1kk) + (pq||kZ^1k), \tag{125}$$

$$\tilde{D}_{pq}^1 = \langle p|D^{01} + Z^{1\dagger}F^0 + F^0Z^1|q\rangle. \tag{126}$$

In view of Eqs. 114, 120, 121 and 113 can be rearranged to

$$\begin{aligned} E_{2+}^1 = {}&\gamma_{ji}G_{ij}^1 + \gamma_{ba}G_{ab}^1 + (X_{ia}C_{ai}^1 + c.c.) \\ &+ I_{ji}S_{ij}^1 + I_{ba}S_{ab}^1 + (I_{ia}S_{ai}^1 + c.c.) + E_{2+}^{[1]}, \end{aligned} \tag{127}$$

where

$$X_{ia} = \frac{1}{2}\bar{g}_{ib}^{jk}\bar{t}_{jk}^{ab} - \frac{1}{2}\bar{t}_{ij}^{bc}\bar{g}_{bc}^{aj} + \gamma_{kj}\bar{g}_{ji}^{ka} + \gamma_{cb}\bar{g}_{bi}^{ca}, \tag{128}$$

$$I_{ba} = \frac{1}{2}\bar{g}_{bc}^{ij}\bar{t}_{ij}^{ac}, \tag{129}$$

$$I_{ji} = -\frac{1}{2}\bar{g}_{jk}^{ab}\bar{t}_{ab}^{ik}, \tag{130}$$

$$I_{ia} = \frac{1}{2}\bar{g}_{ib}^{jk}\bar{t}_{jk}^{ab}, \qquad I_{ai} \stackrel{\text{def}}{=} I_{ia}^*. \tag{131}$$

It is clear that Eq. 127 does not depend explicitly on the occupied–occupied $C_{ij}^1$ and virtual–virtual $C_{ab}^1$ coefficients. The virtual-occupied coefficients $C_{ai}^1$ are to be determined by the CPHF equation

$$H_{ai,bj}\begin{pmatrix} C_{bj}^1 \\ C_{bj}^{1*} \end{pmatrix} = \begin{pmatrix} -G_{ai}^1 \\ -G_{ai}^{1*} \end{pmatrix}, \tag{132}$$

where the orbital Hessian is defined as

$$H = \begin{pmatrix} A & B \\ B^* & A^* \end{pmatrix}, \tag{133}$$

$$A_{ai,bj} = \delta_{ij}\delta_{ab}(\epsilon_a^0 - \epsilon_i^0) + \bar{g}_{aj}^{ib}, \tag{134}$$

$$B_{ai,bj} = \bar{g}_{ab}^{ij}. \tag{135}$$

As it is the contractions $X_{ia}C_{ai}^1$ that are really needed in Eq. 127, instead of solving the CPHF (132) for each cartesian component of the perturbations $G^1$, one can invoke the so-called Z-vector equation [66]

$$\begin{aligned} (X_{ia}X_{ia}^*)\begin{pmatrix} C_{ai}^1 \\ C_{ai}^{1*} \end{pmatrix} &= (X_{ia} \quad X_{ia}^*)(H^{-1})_{ai,bj}\begin{pmatrix} -G_{bj}^1 \\ -G_{bj}^{1*} \end{pmatrix} \\ &= (\gamma_{jb} \quad \gamma_{jb}^*)\begin{pmatrix} G_{bj}^1 \\ G_{bj}^{1*} \end{pmatrix}. \end{aligned} \tag{136}$$

Then, only the following single set of equation needs to be solved for the occupied-virtual block of the relaxed density matrix $\gamma$:

$$(\gamma_{jb}\gamma_{jb}^*)H_{bj,ai} = (-X_{ia} \quad -X_{ia}^*). \tag{137}$$

As a result, Eq. 127 takes the following very compact form

$$E_{2+}^1 = \gamma_{ji}G_{ij}^1 + \gamma_{ba}G_{ab}^1 + (\gamma_{ia}G_{ai}^1 + c.c.) + I_{qp}S_{pq}^1 + E_{2+}^{[1]}. \tag{138}$$

Note that all the orbitals and the energy levels $\epsilon_p^0$ involved in Eqs. 113 to 138 are of zeroth order with respect to the considered first perturbation (denoted as 01) but of infinite order with respect to the second perturbation (denoted as 10). Further differentiating $E_{2+}^1$ with respect to the second perturbation gives rise to

$$\begin{aligned} E_{2+}^{11} = {}&\left[(\gamma_{ia}G_{ai}^{11} + \gamma_{ia}^{10}G_{ai}^{01}) + c.c.\right] \\ &+ \gamma_{ji}G_{ij}^{11} + \gamma_{ji}^{10}G_{ij}^{01} + \gamma_{ba}G_{ab}^{11} + \gamma_{ba}^{10}G_{ab}^{01} \\ &+ I_{qp}S_{pq}^{11} + I_{qp}(C_{pr}^{10*}S_{rq}^{01} + S_{pr}^{01}C_{rq}^{10}) + I_{qp}^{10}S_{pq}^{01} + E_{2+}^{[1]}, \end{aligned} \tag{139}$$

where

$$\begin{aligned} G_{pq}^{11} = {}&G_{pq}^{[11]} + C_{rp}^{10*}G_{rq}^{01} + G_{pr}^{01}C_{rq}^{10} + S_{pr}^{01}C_{rq}^{10}(\epsilon_r^{00} - \epsilon_q^{00}) \\ &- S_{pr}^{01}\epsilon_{rq}^{10} + C_{rk}^{10*}(\bar{g}_{pr}^{qk[01]} - S_{rl}^{01}\bar{g}_{pl}^{qk} - S_{lk}^{01}\bar{g}_{pr}^{ql}) \\ &+ C_{rk}^{10}(\bar{g}_{pk}^{qr[01]} - S_{kl}^{01}\bar{g}_{pl}^{qr} - S_{lr}^{01}\bar{g}_{pk}^{ql}) \end{aligned} \tag{140}$$

$$\begin{aligned} G_{pq}^{[11]} = {}&\tilde{D}_{pq}^{11} - S_{pq}^{11}\epsilon_q^{00} - S_{kl}^{11}\bar{g}_{pl}^{qk} - S_{kl}^{01}\bar{g}_{pl}^{qk[10]} \\ &+ (Z^{10}pq||Z^{01}kk) + (pZ^{10}q||Z^{01}kk) + (Z^{10}pq||kZ^{01}k) \\ &+ (pZ^{10}q||kZ^{01}k) + (Z^{01}pq||Z^{10}kk) + (pZ^{01}q||Z^{10}kk) \\ &+ (Z^{01}pq||kZ^{10}k) + (pZ^{01}q||kZ^{10}k) + (pq||Z^{11}kk) \\ &+ (pq||Z^{01}kZ^{10}k) + (pq||Z^{10}kZ^{01}k) + (pq||kZ^{11}k), \end{aligned} \tag{141}$$

$$\begin{aligned} \tilde{D}_{pq}^{11} = {}&\langle p|[(Z^{10\dagger}D^{01} + Z^{01\dagger}D^{10} + Z^{01\dagger}F^{00}Z^{10} + Z^{11\dagger}F^{00}) \\ &+ h.c.] + D_{ext}^{11}|q\rangle, \end{aligned} \tag{142}$$

$$S_{pq}^{11} = \langle p|(Z^{10\dagger}Z^{01} + Z^{11}) + h.c.|q\rangle, \tag{143}$$

$$\begin{aligned} \bar{g}_{pq}^{rs[10]} = {}&(Z^{10}pr||qs) + (pZ^{10}r||qs) + (pr||Z^{10}qs) \\ &+ (pr||qZ^{10}s), \end{aligned} \tag{144}$$

$$\begin{cases} \epsilon_{ij}^{10} = G_{ij}^{10} + C_{ij}^{10}(\epsilon_i^{00} - \epsilon_j^{00}) + (C_{ak}^{10}\bar{g}_{ik}^{ja} + C_{ak}^{10*}\bar{g}_{ia}^{jk}) \\ \epsilon_{ab}^{10} = G_{ab}^{10} + C_{ab}^{10}(\epsilon_a^{00} - \epsilon_b^{00}) + (C_{ci}^{10}\bar{g}_{ai}^{bc} + C_{ci}^{10*}\bar{g}_{ac}^{bi}) \\ \epsilon_{ai}^{10} = 0 \end{cases}$$

$$(145)$$

$$G_{pq}^{10} = \tilde{D}_{pq}^{10} - S_{pq}^{10}\epsilon_q^{00} - S_{kl}^{10}\bar{g}_{pl}^{qk} + (pq||Z^{10}kk) + (pq||kZ^{10}k), \tag{146}$$

$$\tilde{D}_{pq}^{10} = \langle p|D^{10} + Z^{10\dagger}F^{00} + F^{00}Z^{10}|q\rangle, \tag{147}$$

$$S_{pq}^{10} = \langle p|Z^{10\dagger} + Z^{10}|q\rangle, \tag{148}$$

$$G_{pq}^{01} = \tilde{D}_{pq}^{01} - S_{pq}^{01}\epsilon_q^{00} - S_{kl}^{01}\bar{g}_{pl}^{qk} + (pq||Z^{01}kk) + (pq||kZ^{01}k), \tag{149}$$

$$\tilde{D}_{pq}^{01} = \langle p|D^{01} + Z^{01\dagger}F^{00} + F^{00}Z^{01}|q\rangle, \tag{150}$$

$$S_{pq}^{01} = \langle p|Z^{01\dagger} + Z^{01}|q\rangle, \tag{151}$$

$$\bar{g}_{pq}^{rs[01]} = (Z^{01}pr||qs) + (pZ^{01}r||qs) + (pr||Z^{01}qs) + (pr||qZ^{01}s). \tag{152}$$

Just like the first-order derivative (127), it can be proven that the second-order derivative (139) also does not depend explicitly on the occupied–occupied $C_{ij}^{10}$ and virtual–virtual $C_{ab}^{10}$ coefficients. Therefore, without loss of generality, one can invoke the intermediation normalization condition

$$C_{ij}^{10} = -\langle i|Z^{10}|j\rangle, \tag{153}$$

$$C_{ab}^{10} = -\langle a|Z^{10}|b\rangle \tag{154}$$

so as to simply subsequent derivations. The virtual-occupied coefficients $C_{ai}^{10}$ are to be determined by the CPHF equation

$$\begin{pmatrix} A^{00} & B^{00} \\ B^{00*} & A^{00*} \end{pmatrix} \begin{pmatrix} C_{bj}^{10} \\ C_{bj}^{10*} \end{pmatrix} = \begin{pmatrix} -G_{ai}^{10} \\ -G_{ai}^{10*} \end{pmatrix}, \tag{155}$$

$$A_{ai,bj}^{00} = \delta_{ij}\delta_{ab}(\epsilon_b^{00} - \epsilon_j^{00}) + \bar{g}_{aj}^{ib}, \tag{156}$$

$$B_{ai,bj}^{00} = \bar{g}_{ab}^{ij}. \tag{157}$$

The derivatives of the occupied–occupied and virtual–virtual blocks of the relaxed density matrices are

$$\gamma_{ji}^{10} = \frac{1}{2}(\bar{t}_{jk}^{ab10}\bar{t}_{ab}^{ik} + \bar{t}_{jk}^{ab}\bar{t}_{ab}^{ik10}), \tag{158}$$

$$\gamma_{ba}^{10} = -\frac{1}{2}(\bar{t}_{ij}^{ac10}\bar{t}_{bc}^{ij} + \bar{t}_{ij}^{ac}\bar{t}_{bc}^{ij10}), \tag{159}$$

where

$$\bar{t}_{ab}^{ij10} = (\bar{g}_{ab}^{ij10} + \bar{t}_{cb}^{ij}\epsilon_{ac}^{10} + \bar{t}_{ac}^{ij}\epsilon_{bc}^{10} - \bar{t}_{ab}^{kj}\epsilon_{ki}^{10} - \bar{t}_{ab}^{ik}\epsilon_{kj}^{10})/\Delta_{ab}^{ij}, \tag{160}$$

$$\bar{g}_{pq}^{rs10} = \bar{g}_{pq}^{rs[10]} + \bar{g}_{tq}^{rs}C_{tp}^{10*} + \bar{g}_{pt}^{rs}C_{tq}^{10*} + \bar{g}_{pq}^{ts}C_{tr}^{10} + \bar{g}_{pq}^{rt}C_{ts}^{10}. \tag{161}$$

The equation for the derivative of the virtual-occupied block of the relaxed density matrix is

$$\gamma_{jb}^{10}A_{bj,ai}^{00} + \gamma_{jb}^{10*}B_{bj,ai}^{00*} = -X_{ia}^{10} - \gamma_{jb}A_{bj,ai}^{10} - \gamma_{jb}^{*}B_{bj,ai}^{10*}, \tag{162}$$

where

$$A_{ai,bj}^{10} = \delta_{ij}\epsilon_{ab}^{10} - \delta_{ab}\epsilon_{ji}^{10} + \bar{g}_{aj}^{ib10}, \qquad B_{ai,bj}^{10} = \bar{g}_{ab}^{ij10}, \tag{163}$$

$$X_{ia}^{10} = \frac{1}{2}(\bar{g}_{ib}^{jk10}\bar{t}_{jk}^{ab} + \bar{g}_{ib}^{jk}\bar{t}_{jk}^{ab10}) - \frac{1}{2}(\bar{t}_{ij}^{bc}\bar{g}_{bc}^{aj10} + \bar{t}_{ij}^{bc10}\bar{g}_{bc}^{aj})$$
$$+ \gamma_{kj}^{10}\bar{g}_{ji}^{ka} + \gamma_{kj}\bar{g}_{ji}^{ka10} + \gamma_{cb}^{10}\bar{g}_{bi}^{ca} + \gamma_{cb}\bar{g}_{bi}^{ca10}. \tag{164}$$

The derivatives of $I_{qp}$ read

$$I_{ba}^{10} = \frac{1}{2}(\bar{g}_{bc}^{ij}\bar{t}_{ij}^{ac10} + \bar{g}_{bc}^{ij10}\bar{t}_{ij}^{ac}), \tag{165}$$

$$I_{ji}^{10} = -\frac{1}{2}(\bar{g}_{jk}^{ab10}\bar{t}_{ab}^{ik} + \bar{g}_{jk}^{ab}\bar{t}_{ab}^{ik10}), \tag{166}$$

$$I_{ia}^{10} = \frac{1}{2}(\bar{g}_{ib}^{jk}\bar{t}_{jk}^{ab10} + \bar{g}_{ib}^{jk10}\bar{t}_{jk}^{ab}). \tag{167}$$

Finally, the last term of Eq. 139 is given by

$$E_{2+}^{[11]} = \frac{1}{4}\left[2(C_{pi}^{10*}\bar{g}_{pj}^{ab[01]} + C_{pa}^{10}\bar{g}_{ij}^{pb[01]})\bar{t}_{ab}^{ij} + \bar{g}_{ij}^{ab[11]}\bar{t}_{ab}^{ij} + \bar{g}_{ij}^{ab[01]}\bar{t}_{ab}^{ij10}\right] + c.c., \tag{168}$$

where

$$\bar{g}_{pq}^{rs[11]} = (Z^{11}pr||qs) + (Z^{10}pZ^{01}r||qs) + (Z^{10}pr||Z^{01}qs)$$
$$+ (Z^{10}pr||qZ^{01}s) + (Z^{01}pZ^{10}r||qs) + (pZ^{11}r||qs)$$
$$+ (pZ^{10}r||Z^{01}qs) + (pZ^{10}r||qZ^{01}s) + (Z^{01}pr||Z^{10}qs)$$
$$+ (pZ^{01}r||Z^{10}qs) + (pr||Z^{11}qs) + (pr||Z^{10}qZ^{01}s)$$
$$+ (Z^{01}pr||qZ^{10}s) + (pZ^{01}r||qZ^{10}s)$$
$$+ (pr||Z^{01}qZ^{10}s) + (pr||qZ^{11}s). \tag{169}$$

Note that all the orbitals involved in Eqs. 139 to 169 refer to the zeroth-order ones of positive energy.

So far, the derivations are completely general and apply to all kinds of external perturbations. For nuclear shieldings, the following specifications ought to be made: (1) The nuclear magnetic moment $\vec{\mu}_K$ and the external magnetic field $\vec{B}^{10}$ are to be taken as the first and second perturbations, respectively. (2) As the Dirac operator is only linear with respect to the vector potential, the $D_{ext}^{11}$ operator in (142) vanishes. (3) The $Z^{11}$ operator is not needed and can therefore be set to zero in all the above equations. (4) The various forms of the $Z^{01}$ operator, defined similarly as the $Z_m^{10}$ ones presented in Sect. 2, are only needed in the symmetric formulation where the nuclear magnetic moment and the external magnetic field are treated on an equal footing, but are not needed (i.e., $Z^{01} = 0$) in the asymmetric formulation where the external magnetic field is considered as the primary perturbation. In the former case, it is the RMB $Z_{m,RMB}^{01}$ and the DMB $Z_{m,DMB}^{01}$ operators that are particularly recommended. For the other options (ODA, SD, EFUT), the matrix elements of $\tilde{D}^{01}$ (150) must be evaluated as

$$\tilde{D}_{pq}^{01} = \langle p|D^{01} + Z^{01\dagger}\epsilon_q^{00} + \epsilon_p^{00}Z^{01}|q\rangle \tag{170}$$

in order to avoid possible singularities [2]. This is equivalent to assuming that the unperturbed problem can be solved exactly, viz.,

$$F^{00}\psi_p^{00} = \epsilon_p^{00}\psi_p^{00}. \tag{171}$$

It has been shown [2] that this is a very good approximation and can meanwhile simplify the implementations greatly. Note, however, that the same trick (170) cannot be applied to the matrix elements of $\tilde{D}^{10}$ (147). Otherwise, the gauge invariance would be lost even with the GIAO [24] (5). As discussed in Sect. 4, there are no direct contributions from the Coulomb interaction, such that only the exchange parts of the two-electron integrals are to be retained in all the equations. If wanted, the direct and exchange contributions of the Gaunt interaction can be inserted in due places (6). In view of the following relationships

$$C_{\bar{a}\bar{i}}^{01} = -C_{ai}^{01*}, \tag{172}$$

$$C_{\bar{a}\bar{i}}^{10} = -C_{ai}^{10*}, \tag{173}$$

$$\gamma_{\bar{i}\bar{a}} = \gamma_{ia}^*, \tag{174}$$

$$\gamma_{\bar{i}\bar{a}}^{10} = -\gamma_{ia}^{10*}, \tag{175}$$

the coupled-perturbed equations (132), (137), (155), and (162) can, at the limit of $\vec{B}^{10} = 0$, be reduced, respectively, to

$$(\epsilon_a^{00} - \epsilon_i^{00})C_{ai}^{01} + (g_{a\bar{b}}^{\bar{j}i} - g_{aj}^{bi})C_{bj}^{01} = -G_{ai}^{01}, \tag{176}$$

$$(\epsilon_a^{00} - \epsilon_i^{00})\gamma_{ia} + (\bar{g}_{ij}^{a\bar{b}} + \bar{g}_{ib}^{aj})\gamma_{jb} = -X_{ia}, \tag{177}$$

$$(\epsilon_a^{00} - \epsilon_i^{00})C_{ai}^{10} + (g_{a\bar{b}}^{\bar{j}i} - g_{aj}^{bi})C_{bj}^{10} = -G_{ai}^{10}, \tag{178}$$

$$\begin{aligned}(\epsilon_a^{00} - \epsilon_i^{00})\gamma_{ia}^{10} &+ (\bar{g}_{ji}^{a\bar{b}} - g_{bi}^{aj})\gamma_{jb}^{10} \\ &= -X_{ia}^{10} - \gamma_{jb}(\delta_{ij}\epsilon_{ab}^{10} - \delta_{ab}\epsilon_{ji}^{10} + \bar{g}_{ij}^{a\bar{b}10} + \bar{g}_{ib}^{aj10}),\end{aligned} \tag{179}$$

whose dimensions are just half of the original ones. Note, however, that the implicit summations over $b$ ($j$) should still include all the virtual (occupied) Kramers partners. The virtual orbitals here include both the PES and NES, viz., the residual parts of the (occupied) first-order orbitals are to be expanded in the full basis of zeroth-order PES and NES. Yet, as a good approximation (see 4), the NES can be neglected when solving the CPHF iteratively. One then has a strict no-pair approximation for the correlation effects on nuclear shieldings, which is computationally very much the same as its two-component counterpart. That is, the contributions of

NES to nuclear shieldings are to be accounted for only at the DHF level. Equation 176 is of course not really needed and listed here just for completeness.

Finally, it deserves to be pointed out that the nonrelativistic GIAO-MP2 expressions for nuclear shieldings can readily be deduced from the above general expressions by the following setups:

$$Z^{01} = 0, \tag{180}$$

$$Z^{11} = 0, \tag{181}$$

$$Z^{10} = -i\Lambda_\mu^g, \tag{182}$$

$$D^{01} = \frac{1}{c^2}\left(\frac{\vec{l}_K}{r_K^3} + \frac{3\vec{r}_K(\vec{s}\cdot\vec{r}_K) - \vec{s}r_K^2}{r_K^5} + \frac{8\pi\vec{s}}{3}\delta(\vec{r}_K)\right)\cdot\vec{\mu}_K, \tag{183}$$

$$D^{10} = \left(\frac{1}{2}\vec{l} + \vec{s}\right)\cdot\vec{B}, \tag{184}$$

$$D_{ext}^{11} = \vec{A}_K^{01}\cdot\vec{A}^{10}, \tag{185}$$

$$\gamma_{ai}^{def} = \gamma_{ia}^*, \tag{186}$$

$$\gamma_{ai}^{10\,def} = \gamma_{ia}^{10*}. \tag{187}$$

In terms of these identities we have

$$\begin{aligned}G_{pq}^{11} &= \langle p|(Z^{10\dagger}D^{01} + D^{01}Z^{10} + D_{ext}^{11})|q\rangle + C_{rp}^{10*}D_{rq}^{01} \\ &\quad + D_{pr}^{01}C_{rq}^{10},\end{aligned} \tag{188}$$

$$G_{pq}^{10} = \tilde{D}_{pq}^{10} - S_{pq}^{10}\epsilon_q^{00} - S_{kl}^{10}\bar{g}_{pl}^{qk} + (pq||Z^{10}kk) + (pq||kZ^{10}k), \tag{189}$$

$$\tilde{D}_{pq}^{10} = \langle p|D^{10} + Z^{10\dagger}F^{00} + F^{00}Z^{10}|q\rangle, \tag{190}$$

$$S_{pq}^{10} = \langle p|Z^{10\dagger} + Z^{10}|q\rangle, \tag{191}$$

$$G_{pq}^{01} = D_{pq}^{01}, \tag{192}$$

$$S_{pq}^{01} = 0. \tag{193}$$

The first-order $E_{2+}^1$ (113) and second-order $E_{2+}^{11}$ (139) derivatives are then reduced to

$$E_{2+}^1 = \gamma_{qp}D_{pq}^{01}, \tag{194}$$

$$E_{2+}^{11} = \gamma_{qp}G_{pq}^{11} + \gamma_{qp}^{10}D_{pq}^{01}, \tag{195}$$

which are in agreement with the previous results [67]. The coupled-perturbed equations (176) to (179) remain unchanged.

### 5.2 $j$-MP2: nuclear shielding in terms of MP2 current

At variance with the previous partitioning of the full Hamiltonian (102), one can also think of the following partitioning of the no-pair Hamiltonian $H_+$

$$H_+ = \tilde{H}_0 + \tilde{V}, \tag{196}$$

$$\tilde{H}_0 = \Lambda^{++} \sum_i^N h(i) \Lambda^{++}, \quad h = D^{00} + D^{10} + V_{HF}[\vec{B}^{10}], \tag{197}$$

$$\tilde{V} = \Lambda^{++} \left\{ \frac{1}{2} \sum_{ij}^N \frac{1}{r_{ij}} - \sum_i^N V_{HF}[\vec{B}^{10}](i) \right\} \Lambda^{++}, \tag{198}$$

$$\Lambda^{++} = \prod_{i=1}^N \Lambda^+(i), \quad \Lambda^+ = \sum_{p(\varepsilon_p > 0)} |\psi_p(1)\rangle \langle \psi_p(1)|. \tag{199}$$

The distinction between the present and the previous partitionings of the Hamiltonian resides in that the perturbation $D^{01}$ is not considered in the DHF calculation but is to be treated later on. To evaluate the current $\vec{j}(\vec{B}^{10})$ for a given finite magnetic field $\vec{B}^{10}$, one would need both the first and second order wave functions in the fluctuation potential $\tilde{V}$, viz. $\Psi_0^{(1)}$ and $\Psi_0^{(2)}$. Following the Rayleigh-Schrödinger perturbation theory, it is straightforward to obtain

$$|\Psi_0^{(2)}\rangle = |0\rangle T_{00}^{(2)} + \sum_{K \neq 0} |K\rangle T_{K0}^{(2)}, \quad \tilde{H}_0|K\rangle = E_K^{(0)}|K\rangle, \tag{200}$$

$$T_{K0}^{(2)} = \frac{\langle K|\tilde{V} - E_0^{(1)}|\Psi_0^{(1)}\rangle}{E_0^{(0)} - E_K^{(0)}}, \tag{201}$$

$$|\Psi_0^{(1)}\rangle = \sum_{L \neq 0} \frac{|L\rangle \langle L|\tilde{V}|0\rangle}{E_0^{(0)} - E_L^{(0)}}, \quad E_0^{(1)} = \langle 0|\tilde{V}|0\rangle, \tag{202}$$

where the $T_{00}^{(2)}$ amplitude is to be determined by the unitary normalization condition

$$T_{00}^{(2)} + T_{00}^{(2)*} + \langle \Psi_0^{(1)}|\Psi_0^{(1)}\rangle = 0 \rightarrow T_{00}^{(2)} = -\frac{1}{2}\langle \Psi_0^{(1)}|\Psi_0^{(1)}\rangle. \tag{203}$$

As a matter of fact, only the first equality of Eq. 203 is needed. The current density $\vec{j}_{2+}$ is then obtained as

$$\vec{j}_{2+}(\vec{B}^{10}, \vec{r}_0) = \langle \Psi_0^{(1)}| \sum_i^N \hat{j}(i) |\Psi_0^{(1)}\rangle$$

$$+ \left[ \langle 0| \sum_i^N \hat{j}(i) |\Psi_0^{(2)}\rangle + c.c. \right] = \vec{j}_A + \vec{j}_B + \vec{j}_C + \vec{j}_D,$$

$$\hat{j} = c\vec{\alpha}\delta(\vec{r} - \vec{r}_0), \tag{204}$$

where

$$\vec{j}_A = -\frac{1}{2} \sum_{i,j,a,b,c} \left( \frac{J_{ic}\bar{g}_{cj}^{ab}\bar{g}_{ab}^{ij}}{\Delta_{ab}^{ij}\Delta_i^c} + c.c. \right), \quad J_{pq} = \langle p|\hat{j}|q\rangle, \tag{205}$$

$$\vec{j}_B = \frac{1}{2} \sum_{a,b,i,j,k} \left( \frac{J_{ak}\bar{g}_{kb}^{ij}\bar{g}_{ij}^{ab}}{\Delta_{ij}^{ab}\Delta_a^k} + c.c. \right)$$

$$= -\vec{j}_A(occ \rightarrow vir, vir \rightarrow occ), \tag{206}$$

$$\vec{j}_C = -\frac{1}{2} \sum_{i,j,k,a,b} \frac{\bar{g}_{ab}^{jk}J_{ji}\bar{g}_{ik}^{ab}}{\Delta_{ab}^{ik}\Delta_{ab}^{jk}}, \tag{207}$$

$$\vec{j}_D = \frac{1}{2} \sum_{i,j,a,b,c} \frac{\bar{g}_{ij}^{ac}J_{ab}\bar{g}_{bc}^{ij}}{\Delta_{ij}^{ca}\Delta_{ij}^{cb}} = -\vec{j}_C(occ \rightarrow vir, vir \rightarrow occ). \tag{208}$$

Note that $\vec{j}_B$ ($\vec{j}_D$) can be obtained from $\vec{j}_A$ ($\vec{j}_C$) by replacing the occupied (virtual) orbitals with the virtual (occupied) orbitals, followed by a global negative sign. This internal symmetry allows to simplify the subsequent manipulations. Note also that the energy $E_{2+}^j$ calculated as

$$E_{2+}^j = \int \vec{j}_{2+} \cdot A_K^{01} \, d\vec{r} \tag{209}$$

is second order in $\tilde{V}$ but infinite order in $\vec{B}^{10}$, and is the same as $E_{2+}^1$ in Eq. 113 if $Z^1 = 0$ and the coefficients $C_m^{10}$ therein are determined in an uncoupled manner. The current $\vec{j}_{2+}^{10}$, first order with respect to $\vec{B}^{10}$, reads

$$\vec{j}_{2+}^{10} = \sum_k \left. \frac{d\vec{j}_{2+}}{dB_k^{10}} \right|_{\vec{B}^{10}=0} B_k^{10}, \quad k = x, y, z, \tag{210}$$

$$\left. \frac{d\vec{j}_{2+}}{dB_k^{10}} \right|_{\vec{B}^{10}=0} = \sum_{X=A,B,C,D} \left. \frac{d\vec{j}_X}{dB_k^{10}} \right|_{\vec{B}^{10}=0}. \tag{211}$$

The derivatives $\left. \frac{d\vec{j}_X}{dB_k^{10}} \right|_{\vec{B}^{10}=0}$ can be decomposed into five portions, viz.,

$$\left. \frac{d\vec{j}_X}{dB_k^{10}} \right|_{\vec{B}^{10}=0} = (\vec{j}_X)_{Z^{10}} + (\vec{j}_X)_{vo/ov} + (\vec{j}_X)_{no/nv}$$

$$+ (\vec{j}_X)_{oo/vv} + (\vec{j}_X)_\epsilon, \tag{212}$$

which arise, respectively, from the field dependence (denoted by subscript '$Z^{10}$') in the basis set, the transitions between the occupied and virtual PES (denoted by subscript '$vo/ov$'), the transitions between the occupied PES and virtual NES (denoted by subscript '$no/nv$'), the transitions between the PES of the same occupations $n_p$ (denoted by subscript '$oo/vv$'), as well as the dependence on the orbital energies (denoted by subscript '$\epsilon$'). The first term in Eq. 212 can formally be written as

$$(\vec{j}_X)_{Z^{10}} = \hat{f}(X), \quad X = A, B, C, D, \tag{213}$$

where the action of operator $\hat{f}$ is to replace every orbital $p$ in the expression of $\vec{j}_X$ with $Z^{10}p$, one at a time. All the terms are then summed up, e.g.,

$$
(\vec{j}_A)_{Z^{10}} = -\frac{1}{2}\sum_{i,j,a,b,c}\left\{\frac{\langle Z^{10}i|\hat{j}|c\rangle\langle cj||ab\rangle\langle ab||ij\rangle}{\Delta_{ab}^{ij}\Delta_i^c}\right.
$$

$$
+\frac{\langle i|\hat{j}|Z^{10}c\rangle\langle cj||ab\rangle\langle ab||ij\rangle}{\Delta_{ab}^{ij}\Delta_i^c}+\frac{J_{ic}\langle Z^{10}cj||ab\rangle\langle ab||ij\rangle}{\Delta_{ab}^{ij}\Delta_i^c}
$$

$$
+\frac{J_{ic}\langle cZ^{10}j||ab\rangle\langle ab||ij\rangle}{\Delta_{ab}^{ij}\Delta_i^c}+\frac{J_{ic}\langle cj||Z^{10}ab\rangle\langle ab||ij\rangle}{\Delta_{ab}^{ij}\Delta_i^c}
$$

$$
+\frac{J_{ic}\langle cj||aZ^{10}b\rangle\langle ab||ij\rangle}{\Delta_{ab}^{ij}\Delta_i^c}+\frac{J_{ic}\langle cj||ab\rangle\langle Z^{10}ab||ij\rangle}{\Delta_{ab}^{ij}\Delta_i^c}
$$

$$
+\frac{J_{ic}\langle cj||ab\rangle\langle aZ^{10}b||ij\rangle}{\Delta_{ab}^{ij}\Delta_i^c}+\frac{J_{ic}\langle cj||ab\rangle\langle ab||Z^{10}ij\rangle}{\Delta_{ab}^{ij}\Delta_i^c}
$$

$$
\left.+\frac{J_{ic}\langle cj||ab\rangle\langle ab||iZ^{10}j\rangle}{\Delta_{ab}^{ij}\Delta_i^c}\right\}+c.c. \qquad (214)
$$

The second term in Eq. 212 can formally be written as

$$
(\vec{j}_X)_{vo/ov} = \hat{g}(X), \quad X = A, B, C, D, \qquad (215)
$$

where the action of operator $\hat{g}$ is to replace every orbital $p$ in the expression of $\vec{j}_X$ with $\sum_q C_{qp}^{10}$ ($\epsilon_p > 0, \epsilon_q > 0$, $n_p \neq n_q$), one at a time. All the terms are then summed up, e.g.,

$$
(\vec{j}_A)_{vo/ov} = -\frac{1}{2}\sum_{i,j,a,b,c}\left\{\frac{C_{di}^{10*}J_{dc}\langle cj||ab\rangle\langle ab||ij\rangle}{\Delta_{ab}^{ij}\Delta_i^c}\right.
$$

$$
+\frac{C_{mc}^{10}J_{im}\langle cj||ab\rangle\langle ab||ij\rangle}{\Delta_{ab}^{ij}\Delta_i^c}\cdot+\frac{C_{mc}^{10}J_{ic}\langle mj||ab\rangle\langle ab||ij\rangle}{\Delta_{ab}^{ij}\Delta_i^c}
$$

$$
+\frac{C_{dj}^{10*}J_{ic}\langle cd||ab\rangle\langle ab||ij\rangle}{\Delta_{ab}^{ij}\Delta_i^c}+\frac{C_{ma}^{10}J_{ic}\langle cj||mb\rangle\langle ab||ij\rangle}{\Delta_{ab}^{ij}\Delta_i^c}
$$

$$
+\frac{C_{mb}^{10}J_{ic}\langle cj||am\rangle\langle ab||ij\rangle}{\Delta_{ab}^{ij}\Delta_i^c}+\frac{C_{ma}^{10*}J_{ic}\langle cj||ab\rangle\langle mb||ij\rangle}{\Delta_{ab}^{ij}\Delta_i^c}
$$

$$
+\frac{C_{mb}^{10*}J_{ic}\langle cj||ab\rangle\langle am||ij\rangle}{\Delta_{ab}^{ij}\Delta_i^c}+\frac{C_{di}^{10}J_{ic}\langle cj||ab\rangle\langle ab||dj\rangle}{\Delta_{ab}^{ij}\Delta_i^c}
$$

$$
\left.+\frac{C_{dj}^{10}J_{ic}\langle cj||ab\rangle\langle ab||id\rangle}{\Delta_{ab}^{ij}\Delta_i^c}\right\}+c.c. \qquad (216)
$$

The third term in Eq. 212 is defined similarly as the second term,

$$
(\vec{j}_X)_{no/nv} = \hat{\tilde{g}}(X), \quad X = A, B, C, D. \qquad (217)
$$

It is just that every orbital $p$ in the expression of $\vec{j}_X$ is to be replaced with $\sum_{\tilde{a}} C_{\tilde{a}p}^{10}$ ($\epsilon_{\tilde{a}} < 0$), again one at a time. This term should be very small [already neglected in $E_{2+}^{11}$ (139)], for it represents the interplay between the electron correlation within the PES manifold and the orbital rotations to the NES manifold. It should not to be confused with the genuine correlation contributions of NES to the induced current. The last two terms in (212) should be treated together due to cancelations of intermediate terms. Without going into details, the results read

$$
(\vec{j}_A)_{oo/vv} + (\vec{j}_A)_{\epsilon}
$$

$$
= -\left(\frac{1}{2}\frac{[-\Delta_{abc}^{ijm}\tilde{F}_{mi}^{10}+S_{mi}^{10}(\Delta_{ab}^j\epsilon_c^{00}+\epsilon_m^{00}\epsilon_i^{00})]V_i^c\bar{g}_{cj}^{ab}\bar{g}_{ab}^{mj}}{\Delta_i^c\Delta_c^m\Delta_{ab}^{mj}\Delta_{ab}^{ij}}+c.c.\right)
$$

$$
+\left(\frac{1}{2}\frac{(\tilde{F}_{mj}^{10}+S_{mj}^{10}\Delta_i^i)V_i^c\bar{g}_{cj}^{ab}\bar{g}_{ab}^{im}}{\Delta_i^c\Delta_{ab}^{im}\Delta_{ab}^{ij}}+c.c.\right)
$$

$$
-\left(\frac{(\tilde{F}_{da}^{10}-S_{da}^{10}\Delta_b^{ij})V_i^c\bar{g}_{cj}^{db}\bar{g}_{ab}^{ij}}{\Delta_i^c\Delta_{db}^{ij}\Delta_{ab}^{ij}}+c.c.\right)
$$

$$
-\left(\frac{1}{2}\frac{(\tilde{F}_{dc}^{10}-S_{dc}^{10}\epsilon_i^{00})V_i^d\bar{g}_{cj}^{ab}\bar{g}_{ab}^{ij}}{\Delta_i^c\Delta_d^i\Delta_{ab}^{ij}}+c.c.\right), \qquad (218)
$$

$$
(\vec{j}_B)_{oo/vv} + (\vec{j}_B)_{\epsilon} = -\hat{E}\left((\vec{j}_A)_{oo/vv} + (\vec{j}_A)_{\epsilon}\right), \qquad (219)
$$

$$
(\vec{j}_C)_{oo/vv} + (\vec{j}_C)_{\epsilon} = \frac{1}{2}\left(\frac{(\tilde{F}_{mi}^{10}+S_{mi}^{10}\Delta_{ab}^k)\bar{g}_{ik}^{ab}V_j^m\bar{g}_{ab}^{jk}}{\Delta_{ab}^{mk}\Delta_{ab}^{ik}\Delta_{ab}^{jk}}+c.c.\right)
$$

$$
+\frac{[-\Delta_{adbb}^{ijkk}\tilde{F}_{da}^{10}+(\Delta_b^{ik}\Delta_b^{jk}-\epsilon_d^{00}\epsilon_a^{00})^{00}]S_{da}^{10}\bar{g}_{ik}^{db}V_j^i\bar{g}_{ab}^{jk}}{\Delta_{db}^{ik}\Delta_{db}^{jk}\Delta_{ab}^{ik}\Delta_{ab}^{jk}}
$$

$$
+\frac{1}{2}\frac{[\Delta_{aabb}^{ijkm}\tilde{F}_{mk}^{10}+(\Delta_{ab}^i\Delta_{ab}^j-\epsilon_m^{00}\epsilon_k^{00})S_{mk}^{10}]\bar{g}_{ik}^{ab}V_j^i\bar{g}_{ab}^{jm}}{\Delta_{ab}^{im}\Delta_{ab}^{jm}\Delta_{ab}^{ik}\Delta_{ab}^{jk}}, \qquad (220)
$$

$$
(\vec{j}_D)_{oo/vv} + (\vec{j}_D)_{\epsilon} = -\hat{E}\left((\vec{j}_C)_{oo/vv} + (\vec{j}_C)_{\epsilon}\right), \qquad (221)
$$

where the action of operator $\hat{E}$ is to replace the occupied (virtual) orbitals with the virtual (occupied) ones (cf. Eq. 206). Finally, the mixed second-order energy reads

$$
E_{2+}^{11} = \int \vec{j}_{2+}^{10} \cdot \vec{A}_K^{01}\, d\vec{r}. \qquad (222)
$$

Compared with the previous expression for $E_{2+}^{11}$ (139), the present expression (222) in terms of the MP2 current is much simpler. The latter amounts to treating the coefficients $C_{ai}^{01}$ in an uncoupled manner such that the Z-vector equations (177) and (179) need not be invoked. Instead, only Eq. 178 is to be solved iteratively for the coefficients $C_{ai}^{10}$. The two expressions for $E_{2+}^{11}$ arise from two definitions for the nuclear shielding, through, respectively, the second-order derivative of the energy (cf. Eq. 15) and the induced magnetic field produced by the induced current (cf. Eqs. 5 and 14). The latter appears more natural in view of the experimental measurement (It is the local magnetic field that is really measured). The two definitions are identical only for variationally determined wave functions. To the best of our knowledge, the derivation of the MP2 shielding through the current density has not been documented in the literature. The idea can readily be applied to other four-component relativistic correlation schemes but under the no-pair approximation. As stated before, the bound-state QED prescription must be invoked to work out the correlation contributions of NES to the induced current. Yet, the present derivation in terms of the current density does not apply to two-component relativistic theories because of the lack of a uniquely defined two-component current density operator. However, the nonrelativistic GIAO-MP2 expression for $\vec{j}_{2+}$ $(=\vec{j}_{2d}+\vec{j}_{2p})$ is the same as Eq. 204 but with the current density operator defined as

$$\hat{j} = \hat{j}_d + \hat{j}_p, \tag{223}$$

$$\hat{j}_d = \delta(\vec{r} - \vec{r}_0)\vec{A}^{10}, \tag{224}$$

$$\hat{j}_p = -\frac{i}{2}\delta(\vec{r} - \vec{r}_0)\vec{\nabla} + h.c. \tag{225}$$

The two terms represent the respective diamagnetic $\vec{j}_{2d}$ and paramagnetic $\vec{j}_{2p}$ contributions. The first-order diamagnetic term $\vec{j}_{2d}^{10}$ takes the same form as Eq. 204 but evaluated with $\hat{j}_d$ and the zeroth-order orbitals. The first-order paramagnetic term $\vec{j}_{2p}^{10}$ is to be obtained from Eq. 210 but with $\hat{j}_p$ Eq. 225 and further dropping the $(\vec{j}_X)_{no/nv}$ term in Eq. 212. The mixed second-order energy finally reads

$$E_{NRMP2}^{11} = \int (\vec{j}_{2d}^{10} + \vec{j}_{2p}^{10}) \cdot \vec{A}_K^{01} \, d\vec{r}. \tag{226}$$

## 6 Conclusions and outlook

Four-component relativistic theories for magnetic properties have long been plagued by the problem associated with the missing relativistic diamagnetism. The situation has now changed, and thanks to the novel formulations as summarized here in a unified manner. What is essential is the magnetic balance (MB) condition between the large and small components of the Dirac spinors. In terms of the MB, the otherwise missing diamagnetism arises naturally and the heavy demand on the basis sets of high angular momenta is also greatly alleviated. Various ways of incorporating the MB are strictly equivalent and perform rather similarly in practice. The MB can readily be combined with the GIAO for distributed gauge origins at any level of theory for electron correlation. To illustrate this, the no-pair MP2 expressions have been derived in two different ways, one with the derivative technique and the other through the induced current. The latter, considered here for the first time, is much simpler and appears more natural in view of the experimental measurement. Going beyond the no-pair approximation is possible but the caveat is that the QED prescription on the NES must be followed up, even if only the correlation aspect is under concern. Admittedly, the novel formulations of four-component relativistic theories for NMR parameters have emerged

only rather recently such that there have not yet been many applications. Nevertheless, some important observations have already been made in the literature. Among others, what is particularly interesting is that the NES, characterized previously as a nightmare, have become embarrassingly harmless: Only those extremely compact NES are relevant for total shieldings [2, 68] and their effects are virtually canceled out for chemical shifts [24]. The implication is that the NES can be treated in a simplified manner, e.g., by an uncoupled treatment or simply by atomic calculations in the spirit of "from atoms to molecule" [60, 69–70]. Neglecting the NES amounts to a no-pair approximation even at the mean-field level (in the parlance of Fock space or QED), which is appropriate for chemical shifts. However, such an approximation is only possible with the asymmetric approaches (ODA, EFUT, SD, RMB, DMB) where the MB is applied only to the external magnetic field, but not with the symmetric schemes (such as FFUTm) where the MB is applied to both the external and the nuclear magnetic fields. In the latter case, the no-pair approximation turns out to be imbalanced and leads to severe underestimates of total shieldings (and hence relativistic effects) [2, 16], opposite to the Sternheim diamagnetic approximation [46] which tends to severely overestimate total shieldings. One then has a somewhat funny situation here: The "first-order coupling approximation" (FOCA; correct to order $c^{-2}$) obtained by neglecting the NES in EFUT works very well for chemical shifts but the "second-order coupling approximation" (SOCA; correct to order $c^{-4}$) obtained by neglecting the NES in FFUTm fails miserably [2]. If the NES were neglected in the LRT without the MB, the entire diamagnetism would be lost. Another interesting point to be made here is that the X2C counterparts [55] of the four-component theories can readily be formulated in the same way as the X2C Hamiltonians [59, 60] for electronic structure calculations.

It has long been known [4, 72] that NMR shieldings have a simple relationship with nuclear spin-rotation (NSR) couplings. However, this is true only in the nrl. Surprisingly enough, relativistic theories of NSR couplings are still missing, more than 6 decades after the nonrelativistic formulation [72]. Yet, some preliminary progress (Y. Xiao and W. Liu, unpublished) has been achieved, according to which the two types of observes are not directly related in the relativistic regime. The implications of such findings deserve great attention.

What have been discussed so far are concerned with second-order magnetic properties due to weak magnetic fields. How to solve the Dirac equation in the presence of strong magnetic fields remain to be explored. In this regard, the RMB and particularly the DMB anzätze could be very rewarding.

**Acknowledgments** The research of this work was supported by grants from the National Natural Science Foundation of China (Project No. 21033001). W.L. is grateful to Prof. Dr. Ch. van Wüllen for the invitation and stimulating discussions.

# References

1. Zaccari D, Ruizde Azúa MC, Giribet CG (2007) Phys Rev A 76:022105
2. Xiao Y, Liu W, Cheng L, Peng D (2007) J Chem Phys 126:214101
3. Cheng L, Xiao Y, Liu W (2009) J Chem Phys 130:144102
4. Ramsey NF (1950) Phys Rev 78:699
5. Feiock FD, Johnson WR (1968) Phys Rev Lett 21:785
6. Feiock FD, Johnson WR (1969) Phys Rev 187:69
7. Kolb D, Johnson WR, Shorer P (1982) Phys Rev A 26:19
8. Pyykkö P (1977) Chem Phys 22:289
9. Pyykkö P (1983) Chem Phys 74:1
10. Aucar GA, Oddershede J (1993) Int J Quantum Chem 47:425
11. Vaara J, Pyykkö P (2003) J Chem Phys 118:2973
12. Pecul M, Saue T, Ruud K, Rizzo A (2004) J Chem Phys 121:3051
13. Stanton RE, Havriliak S (1984) J Chem Phys 81:1910
14. Kutzelnigg W (2003) Phys Rev A 67:032109
15. Aucar GA, Saue T, Visscher L, Jensen HJAa (1999) J Chem Phys 110:6208
16. Visscher L (2005) Adv Quantum Chem 48:369
17. Sun Q, Liu W, Kutzelnigg W (2011) Theor Chem Acc 129:423
18. Kutzelnigg W (1999) J Comput Chem 20:1199
19. Kutzelnigg W (2004) Calculation of NMR and EPR parameters: theory and applications. In: Kaupp M, Bühl M, Malkin VG (eds) Wiley-VCH, p 43
20. London F (1937) J Phys Rad 8:397
21. Ditchfield R (1972) J Chem Phys 56:5688
22. Ditchfield R (1974) Mol Phys 27:789
23. Wolinski K, Hinton JF, Pulay P (1990) J Am Chem Soc 112:8251
24. Cheng L, Xiao Y, Liu W (2009) J Chem Phys 131:244113
25. Manninen P, Vaara J (2006) J Chem Phys 124:137101
26. Lantto P, Romero RH, Gómez SS, Aucar GA, Vaara J (2006) J Chem Phys 125:184113
27. Manninen P, Ruud K, Lantto P, Vaara J (2005) J Chem Phys 122:114107
28. Manninen P, Ruud K, Lantto P, Vaara J (2006) J Chem Phys 124:149901(E)
29. Maldonado AF, Aucar GA (2009) Phys Chem Chem Phys 11:5615
30. Ishikawa Y, Nakajima T, Hada M, Nakatsuji H (1998) Chem Phys Lett 283:119
31. Hada M, Ishikawa Y, Nakatani J, Nakatsuji H (1999) Chem Phys Lett 310:342
32. Hada M, Fukuda R, Nakatsuji H (2000) Chem Phys Lett 321:452
33. Kato K, Hada M, Fukuda R, Nakatsuji H (2005) Chem Phys Lett 408:150
34. Visscher L, Enevoldsen T, Saue T, Jensen HJAa, Oddershede J (1999) J Comput Chem 20:1262
35. Zhang ZC, Webb GA (1983) J Mol Struct (THEOCHEM) 104:439
36. Quiney HM, Skaane H, Grant IP (1998) Chem Phys Lett 290:473
37. Quiney HM, Skaane H, Grant IP (1999) Adv Quantum Chem 32:1
38. Grant IP, Quiney HM (2000) Int J Quantum Chem 80:283
39. Iliaš M, Saue T, Enevoldsen T, Jensen HJAa (2009) J Chem Phys 131:124119
40. Komorovsky S, Repisky M, Malkina OL, Malkin VG, Ondík IM, Kaupp M (2008) J Chem Phys 128:104101
41. Hamaya S, Fukui H (2010) Bull Chem Soc Jpn 83:635
42. Komorovsky S, Repisky M, Malkina OL, Malkin VG (2010) J Chem Phys 132:154101
43. Aucar GA, Romero RH, Maldonado AF (2010) Int Rev Phys Chem 29:1
44. Xiao Y, Peng D, Liu W (2007) J Chem Phys 126:081101
45. Kutzelnigg W, Liu W (2009) J Chem Phys 131:044129
46. Sternheim MM (1962) Phys Rev 128:676
47. Pyper NC (1983) Chem Phys Lett 96:204
48. Pyper NC (1999) Mol Phys 97:381
49. Pyper NC, Zhang ZC (1999) Mol Phys 97:391
50. Szmytkowski R (2002) Phys Rev A 65:03112
51. Zaccari DG, Ruizde Azúa MC, Melo JI, Giribet CG (2006) J Chem Phys 124:054103
52. Luber S, Malkin Ondík I, Reiher M (2009) Chem Phys 356:205
53. Dyall KG, Fægri K Jr. (2007) Introduction to relativistic quantum chemistry. Oxford University Press, New York
54. Repiský M, Komorovský S, Malkina OL, Malkin VG (2009) Chem Phys 356:236
55. Sun Q, Liu W, Xiao Y, Cheng L (2009) J Chem Phys 131:081101
56. Shabaev VM, Tupitsyn II, Yerokhin VA, Plunien G, Soff G (2004) Phys Rev Lett 93:130405
57. Kutzelnigg W (1984) Int J Quantum Chem 25:107
58. Dyall KG (1994) J Chem Phys 100:2118
59. Kutzelnigg W, Liu W (2005) J Chem Phys 123:241102
60. Liu W (2010) Mol Phys 108:1679
61. Eschrig H, Servedio VDP (1999) J Comput Chem 20:23
62. Wang F, Liu W (2003) J Chin Chem Soc (Taipei) 50:597
63. Kutzelnigg W (2011) Chem Phys (in press), doi:10.1016/j.chemphys.2011.06.001)
64. Salter EA, Trucks GW, Bartlett RJ (1989) J Chem Phys 90:1752
65. Handy NC, Amos RD, Gaw JF, Rice JE, Sirnandiras ES (1985) Chem Phys Lett 120:151
66. Handy NC, Schaefer HF (1984) J Chem Phys 81:5031
67. Gauss J (1992) Chem Phys Lett 191:614
68. Maldonado A, Aucar G (2007) J Chem Phys 127:154115
69. Liu W, Peng D (2006) J Chem Phys 125:044102
70. Liu W, Peng D (2006) J Chem Phys 125:149901(E)
71. Peng D, Liu W, Xiao Y, Cheng L (2007) J Chem Phys 127:104106
72. Wick GC (1948) Phys Rev 73:57

REGULAR ARTICLE

# Exact decoupling of the relativistic Fock operator

Daoling Peng · Markus Reiher

Received: 18 June 2011 / Accepted: 31 August 2011 / Published online: 7 January 2012
© The Author(s) 2011. This article is published with open access at Springerlink.com

**Abstract** It is generally acknowledged that the inclusion of relativistic effects is crucial for the theoretical description of heavy-element-containing molecules. Four-component Dirac-operator-based methods serve as the relativistic reference for molecules and highly accurate results can be obtained—provided that a suitable approximation for the electronic wave function is employed. However, four-component methods applied in a straightforward manner suffer from high computational cost and the presence of pathologic negative-energy solutions. To remove these drawbacks, a relativistic electron-only theory is desirable for which the relativistic Fock operator needs to be exactly decoupled. Recent developments in the field of relativistic two-component methods demonstrated that exact decoupling can be achieved following different strategies. The theoretical formalism of these exact-decoupling approaches is reviewed in this paper followed by a comparison of efficiency and results.

**Keywords** Relativistic electronic structure theory · Fock operator · Douglas–Kroll–Hess method · X2C method · Picture change error

## 1 Introduction

It is a well-established experimental fact that any mathematical description of electromagnetic phenomena involving electrons and atomic nuclei has to obey the principles of special relativity [1]. As a consequence, a fundamental theory for chemistry should be a relativistically correct quantum-mechanical all-electron theory [2]. While a truly Lorentz-covariant many-electron theory is not available—although its basic principles have been cast in the theory of quantum electrodynamics [3]—it turned out that a semi-classical theory that quantizes the matter field only (first quantization) is sufficient if chemical accuracy for energies is sought, i.e., if relative energies shall be calculated with an accuracy of about 1 kJ/mol. For such a first-quantized theory, a relativistic many-electron Hamiltonian operator may be formulated as

$$H = \sum_i h_{\rm D}(i) + \sum_{i<j} g(i,j) + V_{\rm NN}, \qquad (1)$$

where $V_{\rm NN}$ is the repulsion potential energy operator of the nuclei in a molecule, $h_{\rm D}$ Dirac's $4 \times 4$ one-electron operator, and $g(i, j)$ the two-electron interaction operator including the leading Coulomb term plus magnetic and retardation corrections as comprised by the Breit operator [2]. Hence, the electron–electron interaction in Eq. 1 is approximate. An additional approximation invoked in almost all practical cases is the Born–Oppenheimer approximation, i.e., the assumption of clamped nuclei. For the purist we should note that we refrained from embracing the one-electron Dirac Hamiltonian by positive-energy projectors (see below) for the sake of brevity.

Because of the four-dimensional matrix structure of the Dirac Hamiltonian $h_{\rm D}$, the resulting orbital-based electronic structure methods are called four-component methods named after the number of functions that constitute a one-electron state. These one-electron states commonly referred to as orbitals are known as spinors in relativistic theory. A four-component orbital is called a 4-spinor, a

Published as part of the special collection of articles celebrating the 50th anniversary of Theoretical Chemistry Accounts/Theoretica Chimica Acta.

D. Peng · M. Reiher (✉)
Laboratorium für Physikalische Chemie, ETH Zürich,
Wolfgang-Pauli-Strasse 10, 8093 Zurich, Switzerland
e-mail: markus.reiher@phys.chem.ethz.ch

two-component orbital, which may be an eigenvector of a two-dimensional one-electron Hamiltonian, is called a 2-spinor. Unfortunately, jargon occasionally mixes this notation for operators with that for the corresponding orbitals and one speaks of four-component operators although four-dimensional operators are meant.

As the Dirac Hamiltonian includes a kinetic energy term associated with the electron's spin momentum in a natural way, spin–orbit interactions are consistently described. However, this can already be achieved to arbitrary accuracy by a purely two-dimensional Hamiltonian [4]. The four-dimensional structure is, however, the origin of negative-energy states which are interpreted as positronic states in quantum electrodynamics, but which require ad hoc assumptions in a first-quantized relativistic theory to assure stability of matter (Dirac's hole theory)—and, consequently, lead to conceptual problems when applied in chemistry.

From the point of view of numerical results, quantum chemistry based on standard Schrödinger quantum mechanics, the so-called non-relativistic approach, may yield numerical results that do not deviate significantly from a relativistic description. It is then said that the quantity studied is not affected by so-called relativistic effects. For example, most aspects that are studied in the context of organic molecules hardly show relativistic effects. But if we aim for a quantum mechanical theory valid for all chemistry, i.e., for molecules and molecular aggregates that may contain any atom from the periodic table of the elements, a "fully relativistic" four- or two-component approach is mandatory.

Here, it is important to distinguish between methods that are "quasi-relativistic" and thus do not completely reproduce four-component reference results for the same choice of electronic wave function approximation and methods that yield in principle the same result (and in practice results that agree with four-component results to the desired chemical accuracy). While this work is devoted to the latter kind, we should note that the former type of methods is usually split into scalar relativistic (also called one-component) approaches, whose computational cost are basically equivalent to that of the non-relativistic approach, and spin–orbit interactions including (two-component) approaches. Since the inclusion of spin in the Hamiltonian requires invoking the spin operator and hence the two-dimensional Pauli spin matrices, spin–orbit coupling including methods are always two-component and spin itself is no longer a good quantum number. As the scalar-relativistic approaches describe only kinematic relativistic effects, they change only the one-electron kinetic energy operators in the many-electron Hamiltonian of Eq. 1, but cannot account for any effects that are due to electron spin—like the spin–orbit interaction.

The four-component approach for the optimization of orbitals is computationally more demanding than the non-relativistic one by some, not very large constant factor (because of the matrix structure of the one-electron Hamiltonian in the former case). The four components of the orbital (spinor) also affect the scaling behavior of the four-index transformation for the application of subsequent correlation methods like configuration interaction or coupled cluster, but not the correlation methods themselves [5]. Hence, in ab initio correlation calculations that do not optimize the orbitals, four-component methods have basically the same computational cost as two-component methods. However, the four-component optimization of orbitals always automatically involves the optimization of the negative-energy solutions, which require positive-energy projectors if a variational procedure shall be applied to the otherwise unbounded Hamiltonian. One may consider the negative-energy states as pathologic as they cause interpretative problems in semi-classical relativistic theory employed here, "solved" only by Dirac's hole theory. As a consequence, actual calculations require the continuous update of projectors [6]. From an algorithmic point of view, these projectors are usually implicitly defined and clearly depend on the external potential. As a consequence, free-particle projectors proposed by Sucher [7] are not very appropriate as Heully et al. [8] pointed out. However, for the bound solutions, we are interested in, Talman [9, 10] showed that a minimax principle holds, which can be implemented in variational procedures also in the many-electron case [11]. Still, negative-energy states may cause numerical problems because of the choice of the one-particle basis set (usually kinetically balanced atom-centered Gauss-type functions), in which the negative-energy (continuum) solutions are to be represented.

The negative-energy solutions can be considered as superfluous for an electron-only theory that aims at the positive-energy solutions. However, they can be exactly removed by a decoupling of the negative- and positive-energy states. For this, two pathways have been followed in the past decades, namely the so-called elimination of the small component and the unitary transformation approach. The resulting Hamiltonians feature two-dimensional one-electron operators, which allow one to set up two-component methods.

The development of these two-dimensional operators was neither straightforward nor without difficulties. For instance, some featured energy-dependent operators or operators that are difficult to calculate. The early two-component approaches were called "quasi-relativistic" since some terms were discarded or approximated in the desire to obtain efficient methods to implement in computer programs. This has led to very efficient quasi-relativistic methods of which the zeroth-order regular approximation (ZORA) [12–14] and the second-order Douglas–Kroll–Hess approximation (DKH2) [15] are the most prominent examples. Only within the past decade, it has become clear that one may actually formulate "exact" two-component methods which approach four-component reference results to (almost) arbitrary degree of accuracy. For reviews of these developments see Refs. [16–24].

This work attempts a presentation of the "exact" two-component approaches from the perspective of the one-electron equation that determines the orbitals (spinors) in a quantum chemical calculation. This point of view is usually not taken and instead only the Dirac Hamiltonian in an external electrostatic field is considered, which allows one to omit the discussion of how to deal with the electron–electron interaction. It is, however, the treatment of these two-electron terms (often accompanied by additional atom-based (local) approximations) that introduces approximations which are the reason why we put "exact" in quotation marks. It must be stressed that only a derivation of the various "exact" two-component methods from the point of view of the four-component Fock equation allows us to highlight all approximations made within a particular approach and to relate the different approaches to one another. In addition to this review of the formal aspects of "exact" two-component methods, we present a detailed comparison of them on the basis of new numerical results.

## 2 The Fock one-electron equation for the definition of orbitals

By contrast to standard presentations of the subject, we start from the relativistic (four-dimensional) Fock operator rather than from the Dirac Hamiltonian. This choice is made to clearly highlight the differences of "exact" two-component approaches and the corresponding four-component one. Most quantum chemical approaches approximate the many-electron wave function by a direct product of one-particle states. These one-particle states are obtained as eigenvectors of an effective one-electron Hamiltonian, well known as the Fock operator $f$.

The Fock operator comprises an operator for the kinetic energy of an electron and its interaction with external electromagnetic potentials (including those produced by the atomic nuclei) as well as operators that describe the interaction of the electron with other electrons. Especially, the latter ones are not easy to approximate without compromising the resulting accuracy of numerical results. The particular choice for such two-electron operators defines the electronic structure method under consideration, be it Hartree–Fock (HF), multi-configuration self-consistent field (MCSCF) or Kohn–Sham density functional theory (DFT).

All these methods rely on an effective one-electron eigenvalue equation

$$f\psi_i = \epsilon_i \psi_i \tag{2}$$

with orbital energy $\epsilon_i$ and orbital $\psi_i$. For a four-component method, the Fock operator reads in Gaussian units

$$f = h_D + q_e V_{\text{eff}} - q_e \boldsymbol{\alpha} \cdot \boldsymbol{A}_{\text{eff}}. \tag{3}$$

Here, the electromagnetic scalar and vector potentials, $V_{\text{eff}}$ and $\boldsymbol{A}_{\text{eff}}$, respectively, are sums of all electromagnetic potentials that couple to the electron's charge $q_e = -e$. Note that the potential energy operator $q_e V_{\text{eff}} - q_e \boldsymbol{\alpha} \cdot \boldsymbol{A}_{\text{eff}}$ is to be understood as a four-dimensional matrix operator.

The field-free Dirac Hamiltonian is given by

$$h_D = c\boldsymbol{\alpha} \cdot \boldsymbol{p} + (\beta - 1)m_e c^2, \tag{4}$$

where $\boldsymbol{\alpha}$ and $\beta = \text{diag}(1, 1, -1, -1)$ are the Dirac matrices and $c$ is the speed of light. The vector $\boldsymbol{\alpha}$ contains three four-dimensional matrices whose off-diagonal two-dimensional entries are the Pauli spin matrices $\sigma_i$,

$$\alpha_i = \begin{pmatrix} 0 & \sigma_i \\ \sigma_i & 0 \end{pmatrix} \forall i \in \{x, y, z\}. \tag{5}$$

The first term on the right hand side of Eq. 4 is the relativistic kinetic energy operator and the second term is the rest energy operator of the electron, $\beta\, m_e c^2$ ($m_e$ is its rest mass), shifted by its rest energy $-m_e c^2$ in order to have the same zero-energy reference level as in Schrödinger quantum mechanics.

The effective electromagnetic potentials in Eq. 3 depend on how the electron–electron interaction is described. This question has two facettes as it refers to the choice of the interaction operator, i.e., of $g(i, j)$ in Eq. 1, and to the approximation of the *many-electron* wave function (or density). If we neglect all magnetic and retardation effects in $g(i, j)$, we obtain only terms in $V_{\text{eff}}$ but not in $\boldsymbol{A}_{\text{eff}}$, while those effects, cast, e.g., as Breit operators, enter $\boldsymbol{A}_{\text{eff}}$ (but can be written as a contribution to the scalar potential $V_{\text{eff}}$; see the detailed discussion in chapter 8.1 of Ref. [2]). The wave-function approximation then affects the expressions for the electromagnetic potentials in Eq. 3 as the electromagnetic interaction of an electron with another electron depends on the other electron's 4-current, i.e., on its density and current density to be calculated from the wave function.

We may write the effective electrostatic potential in general as

$$V_{\text{eff}} = V_{\text{ext}}\mathbf{1}_4 + V_{\text{ee}}, \tag{6}$$

where $V_{\text{ee}}$ represents the electron–electron interaction whose explicit form depends on the wave function approximation. The external potential $V_{\text{ext}}$ created by the atomic nuclei and any other external electrostatic field is taken as a diagonal matrix operator (i.e., we do not consider magnetic and retardation effects for the electron–nucleus interaction, which is a common approximation in quantum chemistry as it does not compromise chemical accuracy) However, for the sake of brevity, we suppress the four-dimensional unit matrix $\mathbf{1}_4$ in the following.

In the independent-particle model of four-component Hartree–Fock theory, i.e., in Dirac–Hartree–Fock (DHF) theory we obtain

$$V_{\text{eff}}^{\text{DHF}} = V_{\text{ext}} + J - K, \tag{7}$$

where $J$ denotes the classical Coulomb potential

$$J(\boldsymbol{r}) = q_e \sum_{j=1}^{\text{occ}} \int \frac{\psi_j^\dagger(\boldsymbol{r}') \cdot \psi_j(\boldsymbol{r}')}{|\boldsymbol{r} - \boldsymbol{r}'|} \mathrm{d}\boldsymbol{r}' = q_e \int \frac{\rho(\boldsymbol{r}')}{|\boldsymbol{r} - \boldsymbol{r}'|} \mathrm{d}\boldsymbol{r}', \tag{8}$$

and $K$ denotes the exchange potential defined as

$$K\psi_i(\boldsymbol{r}) = q_e \sum_{j=1}^{\text{occ}} \int \frac{\psi_j^\dagger(\boldsymbol{r}') \cdot \psi_i(\boldsymbol{r}')}{|\boldsymbol{r} - \boldsymbol{r}'|} \mathrm{d}\boldsymbol{r}' \psi_j(\boldsymbol{r}) \tag{9}$$

($K$ produces off-diagonal (so-called odd) contributions of the scalar electron–electron Coulomb potential). In the MCSCF case, the two-electron terms have three spinor indices for a given Fock equation [2, 25, 26]. The four-component Kohn–Sham DFT case is similar to DHF theory and the effective electrostatic potential reads

$$V_{\text{eff}}^{\text{DFT}} = V_{\text{ext}} + J + V_{\text{XC}}, \tag{10}$$

where all exchange and electron-correlation terms are then obtained as functional derivatives of a properly approximated exchange–correlation energy contribution $E_{\text{xc}}$,

$$V_{\text{XC}}(\boldsymbol{r}) = \frac{\delta E_{\text{xc}}[\rho(\boldsymbol{r})]}{\delta \rho(\boldsymbol{r})}. \tag{11}$$

If magnetic interactions among the electrons shall be considered in DHF theory or relativistic DFT, the effective vector potential must be considered

$$\boldsymbol{A}_{\text{eff}}(\boldsymbol{r}) = \boldsymbol{A}_{\text{ext}}(\boldsymbol{r}) + \boldsymbol{A}_{\text{G}}(\boldsymbol{r}) + \boldsymbol{A}_{\text{XC}}(\boldsymbol{r}), \tag{12}$$

where $\boldsymbol{A}_{\text{ext}}$ is an external vector potential induced by an external magnetic field. $\boldsymbol{A}_{\text{G}}$ denotes the (unretarded) magnetic interaction of electrons

$$\boldsymbol{A}_{\text{G}}(\boldsymbol{r}) = \frac{q_e}{c} \int \frac{\boldsymbol{j}(\boldsymbol{r}')}{|\boldsymbol{r} - \boldsymbol{r}'|} \mathrm{d}\boldsymbol{r}', \tag{13}$$

which yields the Gaunt interaction and depends on the total current density $\boldsymbol{j}$ of the electrons in the system. In DHF theory $\boldsymbol{A}_{\text{XC}}$ would contain "magnetic" exchange integrals, while it is the relativistic current-density functional derivative in relativistic DFT,

$$\boldsymbol{A}_{\text{XC}}(\boldsymbol{r}) = \frac{\delta E_{\text{xc}}[\boldsymbol{j}(\boldsymbol{r})]}{\delta \boldsymbol{j}(\boldsymbol{r})}. \tag{14}$$

In the relativistic realm, the exchange–correlation energy depends on both the electron density and the current density, therefore the exchange–correlation energy functional $E_{\text{xc}} = E_{\text{xc}}[\rho(\boldsymbol{r}), \boldsymbol{j}(\boldsymbol{r})]$ also appears in the expression for the effective vector potential.

Retardation effects that account for the transmission of electromagnetic fields between the electrons with the finite speed of light introduce additional terms that we here neglect as it has turned out that such terms—approximated, for instance, in the Breit interaction—lead to negligibly small corrections for chemical applications. Note that these terms depend on the choice of gauge for the electromagnetic fields. Also the magnetic Gaunt interaction has only very small contributions to electronic energy differences and is thus usually omitted in calculations. Note that these electronic contributions to the effective vector potential are usually written as contributions to the effective electrostatic potential (see chapter 8.1 of Ref. [2]).

For the decoupling of the negative-energy states, it will make a difference whether the effective potentials contribute to the block-diagonal or to the off-diagonal entries of the Fock operator. Therefore, it is advantageous to focus first on the electrostatic effective potential and to consider the magnetic contributions at a later stage.

The relativistic Fock operator in the absence of external magnetic field, $\boldsymbol{A}_{\text{eff}} = \boldsymbol{0}$, is

$$f = h_D + q_e V_{\text{eff}}. \tag{15}$$

If only a (block)-diagonal electrostatic potential is taken into account, which is an approximation often made in the derivation of two-component methods, the explicit matrix form of the relativistic Fock operator will then read

$$f = \begin{pmatrix} q_e V_{\text{diag}} & c\boldsymbol{\sigma} \cdot \boldsymbol{p} \\ c\boldsymbol{\sigma} \cdot \boldsymbol{p} & q_e V_{\text{diag}} - 2m_e c^2 \end{pmatrix}, \tag{16}$$

where $\boldsymbol{\sigma}$ denotes the vector of Pauli spin matrices. For example, the one-electron Dirac operator and the Fock operator in four-component DFT have a diagonal electrostatic potential, $V_{\text{diag}} = V_{\text{ext}}$ and $V_{\text{diag}} = V_{\text{ext}} + J + V_{\text{XC}}$, respectively. Recall that a diagonal term in the $2 \times 2$ superstructure representation of the relativistic Fock operator is actually a block-diagonal $2 \times 2$ matrix contribution.

The $2 \times 2$ superstructure of the Fock operator is a general feature and allows one to also write the corresponding 4-spinor in split notation, $\psi = (\varphi, \chi)$, with $\varphi$ and $\chi$ as its upper and lower 2-spinor components. The components $\varphi$ and $\chi$ are often called large and small components, respectively, because of their relative size for small external potentials (i.e., for small nuclear charge numbers $Z$). Since "relative size" is somewhat vaguely defined, we should emphasize that it refers to a comparison of the absolute values of both components over the whole spatial domain, which shows only few exceptions (e.g., positions where the large components have nodes). One may use the square root of the integral of the large and of the small component *densities* over the whole spatial domain to define "size". However, this classifications is not valid for very heavy elements (e.g., for nuclear charge numbers approaching $Z = 100$ and higher $Z$'s) and therefore we consistently use the former notation of 'upper' and 'lower' to distinguish the components of the 4-spinor.

# 3 Theory of exact decoupling

The relativistic Fock equation in Eq. 2 has both positive-energy (so-called electronic) and negative-energy (so-called positronic) solutions

$$f\psi_i^{(+)} = \epsilon_i^{(+)}\psi_i^{(+)} \quad \text{and} \quad f\psi_i^{(-)} = \epsilon_i^{(-)}\psi_i^{(-)}. \qquad (17)$$

The negative-energy solutions $\psi_i^{(-)}$ are not needed in molecular calculations which consider electrons and nuclei only. In DHF and relativistic DFT calculations, they are always kept unoccupied and are never used to construct the electron density and current density from which the effective potentials are calculated. Since HF, MCSCF and DFT approaches all utilize the variational principle, the negative-energy states, whose number is finite in molecular calculations only because of the choice of a finite basis set for the representation of the components of the spinor, cause problems for the variational stability. For this reason it is desirable to derive a Fock operator for electrons only.

## 3.1 Two principal options

In principle, there are two conceptually different approaches to obtain an electron-only Fock operator. One is the projection approach and the other one the block-diagonalization by unitary transformation. In the former one, a projection operator is introduced [7] to remove the negative-energy states

$$P^{(+)} = \sum_i |\psi_i^{(+)}\rangle\langle\psi_i^{(+)}|, \qquad (18)$$

(the sum runs over all positive-energy states). The projected Fock operator $f^{(+)}$,

$$f^{(+)} = P^{(+)}fP^{(+)}, \qquad (19)$$

possesses then only the positive-energy spectrum. But this four-component projected Fock operator is not useful in practice because the projection operator requires all information of the positive-energy states which are actually the result we aim for. One may introduce an atomic

approximation to the projection operator to overcome this problem, which is efficient, but would be a first approximation in an exact-decoupling scheme.

The second option is to block diagonalize the four-component Fock operator by a unitary transformation

$$f^{\text{bd}} = UfU^\dagger = \begin{pmatrix} f^{(2+)} & 0 \\ 0 & f^{(2-)} \end{pmatrix}, \qquad (20)$$

where $f^{(2+)}$ and $f^{(2-)}$ possess only the electronic and positronic solutions, respectively:

$$f^{(2+)}\phi_i^{(+)} = \epsilon_i^{(+)}\phi_i^{(+)} \quad \text{and} \quad f^{(2-)}\phi_i^{(-)} = \epsilon_i^{(-)}\phi_i^{(-)}. \qquad (21)$$

Obviously, one can now simply discard the $f^{(2-)}$ operator and employ the two-component electron-only Fock operator $f^{(2+)}$ to perform a relativistic molecular calculation. The computational applicability of the decoupling transformation depends crucially on whether one can derive an efficient and computationally feasible algorithm to compute the unitary transformation $U$ as well as the corresponding Fock operator $f^{(2+)}$. We will discuss the possible options for their calculation in the following section.

## 3.2 The straightforward solution

The Fock equation, Eq. 2, can be rearranged to derive an operator $X$ that relates the large (upper ($U$)) component $\varphi$ and the small (lower ($L$)) component $\chi$ of a 4-spinor,

$$\chi^{(+)} = X\varphi^{(+)}. \qquad (22)$$

Heully et al. [27] gave a closed form expression of the exact decoupling transformation in terms of this key operator $X$,

$$U_X = \begin{pmatrix} \dfrac{1}{\sqrt{1+X^\dagger X}} & \dfrac{1}{\sqrt{1+X^\dagger X}}X^\dagger \\ -\dfrac{1}{\sqrt{1+XX^\dagger}}X & \dfrac{1}{\sqrt{1+XX^\dagger}} \end{pmatrix}, \qquad (23)$$

whose derivation from the Fock equation is straightforward [2]. This transformation has been called a Foldy–Wouthuysen transformation [28], although the $X$-operator has not been introduced in the original paper by Foldy and Wouthuysen, that had a different focus to which we come back later. Note that the operator $X$ is an electron-only operator which connects the large and small components of a positive-energy spinor. No analytical energy-independent closed form of the $X$-operator was discovered until now. However, the energy-dependent form of $X$ can be easily derived from the Fock equation with the Fock operator of Eq. 16 [2],

$$X = \frac{c\boldsymbol{\sigma}\cdot\boldsymbol{p}}{2m_ec^2 - q_eV_{\text{diag}} + \epsilon_i^{(+)}}. \qquad (24)$$

A more general expression can be derived for the general potential $V_{\text{eff}}$,

$$X = \frac{c\boldsymbol{\sigma}\cdot\boldsymbol{p} + q_e V_{\text{eff}}^{LU}}{2m_e c^2 - q_e V_{\text{eff}}^{LL} + \epsilon_i^{(+)}}. \qquad (25)$$

This energy-dependent expression of $X$ then depends on which electronic spinor it acts on and does not lead to eigenvalue equations, which makes it useless for actual calculations.

### 3.3 The sequential solution

We can decompose the overall transformation $U$ into a sequence of unitary transformations

$$U = \cdots U_3 U_2 U_1 U_0. \qquad (26)$$

which is beneficial if the individual $U_i$ are easier to obtain than an expression for the total $U$. Foldy and Wouthuysen [29, 30] were the first to attempt a derivation of such unitary matrices. Their main result was to give a closed form expression for $U^{\text{fpFW}}$ that (block)-diagonalizes the field-free Dirac Hamiltonian,

$$U^{\text{fpFW}} \begin{pmatrix} 0 & c\boldsymbol{\sigma}\cdot\boldsymbol{p} \\ c\boldsymbol{\sigma}\cdot\boldsymbol{p} & -2m_e c^2 \end{pmatrix} U^{\text{fpFW},\dagger}$$
$$= \begin{pmatrix} E_p - m_e c^2 & 0 \\ 0 & -E_p - m_e c^2 \end{pmatrix}, \qquad (27)$$

The scalar relativistic energy $E_p$ is given by $E_p = \sqrt{\boldsymbol{p}^2 c^2 + m^2 c^4}$, which is occasionally also abbreviated as $E_0$ because it is the zeroth-order term of the exactly decoupled operators [without subtraction of the rest energy]. However, any attempt to achieve a sequential decoupling with operators ordered according to the formal expansion parameter $1/c$ failed because of an ill-defined series expansion [4] that yields operators at most to be used in perturbation theory.

As we aim for an exact variationally stable procedure, we must consider the only other formal expansion parameter, namely the potential, which yields a convergent series expansion of a variational one-electron Hamiltonian [4]. The idea for an expansion in terms of the external electrostatic potential was first proposed by Douglas and Kroll [31], but found no application in electronic structure theory until rediscovered by Hess who also turned its low-order approximation into a practical method [15, 32]. The first unitary transformation in this Douglas–Kroll–Hess (DKH) transformation protocol must necessarily [4] be the above-mentioned closed-form Foldy–Wouthuysen transformation, $U_0 = U^{\text{fpFW}}$, if the off-diagonal terms to be eliminated in the Fock operator are given by $c\boldsymbol{\sigma}\cdot\boldsymbol{p}$ as in Eq. 16 and the new expansion parameter, the potential, shall be introduced to first order for the stepwise elimination up to infinite order by subsequent unitary transformations $U_1, U_2, \ldots, U_\infty$.

Just as a historical side remark, we should mention that Nakajima and Hirao [33] were the first to present third-order DKH results based on Hess' original work on second-order DKH, followed by fourth- and fifth-order results by our group [34] and then sixth-order results by van Wüllen [35]. The final step in this direction was then our implementation of the first infinite-order DKH protocol [36] that allowed the explicit symbolic derivation and evaluation of the DKH one-electron Hamiltonian through all orders in the external potential, which has been made available in standard quantum chemistry programs [37, 38]. Its efficiency was significantly increased by Peng and Hirao [39] (see also below).

In addition to the analytical insights into the Douglas–Kroll–Hess approach presented in Ref. [4], we also presented in Ref. [34] a crucial reformulation called the *generalized* DKH transformation. This was necessary as different authors employed different parametrizations of the unitary transformation matrices in terms of an off-diagonal anti-Hermitean operator $W$ to be chosen such that the lowest-order off-diagonal term in the one-electron Hamiltonian is stepwise eliminated. And it was not at all clear whether these different parametrizations yield the same results for finite DKH orders. The generalized DKH transformation clarified these matters by definition of the most general parametrization of a unitary transformation given by a Taylor series expansion,

$$U_k \rightarrow U_k(W_k) = \sum_{i=0}^{\infty} a_{k,i} W_k^i, \qquad (28)$$

where the expansion coefficient $a_{k,i}$ are different for different parameterization schemes. These coefficients must be chosen such that the unitarity of the transformation is guaranteed.

The infinite-order DKH transformation,

$$U_{\text{DKH}} = \prod_{k=\infty}^{0} U_k, \qquad (29)$$

can exactly decouple the relativistic Fock operator of Eq. 16 (we denote the exact sequential DKH decoupling transformation simply as "DKH", while we add a number to it if decoupling is achieved only to a given order in the electrostatic potential). If a sequential decoupling scheme is applied to the block-diagonalization of the relativistic Fock operator, then Douglas–Kroll–Hess theory is a unique approach and no other option to produce closed-form expressions (ordered by increasing powers of the potential) exists [4]. For exact block-diagonalization of the operator in Eq. 16, the full electrostatic potential $V_{\text{diag}}$ must be considered as an expansion parameter. Standard DKH implementations take, however, only the electron–nucleus potential $V_{\text{ext}}$ because this then requires only a modification of the one-electron part of the Fock operator and leaves the electron–electron term $V_{ee}$ untouched. This approximation yields an efficient scheme but, of course, introduces inaccuracies [40–43]. Of course, the exchange interaction in DHF theory as well as vector potential (i.e., magnetic) contributions make the DKH scheme even more involved due to the occurrence of off-diagonal potential terms in the Fock operator. However, to properly discuss the inclusion of magnetic fields in exact decoupling methods is beyond the scope of this work and we may refer the reader to the overview in Ref. [44] instead.

### 3.4 The two-step solution

The most severe problem of the DKH expansion approach when viewed as an exact-decoupling method is that it introduces too many unitary transformations and thus too many operators, which may make its implementation very complicated (this is no problem for the efficient standard low-order approximations like DKH2 or DKH4, but becomes a true challenge for high orders). Therefore, we should step back and reconsider the case if only one transformation is introduced after the free-particle Foldy–Wouthuysen transformation $U_0$,

$$U_{\mathrm{BSS}} = U_1' U_0 \quad \text{with} \quad U_0 = U^{\mathrm{fpFW}}. \tag{30}$$

This idea was first proposed by Barysz et al. [45] and is therefore know as the BSS approach although it has later been called by Barysz and Sadlej the infinite-order two-component (IOTC) method [46]. Their transformation $U_1'$ is a $U_X$-like operator

$$U_1' = \begin{pmatrix} \frac{1}{\sqrt{1+R^\dagger R}} & \frac{1}{\sqrt{1+R^\dagger R}} R^\dagger \\ -\frac{1}{\sqrt{1+RR^\dagger}} R & \frac{1}{\sqrt{1+RR^\dagger}} \end{pmatrix}, \tag{31}$$

where, instead of the original operator $X$ of the (one-step) $U_X$ transformation of Eq. 23, the operator $R$ connects the upper and lower components of the $U_0$-transformed spinor obtained as an eigenstate of the $U_0$-transformed Fock operator

$$U_0 f U_0^\dagger = \begin{pmatrix} F_{UU} & F_{UL} \\ F_{LU} & F_{LL} \end{pmatrix}. \tag{32}$$

Here, we do not write explicit expressions as this is not important for our following discussion. Barysz and co-workers obviously had to face the same problem discussed above for the $U_X$ transformation. I.e., the expression for $U_1'$ contains either unknown operators or impractical energy-dependent operators and no analytical form was obtained. However, the operator $R$ can be obtained a solution of the following equation

$$F_{LL} R = R F_{UU} + R F_{UL} R - F_{LU}, \tag{33}$$

which is easy to derive from the free-particle Foldy–Wouthuysen-transformed Fock equation. Eq. 33 can be iteratively solved with a proper initial guess. Obviously, the iterative equation and iterative schemes for $R$ are not unique. There exist many ways to set up and solve the iterative equations—a simple alternative to Eq. 33 would be

$$R F_{UU} = F_{LL} R - R F_{UL} R + F_{LU}, \tag{34}$$

but they may result in divergent numerical solutions (for a detailed discussion see Ref. [47]).

There are more difficulties associated with this iterative approach. First, if the iterations are carried out explicitly, the analytic expression of $R$ becomes very complicated already after a few iterations (therefore, the iterative equations are always solved with matrix representations of the corresponding operators). Second, the solutions of the iterative equations are not unique, as it may produce an operator connecting the upper and lower components of any solution of the original Fock operator, which need not necessarily be an electronic solution. This means that even if the iteration converges, it may converge to an unwanted solution. However, by properly organizing the iterative solution, the correct operator $R$ can be obtained and the Fock operator is then exactly decoupled by the BSS transformation.

### 3.5 The one-step solution

It should be clear that if the $U_X$-like transformation $U_1'$ can be obtained by an iterative solution, the original transformation $U_X$ also has an iterative solution. The only difference is that the iterative equations are slightly different and may have different convergence behavior. The $U_X$ transformation is indeed obtained within matrix representation via the so-called eXact-2-Component (X2C) approach [47–53]. However, an important characteristic of the X2C approach is that it invokes a non-iterative construction of the key operator $X$ in $U_X$. The drawbacks of the iterative construction method discussed above do not exist in the X2C approach. In this non-iterative construction method, the matrix operator $X$ is obtained from the electronic eigenvectors of the relativistic Fock–Roothaan equation

$$\begin{pmatrix} F_{UU} & F_{UL} \\ F_{LU} & F_{LL} \end{pmatrix} \begin{pmatrix} C_U^{(+)} \\ C_L^{(+)} \end{pmatrix} = \epsilon^{(+)} \begin{pmatrix} C_U^{(+)} \\ C_L^{(+)} \end{pmatrix}, \tag{35}$$

where the components of the 4-spinor are expanded in a one-electron basis set (in the above equation we kept the typical $2 \times 2$ superstructure of the Fock operator). $C_U^{(+)}$ and $C_L^{(+)}$ are the coefficients of the basis set expansions of upper $(U)$ and lower $(L)$ components of the 4-spinor eigenvectors, respectively. The diagonal matrix $\epsilon^{(+)}$

contains all positive-energy eigenvalues. Once the coefficients $C_U^{(+)}$ and $C_L^{(+)}$ have been obtained, the matrix $X$ is simply obtained by

$$X = C_L^{(+)} \left( C_U^{(+)} \right)^{-1}. \tag{36}$$

This "trick" was called "Douglas–Kroll the easy way" by Jensen [48], who introduced it in a talk at the REHE2005 conference and which was then the starting point for the extensive development by Kutzelnigg and Liu [47, 49, 50]. Since the $X$ matrix is directly evaluated from the electronic solutions of the Fock operator, the four-component Fock operator must first be diagonalized. *But this already solves the problem and no additional unitary transformation is required.*

Hence, a two-component electron-only Fock operator is then actually no longer needed to obtain electronic solutions. However, there are two points that must be considered. First, the exact-decoupling approach not only separates the electronic solutions from the positronic solutions but also constructs the electronic *two-component* spinors instead of four-component ones. This is clearly an advantage, but it will also reduce the effort for the four-index transformation if post-DHF correlation methods shall be applied. These issues have been discussed in detail in Ref. [54]. Second, as relativistic many-electron calculations require an external-field no-pair projection one may view this projection to be accomplished by the exact-decoupling approach. However, if actually a four-component calculation must be carried out before · the two-dimensional operator can be evaluated (as in the X2C case), the "projection by two-component approach" is no valid advantage as the four-component variational solution for the 4-spinors already required (implicit) projection to the electronic solutions (in iterative protocols like the self-consistent field algorithm, the projectors may even be optimized implicitly when solving for the electronic 4-spinors; see the discussion in the Introduction and references given there).

Formally, the X2C decoupling transformation is just the $U_X$ transformation

$$U_{\text{X2C}} = U_X, \tag{37}$$

although $U_X$ is not obtained by Eq. 23 but through Eq. 36.

# 4 Algorithmic aspects of "exact" decoupling methods

Complementary to the principles of exact-decoupling methods discussed above, their actual implementation poses additional challenges that we shall discuss in this section. The most decisive insight which connects all decoupling methods is the construction of a so-called kinetically balanced basis set that ensures the correct

non-relativistic limit when the speed of light approaches infinity (in actual electronic structure calculations, the speed of light is then set to a sufficiently high value, e.g., to 1,000,000 Hartree atomic units).

In four-component calculations, basis functions for the small component must be chosen carefully. One cannot simply employ the same basis set for the large and small components. This would lead to variationally unstable results as already observed in the first attempt by Kim [5, 55] and to a wrong non-relativistic limit [56], The correct non-relativistic limit is obtained if the kinetic-balance (KB) condition [28, 57–60] which relates the basis sets for large and small components,

$$\varphi \to \sum_\mu C_{U,\mu} \varphi_\mu, \ \varphi_\mu \in \{\lambda_k\} \tag{38}$$

$$\chi \to \sum_\mu C_{L,\mu} \chi_\mu, \ \chi_\mu \in \{\boldsymbol{\sigma} \cdot \boldsymbol{p} \lambda_k\}, \tag{39}$$

is obeyed. Here, $\{\lambda_k\}$ represents the space spanned by the set of basis functions $\lambda_k$. The KB condition is a natural requirement as it has its origin in the off-diagonal terms of the $2 \times 2$ superstructure of the Fock operator in Eq. 16. Strictly speaking, the requirement in Eq. 39 is a restricted KB condition. The most rigorous KB condition is defined by Eq. 22, but this equation is not useful as no closed-form solution for $X$ exists and the energy-dependent expressions of Eqs. 24 and 25 are, of course, totally impractical for actual calculations. However, any choice of basis set which guarantees the correct non-relativistic limit is sufficient for exact-decoupling approaches. Hence, the one defined in Eq. 39, which fulfills

$$\chi_\mu^S = X_{\text{KB}} \chi_\mu^L, \quad \text{with} \quad X_{\text{KB}} = \frac{1}{2c} \boldsymbol{\sigma} \cdot \boldsymbol{p}, \tag{40}$$

is appropriate. In the case of numerical instabilities associated with Eq. 40, it can be advisable to choose the small-component basis functions normalized.

Equivalent to the KB condition is the idea of transferring the restriction on the small component's basis functions to the relativistic Fock operator of Eq. 15 by

$$\begin{pmatrix} 1 & 0 \\ 0 & \boldsymbol{\sigma} \cdot \boldsymbol{p} \end{pmatrix} f \begin{pmatrix} 1 & 0 \\ 0 & \boldsymbol{\sigma} \cdot \boldsymbol{p} \end{pmatrix} = \begin{pmatrix} q_e V_{\text{eff}} & c\boldsymbol{p}^2 \\ c\boldsymbol{p}^2 & \mathcal{V} - 2m_e c^2 \boldsymbol{p}^2 \end{pmatrix} \tag{41}$$

where the transformation is clearly inspired by the restricted KB condition. where $\mathcal{V} = (\boldsymbol{\sigma} \cdot \boldsymbol{p}) q_e V_{\text{eff}} (\boldsymbol{\sigma} \cdot \boldsymbol{p})$. The idea was proposed by Dyall [61] in the context of the so-called modified Dirac equation

$$\begin{pmatrix} q_e V_{\text{eff}} & c\boldsymbol{p}^2 \\ c\boldsymbol{p}^2 & \mathcal{V} - 2m_e c^2 \boldsymbol{p}^2 \end{pmatrix} \begin{pmatrix} \varphi_i \\ \tilde{\chi}_i \end{pmatrix} = \epsilon_i \begin{pmatrix} 1 & 0 \\ 0 & \boldsymbol{p}^2 \end{pmatrix} \begin{pmatrix} \varphi_i \\ \tilde{\chi}_i \end{pmatrix}, \tag{42}$$

which we should call here the modified Fock equation. Note that the lower component $\widetilde{\chi}_i$ is no longer the small component of the original Dirac spinor. It is called the pseudo-large component. Now, the same basis set can be employed for the large and pseudo-large components of the modified Fock operator.

However, the modified Fock equation, Eq. 42, changes the overlap metric as can be seen on the right hand side. It is therefore more convenient to introduce a unitary transformation to ensure the important KB condition and also to preserve the identity as the metric. This is achieved with the help of a special operator $s$,

$$s = \frac{\boldsymbol{\sigma} \cdot \boldsymbol{p}}{p}, \text{ with } p \equiv \sqrt{\boldsymbol{p}^2}. \tag{43}$$

because its square is the two-dimensional unit matrix, $s^2 = \mathbf{1}$. The KB unitary transformation [56] then reads

$$U_{\text{KB}} = \begin{pmatrix} 1 & 0 \\ 0 & s \end{pmatrix}. \tag{44}$$

and yields the KB-transformed Fock operator

$$\widetilde{f} = U_{\text{KB}} f U_{\text{KB}}^{\dagger} = \begin{pmatrix} q_e V_{\text{eff}} & cp \\ cp & \frac{V}{\boldsymbol{p}^2} - 2m_e c^2 \end{pmatrix}. \tag{45}$$

Now, the corresponding transformed Fock equation does no longer need a non-identity metric. In the following discussion, we denote the form $\widetilde{O}$ with a tilde on top of the symbol as the KB-transformed operator of an original operator $O$

$$\widetilde{O} = U_{\text{KB}} O U_{\text{KB}}^{\dagger}. \tag{46}$$

It is crucial to employ the KB-transformed operators in the implementation of exact-decoupling methods.

The operator $s$ in Eq. 43 is a quite special operator. The calculation of the square root of $\boldsymbol{p}^2$ needed for $p$ seems to be difficult to evaluate considering the usual definition of the momentum operator as a differential operator. Moreover, there exists a restriction on the evaluation of operator $p^{-1}$ arising from the KB condition. The operator $s$ must preserve the KB condition such that the space spanned by $\{s\lambda_k\}$ must be equivalent to $\{\boldsymbol{\sigma} \cdot \boldsymbol{p}\lambda_k\}$. From the form of operator $s$ we understand that the KB condition thus reduces to requiring that the space spanned by $\{p^{-1}\lambda_k\}$ must be equivalent to $\{\lambda_k\}$. Therefore, $p^{-1}$ must be defined within the basis functions space $\{\lambda_k\}$, which a priori is not a condition trivially fulfilled in position space (only in momentum space it is).

In general, it is easy to calculate the action of operators that are algebraic functions of $\boldsymbol{p}^2$ in the space of eigenfunctions of $\boldsymbol{p}^2$, $\{\theta_i\}$. Then, the operator $\boldsymbol{p}^2$ can be replaced by its eigenvalues

$$\boldsymbol{p}^2 \theta_i = p_i^2 \theta_i \Longrightarrow f(\boldsymbol{p}^2)\theta_i = f(p_i^2)\theta_i. \tag{47}$$

For a finite basis-function space, $\{\lambda_k\}$, Hess [15, 32, 62] suggested that the exact momentum eigenfunctions $\theta_i$ are to be replaced by the eigenfunctions of the matrix representation of $\boldsymbol{p}^2$, $\{\langle \lambda_k | \boldsymbol{p}^2 | \lambda_l \rangle\}$. A transformation into this basis is easily achieved as the non-relativistic kinetic energy matrix, which is available in every quantum chemistry program package, is proportional to $\boldsymbol{p}^2$ and can be diagonalized after multiplication by $-2m_e$. Within this scheme, the KB condition is satisfied since any $p^{-1}\lambda_k$ belongs to the space $\{\lambda_k\}$.

In particular, all unitary transformations applied to the relativistic Fock operator must preserve the KB condition. The explicit form of the KB-transformed free-particle Foldy–Wouthuysen transformation [29] reads

$$\widetilde{U}_0 = \begin{pmatrix} \sqrt{\frac{E_0 + m_e c^2}{2E_0}} & \sqrt{\frac{E_0 - m_e c^2}{2E_0}} \\ -\sqrt{\frac{E_0 - m_e c^2}{2E_0}} & \sqrt{\frac{E_0 + m_e c^2}{2E_0}} \end{pmatrix}. \tag{48}$$

It only consists of operators which are algebraic functions of $\boldsymbol{p}^2$. Therefore, they can be evaluated within the Hess scheme. Obviously, $\widetilde{U}_0$ will not violate the KB condition. For the DKH expansion algorithm to exactly decouple the relativistic Fock operator, the subsequent $U_k$ $(k > 0)$ transformations as well as their components $W_k$ must be evaluated within the KB-transformed space. In other words, one must calculate the matrices of $\widetilde{U}_k$ and $\widetilde{W}_k$ instead of their untransformed forms.

The traditional arbitrary-order DKH approach did not evaluate the KB-transformed $\widetilde{U}_k$, but expanded the final Fock operator in terms of low-level operators [36]. The high-level intermediates, which are very useful to reduce the computational costs, were not used. Instead, the traditional DKH approach then leads to an exponentially scaling algorithm with increasing order of the expansion, which can makes it hard to approach infinite-order results in practice. By contrast, the DKH method using KB-transformed operators scales only polynomially [39] so that the calculation of infinite-order results is feasible for any element from the periodic table.

As a side remark, we should note that the expansion formulation in terms of low-level operators is not directly evaluated. In order to reduce the number of matrix operators needed to evaluate the Hamiltonian, the resolution of identity operator $s^2 = 1$ is inserted into proper positions so that only a small number of operator matrices is required for the evaluation of the two-component DKH operator. In its scalar-relativistic variant, the identity to be employed reduces to $\boldsymbol{p}^2 / \boldsymbol{p}^2$ and only the matrix representation of $\widetilde{\boldsymbol{p} \cdot V \boldsymbol{p}}$ is needed in addition to $\widetilde{T}$ and $\widetilde{V}$ (see Sect. 5.2). We found the insertion approach to be numerically equivalent to the approach using the KB-transformed operators

In the BSS approach, the operator $R$ is also not directly evaluated. In fact, the BSS approach proposes a set of iteration equations for an operator $Y$ to replace $R$. The operator $Y$ is defined as

$$Y = sR = \frac{\boldsymbol{\sigma} \cdot \boldsymbol{p}}{p} R. \tag{49}$$

This definition turned to be the most crucial point for the implementation of the BSS method in a computer program. However, the reason why this step was invoked appears to be mostly historical from our current perspective—namely, it paralleled the procedure of the DKH approach [17], namely to avoid the evaluation of matrix of operators which include an odd number of $\boldsymbol{\sigma} \cdot \boldsymbol{p}$ operators. By contrast, here, we start from a more fundamental point of view, namely from the KB condition and the non-relativistic limit, whose true importance has not been recognized in previous work on the BSS approach.

Not surprisingly, in implementations of the X2C method the $X$ matrix operator is evaluated from the $C_L^{(+)}$ matrix which consists of basis set expansion coefficients of the pseudo-large components instead of the small components. This means that the $X$ matrix is evaluated in the KB-transformed basis-function space.

Exact-decoupling methods have been given many names by different authors mostly for historical reasons. Actually, there exist only three variants of exact-decoupling methods up to now. These three variants and their main protagonists shall be briefly reviewed in the following three subsections. In Sect. 6, we shall demonstrate how efficient the three variants of exact-decoupling methods are. We will then also demonstrate that, if properly implemented, all three variants are almost equally efficient.

## 4.1 One-step transformation

If the exact-decoupling method is algorithmically achieved by only one unitary transformation, we shall call it a "one-step transformation". On formal grounds this notation may be ambiguous as the single unitary transformation can be decomposed or combined from more than one unitary transformation. This is the reason why we write "algorithmically achieved" in order to clearly state that it is a matter of implementation into a computer program. Currently, the only example for a 'one-step transformation' is the $U_X$-transformation, but there might exist other analytical expressions that can be implemented in a single step.

The X2C method implements a one-step transformation. However, the discussion of the X2C method usually starts from the modified Dirac equation proposed by Dyall [61], but the modified Dirac equation method is equivalent to the four-component method with KB-transformed basis functions. Very closely related to this method is the normalized-elimination of the small component (NESC) method [63–67] also proposed by Dyall, which is an electron-only method, but with eigenfunctions expressed in terms of the large components of electronic 4-spinors. In one NESC paper [66], Dyall discussed the transformation (with a renormalization matrix) to pure two-component wave functions, and this version of the NESC method has almost all characteristics of the current X2C method except for the construction scheme of the $X$ matrix. Dyall employed an energy- or eigenfunction-dependent form to evaluate the $X$ matrix iteratively [68], although a non-iterative scheme as in X2C could also be formulated in the NESC framework. It must be noted that when the essential ideas of the X2C approach had been worked out in 2005, it turned out that Filatov and co-workers [68–70] had actually come to similar conclusions considering an iterative NESC approach [71, 72].

In a series of papers by Kutzelnigg and Liu [47, 49, 50] the iterative way to construct the $X$ matrix was discussed in detail. These authors suggested the non-iterative construction scheme for many-electron calculations. Later, the non-iterative construction approach was implemented into the BDF program [51, 52] by Peng and Liu for molecular calculations. The method was first called XQR (exact quasi-relativistic) or infinite-order quasi-relativistic method. This XQR method is a pure two-component method employing Dyall's renormalization matrix. A one-step transformation method named IOTC (infinite-order two-component) was implemented into the DIRAC program [53] by Iliaš and Saue. [Note that the name IOTC has also been used for other exact-decoupling approaches (see Sect. 3.4 above) and it may easily cause misunderstandings.] Iliaš and Saue adopted almost the same algorithm as the one implemented in BDF except for Dyall's renormalization matrix (and they used a numerically more stable expression for the evaluation of the $X$ operator). In their implementation, the basis functions are first converted into an orthonormalized set and every matrix is then evaluated within the orthonormalized set, while the implementation in the BDF program uses the matrices expressed in unnormalized basis functions. Dyall's renormalization matrix turned out to be problematic [73] and a new implementation in the BDF program fixed this problem using a new renormalization matrix, which made the BDF implementation equivalent to the IOTC method of the DIRAC program. Finally, Liu and co-workers found the name XQR to be not suitable to describe this approach. The acronym "X2C" is now commonly used as the name for the one-step exact-decoupling transformation approach as a result from extensive discussions of Jensen, Kutzelnigg, Liu, Saue, and Visscher at the DFT-2007 conference in Amsterdam in August 2007. so far. Unfortunately, any exact-decoupling method may be called an exact two-component method or infinite-order two-component method. However, the acronym X2C has been used only for this special one-step transformation algorithm. The current X2C method has the following features : (1) an $U_X$-type transformation is employed, (2) $X$ is defined within a KB-transformed basis, (3) $X$ is non-iteratively constructed.

## 4.2 Two-step transformation

The first transformation of a two-step method is the free-particle Foldy–Wouthuysen transformation which was considered necessary for the BSS approach [4]. As discussed above, the second step is then the $U_X$-type transformation as first proposed by Barysz et al. [45]. It was the first approach proposed to exactly decouple the relativistic Dirac Hamiltonian in an external electrostatic potential operator. The non-iterative construction of the $Y$-operator has been discussed in a paper by Kędziera and Barysz [74] but was already mentioned in the talk by Jensen [48]. Almost all other calculations published employed the iterative scheme. One exception that used the prescription of Jensen was given in Ref. [75] for calculations on PbO. The scheme is usually called the "BSS approach" [4, 45, 46], but later Barysz et al. preferred the abbreviation "IOTC". Other names for their approach, which have been used, are IOFW (infinite-order Foldy–Wouthuysen) [76, 77] and IODK (infinite-order Douglas–Kroll) [43, 78]. Clearly, the adjective "infinite-order" is not appropriate to describe the latest version of the BSS approach, since this approach uses either an iterative or non-iterative algorithm (like the X2C method) to achieve exact decoupling. A notion refering to an "order" is rooted in the history of this approach, which was proposed in 1997 as an attempt to decouple to a certain order in $1/c$ [4, 45]. As the attribute "infinite-order" is no longer suitable and we continue to simply use the name "BSS method".

The conventional BSS method has the following characteristics : (1) the free-particle Foldy–Wouthuysen transformation is first applied to the Fock operator, (2) a $U_X$-type transformation is employed in the second step, (3) the $Y$ operator is either iteratively or non-iteratively constructed.

However, it appears that the two-step transformation has no advantages over the one-step transformation and hence the X2C scheme can be used instead.

## 4.3 Expansion of the transformation

The DKH method is the only one of this category. In the past, the acronyms DKH (or DK) referred to Hess' original truncated DKH2 method (DKH to 2nd order in the electrostatic potential). However, one should clearly distinguish finite-order from infinite-order DKH results. Also, the still

used abbreviation "DK" for "Douglas–Kroll" should be avoided in favor of "DKH" to highlight Hess' work without which the suggestion by Douglas and Kroll would probably not been known (apart from the fact that it was Hess who demonstrated how to employ the DK transformation in actual calculations). We should note that different parameterizations of $U_k$, i.e., different sets of $a_{k,i}$ expansion parameters, give different exactly decoupled Fock operators. The infinite-order DKH Fock operators are therefore not unique. However, at infinite-order, the results obtained for expectation values are, of course, the same, independent of the chosen parametrization. Only the infinite-order DKH spinors differ by a unitary transformation from one another.

The DKH method has the following characteristics : (1) the free-particle Foldy–Wouthuysen transformation is applied to the relativistic Fock operator, (2) the electrostatic potential is used as an order parameter for an order-by-order expansion of the relativistic Fock operator, (3) the $U_k$ transformation matrices are parametrized in terms of off-diagonal anti-Hermitian $W_k$ of $k$-th order in the electrostatic potential.

## 5 Approximations involved in many-electron calculations

So far, we have discussed the three existing variants for exact-decoupling methods. In principle, they are all exact two-component methods employing the full electrostatic potential $V_{\text{eff}}$. However, in practice, approximations are introduced in order to increase the efficiency (ideally without compromising the accuracy). The discussion to follow now is independent of how the exact-decoupling transformation $U$ is obtained and thus holds for all exact-decoupling methods.

### 5.1 The cumbersome two-electron terms

In many-electron calculations, the effective electrostatic potential contains two terms

$$V_{\text{eff}} = V_{\text{ext}} + V_{ee}[\{\psi_i^{(+)}\}], \tag{50}$$

i.e. the external potential and the effective potential from electron–electron interactions which depends on the positive-energy spinors that enter the expression for the total electronic wave function (cf. Sect. 2) Therefore, the exact-decoupling transformations must be updated if the spinors have changed, e.g., upon their optimization in a self-consistent field procedure (or when the positions of the nuclei, i.e., the molecular structure is changed). In general, this change of the exact-decoupling transformation has been shown to be small [54, 79]. An exact-decoupling transformation constructed only for the external electrostatic potential, $V_{\text{eff}} \rightarrow V_{\text{ext}}$, turned out to be an excellent approximation for molecular many-electron calculation. Within this approximation, the exact two-component Fock operator is approximated as

$$f^{(2+)} \approx \left( U[V_{\text{ext}}] \left( h_D + q_e V_{\text{ext}} + q_e V_{\text{ee}}[\psi_i^{(+)}] \right) U[V_{\text{ext}}]^\dagger \right)^{++},$$
$$(51)$$

where $(\cdots)^{++}$ denotes the upper-left part of the transformed four-dimensional operator. Formally, the choice of $U = U[V_{\text{eff}}]$ for exact decoupling preserves equivalency to the four-component approach, while $U = U[V_{\text{ext}}]$ is equivalent to the external-field no-pair projection approximation.

Untouched by this approximation, the untransformed electron–electron interaction potential operator, $V_{\text{ee}}[\{\psi_i^{(+)}\}]$, still depends on all occupied positive-energy 4-spinors in the many-electron case. To exploit the advantages of the two-component approach, we need to compute it from the 2-spinors $\phi_i^{(+)}$. There exist two approaches to achieve this. One is the back-transformation approach. In an exact-decoupling method, we have available both the exact-decoupling transformation and the 2-spinor. It is then easy to back-transform the two-component orbital to a four-component one:

$$\psi_i^{(+)} = U^{-1} \begin{pmatrix} \phi_i^{(+)} \\ 0 \end{pmatrix} = U^\dagger \begin{pmatrix} \phi_i^{(+)} \\ 0 \end{pmatrix}. \tag{52}$$

However, this approach is only useful to calculate one-electron properties, but it is useless for calculating two-electron integrals, because the former requires two times a back-transformation, while latter requires four. If the back-transformation is applied four times, the computational costs will be higher than those of the corresponding four-component calculation.

If we directly replace the four-component spinor $\psi_i^{(+)}$ by its corresponding two-component spinor $\phi_i^{(+)}$ in the calculation of the $V_{\text{ee}}$ operator, we introduce a picture change error (see below). To correct for this error, we may add a correction term, $V_{\text{cor}}$ [42]. The correction term can be added either before the transformation is carried out,

$$f^{(2+)} \approx (U[V_{\text{ext}}] \, (h_D + V_{\text{ext}} + V_{\text{ee}}[\phi_i^{(+)}]$$
$$+ \, V_{\text{cor}}) \, U[V_{\text{ext}}]^\dagger)^{++}, \tag{53}$$

or as an a posteriori correction

$$f^{(2+)} \approx (U[V_{\text{ext}}] \, (h_D + V_{\text{ext}}) \, U[V_{\text{ext}}]^\dagger)^{++}$$
$$+ \, V_{\text{ee}}[\phi_i^{(+)}] + V_{\text{cor}}. \tag{54}$$

Note that the correction term $V_{\text{cor}}$ is not the same in both equations.

Atomic mean field (AMFI) [80] and screened spin–orbit (SNSO) [81] approaches have been proposed to correct for

two-electron picture change errors, but so far they have only been applied to quasi-relativistic methods. An extension to exact-decoupling methods would be most desirable. However, we should emphasize that in many calculations the bare (untransformed) electron–electron interaction operator is employed in the two-component Fock operator

$$f^{(2+)} \approx \left( U[V_{\text{ext}}](h_D + V_{\text{ext}}) U[V_{\text{ext}}]^\dagger \right)^{++} + V_{\text{ee}}[\phi_i^{(+)}]. \tag{55}$$

It is already known for a long time from quasi-relativistic calculations that this bare-potential approximation provides reliable results for valence electron properties (this is particularly true (see e.g., Refs. [82, 83]) for the scalar-relativistic variant introduced in the next section, but may be different for the truely two-component method including spin–orbit splitting [84]; but see also the numerical results section below). However, it might be not good enough for core-electron properties and gives large errors for the spin–orbit splitting of high-angular-momentum orbitals [42, 81].

### 5.2 Scalar-relativistic approximations

There exist scalar versions of the exact-decoupling methods, which have the huge advantage that they can be easily interfaced with a standard non-relativistic quantum chemistry program package. These scalar-relativistic versions allow for an efficient description of kinematic relativistic effects. However, scalar exact-decoupling is not uniquely defined, because there is no unique definition of the scalar full-relativistic Fock equation [85]. There exist two principal ways to obtain a scalar exact-decoupling Fock operator.

One is the "a priori" approach, where a scalar fully relativistic Fock equation is defined first, then follows the same steps as in the exact decoupling of the two-dimensional operator to obtain a scalar electron-only Fock operator. The commonly used scalar fully relativistic Fock equation is obtained by replacing the two-component $\mathcal{V}$ term in the Fock equation, Eq. 42, by a scalar operator $(\boldsymbol{p} \cdot V\boldsymbol{p})$. The scalar exact-decoupling method would then provide the same eigenvalues. Another option is the "a posteriori" approach. Once the two-component electron-only Fock operator has been obtained, the spin-dependent terms are discarded (by virtue of Dirac's relation) to obtain the scalar version.

### 5.3 Local approximations

Because of the fact that all operators in molecular electronic structure calculations are evaluated in a one-electron basis set, the matrix operations required by the exact-decoupling methods require a computational effort that scales with the molecular size rather than with the number of heavy atoms, which contribute most to the numerical relativistic effect. For large molecules, efficient local approximations are required. Naturally, one may restrict the unitary transformation to those matrix elements of basis functions that are located at a heavy atom (see e.g., the work by Peralta and Scuseria [86, 87]). A more systematic analysis has been provided by Thar and Kirchner [88]. The main conclusion from these studies is that the unitary transformation can be restricted to atom-same-atom diagonal blocks of the relativistic Fock operator without loss of numerical accuracy. However, a more rigorous localization scheme, which would also include all relevant atom-other-atom off-diagonal blocks of the relativistic Fock operator, is desired.

## 5.4 Transformed expectation values

To correctly evaluate an expectation value of a molecular property operator $P$ from a two-component wave function the operator $P$ must be transformed as well

$$
\begin{aligned}
\langle \psi_i^{(+)}|P|\psi_i^{(+)}\rangle &= \langle U\psi_i^{(+)}|UPU^\dagger|U\psi_i^{(+)}\rangle \\
&= \langle \phi_i^{(+)}|(UPU^\dagger)^{++}|\phi_i^{(+)}\rangle \\
&= \langle U^\dagger\phi_i^{(+)}|P|U^\dagger\phi_i^{(+)}\rangle.
\end{aligned}
\tag{56}
$$

If the transformation of the property operator is neglected, a picture change error [89, 90] is introduced, whose magnitude depends on the type of property considered [91–97]. In general, the picture change error is large for core properties. If the two-component result does not match the four-component reference even though the property operator has been properly transformed, then this is most likely because of other approximations discussed in this section.

## 6 Numerical comparison of the three exact-decoupling variants

For a detailed numerical one-to-one comparison of the three exact-decoupling methods, (infinite-order) DKH, X2C, and BSS, we have implemented them into the MOLCAS program (see Sect. 7) However, only scalar-relativistic versions of these methods are currently available in MOLCAS. For this reason, the explicit order does not need to be denoted. To also shed light on spin–orbit effects, we carried out atomic calculations for the radon atom as an example (see Sect. 7).

### 6.1 Scaling behavior

The efficiency of the three decoupling methods are compared in calculations of a test one-electron atomic system

$(Rn^{85+})$. 100, 200, and 300 even-tempered Gaussian basis functions were used. Since a 64-bit calculation turned out to be too fast for this test, all calculations were carried out with 128-bit precision (this also cures the failure of the diagonalization routine for large basis sets in 64-bit precision). The calculations were performed on the Opteron 250 CPU. The evaluation of one-electron integrals has not been included in the measurement of the CPU time, only the transformation steps have been counted. Since the calculations are dominated by matrix multiplications and diagonalization, all methods are of the order $\mathcal{O}(aN^2 + bN^3)$ where $N$ denotes the number of basis functions. The formal scaling analysis is confirmed by the data given in Table 1, where the ratios are $7.8 \approx 2^3$ and $26.2 \approx 3^3$ for increasing the basis set from $N = 100$ to $N = 200$ and $N = 300$. This demonstrates that the diagonalization dominates the computational effort, i.e., that the prefactor $a$ is rather small. The ratio between different methods is then a constant factor. The approximate DKH2 method is the fastest one, but it does, of course, not achieve *exact* decoupling. The computation time of the BSS method is almost the same as (still approximate) DKH8, while X2C is a little bit faster than BSS since the additional free-particle FW transformation is missing in the X2C approach.

### 6.2 The hydrogen-like $Rn^{85+}$ heavy ion

Table 2 presents results for the $1s$ state of the one-electron heavy ion $Rn^{85+}$ obtained with the different variants of exact-decoupling methods as well as the four-component Dirac equation solution denoted as DEQ. As explained above, we will use the abbreviation 'DKH' for results converged with respect to the order in the external potential. We employed a 35th order DKH scheme, i.e., DKH35, in all cases presented here and below, which yields results

**Table 1** Computation times (in seconds) of various decoupling methods (for $Rn^{85+}$)

|        | $N = 100$ | $N = 200$     | $N = 300$        |
|--------|-----------|---------------|------------------|
| DKH2   | 1.8       | 12.6 (7.1)    | 41.5 (23.5)      |
| DKH4   | 2.6       | 19.3 (7.3)    | 63.9 (24.3)      |
| DKH6   | 4.4       | 33.2 (7.5)    | 110.8 (25.0)     |
| X2C    | 7.0       | 54.2 (7.7)    | 181.1 (25.9)     |
| DKH8   | 7.4       | 57.1 (7.8)    | 191.0 (26.0)     |
| BSS    | 7.5       | 58.3 (7.8)    | 194.5 (26.0)     |
| DKH10  | 12.6      | 97.8 (7.7)    | 326.0 (25.8)     |
| DKH12  | 19.3      | 151.1 (7.8)   | 505.6 (26.2)     |
| DKH14  | 29.5      | 230.3 (7.8)   | 771.4 (26.1)     |
| DKH16  | 41.3      | 322.4 (7.8)   | 1,082.3 (26.2)   |

$N$ the number of basis functions. Time ratios with respect to $N = 100$ are given in parentheses

**Table 2** Energy eigenvalue and expectation values of *untransformed* operators $\{r^n, n = -2, -1, 1, 2\}$ of the $1s$ state in the one-electron heavy ion $\mathrm{Rn}^{85+}$

|          | DEQ            | DKH            | X2C            | BSS            |
|----------|----------------|----------------|----------------|----------------|
| $\epsilon_i$ | −4,154.6625406 | −4,154.6625406 | −4,154.6625406 | −4,154.6625406 |
| $r^{-2}$ | 31,523.20      | 50,897.75      | 51,129.21      | 51,391.38      |
| $r^{-1}$ | 110.1254       | 126.3773       | 125.1675       | 126.4942       |
| $r$      | 1.488335E−02   | 1.389296E−02   | 1.396132E−02   | 1.389185E−02   |
| $r^2$    | 3.080482E−04   | 2.761091E−04   | 2.779321E−04   | 2.760894E−04   |

The energy eigenvalue $\epsilon_i$ is obtained to be exactly the same for all methods. However, note that the $\{r^n, n = -2, -1, 1, 2\}$ are affected by picture change errors to highlight the fact that two-component eigenfunctions are obtained that are identical only up to another unitary transformation. All data are in Hartree atomic units

well converged to the infinite-order result. All eigenvalues are identical to the original DEQ (in the same finite basis set) as one would have expected for exact-decoupling methods. Of course, this is possible because we do not invoke any approximations for the electron–electron interaction as this is absent in a one-electron system. The equivalence of the energy eigenvalues is, of course, not a surprising result since we have already discussed the equivalence of the three variants for exact decoupling for any finite basis set.

In order to study the difference of eigenfunctions, we report picture change error affected expectation values of radial momenta $\{r^n, n = -2, -1, 1, 2\}$. If we would properly transform the operators, all expectation values would be the same as they should be. Instead, we evaluate the expectation value as an integral of the squared two-component eigenfunction multiplied by the proper power of $r$, $r^n$. The lower the power $n$ the more weight is given to the core part of the squared eigenfunction. From the data in Table 2, we can see that there exist discrepancies among different exact-decoupling methods which indicates that their eigenfunctions are indeed different. The fact that the differences observed for different operators $r^n$ are of the same order indicates that discrepancies exist in the whole range of the two-component eigenfunctions. Also, the large difference between the four-component and two-component results shows that the picture change error is not negligible.

Table 3 reports additional (picture change error affected) expectation values for the operator $r^{-1}$ for different states of $\mathrm{Rn}^{85+}$. It can be seen that larger deviations from the DEQ reference result are found for core orbitals. The DKH results are found to be very close to BSS results, especially for the outer core orbitals, which may be due to the free-particle Foldy–Wouthuysen transformation that is the first step in both schemes. *We should stress again that, if the operators $r^n$ would have been properly transformed or if the two-component eigenstates were back-transformed to the four-component picture, the exact decoupling methods yield expectation values are identical to the DEQ*

**Table 3** Picture change error affected expectation values of the operator $r^{-1}$ for ground and excited states of the $\mathrm{Rn}^{85+}$ hydrogen-like ion

|            | DEQ       | DKH       | X2C       | BSS       |
|------------|-----------|-----------|-----------|-----------|
| $1s_{1/2}$ | 110.12545 | 126.37731 | 125.16747 | 126.49422 |
| $1p_{1/2}$ | 29.275384 | 28.034662 | 27.975650 | 28.036643 |
| $1p_{3/2}$ | 22.643647 | 23.005977 | 22.989564 | 23.006007 |
| $1d_{3/2}$ | 10.237159 | 10.166282 | 10.165467 | 10.166282 |
| $1d_{5/2}$ | 9.7717553 | 9.8145082 | 9.8138649 | 9.8145083 |
| $1f_{5/2}$ | 5.5349079 | 5.5216606 | 5.5216008 | 5.5216606 |
| $1f_{7/2}$ | 5.4423748 | 5.4519652 | 5.4519099 | 5.4519652 |

All values are in Hartree atomic units

*reference which is the reason, why we did not report them in Tables 2 and 3.*

This last point cannot be overemphasized. Within the finite basis set used, we obtain *exactly* the same energies and expectation values for all states and methods considered. It turned out that basis set convergence of the $r^{-1}$ and $r^{-2}$ momenta for one-electron heavy ions is much better than first observed in Ref. [91]. The persistent (small) deviation of high-order DKH results from the *exact* DEQ reference for these operators was believed in Ref. [91] to be an artifact of the finite basis set used. However, we discovered a bug in the original implementation of Ref. [91], which caused a wrong prefactor in front of one of the commutators of the third-order DKH property operator expression. As a consequence, the comparatively small deviation from the DEQ result completely vanishes upon correction of the erroneous prefactor indicating again the excellent convergence of the DKH$n$ series.

6.3 A many-electron case: the Rn atom

In Table 4, we present results of DFT calculations of the ground state of the neutral radon atom. The orbital energies of exact decoupling methods now differ from each other and from the four-component results due to the picture

**Table 4** Total and selected B3LYP as well as Hartree–Fock orbital energies of the neutral Rn atom

| | 4c | DKH + SNSO | DKH | X2C | BSS |
|---|---|---|---|---|---|
| **B3LYP results** | | | | | |
| Total | −23,611.4636 | −23,597.9850 (0.057) | −23,598.9075 (0.053) | −23,599.8921 (0.049) | −23,598.9087 (0.053) |
| $1s_{1/2}$ | −3,619.9163 | −3,612.3095 (0.210) | −3,611.4756 (0.233) | −3,612.2372 (0.212) | −3,611.4679 (0.233) |
| $3s_{1/2}$ | −162.3668 | −162.1464 (0.136) | −162.1294 (0.146) | −162.1402 (0.140) | −162.1294 (0.146) |
| $6s_{1/2}$ | −0.8575 | −0.8568 (0.089) | −0.8568 (0.090) | −0.8568 (0.086) | −0.8568 (0.090) |
| $2p_{1/2}$ | −634.9644 | −633.8152 (0.181) | −635.6840 (0.113) | −635.7755 (0.128) | −635.6833 (0.113) |
| $2p_{3/2}$ | −533.8440 | −533.1097 (0.138) | −532.3910 (0.272) | −532.4396 (0.263) | −532.3909 (0.272) |
| $2p_{SO}$ | 101.1203 | 100.7055 (0.410) | 103.2929 (2.149) | 103.3359 (2.191) | 103.2923 (2.148) |
| $6p_{1/2}$ | −0.4160 | −0.4160 (0.006) | −0.4189 (0.683) | −0.4189 (0.693) | −0.4189 (0.682) |
| $6p_{3/2}$ | −0.2789 | −0.2787 (0.057) | −0.2779 (0.333) | −0.2779 (0.332) | −0.2779 (0.333) |
| $6p_{SO}$ | 0.1372 | 0.1373 (0.133) | 0.1409 (2.747) | 0.1410 (2.779) | 0.1409 (2.746) |
| $5d_{3/2}$ | −1.8692 | −1.8723 (0.166) | −1.8862 (0.907) | −1.8861 (0.901) | −1.8862 (0.907) |
| $5d_{5/2}$ | −1.7045 | −1.7046 (0.008) | −1.6962 (0.484) | −1.6961 (0.492) | −1.6962 (0.484) |
| $5d_{SO}$ | 0.1648 | 0.1677 (1.793) | 0.1900 (15.305) | 0.1900 (15.310) | 0.1900 (15.305) |
| $4f_{5/2}$ | −8.2980 | −8.3035 (0.066) | −8.3759 (0.938) | −8.3752 (0.931) | −8.3759 (0.938) |
| $4f_{7/2}$ | −8.0377 | −8.0455 (0.097) | −7.9930 (0.556) | −7.9924 (0.563) | −7.9930 (0.556) |
| $4f_{SO}$ | 0.2603 | 0.2580 (0.877) | 0.3828 (47.080) | 0.3828 (47.077) | 0.3828 (47.080) |
| MAE | | 0.293 | 0.397 | 0.377 | 0.397 |
| MRE (%) | | 0.088 | 0.392 | 0.392 | 0.392 |
| **Hartree–Fock results** | | | | | |
| Total | −23,602.1044 | −23,591.3125 (0.046) | −23,592.1434 (0.042) | −23,593.2115 (0.038) | −23,592.1283 (0.042) |
| $1s_{1/2}$ | −3,641.1973 | −3,635.6614 (0.152) | −3,634.9353 (0.172) | −3,635.5917 (0.154) | −3,634.9235 (0.172) |
| $3s_{1/2}$ | −166.8331 | −166.6451 (0.113) | −166.6291 (0.122) | −166.6378 (0.117) | −166.6289 (0.122) |
| $6s_{1/2}$ | −1.0714 | −1.0704 (0.094) | −1.0704 (0.091) | −1.0704 (0.090) | −1.0704 (0.091) |
| $2p_{1/2}$ | −642.3301 | −642.0151 (0.049) | −643.9010 (0.245) | −643.9966 (0.259) | −643.9001 (0.244) |
| $2p_{3/2}$ | −541.1023 | −540.5336 (0.105) | −539.8050 (0.240) | −539.8547 (0.231) | −539.8049 (0.240) |
| $2p_{SO}$ | 101.2278 | 101.4815 (0.251) | 104.0960 (2.833) | 104.1419 (2.879) | 104.0952 (2.833) |
| $6p_{1/2}$ | −0.5403 | −0.5411 (0.151) | −0.5440 (0.688) | −0.5441 (0.700) | −0.5440 (0.688) |
| $6p_{3/2}$ | −0.3840 | −0.3836 (0.089) | −0.3827 (0.319) | −0.3827 (0.318) | −0.3827 (0.319) |
| $6p_{SO}$ | 0.1563 | 0.1575 (0.742) | 0.1613 (3.162) | 0.1613 (3.200) | 0.1613 (3.161) |
| $5d_{3/2}$ | −2.1897 | −2.1940 (0.197) | −2.2085 (0.863) | −2.2084 (0.857) | −2.2085 (0.863) |
| $5d_{5/2}$ | −2.0165 | −2.0156 (0.047) | −2.0066 (0.491) | −2.0065 (0.498) | −2.0066 (0.491) |
| $5d_{SO}$ | 0.1731 | 0.1784 (3.037) | 0.2019 (16.634) | 0.2019 (16.640) | 0.2019 (16.634) |
| $4f_{5/2}$ | −9.1939 | −9.2009 (0.076) | −9.2763 (0.896) | −9.2757 (0.889) | −9.2763 (0.896) |
| $4f_{7/2}$ | −8.9282 | −8.9313 (0.034) | −8.8764 (0.580) | −8.8758 (0.587) | −8.8764 (0.580) |
| $4f_{SO}$ | 0.2657 | 0.2696 (1.481) | 0.3999 (50.506) | 0.3999 (50.503) | 0.3999 (50.506) |
| MAE | | 0.203 | 0.363 | 0.346 | 0.363 |
| MRE (%) | | 0.077 | 0.396 | 0.395 | 0.396 |

The relative errors (in %) with respect to the four-component reference values are presented in parentheses. The mean absolute error (MAE) and mean relative error (MRE) are indicated for the complete set of occupied orbitals. The *SO* (spin–orbit splitting) entry is the energy difference between orbitals $j = l + 1/2$ and $j = l - 1/2$. All values are in Hartree atomic units

change error in the two-electron Coulomb interaction term. For the mean absolute errors (MAEs) of the complete set of occupied electron orbitals, the results of DKH, X2C, and BSS calculations are 0.397, 0.377 and 0.397 Hartree for B3LYP (and 0.363, 0.346 and 0.363 Hartree for Hartree-Fock), respectively. These deviations are basically the same. All variants have the same mean relative errors (MREs) for all occupied orbitals with a value of 0.397% (B3LYP). The differences of the total energies are also at the same level. It is evident that, for the calculation of many-electron systems without a proper inclusion of the two-electron interaction operator, current exact-decoupling methods provide the same accuracy for electronic energies, but do not fully reproduce the reference result.

Neglecting the picture change of the two-electron interaction operator also leads to large errors for the spin–orbit splitting, especially for orbitals of high-angular momentum. As we can see from Table 4, the relative errors of spin–orbit splitting of the $6p$, $5d$, and $4f$ orbitals are roughly 2.7, 15.3, and 47.1% for B3LYP (and 3.2, 16.6, and 50.5% for Hartree–Fock), respectively, when compared with the four-component reference. The spin–orbit splitting of the $f$ shell has an error larger than 50% for B3LYP.

The SNSO [81] approach proposed by Boettger is a simple method to correct the two-electron picture change error and can be applied for atomic systems. We employed the same parameters for our SNSO infinite-order DKH calculation as Boettger did for the DKH2 approximation (it is obvious that such parameters are not very suitable for the exact-decoupling methods and we will improve on it in future work). We find that the DKH method with SNSO correction significantly reduces the relative errors of spin–orbit couplings, especially for the high-angular-momentum orbitals. The relative error of $6p$, $5d$, and $4f$ orbitals decreases to 0.1, 1.8, and 0.9% for B3LYP (and 0.7, 3.0, and 1.5% for Hartree–Fock), respectively. The MAE and MRE are also improved. The MRE of all occupied orbitals is decreased to 0.088% in the case of B3LYP and to 0.077% in the case of Hartree–Fock. However, the total energy is not improved because the contribution of high-angular momenta is too small.

The SNSO correction approach includes only the correction of spin–orbit terms and thus no scalar-relativistic terms are involved (by contrast to the approach of van Wüllen [42] mentioned earlier). The error in total electronic energy is mainly determined by the innermost $1s$ orbital, which accounts for 56% of the total energy error. If a scalar-relativistic picture change correction term can also be included to account for the deficiencies in the approximate treatment of the electron–electron interaction, we may expect that it will improve the total electronic energy as well as the inner core orbitals. If one is only interested in energy differences, as is usually the case in chemical applications, which may mainly come from the valence orbitals, a spin–orbit picture change correction scheme may be already good enough. The SNSO approach combined with the X2C or BSS methods yields the same results as for DKH. We therefore reported only the SNSO-DKH results in Table 4.

The analysis of energies does not provide a complete picture of the accuracy of the exact-decoupling methods. We therefore also report results of other properties. Again, we utilize data for the expectation values of operator $r^{-1}$ as listed in Table 5. The leftmost column shows the results without picture change correction. This leads to 5.2% relative error of the total quantities. The error mainly stems from the $1s$ orbital, which has 14.5% relative error. Even for the MRE of all occupied orbitals, it is 1.14% for B3LYP (and 1.13% for Hartree–Fock) and obviously not negligible.

Clearly, in the step of calculating the one-electron expectation values, the picture change correction, i.e., the unitary transformation of the $R^{-1} \equiv \sum_i r_i^{-1}$ operator, is necessary. As we can see from Table 5, the picture change correction reduces the relative error of total expectation value from 5.2 to 0.1%. The MAE and MRE are also reduced from 0.4857 and 1.144% to 0.0101 and 0.156% for B3LYP (and 0.4838 and 1.132% to 0.0069 and 0.051% for Hartree–Fock), respectively. There errors are close to or even better than those obtained for energies. Note that the picture change error not only affects the operator $R^{-1}$, but also the two-component spinors through the neglect of the transformation of $V_{ee}$ in the optimization of the spinors. Adding the SNSO correction term improves further the accuracy of expectation values. However, the total expectation value is not improved for the same reason as discussed above. We even observe that the results for $2p_{1/2}$ become worse, although $2p_{3/2}$ turns out to be more accurate. For high-angular momentum orbitals $d$ and $f$ all results are improved.

### 6.4 Contact densities

Contact densities are most sensitive to the proper set-up of transformation of operators [93, 94, 98–100]. Picture change affected results are dramatically wrong. Especially, the contact density of $p_{1/2}$ would be zero without a proper treatment of the picture change. Our results for the contact density of the Rn atom in Table 6 show that all exact decoupling methods provide results of the same accuracy. Not unexpectedly, the SNSO correction does not improve the accuracy of the contact densities as it has been designed to correct the spin–orbit splitting, while the contact density is dominated by the $1s$ orbital, which has a contribution of more than 80%.

Since the RnH molecule considered for an illustrative calculation of relative contact densities with the Rn atom as reference is somewhat artificial (we unexpectedly obtained with Hartree–Fock and with B3LYP an increased contact density for the Rn contact density in the molecule; as can be seen for Hartree–Fock in Table 7), we also studied contact densities and contact-density shifts for the heavy water analog of the sixth period of the periodic table, i.e., for $PoH_2$, for which we obtained negative shifts. The most important result from Table 7 is that DKH and X2C yield the same results for the contact-density shift and very similar absolute contact densities.

**Table 5** B3LYP and Hartree-Fock expectation values of the $r^{-1}$ operator for selected orbitals of the neutral Rn atom as well as for the total operator, $R^{-1} \equiv \sum_i r_i^{-1}$

| | 4c | DKH + SNSO | DKH | X2C | DKH [pce] |
|---|---|---|---|---|---|
| **B3LYP results** | | | | | |
| Total | 699.505703 | 698.833118 (0.096) | 698.944382 (0.080) | 698.991685 (0.073) | 735.502906 (5.146) |
| $1s_{1/2}$ | 109.179301 | 109.071618 (0.099) | 109.059783 (0.109) | 109.070963 (0.099) | 125.051113 (14.537) |
| $3s_{1/2}$ | 9.969738 | 9.962444 (0.073) | 9.961970 (0.078) | 9.962248 (0.075) | 10.466635 (4.984) |
| $6s_{1/2}$ | 0.695418 | 0.695001 (0.060) | 0.694979 (0.063) | 0.694997 (0.060) | 0.699724 (0.619) |
| $2p_{1/2}$ | 27.194679 | 27.093849 (0.371) | 27.200967 (0.023) | 27.205183 (0.039) | 26.138550 (3.884) |
| $2p_{3/2}$ | 21.081260 | 21.072395 (0.042) | 21.043522 (0.179) | 21.045230 (0.171) | 21.347845 (1.265) |
| $6p_{1/2}$ | 0.579729 | 0.579411 (0.055) | 0.580965 (0.213) | 0.581003 (0.220) | 0.579486 (0.042) |
| $6p_{3/2}$ | 0.492555 | 0.492325 (0.047) | 0.491740 (0.166) | 0.491751 (0.163) | 0.492103 (0.092) |
| $5d_{3/2}$ | 1.247426 | 1.247309 (0.009) | 1.250741 (0.266) | 1.250737 (0.265) | 1.249353 (0.154) |
| $5d_{5/2}$ | 1.202089 | 1.202033 (0.005) | 1.199814 (0.189) | 1.199806 (0.190) | 1.200621 (0.122) |
| $4f_{5/2}$ | 2.974877 | 2.972309 (0.086) | 2.982564 (0.258) | 2.982533 (0.257) | 2.979985 (0.172) |
| $4f_{7/2}$ | 2.929191 | 2.930927 (0.059) | 2.923331 (0.200) | 2.923299 (0.201) | 2.925170 (0.137) |
| MAE | | 0.0083 | 0.0101 | 0.0099 | 0.4857 |
| MRE (%) | | 0.067 | 0.156 | 0.155 | 1.144 |
| **Hartree–Fock results** | | | | | |
| Total | 699.258976 | 698.707959 (0.079) | 698.819315 (0.063) | 698.865968 (0.056) | 735.341956 (5.160) |
| $1s_{1/2}$ | 109.178769 | 109.059584 (0.109) | 109.048561 (0.119) | 109.058925 (0.110) | 125.027648 (14.516) |
| $3s_{1/2}$ | 9.980194 | 9.972547 (0.077) | 9.972099 (0.081) | 9.972330 (0.079) | 10.475999 (4.968) |
| $6s_{1/2}$ | 0.676254 | 0.675837 (0.062) | 0.675876 (0.056) | 0.675890 (0.054) | 0.680187 (0.582) |
| $2p_{1/2}$ | 27.149872 | 27.085486 (0.237) | 27.192072 (0.155) | 27.196654 (0.172) | 26.131226 (3.752) |
| $2p_{3/2}$ | 21.071816 | 21.066569 (0.025) | 21.037648 (0.162) | 21.039426 (0.154) | 21.341624 (1.280) |
| $6p_{1/2}$ | 0.574960 | 0.575095 (0.023) | 0.576472 (0.263) | 0.576515 (0.270) | 0.575063 (0.018) |
| $6p_{3/2}$ | 0.492764 | 0.492489 (0.056) | 0.491984 (0.158) | 0.491998 (0.156) | 0.492334 (0.087) |
| $5d_{3/2}$ | 1.243663 | 1.243946 (0.023) | 1.247251 (0.288) | 1.247249 (0.288) | 1.245899 (0.180) |
| $5d_{5/2}$ | 1.198849 | 1.198650 (0.017) | 1.196512 (0.195) | 1.196505 (0.196) | 1.197296 (0.130) |
| $4f_{5/2}$ | 2.975837 | 2.973897 (0.065) | 2.984120 (0.278) | 2.984090 (0.277) | 2.981547 (0.192) |
| $4f_{7/2}$ | 2.930898 | 2.932343 (0.049) | 2.924769 (0.209) | 2.924738 (0.210) | 2.926603 (0.147) |
| MAE | | 0.0069 | 0.0117 | 0.0114 | 0.4838 |
| MRE (%) | | 0.051 | 0.170 | 0.170 | 1.132 |

The relative errors (in %) with respect to the four-component data are presented in parentheses. The mean absolute error (MAE) and mean relative error (MRE) are indicated for the complete set of occupied orbitals. 'pce' denotes picture-change affected results. All values are in Hartree atomic units

## 7 Conclusion and outlook

In this paper, we reviewed the current status of exact-decoupling methods applied to the relativistic Fock operator. Three different approaches—DKH, X2C, and BSS—exist for this purpose and they are all intimately related. In addition to this discussion of all important formal aspects, we then set out to provide numerical results which are obtained on the same basis (same program platform, same basis set, same electronic structure method). The main results of this study, of which some have already been obtained in previous work as cited above, may be summarized as follows.

(a) The exact decoupling of the relativistic Fock operator can be achieved with either DKH, X2C, or BSS. The iteration scheme within the X2C and BSS approaches may suffer from convergence problems, but can be cured by the non-iterative scheme.

(b) DKH, X2C, and BSS calculations in a finite basis set produce the same eigenvalues as the four-component reference (in the same KB basis set). However, their eigenfunctions may differ (but are related by to one another by a unitary transformation).

(c) In many-electron calculations, the exact decoupling transformation is usually carried out with the external potential only, thus introducing an approximation. Employing untransformed two-electron potentials is computationally very beneficial, but also introduces a

**Table 6** B3LYP and Hartree–Fock contact densities and contributions from individual orbitals of the neutral Rn atom in Hartree atomic units

| | 4c | DKH + SNSO | DKH | X2C | DKH [pce] |
|---|---|---|---|---|---|
| **B3LYP results** | | | | | |
| Total | 3,813,828.11 | 3,812,599.15 (0.032) | 3,814,303.57 (0.012) | 3,815,690.02 (0.049) | 11,155,986.64 (192.514) |
| $1s_{1/2}$ | 1,550,219.61 | 1,550,969.85 (0.048) | 1,550,367.62 (0.010) | 1,550,950.37 (0.047) | 4,607,441.57 (197.212) |
| $2s_{1/2}$ | 248,537.74 | 248,697.85 (0.064) | 248,617.89 (0.032) | 248,689.67 (0.061) | 739,534.03 (197.554) |
| $3s_{1/2}$ | 58,180.21 | 58,241.41 (0.105) | 58,224.13 (0.075) | 58,239.10 (0.101) | 173,192.71 (197.683) |
| $4s_{1/2}$ | 15,342.95 | 15,361.35 (0.120) | 15,356.94 (0.091) | 15,360.79 (0.116) | 45,680.35 (197.729) |
| $5s_{1/2}$ | 3,527.57 | 3,532.14 (0.129) | 3,531.22 (0.103) | 3,532.09 (0.128) | 10,503.85 (197.764) |
| $6s_{1/2}$ | 551.01 | 551.73 (0.131) | 551.61 (0.109) | 551.74 (0.133) | 1,640.81 (197.782) |
| $2p_{1/2}$ | 22,656.27 | 21,456.97 (5.293) | 22,615.67 (0.179) | 22,630.44 (0.114) | 0.00 (100.000) |
| $3p_{1/2}$ | 5,966.57 | 5,655.97 (5.206) | 5,956.62 (0.167) | 5,959.76 (0.114) | 0.00 (100.000) |
| $4p_{1/2}$ | 1,561.98 | 1,481.19 (5.173) | 1,560.03 (0.125) | 1,560.82 (0.075) | 0.00 (100.000) |
| $5p_{1/2}$ | 331.14 | 314.11 (5.141) | 330.97 (0.051) | 331.13 (0.001) | 0.00 (100.000) |
| $6p_{1/2}$ | 38.99 | 37.00 (5.100) | 39.08 (0.222) | 39.10 (0.274) | 0.00 (100.000) |
| **Hartree–Fock results** | | | | | |
| Total | 3,802,285.57 | 3,789,453.14 (0.337) | 3,791,548.74 (0.282) | 3,792,519.57 (0.257) | 11,094,058.60 (191.773) |
| $1s_{1/2}$ | 1,546,350.48 | 1,542,225.22 (0.267) | 1,541,792.70 (0.295) | 1,542,205.45 (0.268) | 4,583,791.56 (196.426) |
| $2s_{1/2}$ | 247,684.15 | 246,976.59 (0.286) | 246,934.60 (0.303) | 246,966.92 (0.290) | 734,830.91 (196.681) |
| $3s_{1/2}$ | 57,987.93 | 57,844.73 (0.247) | 57,836.71 (0.261) | 57,842.34 (0.251) | 172,111.81 (196.806) |
| $4s_{1/2}$ | 15,121.89 | 15,087.81 (0.225) | 15,086.38 (0.235) | 15,087.72 (0.226) | 44,894.20 (196.882) |
| $5s_{1/2}$ | 3,339.99 | 3,332.53 (0.223) | 3,332.59 (0.222) | 3,332.88 (0.213) | 9,917.13 (196.921) |
| $6s_{1/2}$ | 499.71 | 498.47 (0.248) | 498.58 (0.225) | 498.62 (0.217) | 1,483.69 (196.912) |
| $2p_{1/2}$ | 22,383.11 | 21,341.25 (4.655) | 22,482.12 (0.442) | 22,507.49 (0.556) | 0.00 (100.000) |
| $3p_{1/2}$ | 5,898.06 | 5,627.65 (4.585) | 5,923.69 (0.435) | 5,929.55 (0.534) | 0.00 (100.000) |
| $4p_{1/2}$ | 1,528.81 | 1,459.34 (4.544) | 1,536.26 (0.487) | 1,537.73 (0.583) | 0.00 (100.000) |
| $5p_{1/2}$ | 311.92 | 297.89 (4.499) | 313.70 (0.570) | 314.00 (0.667) | 0.00 (100.000) |
| $6p_{1/2}$ | 36.74 | 35.11 (4.429) | 37.04 (0.818) | 37.08 (0.920) | 0.00 (100.000) |

The relative errors (in %) with respect to the four-component values are presented in parentheses. 'pce' denotes picture-change affected results

**Table 7** Hartree–Fock contact densities for Rn and Po atoms and for the molecules RnH (bond distance 108.7 pm))and PoH$_2$ (Po–H distance 177.3 pm)

| Method | Rn | RnH | $\Delta$RnH | Po | PoH$_2$ | $\Delta$PoH$_2$ |
|---|---|---|---|---|---|---|
| sfDKH [pce] | 11,097,679.77 | 11,097,810.30 | 130.53 | 9,448,555.63 | 9,448,442.35 | −113.28 |
| sfDKH | 3,744,321.59 | 3,744,366.39 | 44.80 | 3,197,017.33 | 3,196,979.51 | −37.82 |
| sfX2C | 3,745,217.87 | 3,745,262.68 | 44.81 | 3,197,772.34 | 3,197,734.51 | −37.83 |

Difference densities are denoted as $\Delta$RnH = RnH − Rn, $\Delta$PoH$_2$ = PoH$_2$ − Po. 'sf' denotes the scalar-relativistic spin-free approximations of the exact-decoupling methods. 'pce' denotes picture-change-affected results

picture change error, which may be compensated by an effective correction term.

(d) The SNSO correction of the two-electron picture change error improves significantly the accuracy of the spin–orbit splitting. Since no scalar-relativistic correction is included in this ansatz, total expectation values are not improved.

(e) Picture change corrections of property operators are mandatory—especially for core properties such as the contact density.

The discussion in this work has highlighted various directions of future developments for exact-decoupling methods. The basic theory is well established, but a couple of practical issues for actual calculations are still to be solved. Examples are:

(a) The development of an intelligent infinite-order DKH method, which automatically truncates the expansion at desired accuracy without the pre-input of the desired order.

(b) The efficiency of transformations of property operators should be improved. This is most apparent for the position-dependent density operator.

(c) Two-electron picture change correction schemes for exact decoupling methods require more work. Developments could be based on ideas of the AMFI and SNSO approaches.

**Acknowledgments** This work has been financially supported by ETH Zürich and by SNF Grant No. 200020-132542/1.

**Open Access** This article is distributed under the terms of the Creative Commons Attribution Noncommercial License which permits any noncommercial use, distribution, and reproduction in any medium, provided the original author(s) and source are credited.

## Appendix: Computational methodology

For the molecular calculations presented in this paper, we have been implemented the scalar-relativistic polynomial-cost DKH algorithm as well as X2C and BSS into the Molcas programme package [37] by docking the module described in Ref. [39] to the existing interface to the exponentially scaling DKH module [36]. Calculations with truly two-component versions are performed with the atomic program presented in Ref. [39], where also the four-component approach is available. Dyall's TZ basis set was employed for Rn and Po in uncontracted form yielding a (30s26p17d11f) primitive basis for both. For H, Dunning's aug-cc-pVTZ basis set was used (6s3p2d)/(4s3p2d) [101]. The (finite) nuclear charge distribution was modeled by a Gaussian distribution [102]. In the DFT calculations we applied the B3LYP hybrid density functional [103–105]. Different Rn–H distance were tested for RnH (from 171.6 to 108.7 pm) and the contact density turned out to be always positive. For $PoH_2$ we chose Po–H bond lengths of 177.3 pm and an angle of 89.6°.

## References

1. Dirac PAM (1958) The principles of quantum mechanics, 4th edn. Oxford University Press, Oxford
2. Reiher M, Wolf A (2009) Relativistic quantum chemistry. Wiley-VCH, Weinheim
3. Akhiezer AI, Berestetskii VB (1965) Quantum electrodynamics, vol XI of interscience monographs and texts in physics and astronomy. Interscience Publishers, New York
4. Reiher M, Wolf A (2004) J Chem Phys 121:2037
5. Reiher M, Hinze J (2003) Four-component ab initio methods for electronic structure calculations of atoms, molecules, and solids, volume relativistic effects in heavy-element chemistry and physics. Wiley, Chichester, pp 61–88
6. Mittleman MH (1981) Phys Rev A 24:1167
7. Sucher J (1980) Phys Rev A 22:348
8. Heully J-L, Lindgren I, Lindroth E, Mårtensson-Pendrill A-M (1986) Phys Rev A 19:4426
9. Talman JD (1986) Phys Rev Lett 57:1091
10. LaJohn L, Talman JD (1992) Chem Phys Lett 189:383
11. Thyssen J, Fleig T, Jensen HJA (2008) J Chem Phys 129:034109
12. Chang C, Pelissier M, Durand P (1986) Phys Scr 34:394
13. van Lenthe E, Baerends EJ, Snijders JG (1993) J Chem Phys 99:4597
14. van Lenthe E, Baerends EJ, Snijders JG (1994) J Chem Phys 101:9783
15. Hess BA (1986) Phys Rev A 33:3742
16. Hess BA, Marian CM (2000) In: Jensen P, Bunker PR (eds) Computational molecular spectroscopy. Wiley, Chichester, p 169
17. Barysz M, Sadlej AJ (2001) J Mol Struct (THEOCHEM) 573:181
18. Wolf A, Reiher M, Hess BA (2002) Two-component methods and the generalized Douglas–Kroll transformation, theoretical and computational chemistry. Elsevier, Amsterdam, pp 622–663
19. Wolf A, Reiher M, Hess BA (2004) Transgressing theory boundaries: the generalized Douglas–Kroll transformation. World Scientific, Singapore
20. Reiher M (2006) Theor Chem Acc 116:241
21. Liu W (2010) Mol Phys 108:1679
22. Barysz M (2010) Two-component relativistic theories, vol 10 of challenges and advances in computational chemistry and physics. Springer, Dordrecht
23. Reiher M (2011) WIREs computational molecular science. doi:10.1002/wcms.67
24. Nakajima T, Hirao K (2011) Chem Rev doi:10.1021/cr200040s
25. Hinze J, Biegler-König F (1990) Numerical relativistic and non-relativistic MCSCF for atoms and molecules. Elsevier, Amsterdam, pp 405–446
26. Reiher M, Hinze J (1999) J Phys B: At Mol Opt Phys 32:5489
27. Heully J-L, Lindgren I, Lindroth E, Lundqvist S, Mårtensson-Pendrill A-M (1986) J Phys B 19:2799
28. Kutzelnigg W (1997) Chem Phys 225:203
29. Foldy LL, Wouthuysen SA (1950) Phys Rev 78:29
30. Tani S (1951) Prog Theor Phys 6:267
31. Douglas M, Kroll NM (1974) Ann Phys (NY) 82:89
32. Hess BA (1985) Phys Rev A 32:756
33. Nakajima T, Hirao K (2000) J Chem Phys 113:7786
34. Wolf A, Reiher M, Hess BA (2002) J Chem Phys 117:9215
35. van Wüllen C (2004) J Chem Phys 120:7307
36. Reiher M, Wolf A (2004) J Chem Phys 121:10945
37. Aquilante F et al (2009) J Comput Chem 31:224
38. Werner H-J et al (2009) Molpro, version 2009.1, a package of ab initio programs, see http://www.molpro.net
39. Peng D, Hirao K (2009) J Chem Phys 130:044102
40. Samzow R, Hess BA, Jansen G (1992) J Chem Phys 96:1227
41. Park C, Almlöf JE (1994) Chem Phys Lett 231:269
42. van Wüllen C, Michauk C (2005) J Chem Phys 123:204113
43. Seino J, Hada M (2008) Chem Phys Lett 461:327
44. Luber S, Malkin Ondik I, Reiher M (2009) Chem Phys 356:205
45. Barysz M, Sadlej AJ, Snijders JG (1997) Int J Quantum Chem 65:225
46. Barysz M, Sadlej AJ (2002) J Chem Phys 116:2696
47. Kutzelnigg W, Liu W (2005) J Chem Phys 123:241102
48. Jensen HJA (2005, April) Talk on conference on relativistic effects in heavy elements—REHE 2005, Mülheim
49. Kutzelnigg W, Liu W (2006) Mol Phys 104:2225
50. Liu W, Kutzelnigg W (2007) J Chem Phys 126:114107
51. Liu W, Peng D (2006) J Chem Phys 125:044102
52. Peng D, Liu D, Xiao Y, Cheng L (2007) J Chem Phys 127:104106

53. Iliaš M, Saue T (2007) J Chem Phys 126:064102
54. Sikkema J, Visscher L, Saue T, Ilias M (2009) J Chem Phys 131:124116
55. Kim Y-K (1967) Phys Rev 154:17
56. Kutzelnigg W (1984) Int J Quantum Chem 25:107
57. Lee YS, McLean AD (1982) J Chem Phys 76:735
58. Ishikawa Y, Binning R, Sando K (1983) Chem Phys Lett 101:111
59. Stanton R, Havriliak S (1984) J Chem Phys 81:1910
60. Dyall KG, Grant IP, Wilson S (1984) J Phys B At Mol Phys 17:493
61. Dyall KG (1994) J Chem Phys 100:2118
62. Jansen G, Hess BA (1989) Phys Rev A 39:6016
63. Dyall KG (1997) J Chem Phys 106:9618
64. Dyall KG (1998) J Chem Phys 109:4201
65. Dyall KG, Enevoldsen T (1999) J Chem Phys 111:10000
66. Dyall KG (2001) J Chem Phys 115:9136
67. Dyall KG (2002) J Comput Chem 23:786
68. Filatov M, Dyall KG (2007) Theor Chem Acc 117:333
69. Filatov M, Cremer D (2003) J Chem Phys 119:11526
70. Filatov M, Cremer D (2005) J Chem Phys 122:064104
71. Filatov M (2006) J Chem Phys 125:107101
72. Kutzelnigg W, Liu W (2006) J Chem Phys 125:107102
73. Liu W, Peng D (2009) J Chem Phys 131:031104
74. Kędziera D, Barysz M (2007) Chem Phys Lett 446:176
75. Ilias M, Jensen HJA, Kellö V, Roos BO, Urban M (2005) Chem Phys Lett 408:210
76. Fukui H, Baba T, Shiraishi Y, Imanishi S, Kudo K, Mori K, Shimoji M (2004) Mol Phys 102:641
77. Seino J, Hada M (2007) Chem Phys Lett 442:134
78. Seino J, Uesugi W, Hada M (2010) J Chem Phys 132:164108
79. Peng D unpublished results
80. Hess BA, Marian CM, Wahlgren U, Gropen O (1996) Chem Phys Lett 251:365
81. Boettger JC (2000) Phys Rev B 62:7809
82. Hess BA, Kaldor U (2000) J Chem Phys 112:1809
83. Wolf A, Reiher M, Hess BA (2004) J Chem Phys 120:8624
84. Mayer M, Krüger S, Rösch N (2001) J Chem Phys 115:4411
85. Visscher L, van Lenthe E (1999) Chem Phys Lett 306:357
86. Peralta JE, Scuseria GE (2004) J Chem Phys 120:5875
87. Peralta JE, Uddin J, Scuseria GE (2005) J Chem Phys 122:084108
88. Thar J, Kirchner B (2009) J Chem Phys 130:124103
89. Baerends EJ, Schwarz WHE, Schwerdtfeger P, Snijders JG (1990) J Phys B At Mol Phys 23:3225
90. Kellö V, Sadlej AJ (1998) Int J Quantum Chem 68:159
91. Wolf A, Reiher M (2006) J Chem Phys 124:064103
92. Mastalerz R, Barone G, Lindh R, Reiher M (2007) J Chem Phys 127:074105
93. Mastalerz R, Lindh R, Reiher M (2008) Chem Phys Lett 465:157
94. Mastalerz R, Widmark P-O, Roos BO, Lindh R, Reiher M (2010) J Chem Phys 133:144111
95. Pernpointner M, Schwerdtfeger P (1998) Chem Phys Lett 295:347
96. Dyall KG (2000) Int J Quantum Chem 78:412
97. Kellö V, Sadlej AJ (2001) J Mol Struct (THEOCHEM) 547:35
98. Reiher M (2007) Faraday Discuss 135:97
99. Fux S, Reiher M (2011) Struct Bonding. doi: 10.1007/430_2010_37
100. Knecht S, Fux S, van Meer R, Visscher L, Reiher M, Saue T (2011) Theor Chem Acc 129:631
101. Dunning TH Jr (1989) J Chem Phys 90:1007
102. Visscher L, Dyall KG (1997) At Data Nucl Data Tables 67:207
103. Becke AD (1993) J Chem Phys 98:5648
104. Lee C, Yang W, Parr RG (1988) Phys Rev B 37:785
105. Stephens PJ, Devlin FJ, Chabalowski CF, Frisch MJ (1994) J Phys Chem 98:11623